Crop Production

Crop Production

Evolution, History, and Technology

C. Wayne Smith
Department of Soil & Crop Science
Texas A&M University

John Wiley & Sons, Inc.

NEW YORK / CHICHESTER / BRISBANE / TORONTO / SINGAPORE

Copyright © 1995 by John Wiley & Sons, Inc.

All rights reserved. Published simultaneously in Canada.

This publication is designed to provide accurate and
authoritative information in regard to the subject
matter covered. It is sold with the understanding that
the publisher is not engaged in rendering legal, accounting,
or other professional services. If legal advice or other
expert assistance is required, the services of a competent
professional person should be sought.

Library of Congress Cataloging in Publication Data:
Smith, C. Wayne.
    Crop production : evolution, history, and technology / C. Wayne
Smith.
        p.   cm.
    Includes bibliographical references and index.
    ISBN 0-471-07972-3
    1. Field crops—United States.   I. Title.
    SB187.U6S54   1995
    633'.00973—dc20   95-12070

10 9 8 7 6 5 4 3

Preface

CROP PRODUCTION, EVOLUTION, HISTORY, AND TECHNOLOGY

Jean Henry Casimir Fabre, the French Naturalist of the nineteenth century is quoted as saying, "History . . . celebrates the battlefields whereon we meet our death, but scorns to speak of the plowed fields whereby we thrive; it knows the names of the king's bastards but cannot tell us the origin of wheat. That is the way of human folly." This statement remains true for most of us today. We in the agricultural university teaching profession often get caught up in teaching basic scientific facts, chemical formulas, equations to explain physical phenomena, the intricacies of the chromosome, and, today, even the intricacies of individual genes. Students learn how nitrogen molecules are assimilated into amino acids, how those amino acids are assimilated into protein, and how proteins are assimilated into cell structures. And this should be taught. It is the basic science that forms the critical mass for new technology that will keep the American farmer the most efficient producer of food and fiber in the world.

But those who would be learned in agriculture must understand the production practices and strategies that make our bountiful harvests possible. They must also be acquainted with marketing aspects of row crops and with the conversion of raw agriculture commodities into the products found at the consumer level. Such information has to be a part of any book on production agriculture, and it is so with this one. Each chapter is written to provide insights into production and marketing aspects that are unique to each of the major grain, fiber, and oilseed crops produced in the United States. Aspects such as choosing the right cultivar, tillage, planting rates, rotations, fertility, and so on are considered.

But agriculture is more than basic science; it is more than production practices and marketing strategies. Agriculture is evolution, history, and sociology. It is woven into the fabric of humankind and our existence. Individual people made invaluable contributions, and they should be recognized, and we should acknowledge those contributions whenever possible. Such men as Henry A. Wallace who was not only the Vice President of the United States but also developed the first commercial hybrid corn sold in Iowa; or David Fife who found and selected a single wheat plant that gave rise to the millions of acres and the vast industry surrounding hard red spring wheat; or Walter Burling who smuggled seed of a productive cotton biotype out of Mexico in 1806; or John Thurber who was responsible for introducing a cultivar of rice into the United States in 1690 that would not be replaced by better cultivars for almost 200 years.

Agriculture is events, some major and some apparently very minor, that gave impetus to the demand and production of U.S. row crops. Such events as the invention of the "penny-in-the-slot" machine that would help make roasted, in-shell peanuts a popular snack around the turn of this century. Or a blacksmith by the name of John Deere who would recognize that a cast iron plow covered with a steel plate would slice through the virgin earth more easily and without soil adhering to it. Or the rise to power of a madman named Adolf Hitler who would disrupt world trade of soybean oil and set the stage for that crop to become one of the highest dollar value crops grown in the United States today.

This text was written to bring together all of these aspects of eight of our major grain, fiber, and oilseed crops. There is a life intrinsic in every crop, in its evolution, its history, and its production. I believe that those who are truly students of agriculture will find that to be so. It is my hope that those who read this book will find it informative and encouraging.

C. WAYNE SMITH

Acknowledgments

Appreciation and acknowledgment is given to the following individuals who made contributions to this book and made editorial suggestions during its development:

Dr. A. J. Bockholt, Associate Professor, Department of Soil and Crop Sciences, Texas A&M University, College Station, Texas.

Dr. L. G. Heatherly, Research Agronomist, USDA-ARS, Stoneville, Mississippi.

Dr. A. D. Klosterboer, Professor and Extension Specialist, Texas A&M University Research and Extension Center, Beaumont, Texas.

Dr. T. D. Miller, Professor and Extension Specialist, Department of Soil and Crop Sciences, Texas A&M University, College Station, Texas.

Dr. K. F. Schertz, Research Geneticist, USDA-ARS, College Station, Texas.

Dr. O. D. Smith, Professor, Department of Soil and Crop Sciences, Texas A&M University, College Station, Texas.

Mr. J. K. Walker, Professor Emeritus, Entomology Department, Texas A&M University, College Station, Texas.

Dr. W. D. Worrall, Associate Professor, Texas A&M University Research and Extension Center, Vernon, Texas.

The author appreciates the contributions of Dawn Deno, who typed most of the manuscript; Linn White, Staci Frerich, and Kristen Kurten, who helped type and proof the manuscript; and Bobby Bredthauer and Brad Albright, who made invaluable contributions toward the completion of this work.

This book is dedicated to
my wife, Elizabeth,
and to our daughters
Amy and Karen

Contents

4. Barley   174

5. Rice   220

Crop Production

1

Corn (Zea mays L.)

INTRODUCTION

Corn, or maize (from the Arawak Amerindians name, maiz) has been called the gift of the Amerindians to the world, as well it should be. Corn was the most basic and widespread food in the Americas during pre-Columbian times. Columbus took grains of corn with him on his return to Spain, as he most likely did on return trips of later missions in 1493–1496, 1498–1500, and 1502–1504. Corn production spread quickly around the world, particularly along the trade routes of the Portuguese; Italy and southern France by 1494; Egypt by 1517; northern Spain and all of Portugal by 1525; northern Europe by 1571; Balkan Peninsula by 1575; and Africa, India, and parts of the Far East during the 16th century. By one account, corn reached the Philippine Islands as an introduced cultigen before Magellan in 1521.

ORIGIN

The origin of maize has been an item of considerable debate over the past 50 years. Unlike many of the cereal grains such as wheat and barley that evolved and were selected as food crops in the Old World, no wild forms of corn have been found. Scientists can readily find intermediate, if not the actual ancestral forms of Asian cereal grains growing wild today, and can therefore develop models for the evolution, domestication, and cultivation of these grains. This is not so for corn. To date, no feral plant has been found having a reproductive structure remotely similar to the corn ear. In fact, corn is unique in that this highly productive plant could not survive

1

under natural conditions. Such survival requires some mechanism for seed dispersal —for example, a shattering inflorescence, morphological structures such as barbs for entanglement in animal fur, or structures and size that allow for movement and dispersal by air current. Corn has none of these. Mature grains are confined in a modified leaf and stem structure that will drop to the ground to either decay or germinate to produce a clump of plants that may not produce offspring because of interplant competition. Even if eaten as whole grains by an animal, the grain is unprotected (that is, not enclosed in a hard fruit case) and will most likely be digested and not excreted as a whole grain to germinate, grow, and reproduce. Unaided by humans, corn would be extinct in a few generations.

There are two theories about the origin of corn, and because the literature supporting those theories provides interesting insights into the scientific method of observations and hypothesizing, the student of U.S. row crops should be familiar with not only the theories but also with some of the history, dialogue, and major players in this debate since 1939.

P. C. Mangelsdorf and R. C. Graves published a lengthy account in 1939 about the origin of modern corn. This account speculated that the Asiatic migrants of, say, 25,000 years ago encountered such an abundance of game and fish in North America that the "invention" of agriculture was not necessary. However, when these migrants reached the lowlands of northern South America, they encountered a scarcity of game but an abundance of excellent food crops: manioc (edible, starchy root of cassava), sweet potato, and pod corn. These circumstances lead to a semi-sedentary life style where farming supplemented hunting and gathering.

This "pod corn" had small, hard, stony seeds that were enclosed in hard, protective glumes. It probably was not considered a good food source until an "accidental (?) application" of heat caused the seeds to burst or pop, exposing the endosperm for easy digestion. As these aborigines migrated into the Andes Mountain range, they carried the wild pod corn with them, it being the best cereal grain available. Here they settled into a more restricted life style where hunting and gathering supplemented farming rather than vice versa. Somewhere and sometime during the adaptation to farming, humans discovered a mutant pod corn plant in which the grains were not enclosed in hard glumes, thus being consumable as popped or unpopped grains.

Eventually, the food crops of South America—sweet potatoes, peanuts, peppers, manioc, and selected pod corn—were introduced into Central America, where beans and squash were the basis for a well-established agriculture. Here, at some point in history, pod corn outcrossed with a wild grass, *Tripsacum* spp., giving rise to a new genus of plants, originally placed into the genus *Euchlaena,* later to be classified as *Zea mexicana* and commonly called teosinte. Subsequent, repeated intermating of teosinte, pod corn, *Tripsacum,* and resulting progeny gave rise to the great diversity of *Zea mays:* popcorn, dent, flint, and flour corns. These newly evolved types spread in all directions from central Mexico, intermating and producing new populations from which Amerindians on two continents selected increasingly productive and diverse biotypes.

This theory of corn's evolution, Mangelsdorf's pod corn-*Tripsacum*-teosinte, or

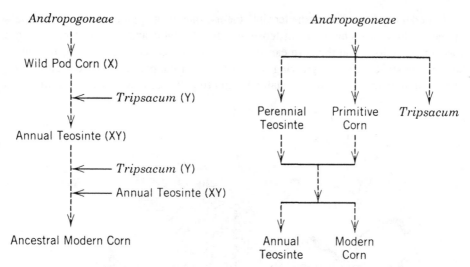

Figure 1.1   *Genealogy of modern corn according to Mangelsdorf. The tripartite theory includ-ing* Tripsacum *(left) was modified in 1986 (right) after the discovery in 1979 of a diploid perennial teosinte.*

tripartite, hypothesis was discarded after electron microscopy of pollen grain struc-ture of corn-*Tripsacum* hybrids and teosinte apparently excluded *Tripsacum* as a possible parent of teosinte. However, with the discovery of a diploid, perennial teosinte in 1979, Mangelsdorf revised his theory of the evolution of modern corn, proposing that the newly discovered perennial teosinte was the missing link between modern corn and wild pod corn (Figure 1.1). (Previously documented perennial teosinte was tetraploid and would not hybridize with modern corn, a diploid.)

The second and most widely accepted theory was proposed by W. C. Galinat of the University of Massachusetts, H. H. Iltis, Professor and Herbarium Director at the University of Wisconsin, and C. W. Beadle, Professor at the University of Chicago, retired. This theory contends that teosinte, a wild grass that still flourishes in Mexico and Guatemala, is the wild progenitor of modern corn. A number of biotypes of teosinte can be found today, both annual and perennial plant habits, several races and species, along with two ploidy levels. Supporters of this widely accepted theory cite a number of factors as evidence. These include (1) the ease of and frequency of hybridization of corn and annual teosinte under natural conditions; (2) both are diploid with n = 10; (3) chromosome structure similarity; (4) plant morphological similarities; (5) overlapping size ranges of pollen; (6) isozyme and DNA similarities; and (7) archaeological evidence.

This hypothesis mandates the evolution of corn in Central America, specifically Mexico and maybe Guatemala, and not South America as proposed by the early tripartite theory. It also mandates a rather abrupt morphological mutation (or evolu-tion) of teosinte fruiting habit to the corn fruiting habit, a change from being adapted to survival in the wild to being almost certainly unadapted to survival without the assistance of humankind.

To understand and accept the teosinte theory, one must appreciate the magnitude of the evolutionary change from teosinte's growth habit and fruiting structure to corn's morphology and the corn ear, or female inflorescence (Figure 1.2). Teosinte has several stalks or tillers growing from the base of the plant, while corn has only one, although there can occasionally be tillering in corn. Seeds of teosinte are

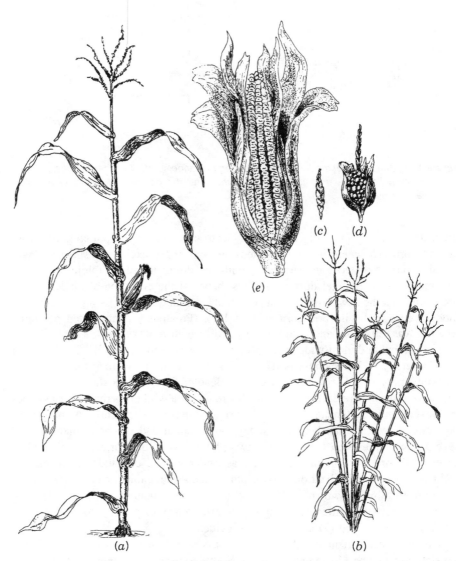

Figure 1.2 *Comparison of teosinte and modern corn. (a) modern corn plant; (b) teosinte plant with many tillers; (c) teosinte spike or ear; (d) the first corn ear as proposed by W. C. Galinat from the oldest remains from Tehuacan, Mexico; and (e) modern corn ear. (Drawing by Mike Hodnett)*

singular and each is enclosed in a hard fruit case, or glumes. There are usually six to ten seeds per female inflorescence arranged in single row, or spike. The teosinte spike has a shattering rachis (analogous to the cob of corn) for seed dispersal, a necessary trait for natural survival of the species. Seeds of the modern corn plant are enclosed in a number of modified leaves, called husks or shucks, and are arranged in several rows along a non-shattering rachis or cob. As already noted, this arrangement makes corn unfit for natural survival.

There are two schools of thought about how the corn ear evolved. One school of thought defended by George Beadle is that only two mutations are required to convert annual teosinte to a corn phenotype. Phenologically, these mutations are (1) from teosinte's shattering spike to corn's non-shattering female spike, so that the kernels would not be scattered and lost when gathered by early humans; and (2) from grains being enclosed in hard fruit cases to being enclosed by soft glumes so that the grains could be easily thrashed and digested. Although these appear to be major changes, they are similar to those that had to occur in any cereal grass (for example, wheat and barley) to become domesticated. It is the uniqueness of the modern corn ear that makes corn appear to be an almost unbelievable transformation. Beadle has supported this hypothesis by evaluating progeny from crosses of a primitive race of Mexican corn, called Chapalote, with a corn-like American teosinte, Chalcote. These experiments indicated that only five major genes with independent inheritance could account for the fruiting differences between teosinte and corn, confirming the possible evolution of annual teosinte to modern corn. Teosinte and corn freely hybridize in the wild and intermediates are fully fertile, supporting the conclusions that intermating of corn with teosinte and with corn-teosinte offspring would, or does, create a large amount of genetic variability from which early humans could have selected, even though multiple selections probably were required, a non-shattering, productive ear on a single-stalk phenotype.

MOVEMENT FROM ORIGIN

The place of evolution of corn appears firmly fixed today as central and southern Mexico and maybe northern Guatemala. The possibility of multiple sites of origin within the range of feral teosinte is a distinct possibility (Figure 1.3). It would seem highly improbable, as indeed it would with all domesticated crop plants, that humans came along at just the right time to find perhaps the greatest discovery of our existence. And, the traits that make corn such a perfect cereal grass (i.e., a large multi-rowed ear, ease of shelling, and a non-shattering rachis) also mandate that it was human-selected.

That humans selected teosinte-corn ears with multiple rows of grain and grains not enclosed in hard fruit cases also seems obvious. And, as noted earlier, the evolution of these traits must be human-assisted because they have no natural survival value. Galinat summarized these first, deliberate selections by humans in the origin of corn as

1. Paired female spikelets in corn in comparison with solitary ones in teosinte.
2. A many-ranked central spike in corn in comparison with a two-ranked one in teosinte. [Note that two-ranked equals four-rowed, author.]
3. Non-shattering rachis (cob) in corn in comparison with shattering one in teosinte.

Figure 1.3 *Origin, evolution, and movement of corn. Shading shows area of feral teosinte distribution today. Archaeological dating of corn culture: (1) Tehuacan Caves, Puebala, Mexico, 5000 B.C.; (2) Ecuador, before 3000 B.C.; (3) Peru, before 3000 B.C.; (4) northern Chile, 2700 B.C.; (5) northeast Tamaulipus, Mexico, 2500 B.C.; (6) southern New Mexico, USA, 1200 B.C.; (7) Venezuela, 400 B.C.; (8) Tennessee, USA, 200 A.D.; (9) New England states, USA, 1400 A.D.; (10) U.S. Corn Belt, 1880 A.D. Source: Shroyer and Hickman, 1988. (Drawing by Mike Hodnett)*

Archaeological and botanical evidence suggest that corn was domesticated and existed by 5000 B.C. in central Mexico. Plant remains, particularly cobs, found in the Tehuacan caves in the state of Puebla, Mexico, are the earliest dated remains. These cobs were from a very primitive race of corn (perhaps better identified as an intermediate between teosinte and all extant corn races). These cobs are similar to teosinte cobs in that they are soft, very small—about 1 inch long—and slender, and the grains would have been very small and hard, probably prepared by popping. But the plants that produced these tiny cobs were more corn-like than teosinte-like in that the cobs were non-shattering and had either four or eight rows of grains (also called two- or four-ranked). These cobs also are similar to the most primitive, stable biotypes, or races, of corn grown in western Mexico and the Yucatan Peninsula in recent times. These races do not compare in yield potential with modern corn, but may be planted even today as a reliable and quick-maturing crop of pop-type grain. Kernels are sometimes brownish, thus resembling teosinte kernels.

From central and southern Mexico, corn culture spread in all directions. Archaeological excavations have documented corn culture as far south as northern Chile by 2700 B.C. and as far north as New Mexico by 1200 B.C. The oldest remains north of the tropics were discovered at the La Perra rockshelter in northeast Tamaulipus, Mexico, and dated to 2500 B.C.

It is obvious from these dates that the acceptance of corn and its culture did not happen quickly, nor did it spread outside the tropics very quickly. One can only speculate about how long it would take a culture of even a settled agriculture to accept a new crop, especially one that originally had to be popped to be easily digested. But we can say with some confidence that taking a photoperiodic plant out of the tropics and selecting for a day-neutral type that could reproduce under the long day conditions prevailing outside the tropics would take many years and many generations. The lag time from domestication of a plant and its wide cultivation has been estimated to be from 2,000 to 6,000 years.

Once introduced into the Old World by Columbus in 1493, corn culture spread to most of the eastern hemisphere within 100 years. It's spread and acceptance occurred so rapidly that later systematic botanists would suggest that corn evolved in Asia and predated its introduction in the 1500s. A discovery of waxy corn in western China in 1909 and reports in the 1940s that open pollinated biotypes of corn in China did not resemble Mexican and Central American races, but did resemble South American races, suggested at least an introduction before Columbus. Stone carvings from 12th and 13th century temples in southern India depict what appears to be ears of corn. However, they also resemble the fruit of a *Pandanus* spp., or screw pine, native to that part of the world. Also, the carvings may not be as old as the temples. These reports have been countered sufficiently to conclude that corn did not originate in Asia, nor was there a pre-Columbian introduction.

RACES OF CORN

By the time Columbus discovered the Americas in 1492, corn had evolved into an amazing number of distinct, stable biotypes. Many races resulted from natural selection pressure as humans planted corn in an ever-expanding circumference from its points of origin. Other races were the result of human selection of biotypes that fit certain, arbitrary needs. For example, even into near modern times the Apache Amerindians often planted corn in the spring before they departed for a summer of hunting and gathering, returning later to gather whatever mother nature provided (somewhat analogous to the decrue system of farming—see Grain Sorghum). This procedure would select for hardy, drought-tolerant, competitive biotypes. Other human-directed evolution was the selection for differing rates of maturity, higher yield, and different types of grains. At the time of Columbus's first voyage, Amerindians had biotypes preferred for boiling, roasting, popping, milling for tortillas, and brewing. We often refer to an ear of corn having a number of different colored grains and used for harvest festivals or Halloween decorations as "Indian Corn." This is a misnomer as the Amerindians probably took pride in the purity of their land-race varieties. The multicolored grains on the same ear result from pollination by a number of different pollen parents, a phenomenon known as the "xenia" effect. The Amerindians would have used this phenomenon as an aid to selecting only pure seed for next year's crop. Other isolating mechanisms would include natural barriers, milpas, or strip farming in forest with trees between plantings of races, and, of course, the tendency of pollen produced on the same plant as the ear to grow faster than foreign pollen.

This movement from origin, cultivation outside of the tropics, and human-directed evolution gave rise to an enormous diversity. Not only are there six basic groups—pod, flint, dent, floury, pop, and sweet (Pod corn has a female inflorescence similar to the corn ear as we know it except that each kernel is enclosed in glumes. Pod corn will not be considered further in this text.); these basic groups are further divided into races. The number of races in Latin America alone has been estimated to be as many as 250. (A race is a group of individuals with enough characteristics in common to permit their recognition as a group within the larger classification of species.) The first large-scale efforts to scientifically categorize the diversity of corn followed the publication of Vavilov's theory that the area of greatest phenotypic diversity in a species corresponds to its center of origin. Early efforts at classification of corn races were based on kernel and ear parameters, but more recent efforts have expanded the list of parameters to include reproductive structures, especially tassel morphology. Although it is beyond the scope of this text, the student of agriculture should be aware of the names and geographical locations of some of the races of maize in the western hemisphere (Table 1.1). Twenty characters were used to classify candidates as to race. These included ear pointing, frequency of endosperm type (i.e., dent, flint, pop, or floury), frequency of kernel color (i.e., white, yellow, or other), cob and kernel characters, and altitude, latitude, and longitude where growing.

Corn is commonly classified into five groupings based on the amount, quality, and arrangement patterns of kernel endosperm (Figure 1.4). These "type" classifications have been in use for such a length of time that their origin is not obvious.

1. Dent: Kernels have very hard, vitreous, horny endosperm at the sides and back, but the central core is soft and floury. This central core extends

TABLE 1.1   A Few of the Numerous Races of Corn in the Western Hemisphere.

Race—Country	Race—Country
Chapalote—Mex.[1]	Oke—Arg.
Harinoso de Ocho—Mex.	Bola Blanca—Arg.
Tuxpeno—Mex.	Kulli—Bol.
Pepitilla—Mex.	Cholito—Bol.
Olotillo—Mex.	Harinoso Tarapaqueno—Chi.
Tabloncillo—Mex.	Choclero—Chi.
White Pop—Cuba	Capio Negro—Chi.
White Dent—Cuba	Dulce Evergreen—Chi.
Sabanero—Col.	Sabanero—Ecu.
Yucatan—Col.	Chillo—Ecu.
Cacao—Col.	Bl. Harinoso Dent—Ecu.
Chococeno—Col.	Negrito—Ven.
Avati Pichinga—Para.	Puya—Ven.
Moroti Guapi—Para.	Nal Tel[2]—Mex./Ecu.
Lenha—Bra.	Cateto—Arg./Uru./Bra.
Caingang—Bra.	Cuban Flint—Cuba/Col./Ven.
Capia Blanco—Arg.	Chullpi[3]—Peru

Source: Adapted from Goodman and Bird, 1977.

1. Mex. = Mexico; Col. = Columbia; Para. = Paraguay; Bra. = Brazil; Arg. = Argentina; Bol. = Bolivia; Chi. = Chile; Ecu. = Ecuador; Ven. = Venezuela; Uru. = Uruguay.
2. May be the most primitive extant race.
3. May be the most primitive extant sweet corn.

to the top, or crown, of the kernel and collapses upon drying, resulting in the distinctive indentation. Dent-type corn dominates U.S. production.

2. Flint: Kernels have very hard, thick, vitreous endosperm that surrounds a small, granular endosperm center. These kernels are round and smooth, with no indentation of the crown at maturity. Almost no flint corn is produced in the United States.

3. Floury: Kernels contain soft, essentially non-vitreous endosperm. These kernels can be easily ground and were used extensively by the Amerindians. Floury corn is perhaps the oldest cultivated corn, and kernels have been found in graves of ancient Aztecs and Incas. Kernels shrink more or less uniformly and therefore no denting occurs. Almost no floury corn is grown in the United States, but it continues to be grown today in the Andean region of South America.

4. Pop: Pop corn is an extreme form of flint corn. Kernels are small, rounded and very hard. Pop corn is grown as a specialty crop in the United States and other countries around the world.

5. Sweet: Kernels are translucent and dry to a wrinkled condition. Sweet corn differs from dent by only one recessive gene (su) that prevents the conversion of sugar to starch. Because sweetness is conditioned by a

Hard (flinty):

Soft and Hard:

Soft (granular):

Sugar (glassy):

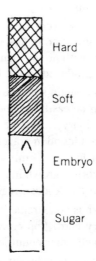

Hard

Soft

Embryo

Sugar

Figure 1.4 *Illustration of the classes of corn based on endosperm hardness. (Drawing by Mike Hodnett)*

recessive gene, plants must be isolated from other corns as pollination of the sweet corn ear by a non-sweet corn pollen would result in nonsweet kernels—again, the xenia effect. Sweet corn ears are harvested 18 to 20 days after pollination and consumed as fresh, canned, or frozen food in the United States.

TAXONOMIC CLASSIFICATION

There are four wild species of the genus *Zea* found in Mexico and northern Central America. These include two perennials, *Z. perennis* (2n = 40) and *Z. diploperennis* (2n = 20), and two annuals, *Z. luxurians* (2n = 20) and *Z. mexicanna* (2n = 20). Corn (*Z. mays* (2n = 20)) evolved from *Z. mexicanna,* common name, teosinte. *Zea mays* is commonly called corn in the United States, but referred to as maize in most other parts of the world.

CULTIVAR DEVELOPMENT IN UNITED STATES

Because corn is native to the Americas and was growing in present-day continental United States by the time of Columbus's first voyage, early cultivar development is not easy to document. We know that pilgrims and emigrants found corn culture already established when they arrived. We suspect that immigrants arriving in the 1600s expected to find corn, because corn had been introduced into most of the Old World before 1600, and it may have been widely known that it came from the New World.

When explorers landed to survey and conquer a new world, they found "pegatowr" growing and sustaining life in present-day Virginia and North Carolina. In the New England states, "ewachimneash" and "ohnasta" were grown by the Iroquois Amerindians and other tribes. As already noted, the Arawak Amerindians encountered by Columbus's men on the island of Cuba on November 5, 1492, called this grain maiz or mahiz.

It was the Amerindians that selected and developed *Z. mays,* and they did not

---

Popcorn is the extreme case of flint endosperm; flint corn kernels are composed mostly of flint endosperm.

Dent corn kernels are composed of a soft central core of endosperm surrounded by hard endosperm. Upon drying, the soft core endosperm collapses or contracts to a greater degree than the hard endosperm, resulting in a characteristic indentation at the top of the kernel.

Flour corn endosperm is soft and easily ground. Kernels may have a slight dent at the crown.

Sweet corn endosperm contains more sugar than regular corn endosperm resulting in uneven drying and therefore kernels have a wrinkled, but glassy appearance.

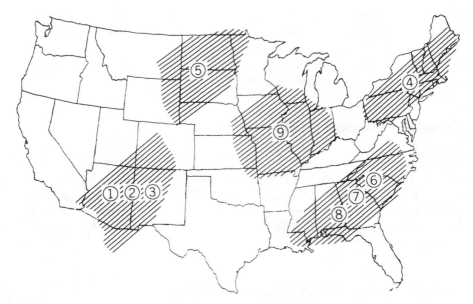

Figure 1.5 *Races of corn and approximate dates of establishment where known in present day United States: (1) Pima-Papago—1500 B.C.; (2) Southwestern 12-row; (3) Southwestern Semidents; (4) Northern Flints and Flours—before 1400 A.D.; (5) Great Plains Flint and Flours—1400 A.D.; (6) Southern Dents—1500 A.D.; (7) Southeastern Flints and Flours; (8) Derived Southern Dents; (9) Corn Belt Dents—1880. (Drawing by Mike Hodnett)*

keep records of such accomplishments. However, archaeology has yielded insights into the spread of corn culture into the continental United States and, at the same time, identified the kind of corn grown in different regions (Figure 1.5). Movement to and adoption of corn culture in the Eastern United States postdated that in the Southwest. This is intuitively correct because corn evolved in Mexico. Scientists still do not know the time of introduction into some parts of the United States, but it seems evident that corn was grown in the Southwest, Great Plains, Northeast and Atlantic coast before colonial times.

Corn had been diversified into a number of racial complexes by 1600 A.D. (Table 1.2). These were selected by the Amerindians for food and ceremonial purposes. The non-Amerindian conquerors simply came in and took over the human-directed evolution process. Although the Amerindians' diet relied heavily on corn, non-Amerindians probably ate corn as a last resort and looked on it more as a feed grain for animals. Nevertheless, the races of corn in the United States by 1600, and those since developed, can be classified into nine races.

The Pima-Papago race is apparently the same or very similar to archaeological remains of corn grown by the Hohokam Amerindians. Apparently, the Hohokam were a village-dwelling people with a well developed irrigation system in the Gila and Salt River Valleys of the U.S. Southwest. Archaeological records indicate that their culture flourished from about 300 B.C. and then began to decline (reasons unknown) about 1200 A.D. The Spanish found the area inhabited by the Pima-

Table 1.2   Description of U.S. Races of *Zea mays*.

Race	Origin	Ear	Endosperm
Pima-Papago[1]	Chapalote from Mexico and perhaps introgression of 2 other races.	5″ × 1″; strongly compressed at base and gently tapering to tip; 10–16 rows of grain with 10 normal.	Flint or flour, with flour common.
Southwestern[1,2] Semident	Tuxpeno introgressed into flint and flour types preceding it.	Similar to Southwestern 12-Row; 14–16 rows of kernels; slight dent.	Mostly flour.
Southwestern 12-row	Origin unknown; similar phenology to Northern Flints, probably common ancestor.	Enlarged at base and attached to large indurated shank; kernels are wide; 12–14 straight rows.	Flint and flour.
Great Plains Flints and Flours	Mixture of Northern Flints and Southwestern types.	Crescent-shaped kernels; 8–10 rows with low frequency of 12–14 rows.	Flint and flour.
Northern Flints	Unknown; suggested from Mexican race Harinoso de Ocho; the U.S. Southwest or Guatemala.	8 to 10 rows of crescent-shaped kernels; enlarged at base; shanks long and thick.	Primarily flint but some flour types.
Southern Dents	Mexico; perhaps from Mexican race Tuxpeno.	8 to 26 rows of well dented kernels.	Floury core to cap, surrounded by flinty.
Southeastern:			
Caribbean Flint	West Indies islands; eastern South America.	12–14 rows; compressed base; gently tapered.	Flinty.
Cherokee Flour	Harinoso de Ocho; northwestern Mexico.	7–10 inches long with 8–10 rows of wide, crescent-shaped kernels.	Flour.
Derived Southern Dents	Hybridization of Southern Dents, Cornbelt Dents and Southeastern Flints.	Somewhat less dent than Southern Dents.	Floury core to cap surrounded by flinty.
Cornbelt Dents	Southern Dent and Northern Flint hybridization.	Slightly tapered or cylindrical; 14–22 rows.	Floury core to cap surrounded by flinty.

1. Southwest races may have genes from corn types existing in the area before 1500 B.C., plus types from the Mexican Plateau and eastern North America introduced about 1200–1300 A.D.
2. May not be a distinct race, but a mixture of at least two races.

Papago people. The Pima-Papago corn race also is similar to the remains of corn grown by the "Basketmaker" group of Amerindians that settled around the four-corners area of Utah, Colorado, Arizona and New Mexico.

Although the oldest remains of corn found in southern New Mexico is dated at about 1200 B.C., evidence suggests an established, distinct race by the time of the Spanish Conquest. The earliest corn into the area was most likely the teosinte-like Mexican corn race, Chapalote, which apparently, along with introgression from at least two other races, formed the basis of the Pima-Papago race. Some suggest that the race evolved after 300 A.D., while other scientists believe that it evolved much earlier and had spread to Oklahoma, Arkansas, Illinois and Georgia by 100 A.D. This race is represented today by the Kokoma cultivar of the Hopi Amerindian tribe.

Southwestern Semident is a complex evolved from the introgression of Mexican race, Tuxpeno, a dent type, into the flint and flour types already existing in the area. Most agree that the dent parent(s) came from Mexico, but it has been suggested that the genes for the kernel indentation came from prehistoric dent types adapted to Utah and Colorado. This race was found predominately in the four-corners area of Arizona, New Mexico, Colorado and Utah. Some authorities do not recognize the Southwestern Semident as a distinct race.

Plants of the Southwestern 12-row race are similar morphologically to the Northern Flints, but there appears to be no evidence to support the thesis that this race evolved from the introduction of eastern types. However, there is sufficient evidence to suggest a common origin. Southwestern 12-row is characterized by short plants, highly tillered, having ears with an enlarged base with wide, slightly crescent-shaped kernels arranged in 12 to 14 straight rows.

The Great Plains Flints and Flours evolved from mixing Northern Flints (the dominant race grown east of the Mississippi Valley and north of Virginia by 1600 A.D.) and biotypes from the Southwest. Data from archaeological digs in South Dakota indicate that the race was established there by the mid-1400s. This race is obviously composed of both flint and flour endosperm types, with ears having eight to ten rows of kernels similar to the Northern Flints. However, ears with 12 to 14 rows of small, square kernels persist. Representative cultivars are Pawnee White, Omaha Flour, Ponca Red Flour, Winnebago White, Winnebago Blue, Dakota Square, Arikara Flints, and Rhee Flint.

Northern Flints were grown widely in the northeast quadrant of the continental United States and reaching as far south as Virginia in pre-Columbian times. Authorities speculate that the Northern Flints originated from Harinoso de Ocho, because both races have eight-row ears. Others suggest a link between the Northern Flints and the San Marceno and Serrano races from Guatemala, all three races having similar ear morphology. The Northern Flints are characterized by relatively short, highly tillered stalks frequently having two ears per main culm. Ears tend to have long, thick shanks, enlarged butt, and eight to ten rows of large, crescent-shaped kernels. Some flour endosperm types persist. Representative cultivars are Longfellow, Smut Nose, Canada Flint, Thompson Flint, and New York Council Flour Corn.

The Southern Dent racial complex was grown during colonial times in the southeastern states as far north as Virginia and west to central Texas. Little is known about this race in pre-Colonial times, nor is it known by what route it arrived in the Southeast. Many cultivars of Southern Dent are similar to races still grown in Mexico, such as Tuxpeno, Pepitilla, Tabloncillo, and Olotillo. A few of the more common old cultivars of the Southern Dent race are Gourdseed, Shoepeg, Tuxpan, Hickory King, Jellicorse, and Mexican June.

The Southeastern Flints and Flours are subdivided into two subraces, Caribbean Flints and Cherokee Flour. As the name implies, the Caribbean Flints grown in the Southeast United States were nearly identical to flint corn grown in the West Indies Islands and similar to the Cuban Flint race and the Cateto race of Argentina, Uruguay, and Brazil. The Caribbean Flints have 12 to 14 rows of kernels and ears that taper gently toward the tip from a slightly compressed butt. Old cultivars (perhaps better described as biotypes) include Creole Flint, Cuban Flint, Spanish, and Jarvis Golden Prolific.

The Cherokee Flour subrace distribution is currently limited to the eastern bank of the Cherokee River at Qualla, North Carolina. Cherokee Flour plants are tall, producing none to few tillers, and late-maturing with high ears. Ears are seven to ten inches long with eight to ten rows of kernels. The subrace is still used by the Cherokee Amerindians for corn meal and flour. There is apparently no evidence that this subrace was ever grown by non-Amerindian people.

Derived Southern Dents appear to have resulted from the hybridization of Southern Dents, Southeastern Flints, and Corn Belt Dents. This racial complex is characterized by less indentation of the kernel cap and increased yield potential relative to Southern Dents. Some of the more well-known cultivars were Caraway's Prolific, Giant Yellow Dent, Southern Snowflake, Horsetooth, Whatley's Prolific, and Latham's Double.

The Corn Belt Dents were developed by hybridization of Southern Dents and Northern Flints. This mixing occurred during the late 1800s and gave rise to the most productive and widely grown race before the advent of hybrid corn. Corn Belt Dents are easily distinguished from other races: strong arching leaves; heavy, many-branched male inflorescence or tassel; slightly tapered ears with 14 to 22 rows of dented kernels; mostly red cobs; and predominantly yellow endosperm although some white endosperm cultivars have been developed.

It is apparent from the foregoing discussion that a great diversity of corn genotypes existed in the United States before occupation by non-Amerindians. This diversity was and is recognizable by plant, ear morphology, kernel morphology, and endosperm type. However, although a thorough discussion is beyond the scope of this text, scientists are now able to use cytological and molecular techniques to refine their classifications. Although simple by today's technology, cytological examination of chromosomes has helped to identify and distinguish races, especially the evaluation of chromosome knobs. For example, the Southwestern 12-row racial complex is similar morphologically to the Northern Flints, but the Southwestern 12-row has many chromosome knobs of a distinct pattern while the Northern Flint types have fewer than five knobs.

DEVELOPMENT OF MODERN CULTIVARS AND HYBRIDS

The Amerindians of present-day continental United States apparently were so concerned with "varietal" purity that they gave little or no attention to hybridization (natural or artificial) of two cultivars followed by selection of better cultivars. These Amerindians had all of the germplasm subsequently used by European settlers to produce improved cultivars. That the Amerindians knew nothing about the sex of plants may be accurate, but they did know the effects of planting two types too closely, that being the loss of phenotype purity. It appears then that the Amerindian simply did not select directly for higher yielding progeny, but rather, as his counterpart in the Old World, accepted at face value what mother nature provided.

But as the fear of being ostracized by the community, excommunicated, or worse, from the church, declined in the 18th century, Europeans began to speak out about their ideas and test their theories. Cotton Mather, the famous Puritan who approved of the persecution of "witches," recorded the effects of wind direction on the distribution of colored grains on plants that should have produced only yellow grains. Eight years later in 1724, Paul Dudley would refer to this phenomena as a "wonderful copulation." In 1727, James Logan conducted perhaps the only scientific experiment with corn in the 18th century, proving that both the tassel and silks were necessary for kernel development.

Practical agriculture took root in the 19th century and farmer-seedsmen began to intervene in natural evolution. John Lorain of Philipsburg, Pennsylvania, published *Nature and Reason Harmonized in the Practice of Husbandry* in which he notes, ". . . results seem to determine that, if nature be judiciously directed by art, such mixtures as are best suited for the purpose of farmers, in every climate of this county where corn is grown, may be introduced." By mixture, Lorain meant the selection of improved cultivars from planting Gourdseed, a Southern Dent, and a Northern Flint in proximity and making selections in subsequent generations of the resulting progeny. One must keep in mind the high degree of outcrossing that occurs in corn. It is such that interplanting of two or more types will result in hybridizations occurring every year, creating a plethora of new genotypes.

Joseph Cooper, a New Jersey farmer, published an account of his corn breeding endeavor in an 1808 Philadelphia Agricultural Society publication. Cooper wrote of his experience from interplanting Guinea corn (not to be confused with grain sorghum by the same name) and his regular flint cultivar, followed by selection of improved types. Guinea corn was thought to have originated/evolved in equatorial West Africa and introduced through the slave trade. Today we know that it was a reintroduction to the New World, as it had to have originated in the Americas. Apparently, the only characteristic of this Guinea corn that was of interest to Cooper was its multiple ears per stalk.

In 1843, Mr. Hendricksen of Middletown, Ohio, reported his seed development program in the American Agriculturist, which was identical to that of John Lorain. Hendricksen used five parents in his mixture, including Gourdseed, a Northern Flint, and Virginia Yellow. During harvest, he then kept ears from desirable plants for the next season's planting seed.

Probably the most important cultivar developed before hybrids was Reid's Yellow Dent—perhaps the ultimate serendipitous event. In 1847, Robert Reid left Ohio and moved to Illinois, bringing Gordon Hopkins Gourdseed corn that had been grown for generations by the Gordon Hopkins family of Virginia. Apparently because of poor germination, poor seedling vigor, or poor performance of Gordon Hopkins Gourdseed the previous year, Reid replanted some missing hills with Little Yellow Flint, a flint that had been grown for centuries by Amerindians of the Northeast. Over the course of several years, Robert and his son James continued to select the best ears from the best plants from the resulting natural hybridizations. By 1893, the Reids had established the Reid's Yellow Dent cultivar, perhaps the greatest single stride in productivity until the development of hybrids. This cultivar made such an impression at the 1893 Chicago Fair that it also became known as World's Fair Corn; others dubbed it Corn Belt Dent (although it and its successors are called this generically as a race). Reid's Yellow Dent became the standard by which all other cultivars grown in the North-Central United States (the Corn Belt) were compared.

Reid's Yellow Dent was the basis for Funk's Yellow Dent, a source of many of today's inbred lines. But probably the only improved strain of Reid's Yellow Dent was developed by George Krug of Woolford County, Illinois. Krug combined a Nebraska strain of Reid's Yellow Dent with Iowa Gold Mine and selected the highest-yielding, non-hybrid yellow corn ever grown in the Corn Belt. At a time when everyone else was selecting for the prettiest corn ears, those that were long, gently tapering with straight rows of kernels so that they could win "best of fair," Krug was only interested in yield. One should understand that to win "best of fair" was the surest way to ensure brisk seed sales. Krug never entered his corn in county or state fairs. His corn was variable in ear and kernel, and in a three-year yield contest in the early 1920s, Krug's corn outyielded the best show line of Reid's Yellow Dent by ten bushels per acre. Once introduced to competition, or seedsmen, Krug began selecting for uniformity and a concomitant loss in yield potential resulted. However, some of Krug's high-yield factors were saved for posterity by Lester Pfister who began developing inbred lines of Krug for hybrid production in 1922. Krug himself became known as the Corn King, but he never wanted and did not gain financially from his success.

The other open-pollinated Corn Belt Dent corn king was a Mennonite elder from Lancaster County, Pennsylvania, named Isaac Hershey. Perhaps it was his "no-frills" Mennonite background that lead Hershey to select for yield superiority over uniformity and looks, just as George Krug had done in Illinois. Hershey's final product was the culmination of growing a mixture of a late-maturing, large-eared corn with an early flint type, followed by adding six other cultivars to the mixture over the course of several years. In 1910, he stopped mixing and began selecting for early maturity. The result was Lancaster Sure Crop. Hershey selected for disease resistance, ease of breaking the ear away from the shank, and for long ears, both smooth and rough. Lancaster Sure Crop became famous throughout the Northeastern states. He, too, then began to try and please seed dealers by selecting for uniformity (seed dealers and developers wanted to win beauty contests as a means of advertisement) and yield potential suffered.

Lancaster Sure Crop became a major contributor to today's hybrids through the efforts of F. D. Richey at the Iowa State Agricultural Experiment Station. Today, Cornbelt hybrids usually have one inbred parent that is derived from Lancaster Sure Crop as the flint component and one inbred parent that is a descendent of Reid's Yellow Dent as the dent component. Is it a lack of wide experimentation or serendipitous that today's high-yielding U.S. hybrids are derived from only two racial complexes?

The first controlled hybridizations directed at harnessing hybrid vigor, or heterosis, were done by W. J. Beal in 1877 at Michigan State University, then known as Michigan Agricultural College. Beal would later lay the foundation of hybrid development by showing increased yields as a result of forced cross pollination between two dissimilar open pollinated cultivars.

George Harrison Shull contributed the next insight into hybrid corn. After only two generations of inbreeding, or self fertilization, Shull correctly identified the cause of what was to become known as "inbreeding depression" and he identified the results of the hybridization of two unlike inbreds—that is, heterosis. Shull reported in 1908 that an ordinary field of corn consisted of a series of complex hybrids, that the decline in plant vigor and yield from self fertilization was due to a gradual increase in homozygosity, and proposed, although he had no interest in the commercialization of science, that corn breeders should strive to identify and maintain the best hybrid combinations rather than trying to produce the best "pure line."

E. M. East was a contemporary of Shull and their work followed parallel lines of thought and results. He too was more interested in science than commercialism, but he recognized the practical implications of his work. East, as did Shull, interacted with seedsmen to influence their acceptance of his ideas in hybrid corn. Some of the inbreds developed by East were used by breeders in developing commercial hybrids.

The commercial visions of two men serve to demonstrate the "can do" of the American agriculturalist. Lester Pfister (noted earlier) had to quit school when his father died, and took a job testing the yield potential of different corn strains. Pfister began to develop inbreds of Krug's corn in 1922. Neighbors ridiculed and laughed at this young man putting paper bags over corn tassels, reducing productivity from 388 ears (and therefore lines of Krug) to only 4. The next five years were devoted to crossing experiments, concomitant with going deeply into debt, with foreclosure apparent in 1933. That year Pfister finally found a superior double cross hybrid that outyielded Krug so substantially that within five years Pfister was grossing one million dollars a year. He was named "the outstanding corn breeder of the world" by *Life* and *Reader's Digest*.

Similarly, Henry A. Wallace, later to be Secretary of Agriculture, Vice President of the United States under F. D. Roosevelt, and Secretary of Commerce under Harry Truman, began the Hi-Bred Corn Company with a number of friends in 1926. The company struggled through 1932 but gained solid financial footing thereafter. Hi-Bred Corn Company later became Pioneer Hi-Bred International, Inc.

Henry Wallace also developed the first commercial hybrid corn produced and sold in Iowa. Wallace contracted with George Kurtzweil to produce Copper Cross hybrid in 1923, and gave Kurtzweil enough inbred material to plant a one-acre plot of East Leaming as the female parent and a meager supply of Bloody Butcher as the pollen parent. About 15 bushels of F_1 hybrid seed were produced and sold the next year for $1 per pound. Thus, Copper Cross became the first commercial hybrid corn developed and sold in the U.S. Cornbelt.

An interesting footnote of history is that all of the detasseling of the first hybrid seed corn in Iowa was done entirely by a woman, George's sister, Ruth. Lester Pfister later developed a detasseling machine, and male sterility would later make detasseling unnecessary, but all hybrid corn production during the early days of hybrids had to be hand-detasseled, a job that employed many teenagers after school and during their summer months.

The first hybrids of corn that were commercial successes were double cross hybrids. That is, they were the result of crossing two hybrids to produce seed for sale—i.e., (inbred A × inbred B) × (inbred C × inbred D). It was known early on that the highest degree of hybrid vigor, or heterosis, and therefore higher yields came from single crosses—i.e., inbred A × inbred B. However, the detrimental effects of inbreeding depression in the first inbreds were such that too few seeds were produced on the female inbred to be profitable. Far greater numbers of seed were produced when two single cross hybrids were crossed, thus providing for greater sales and profits for the seedsmen plus increased yields for farmers.

In the late 1950s, U.S. production of single cross hybrids began to increase with a concomitant decrease in the production and use of double cross hybrids. The single most important factor in this shift was the development of improved inbreds that would produce greater F_1 seed yields when hybridized. Today, almost all commercial corn hybrids in the United States are single cross hybrids. One percent of the U.S. Cornbelt was planted to hybrid seed in 1933; 78 percent by 1943; and almost 100 percent by 1965.

HISTORICAL EVENTS

From a historical perspective, just the settling of the New World by Europeans made corn a major world commodity. Although Europeans consumed grains such as wheat, barley, and rye, they also consumed grain through their domesticated livestock, fish, and poultry (chicken and geese). The Amerindian diet consisted of plant products, supplemented by hunting and fishing. Therefore, corn was a food to the Amerindian, but the conquering Europeans were not interested in this strange new grain as a food; rather, they consumed corn through their domesticated animals when possible, and as the colonists became more established and affluent, they demanded more meat, thereby increasing demand for corn.

The success of Reid's Yellow Dent may be partially attributable to its development and spread coinciding with white settlers taking major corn-growing territory from Amerindians in the northeast at the close of the Black Hawk War in 1832. The opening of the Erie Canal in 1825 also created new lands for settlement on which to produce greater numbers of animals for an ever-increasing non-Amerindian population. Railroads made transport of livestock, especially hogs, to slaughter and packing centers such as Chicago and Cincinnati quick and easy. Transport of slaughtered animal products to major metropolitan areas for consumption was made easy by railroads and later by modern highways and trucks. Greater demand for animal products by an enlarging and evermore affluent society mandated more corn as feed, and in the 20th century more corn for new products brought about by modern chemistry.

The development and commercialization of hybrid corn was a significant event in the development of the U.S., and world, corn industry. One must realize that hybrid corn production began during the worst economic depression that the United States and the modern world has ever experienced. That in 1933, only 1 percent of the U.S. Cornbelt was planted to hybrids and, within ten years, 78 percent of the acreage was planted to hybrids is remarkable. Prices received were only about 60 percent of late 1920s prices, and hybrid seeds were more expensive. But higher yields, uniform ear height, and uniform maturity attracted forward-thinking farmers. U.S. yields averaged 22.6 bushels in 1933, rose to 32.2 bushels by 1943, a 30 percent increase in only ten years, and averaged 118.5 bushels by 1990, more than a four-fold increase. It also should be noted at this point that the performance of hybrids during the dust bowl years of 1934 and 1936 gained hybrids a reputation for being more drought-tolerant then open pollinated cultivars, further driving the acceptance of hybrid corn by the American farmer.

DEVELOPMENT OF U.S. INDUSTRY

Except for possibly sweet corn and popcorn, the early Europeans settling the present day continental United States used corn as a feed for their domesticated animals: swine, chicken, and cattle. Less affluent people, especially those living in the southern states, used flour and dent corn as a source of meal and flour for breads. While it is beyond the intent of this text, there seems to be ample evidence that a large proportion of tenant farm families in the southeast United States subsisted on fatback (pork), cornmeal, and molasses, resulting in widespread pellagra, a disease caused by poor amino acid balance in human diets. However, corn meal continues to be used in a variety of breads and corn flour is consumed as tortillas, particularly in the southwest U.S., tostadas, tortilla chips, corn chips, etc. Corn also is consumed today as hominy, pearled corn kernels cooked in lye solution, and grits. The per capita consumption of dry, processed corn decreased from 100 pounds in 1899 to only 7 pounds by 1970.

Popcorn is a popular snack food in the United States, the average citizen consuming 56 quarts of popped popcorn. During World War II, the Americans remaining at home drastically reduced their consumption of candies so that these confectioneries could be available for their armed forces. These Americans turned to popcorn as their snack food. Also, in the 1930s, 1940s, and 1950s, no trip to the movies would have been complete without popcorn. However, commercialized popcorn had its start long before the second world war, when Charles Cretors of Chicago developed a steam-driven machine in 1885 that mass-produced popped popcorn. Vendors sold popcorn from the streets of Chicago in these hand-propelled contraptions, later to be horse drawn and then motorized with the advent of the automobile. Within a few years, Frederick W. Ruckheim had added molasses and called his "invention" Cracker Jack. (Actually the Amerindian had invented sweetened popcorn hundreds of years before Mr. Ruckheim.) He and his brother added peanuts to their sweetened popcorn and Cracker Jack became an instant hit at the 1893 World's Columbian Exposition in Chicago. Another midwesterner would propel popcorn into the "gourmet" world, peddling his improved popcorn with higher popping volume from the trunk of his car during the 1960s until his company was bought by Hunt-Wesson in 1974. The buy-out was followed by a $6,000,000 ad campaign that made Orville Redenbacher, a former Agricultural County Agent in Indiana, a household name.

Corn is the most popular feed grain produced in the United States and is fed primarily to hogs and cattle, with some going to chickens and sheep. The average production for 1988–1990 was $6,796 \times 10^6$ bushels ($380,562 \times 10^6$ pounds at 56 pounds/bushels), while average U.S. production of grain sorghum and oats for the same time period was 588×10^6 bushels ($33,516 \times 10^6$ pounds at 57 pounds/bushels) and 316×10^6 bushels of oats ($10,115 \times 10^6$ pounds at 32 pounds/bushels), respectively. About 45 percent of the corn produced in the United States is fed to livestock on the farm where it is produced, leaving about 55 percent to be sold into feed, food, and manufacturing markets. The mixed feed market is the largest manufacturing market for corn, but an increasing amount is processed through dry and wet milling industries for food and industrial products.

Dry milling of corn until more modern times was essentially the same as milling wheat or other grains—i.e., crushing between two grindstones. This resulted in a product with poor shelf life because the oil in the germ would soon become rancid. Today, dry milling of corn consists of (1) cleaning, (2) conditioning, (3) degerming, (4) drying, (5) cooling, (6) grading, (7) grinding, and (8) sifting and classifying. Dry milling of dent corn separates the kernel into germ (embryo), bran (pericarp or seed coat), and endosperm (Table 1.3). Unlike wheat that is fractionated by use of a series of rollers, corn is broken by attrition mills and later by entoleters. Attrition, meaning to rub away or wear down, mills vary in design but essentially consist of an inner cone that rotates inside a larger housing. The inner rotor and the inside of the housing have knob-like projections that strike and rub the tempered (i.e., having moisture added) corn kernels, thereby breaking the kernels into germ, seedcoat, and endosperm.

TABLE 1.3 Composition as Percentages of Whole and Fractionated Corn Kernels.

Fraction	Kernel	Starch	Sugar	Protein	Oil	Total excluding kernel
Kernel	100.0	71.5	2.0	10.3	4.8	88.6
Endosperm[1]	81.9	86.4	0.6	9.4	0.8	97.2
Germ	11.9	8.2	10.8	18.8	34.5	72.3
Bran	5.3	7.3	0.3	3.7	1.0	12.3

Source: Adapted from Earle et al., 1946.

1. Composed of starchy and aleurone fractions.

TABLE 1.4 Products Made from Corn.

Milling Process	Product	Uses		
		Food	Industrial	Feed
Dry	Endosperm	Breakfast cereals Mixes (cookie, etc.) Grits Snack foods Breads Baby foods	Gypsum Plywood Particle board, etc. Pharmaceuticals Charcoal binders Paper Corrugating Oil well drilling fluid	Prepared pet feed
Wet	Starch	Baking powder Confections Desserts Gravies Sauces	Adhesives Ceramics Inks Metal castings Paints Textile sizing	Corn gluten meal
Wet	Sugar	Baby foods Baked foods Canned fruits Condensed milk Frozen fruits	Chemicals Pharmaceuticals	Mixed rations
Wet	Syrup	Bakery products Confections Sauces Frozen fruits Frozen desserts Ice cream Jellies Table syrup Soft drinks	Leather tanning Paper Pharmaceuticals Textiles Tobacco curing	
Wet	Oil	Bakery products Condiments Confections Margarine Salad dressings	Leather tanning Paints Soaps Synthetic rubber Varnish	Corn oil meal

Source: Adapted from Chicago Board of Trade, 1977; Watson, 1988; Hobbs et al., 1986.

The bran is used in feed stuffs while oil is extracted from the germ. Larger endosperm pieces are further processed to produce pearled (polished to remove rough edges) hominy to be used as hominy or corn flakes. Smaller pieces of endosperm are sized and used in the brewing industry, or consumed as grits, corn meal, or corn flour (Table 1.4).

Corn is the number-one feed grain for hogs, cattle, poultry, and for sheep if priced competitively and if available. Corn is adequate for horses and mules, but tradition, not science, dictates oat as the number-one grain for these animals. Dairy cattle are fed ground formulations of corn and other grains supplemented with protein concentrates and other additives. Corn meets most needs of livestock and

TABLE 1.5 Feed Value of Several Feed Stuffs Compared with Corn for Swine.

Ingredient[1]	Feeding Value[2]
Corn, yellow	100
Corn, high lysine	100–105
Alfalfa hay, field cured	30–40
Barley	85–100
Grain sorghum	80–100
Oats	80–100
Rye	90
Triticale	90–95
Wheat, hard	100–105

Source: Adapted from Reese et al., 1992; Holden et al., 1984.

1. Air dried.
2. As a percentage of a complete diet.

poultry because of its high-starch and low-fiber content, making it one of the most concentrated natural sources of energy. Corn is the feed against which all others are measured (Table 1.5). In 1980/81, 37 percent of the corn grain fed to livestock in the United States was consumed by hogs, with 23 percent fed to beef cattle and 17 percent consumed by dairy cattle. Consumption by chickens and turkeys accounted for 22 percent. As it is essentially an energy feed, it is often supplemented with protein, minerals, and vitamins in mixtures or as additional feed stuff.

Other products of the dry milling industry are breakfast cereals, snack foods, breads, and baby foods. Dry milled corn also is used in the manufacture of such items as plywood, paper, and pharmaceuticals.

The wet milling industry dates only to the early 1800s and only to 1831 in the United States. The Napoleonic Wars of the early 1800s resulted in a blockade of Europe by the British, cutting off supplies of sugar cane from the British West Indies. Napoleon offered a reward of 100,000 francs for anyone finding an alternative sweetener. J. L. Proust claimed the price by devising a method of extracting a sweet substance from unfermented grape juice, later to be identified as dextrose.

In 1811, a Russian chemist, G. S. C. Kirchoff, searching for improved materials for the manufacture of porcelain, accidentally overcooked a mixture of potato starch and sulfuric acid. The result was a sweet-tasting syrup, a fortuitous event earning him a lifetime pension from the Russian emperor. Chemists later determined that the sweet taste was caused by the hydrolytic conversion of starch into its basic unit, dextrose or crystalline glucose, the same as with Proust's unfermented grape juice sweetener.

The starch hydrolysis industry began in the United States in 1842 with the opening of a mill in Sackets Harbor, New York, to convert potato starch into syrup. The first corn syrup in the United States was produced in Buffalo, New York, in 1866. A major breakthrough occurred in 1957 with the patenting of an enzymatic process to covert dextrose to fructose, a 6-carbon sugar that is sweeter tasting than sucrose, or table sugar, a 12-carbon compound (Table 1.6). This provided food

TABLE 1.6 Composition and Relative Sweetness of Nutritive Sweeteners.

Sweetener	Chemical Formula	Percent Fructose	Percent Dextrose	Relative Sweetness
Sucrose	$C_{12}H_{22}O_{11}$	0	0	1.0
Dextrose	$C_6H_{12}O_6$	0	100	0.7
Fructose	$C_6H_{12}O_6$	100	0	1.5–1.7
42% HFCS[1]	——	42	52	1
55% HFCS	——	55	40	1+
90% HFCS	——	90	<10	1.2–1.6

Source: Adapted from Hobbs et al., 1986. With permission from Dr. Hobbs, Cereal Foods World 31:852–865. 1986, American Assoc.of Cereal Chemists.

1. HFCS: High fructose corn syrup.

manufactures with a means whereby they could reduce calories while maintaining sweetness. The first large-scale production of 90 percent high-fructose corn syrup (HFCS) began in 1978 and by 1984, 100 percent of the sweetener in Coca Cola and Pepsi Cola was HFCS. High fructose corn syrup is used also in bakery products, canned and processed foods, dairy products and, of course, confectioneries.

The wet milling process begins with steeping clean corn in water for 30 to 35 hours at 117 to 127°F to soften the kernel. Steeped corn then is cracked by use of an attrition mill to separate the germ (embryo) and release about one-half of the endosperm (prime starch fraction). The germ is about 40 percent oil (dry basis) and is separated from the prime starch and other fractions by flotation. The germ will then be processed for oil recovery by means of heavy mechanical expellers with the residue used as germ meal or mixed with gluten and fiber to produce corn gluten feed (both high-protein feeds). The expelled oil will be further refined to produce edible oils.

Once the prime starch is isolated, the remaining bound starch is saccharified or used for fermentation to ethyl alcohol. The prime starch will be further refined to produce fine-textured starch and gluten. Purified starch comprises about 68 percent of corn's dry weight. Approximately 25 percent of the purified starch is used as starch in a number of products, while 75 percent is sold as hydrolyzed products— i.e., syrups and dextrose as noted previously (Table 1.4).

U.S. AND WORLD PRODUCTION

The European population in the United States expanded westward into the heartland with the ending of Amerindian hostilities in the first half of the 19th century and with the opening of the Erie Canal in 1825. Corn production concomitantly increased to feed increasing numbers of cattle and hogs for meat demanded by the increasing affluent and expanding population (Table 1.7). Corn production expanded in southern states to feed increasing numbers of mules and horses needed for cotton and tobacco production. Acreage devoted to corn increased steadily through the 1920s until the Great Depression of the 1930s began to decrease the affluence of

TABLE 1.7 Harvested Acreage, Average Yield, Prices Received and Disposition of Corn Produced for Grain in the United States, 1866–1990.

Year	Harvested Acres	Average Yield	Total Production	Prices Received	Disposition[5]	
					Domestic	Exported
	(×1000)	(bu.[1]/ac.)	(×1,000,000 bu.)	(cents/bu.)	(×1,000,000 bu.)	
1866	30,017	24.3	731[2]	65.7	715	16
1870	38,388	29.3	1,125	52.1	1,107	18
1875	52,446	27.7	1,450	41.9	1,387	63
1880	62,545	27.3	1,708	39.0	1,624	83
1885	71,854	28.6	2,058	32.2	1,996	62
1890	74,785	22.1	1,650	49.6	1,624	26
1895	90,479	28.0	2,535	25.2	2,417	118
1900	94,852	28.1	2,662	35.0	2,509	153
1905	95,746	30.9	2,954	40.6	2,839	115
1910	102,267	27.9	2,853	51.6	2,788	65
1915	100,623	28.1	2,829	68.0	2,777	52
1920	101,359	30.3	2,695	61.8[3]	2,580	115
1925	101,331	27.6	2,382	75.0	2,357	25
1930	101,465	20.5	1,757	59.6	1,754	3
1935	95,804	24.0	2,015	65.5[4]	2,014	1
1940	76,443	28.9	2,207	61.8	2,192	15
1945	77,928	33.1	2,577	123.0	2,555	22
1950	72,398	38.2	2,764	152.0	2,654	110[6]
1955	68,462	42.0	2,873	135.0	2,764	109
1960	71,422	54.7	3,907	100.0	3,630	277
1965	55,332	73.8	4,084	116.0	3,411	673
1970	57,358	72.4	4,152	133.0	3,635	517
1975	67,625	86.4	5,841	254.0	4,130	1,711
1980	72,961	91.0	6,639	312.0	4,248	2,391
1985	75,209	118.0	8,875	223.0	7,648	1,227
1990	66,952	118.5	7,934	228.0	6,209	1,725

Source: Adapted from USDA Agricultural Statistics 1936, and following years.

1. One bu. of U.S. No. 1 grade corn = 56 lbs.
2. Total production calculated in grain equivalents on entire harvested acreage until 1919, and reported as actual grain thereafter.
3. The average price received was $1.415/bu. for the previous 4 years.
4. The average price received was $0.3185/bu. for 1931 and 1932 because of overproduction and depressed world markets during the Great Depression.
5. Domestic consumption calculated as total production minus domestic exports, i.e., it does not include imports or carryover from previous years.
6. Includes shipments under the Army Civilian Supply Program begun in 1946 and continuing into later years.

Americans, and urbanites began to decrease the amount of meat they consumed. The 1930s also saw the beginning of the agricultural machine age, decreasing the amount of corn needed for animal power.

Acres devoted to corn reached over 100,000,000 by 1910, but have declined since 1930. Yield per acre, however, remained static at about 30 bushels/acre until the acceptance of hybrid cultivars and commercial fertilizer, both events occurring in the 1930s and 1940s on a widespread scale. By 1990, the United States averaged

TABLE 1.8 Production Statistics for Corn for Grain by States, 1990.

State	Harvested Acres	Average Yield	Production
	(×1,000 ac.)	(bu.[1]/ac.)	(×1,000 bu.)
Alabama	240	58	13,920
Arizona	7	160	1,120
Arkansas	73	95	6,935
California	160	160	25,600
Colorado	830	155	128,650
Delaware	172	115	19,780
Florida	75	71	5,325
Georgia	550	68	37,400
Idaho	30	130	3,900
Illinois	10,400	127	1,320,800
Indiana	5,450	129	703,050
Iowa	12,400	126	1,562,400
Kansas	1,450	130	188,500
Kentucky	1,200	100	120,000
Louisiana	186	116	21,576
Maryland	450	118	53,100
Michigan	2,070	115	238,050
Minnesota	6,150	124	762,600
Mississippi	140	80	11,200
Missouri	1,960	105	205,800
Montana	9	95	855
Nebraska	7,300	128	934,400
New Jersey	75	118	8,850
New Mexico	55	145	7,975
New York	620	98	60,760
North Carolina	1,070	68	72,760
North Dakota	460	80	36,800
Ohio	3,450	121	417,450
Oklahoma	88	114	10,032
Oregon	18	150	2,700
Pennsylvania	970	113	109,610
South Carolina	320	48	15,360
South Dakota	3,000	78	234,000
Tennessee	510	86	43,860
Texas	1,450	90	130,500
Utah	19	140	2,660
Virginia	365	100	36,500
Washington	80	175	14,000
West Virginia	50	105	5,250
Wisconsin	3,000	118	354,000
Wyoming	50	120	6,000
U.S.	66,952	118.5	7,934,028

Source: Adapted from USDA Agricultural Statistics, 1991.

1. One bu. of U.S. No. one grade corn = 56 lbs.

over 118 bushels/acre, with a range of 48 bushels/acre in South Carolina to 175 bushels/acre in Washington (Table 1.8). Total production in 1990 was 7,934 × 10⁶ bushels (444,304 × 10⁶ pounds). Other production statistics are shown in Tables 1.7 and 1.8.

The United States far outproduces every other country today. The United States, in 1990, produced 7,934 × 10⁶ bushels (444,304 × 10⁶ pounds) while the second leading producer, the People's Republic of China, produced only 3,811 × 10⁶ bushels (213,416 × 10⁶ pounds). The United States produced almost 42 percent of the total world production of corn for the 1990/91 production year (Table 1.9). Total world production was 18,741 × 10⁶ bushels (1,049,496 × 10⁶ pounds) from 112 countries.

CORN FOR FORAGE

The entire corn plant may be harvested and used as feed directly as green chop or ensiled to be fed later. Whole plant corn, either green chop or silage, surpasses all other forage crops in dry matter yield and total digestible nutrients per unit land area. Corn should be harvested for silage when the grain is in the medium to hard dough stage of development (about 35 percent moisture), and the entire plant is between 30 and 50 percent dry matter (i.e., 50 to 70 percent moisture) to obtain maximum dry matter yields per unit land area. Silage is fed usually to dairy and beef cattle, with special handling precautions to prevent off flavor in milk. As in the case of grain, the north central states outproduce other regions of the United States (Table 1.10).

PLANT MORPHOLOGY

The corn plant is a tall-growing, monecious monocot with broad, horizontal leaves borne alternately along the length of the main culm. Normally, the main culm will have about 20 leaves. These leaves are composed of long, overlapping sheaths giving rise to wide, arching blades that are, in general, perpendicular to the culm. Some cultivars that have been developed for tolerance to excessive plant densities have blades that are more erect. U.S. hybrids usually have a single main culm, or stalk, but occasionally basal branches, called tillers, can be found, especially in less than optimum plant densities (Figure 1.2).

Corn is the only major cereal grass having unisexual flowers, with both male, tassel, and female, ear, on the same plant—that is, a monecious plant. The tassel is a many-branched panicle that terminates the main culm. The main axis of the panicle bears four rows of spikelets, while branches of the main axis have two rows of spikelets. Spikelets are arranged in pairs, one pedicellate and one sessile, and each is composed of two florets, or flowers. Each floret has three anthers, each estimated to produce 2,500 pollen grains, and an average tassel has 10,000 anthers, thus producing 25,000,000 pollen grains. The ear will have, say, 1,000 ovules,

TABLE 1.9 Production Statistics for Major Corn-Producing Countries in 1990/91.

Country/region	Harvested Acres (×1,000 ac.)	Average Yield (bu.[1]/ac.)	Total Production (×1,000 bu.)
Canada	2,559	110.1	281,700
Guatemala	1,941	25.8	50,145
Mexico	16,302	34.1	554,976
United States	66,925	118.5	7,932,378
N. & C. America	90,834	97.8	8,891,503
Argentina	4,817	62.1	299,136
Brazil	31,863	29.3	932,832
Colombia	2,030	24.7	50,145
Venezuela	1,186	33.5	39,754
S. America	43,467	33.1	1,439,356
France	3,952	94.6	373,920
Germany	566	108.0	61,087
Greece	400	142.6	57,072
Italy	1,897	121.7	230,807
Spain	1,112	99.1	110,208
Austria	489	130.3	63,763
Bulgaria	988	49.4	48,846
Hungary	2,673	63.6	169,917
Romania	6,079	39.2	238,679
Yugoslavia	5,506	48.1	264,657
Europe	25,043	67.9	1,699,525
Soviet Union	7,040	55.1	388,090
Egypt	2,075	87.1	180,977
Ethiopia	2,841	27.7	78,720
Kenya	4,384	19.8	86,592
Malawi	3,359	18.8	62,976
Nigeria	4,446	13.4	59,827
So. Africa	7,474	43.6	326,688
Tanzania	4,026	23.7	95,645
Zambia	1,729	28.5	49,200
Zimbabwe	2,719	22.9	62,425
Africa	49,783	24.9	1,234,054
People's Republic of China	52,863	72.0	3,810,835
India	14,706	24.2	357,113
Indonesia	7,040	29.0	204,672
N. Korea	1,112	102.6	114,144
Pakistan	2,087	22.3	46,642
Philippines	9,537	21.0	200,815
Thailand	3,335	44.8	149,568
Turkey	1,507	54.8	82,656
Asia	95,977	52.9	5,074,291
Oceania	173	82.4	14,248
World total	312,317	60.1	18,741,067

Source: Adapted from USDA Agricultural Statistics, 1992.

1. One bu. of U.S. No. one grade corn = 56 lbs.

TABLE 1.10   Production Statistics for Corn Silage for States Producing over 1,000,000 Tons in 1990.

State	Harvested Acres	Average Yield	Production
	(×1,000 ac.)	(t./ac.)	(×1,000 t.)
California	210	25.0	5,250
Colorado	117	22.5	2,633
Idaho	68	23.0	1,564
Illinois	130	14.0	1,820
Indiana	100	17.0	1,700
Iowa	300	15.5	4,650
Kansas	120	14.0	1,560
Kentucky	140	15.0	2,100
Maryland	95	14.0	1,330
Michigan	280	14.5	4,060
Minnesota	480	12.0	5,760
Missouri	90	13.0	1,170
Montana	55	19.0	1,045
Nebraska	325	13.5	4,388
New York	580	15.0	8,700
North Carolina	85	12.0	1,020
North Dakota	360	4.0	1,440
Ohio	180	16.0	2,880
Pennsylvania	390	16.0	6,240
South Dakota	375	6.2	2,325
Tennessee	100	16.0	1,600
Texas	85	13.0	1,105
Vermont	76	17.5	1,330
Virginia	160	14.0	2,240
Wisconsin	670	14.0	9,380
U.S.	6,124	14.2	86,844

Source: Adapted from USDA Agricultural Statistics, 1991.

which means that the corn plant produces 25,000 pollen grains for each potential kernel.

The ear terminates a lateral branch, and modern corn hybrids usually have only one but occasionally more than one lateral branch per stalk can be found. These lateral branches are composed of compressed nodes, often called the shank of the ear, that give rise to the shucks, or modified leaves, that enclose the ear. The ear, female inflorescence, is a modified panicle with the main axis thickened and lignified to produce the cob. This axis bears paired spikelets, each containing two florets, one functional and the other usually sterile. This feature then means that the ear will have an even number of rows of grain. Each floret contains an ovary with style and stigma. The styles, called silks, elongate until they exit the shuck, or husk, and pollination terminates their growth. Styles, those at the base of the ear, can be 12 inches long or longer. As the grain enlarges during development, the glumes, also called chaff, of each female floret are not showy and are often overlooked by the novice, as will be the cupule, a remnant structure from it's teosinte ancestor.

STAGES OF GROWTH

The level of knowledge and expertise necessary to properly manage row crop production has increased dramatically over the past several years with the advent of chemical pesticides and growth regulators. These products must be applied in precise amounts and at precise times during the life of the crop plant to be most efficient and effective. The effective producer also must understand the development of the plant in order to apply fertilizer and irrigation water in a timely fashion for most efficient crop use and maximum crop production. To accomplish this, producers and the agricultural support community must be able to communicate effectively relative to the stage of development of individual plants and fields. This process has been helped by the identification of stages of growth of all of the major row crops produced in the United States, including corn.

Stages of growth of the corn plant are keyed to the emergence of leaves before pollination and then depends upon the development of the kernels on the ear. A leaf is considered emerged for staging purposes whenever its collar is clearly visible above the whorl. The major disadvantage of this staging system is that the lower leaves can be torn loose and lost from the plant by the rapid enlargement of the stem once stem elongation begins. However, the lowest leaf remaining on the stalk can be identified by the length of the internode below the leaf's node of attachment. The internodes below the first four leaves never elongate. The internode penultimate to the fifth leaf node elongates about 0.5 inches; the internode penultimate to the sixth leaf node elongates about one inch; two inches for the seventh leaf; and about 3.5 inches for the internode penultimate to the eighth leaf.

After silks emerge from the ear, stages are identified by the stage of kernel development. Stage 6, the stage immediately following the silking stage, is characterized by a full size cob with kernels in the early stage of development, called blisters. Other stages, dough, dent and maturity are identified in Table 1.11. A field is considered to be at a particular stage of growth when at least 50 percent of the plants sampled are at that stage (see Chapter 7, Soybean, for a more detailed discussion of staging fields). Stages of growth of corn are referred to usually by the stage name and not the stage number.

PRODUCTION PRACTICES

Cultivar Choice

All cultivars of corn sold for commercial production in the United States are hybrids. Basically, there are only four types of hybrid cultivars: single-cross, modified single-cross, three-way-cross, and double-cross hybrids. Single-cross hybrids are produced by the hybridization of two genetically unrelated inbred lines. A modified single cross uses the hybrid of two related inbreds as one parent, say $A_1 \times A_2$, and then an unrelated inbred, B, as the male parent. The increase in vigor of the $A_1 \times A_2$ hybrid aids seed production. Final yield obtained by the producer will be only

TABLE 1.11 Stages of Growth in Corn, and Cumulative Growing Degree Days and Calendar Days Required to Reach Successive Stages.

Stage Number	Stage	Stage Description	Growing Degree Days[1]	Calendar Days
0.0	Emergence	Tips of leaves have emerged from the coleoptile.	120	10
0.5	Two-leaf	Two leaves fully emerged with collars visible.	200	17
1.0	Four-leaf	Four leaves fully emerged.	—	—
1.5	Six-leaf	Six leaves fully emerged; tassel initiation.	475	30
2.0	Eight-leaf	Eight leaves fully emerged; tassel developing rapidly.	—	—
2.5	Ten-leaf	Ten leaves fully emerged; tassel developing rapidly; ear shoots developing at 6 to 8 nodes.	740	50
3.0	Twelve-leaf	Twelve leaves fully emerged; ears developing rapidly; number of ovules determined.	—	—
3.5	Fourteen-leaf	Fourteen leaves fully emerged; tassel near full size; one or two ears developing rapidly; silks developing.	1,000	60
4.0	Sixteen-leaf Tasseling	Sixteen leaves fully emerged; ears and silks developing rapidly; tassel emerges.	1,150	75
5.0	Silking	Leaves and tassel fully emerged; elongation of stem has ceased; cob and silks growing rapidly; silks will continue to grow until fertilized.	1,480	75
6.0	Blister	Cob and husks fully developed; starch has begun to accumulate in kernels.	—	85
7.0	Dough	Kernels growing rapidly; consistency like bread dough.	1,925	95
8.0	Beginning dent	A few kernels showing dents.	—	—

(continued)

TABLE 1.11   (*Continued*)

Stage Number	Stage	Stage Description	Growing Degree Days[1]	Calendar Days
9.0	Dent	Kernels fully dented; dry matter accumulation near complete.	2,450	105
10.0	Physiological maturity	Dry matter accumulation is complete; grain moisture is 35%.	2,765	120

Source: Adapted from Illinois Agronomy Handbook, 1991–1992; Bark, 1986; Hanway, 1966.

1. Growing degree days based on a minimum temperature of 50°F and a maximum of 86°F. May be referred to as DD_{50} units.

slightly reduced relative to a single-cross hybrid and better than with a three-way hybrid. Three-way hybrids have a single-cross hybrid as the female parent and an inbred line as the male, or pollen, parent. A double-cross hybrid results from the hybridization of two single-cross hybrids. Essentially all U.S. hybrids have been single-cross or modified single-cross hybrids since the 1960s. Lack of sufficient seed produced on early versions of commercial single-cross hybrids, along with a widespread, but not necessarily accurate, belief that single-cross hybrids were not as widely adapted over years or locations as were double-cross hybrids delayed their acceptance. Although it is true that double-cross hybrids may be genetically more heterozygous and therefore may be theoretically more stable across environments, single crosses may be just as stable. Studies have shown that the genetic yield enhancement from the earliest hybrids until today's hybrids is about 54 pounds of grain per acre per year. Today's hybrids have the genetic potential to yield twice as much as the first commercial hybrids, and when improvements in fertility, pest management, and equipment are factored in, production per unit land area is about four times greater than production prior to 1940 (Table 1.7).

Hybrid selection is one of the most important choices that the corn producer must make each year. It is, hopefully, a decision that he or she will make only once, since once the seed are planted, it will be the cultivar that they will live with all season. Corn performance trials conducted by state agricultural experiment stations are the best source of information obtainable on relative performance of corn hybrids. Such reports commonly provide multiple-year average yields, percent moisture at harvest, percent lodging (both root and stalk), percent ear drop, percent stand, and may provide quality parameters such as percent protein, oil, and starch of kernels. For example, the Iowa State University Extension report for district 3 in Iowa reported these data, plus type of hybrid (i.e., single, double, or three-way) for 1988–90 on 240 hybrids. The producer should be aware that it is possible to buy the same hybrid under different names from different companies. This can happen when more than one company buys inbred lines from state experiment stations, foundation seed companies, or they market public hybrids (usually released by state experiment stations) under their own brand name.

The fact that a hybrid can be marketed under more than one name takes on importance when a producer is trying to spread harvest by planting a number of hybrids having differing maturity dates. Corn hybrids are usually marketed as early, mid- or full-season, terms that can have different meanings depending on location. Also, companies may not market hybrids under common claims. That is, company A may market a hybrid primarily in southern Illinois as a mid-season hybrid, while, assuming this hybrid to be a public release, company B markets the same hybrid as a full-season hybrid, especially if company B markets the same hybrid further north in Illinois.

Maturity may be identified by days to maturity or by growing degree days (GDD), also called degree days or heat units. The latest-maturing hybrid planted for grain should reach maturity at least two weeks before the expected first killing frost.

Other considerations in choosing a cultivar for dry land production would be the timing of expected rainfall and the stage of growth at that time. Corn requires maximum moisture from just before pollination to early dent, and moisture deficit during these growth stages will be more detrimental than if occurring earlier during the growing season (Figure 1.6). Planting hybrids that vary in maturity will hedge against moisture stress at pollination.

Number of days to physiological maturity, defined as the point when the kernel has dried to 35 percent moisture, or GDDs provide the producer with a better indication of maturity differences among hybrids than the terms early, mid- or full-season. GDDs are calculated as $[(Tmax + Tmin)/2] - 50$, where Tmax = the maximum temperature in degrees Fahrenheit for a given 24 hours and Tmin = the minimum for the same 24 hours. The value 50, for 50°F, is subtracted because growth and physiological processes in corn slow to near zero at this temperature. Likewise, temperatures above 86°F do not increase growth and development in corn, so Tmax is never over 86 for calculation purposes. The numbers of GDD units required to reach successive stages of growth are shown in Table 1.11. These values are averages for full-season hybrids and serve only as guides; however, the GDDs required for early season hybrids will be proportional. For example, time to silking will require about half the GDDs as required to reach physiological maturity regardless of total requirement. Put in the context of GDDs, it becomes obvious that the terms full, mid, and early apply to specific geographical locations as indicated for the state of Illinois in Figure 1.7.

Other factors to consider in hybrid selection other than yield potential and maturity are resistance to lodging, insects, and diseases. Again, state experiment station and extension service reports provide the best unbiased information. Multiple-year data from properly designed experiments are preferred.

Date of Planting

Planting may begin February 1 in southern Texas and mid-May along the Canadian borders (Figure 1.8). Full-season hybrids with potential for maximum grain yields should be planted first and earlier-maturing hybrids should be used if workload or inclement weather force the producer into later planting. Producers may spread their

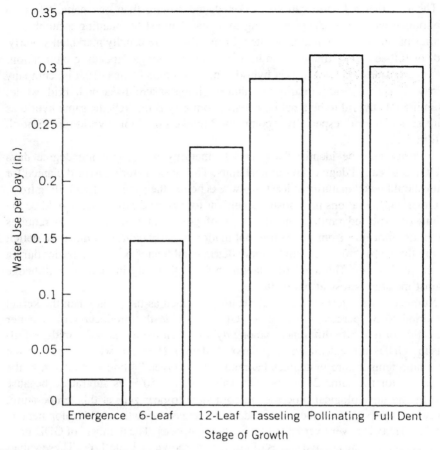

Figure 1.6 *Moisture requirements for corn in Georgia.* Source: *Adapted from Congleton, 1983.*

risk of production shortfall by planting hybrids of mid- and early-season maturity along with full-season hybrids during the recommended planting date range. This practice will spread the time of pollination, thereby avoiding the risk of having all of one's crop at pollination during a drought period. Planting more than one maturity group will also spread harvest, allowing for more economical use of equipment and facilities.

Although water is the most abundant compound on earth, a deficit during the growing season can have devastating affects. The water use per day by corn increases rapidly from about 30 days before silking, peaks during fertilization and early grain fill, and declines thereafter. Moisture deficit is most damaging during the silking to early grain fill period, causing up to 8 percent yield reduction per day of stress (Table 1.12). Producers should plant as early as possible during the recommended planting date range to avoid summer droughts when irrigation is not available.

Calendar day is only a rough guide to when to begin planting corn. Soil condition

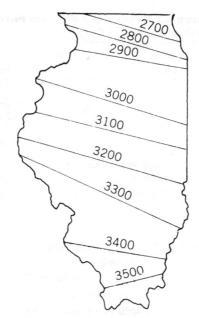

Figure 1.7   *Average number of growing degree days accumulated from May 1 through September 30 for the state of Illinois, 1951–1980.* Source: *Adapted from Illinois Agronomy Handbook, 1991–1992. (Drawing by Mike Hodnett)*

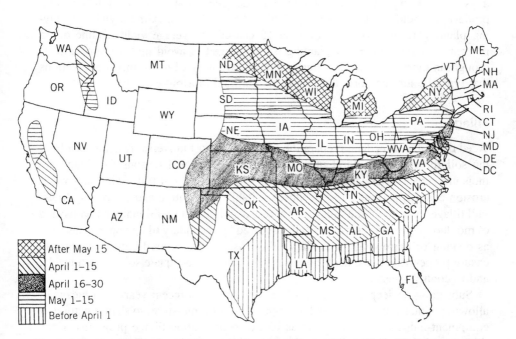

Figure 1.8   *Usual beginning planting dates for corn in the continental United States according to the USDA-SRS. (Redrawn by Mike Hodnett)*

TABLE 1.12   Approximate Minimum and Maximum Percent Yield Loss per Day of Moisture Stress[1] in Corn.

Days Before and After Silking	Yield Loss (%) from 1 Day of Moisture Stress
30–2 days before	2.2–3.6
At silking	2.8–8.0
10 days after	3.0–7.0
30 days after	3.0–5.5
Until maturity	Declines to 0

Source: Adapted from Shaw, 1988. Reproduced from Corn and Corn Improvement, 1988, American Society of Agronomy.

1. Based on the relationship of available moisture and atmospheric demand.

2. Silking occurs over 7 to 10 days.

is the key factor. Planting should not begin until the soil temperature at planting depth is 50 to 55°F and is expected to remain so or increase over the next five days. This is about April 20 in Iowa, March 25 in southeast Kansas and February 15 in central Florida and south Texas. Corn planting should be delayed if soils are too wet for proper seedbed preparation and seed placement. Corn planted in soils at 50 to 55°F may take up to 21 days to emerge, while seeds planted in soil at 60 to 65°F may require as few as 8 days.

As a general rule, planting slightly too early is preferable to planting a few days too late, as yield decline is less drastic, although there usually is a range of several days in which to plant for maximum yields. Long-term data from Iowa indicate that producers should plant between April 20 and May 10 for maximum yields. Delaying planting from May 15 until June 15 caused an average yield decline of 35 percent of maximum. Early planting is encouraged to avoid mid- to late-summer drought in most of the United States, to take advantage of lower insect populations that typically are the norm until mid-summer, and to avoid early frost.

Tillage

Weed control and seedbed preparation are the two most universal reasons for tilling the soil for any agricultural plant commodity. Additional reasons for tillage are moisture conservation, preservation or improvement of soil tilth, wind and water erosion control, and preparing seedbeds suitable for planting and cultivating tools. Full tillage for corn production varies across locations but traditionally has consisted of moldboard plowing, spring or fall, followed by secondary tillage operations such as disking or harrowing. Secondary tillage usually is required before planting to ensure proper seed placement, good seed-soil contact, adequate soil oxygen levels, and to control weeds.

Subsoiling or deep chiseling has become popular in recent years as a means of allowing winter rain or snow melt to recharge subsoil moisture, to shatter natural or equipment-induced hardpans, and as part of conservation tillage programs. Deep tillage, such as with subsoilers or chisels, will have maximum effect on soils where compaction layers (hardpans) are likely to form and restrict root penetration and thereby reduce the volume of soil from which water and nutrients can be extracted. Compaction layers are more likely in coarse or sandy soils than in the more clayey soils, and in areas where soils do not experience freezing and thawing cycles during the winter months—i.e., in the southern states having coastal plains.

Conservation tillage has gained in popularity throughout the country as producers have realized its benefits. Conservation tillage is a general term meaning any system that reduces the number of tillage operations performed in the production of any crop. Conservation tillage in corn may mean (1) Chisel plow-disk-plant, (2) disk-plant, (3) strip-till, or (4) no-till. The major emphasis in each of these systems is erosion control, but other benefits include moisture conservation, reduced energy requirement, reduced labor, and reduced equipment wear. Additional benefits include preservation of soil tilth and allowing for production on land too steep for production under full tillage.

No-till usually means planting in one operation by opening a small V-shaped drill into which seeds are dropped and then covered with virtually no disturbance of the previous crop's residue. This tillage system provides for the maximum effects of reduced tillage. In studies conducted in Nebraska, reduction in soil loss on land having a 5 percent slope was greatest for a no-till production scheme. Other conservation tillage systems used with corn and soybeans were much less effective (Table 1.13). Yields, assuming proper management, are competitive with those obtained under conventional tillage on soils that are suitable for no-till production. Soils having poor internal drainage, and therefore poor soil aeration, should not be planted no-till (Table 1.14). Yields with no-till on poorly drained soils are more competitive when planting is delayed and soils have warmed, and in more southern locations, assuming other factors are equal. Previous crop residues not only protect against soil erosion but may act as insulation that can accentuate cool temperatures and delay planting or slow early season growth.

Strip tillage consist of some method of tilling a more or less narrow band in front of the planter unit during planting and leaving middles untilled. Some systems use a sweep or other devise to "plow out" corn or sorghum butts from the previous year. Other systems use a rotary tiller in front of the planter unit. This system appears to

TABLE 1.13 Effects of Previous Crop Residue on Soil Loss Using Varying Degrees of Tillage.

Tillage System	Corn Residue		Soybean Residue	
	Cover	Soil Loss	Cover	Soil Loss
	(%)	(t./ac.)	(%)	(t./ac.)
Conventional[1]	4	10.1	2	14.3
Reduced till[2]	14	7.5	7	11.5
No-till plant	39	3.2	27	5.0

Source: Adapted from Shroyer and Hickman, 1988.

1. Included moldboard plowing and disking.
2. Included chiseling and disking.

TABLE 1.14 Effect of Soil Internal Drainage on Corn Yields Using Conventional and No-Till Tillage Systems in Indiana.

	Tillage System	
Drainage Class	Conventional Tillage[1]	No-till
	---------------- (bu./ac) ----------------	
Moderately well to well drained	108	112
Somewhat to poorly drained	127	97

Source: Adapted from Griffith et al., 1987.
1. Plow, disk twice, plant.

work best when a bed or ridge is left during cultivation of the previous year's crop and the top of this ridge is then tilled during planting. This form of strip tillage may be called "ridge tillage."

Disk-plant and chisel-disk-plant are less severe forms of conservation tillage. These will allow some remaining residue that may reduce soil loss (Table 1.13). Obviously, the amount of total soil loss will depend on when the tillage operations are carried out. Chiseled and disked, and disked alone, just ahead of planting will provide maximum protection against winter erosion but still leaves a near bare and disturbed ground from tillage until the crop develops ground cover.

Regardless of tillage level, seeds should be placed 1 to 2 inches deep under ideal conditions—i.e., seed depth temperature at least 55°F with good moisture, aeration, and good soil-seed contact. Seeds should be planted more shallow if planted early in the planting season when soils are cooler and evaporation is lower, and deeper as soils warm and soil drying becomes more of a problem. Every effort should be made to place seed into moist soil, even planting seeds as deep as 3 inches. Planting seeds deeper than 3.5 inches usually will result in emergence problems and therefore poor stands. It should be noted that planting depth will not affect root depth, as seminal and primary roots serve only to support the plant during emergence and seedling stages. Corn's permanent root system develops from nodes above the seed and at about the same depth regardless of seed depth.

Seeding Rate

The optimum number of plants per acre for maximum grain production varies with yield potential, moisture, tillage, location, and hybrid. Producers should always check the planting seed bag for recommended seeding rates. Optimum rates range from 12,000 kernels per acre (kpa) when planted dryland in northern Kansas, to 30,000+ kpa for some hybrids in more productive environments. Final plant stands normally will be reduced by 10 to 15 percent of kpa.

In general, soils with higher water-holding capacities will support a higher plant population, as of course will irrigated soils. Early planting usually results in more vigorous plants with better root systems that should maximize yield per plant and therefore require fewer plants per acre. Later-planted corn tends to produce smaller plants that are less productive individually, and therefore per acre yields are maximized at higher densities. Where possible, plants should be spaced equidistant to maximize light interception and rhizosphere exploitation, especially in later season. This is accomplished by decreasing the distance between rows and increasing the distance between plants within rows. Planting corn in narrow rows will not increase yields under stress conditions, such as drought or limited fertility, and may not be economical under some other conditions. Seeding rates should be increased about 15 percent above the recommended rate when planting no-till or with extreme reduced tillage production, and by about 10 percent when planted for silage. Producers should calibrate planters to drop the appropriate number of seeds per foot of row rather than using volume or weight to estimate how many seeds are being planted.

Seed size does not influence yield per se, but may influence seedling vigor. Larger corn seed of the same hybrid will produce a more vigorous seedling than smaller seed. Environmental conditions, genetics, and location on the ear affect seed size. Large, flat seeds are produced in the middle of the ear, while small rounds are produced toward the tip and large rounds are found at the ear's base.

Fertility

Corn requires 170 pounds of nitrogen, 35 pounds of phosphorus (as P_2O_5) and 175 pounds of potassium (as K_2O) to produce 150 bushels of grain (Table 1.15). More fertilizer is applied to corn than to any other crop grown in the United States. Current standards of corn production could not be sustained without commercial fertilizer, especially nitrogen, phosphorus, and potassium. Producers should always obtain a soil test from their state agricultural experiment station to determine needed nutrients and soil pH. A soil pH of 5 to about 8 is optimum for corn production. Soils with a pH below 5 will result in Al, Mn, and Fe toxicity, and if the pH drops below 4, P will become limiting because of being precipitated with Al. Alkaline

TABLE 1.15 **Nutrients Required to Produce 150 Bushels (8,400 lbs. at 56 lb./bu.) of Corn.**

Elements	Grain		Stover		Total	
	(lbs./ac.)	(lbs./bu.)	(lbs./ac.)	(lbs./bu.)	(lbs./ac.)	(lbs./bu.)
Nitrogen	115	0.77	55	0.37	170	1.13
Phosphorus (P_2O_5)	28	0.19	7	0.05	35	0.55
Potassium (K_2O)	35	0.23	140	0.93	175	1.40
Calcium	1.3	0.01	35	0.23	36	0.24
Magnesium	10	0.07	29	0.19	39	0.26
Sulfur	11	0.07	8	0.05	19	0.13
Chlorine	4	0.03	68	0.45	72	0.48
Iron	0.1	——[1]	1.8	0.01	1.9	0.01
Manganese	0.05	——	0.25	——	0.30	——
Copper	0.02	——	0.08	——	0.10	——
Zinc	0.17	——	0.17	——	0.34	——
Boron	0.04	——	0.12	——	0.16	——
Molybdenum	0.005	——	0.003	——	0.008	——

Source: Adapted from Barber and Olson, 1968, Changing Patterns in Fertilizer Use, Soil Society of America.

1. Less than 0.005.

soils, pH greater than 7, may present problems of P precipitating with Ca and Zn, and Fe will become toxic.

Optimum fertility not only maximizes yield but also minimizes cost per unit of production and preserves the environment. Too much fertilizer, particularly excess nitrogen, constitutes an economic loss to the producer and may pollute surface and groundwater. There are 18 elements that are necessary for plant health and maximum production. The three primary structural elements, C, O_2, and H, are extracted from the atmosphere. For practical purposes, all other elements are obtained from the soil. Elements, their typical dry weight concentration in plants, and their major functions are shown in Table 1.16.

Inorganic fertilizer may be applied preplant or split with part applied preplant and the remainder applied as a sidedress after the crop is established. However, the producer should be aware that rapid uptake of N, P, and K begins about 25 days post emergence and is 50 to 60 percent complete by the silking stage, mandating that sidedress should be made early, at least by the five-leaf stage of growth, or about 20 days after emergence.

Under more traditional tillage, dry fertilizer is applied topically and then incorporated into the soil by disking or plowing. Fertilizer placement is a major problem with extreme conservation tillage—i.e., no-till. Nitrogen applied to the surface of land with high levels of residue may result in (1) excess leaching in soils with good percolation rates since runoff is reduced on high-residue fields, (2) excessive denitrification because of longer saturation times, (3) nitrogen tie-up by decay microorganisms, and (4) volatilization of ammonia from urea-based fertilizer.

Other nutrients, especially P and K since they are required in large amounts, that must be applied topically with no-till will remain predominately in the upper 3 inches of finer-textured soils. This situation will result in poor root growth at deeper levels causing plants to be more susceptible to drought stress. The obvious solution is to sample soils from stratified zones such as 0–3, 3–6, 6–9, and 9–12 inches deep in order to monitor the situation. When the situation demands action, the producer must consider some sort of tillage to mix fertilizer into and throughout the plow layer of soil.

Several states have reported increased yields or other desirable results such as early plant growth or earlier maturity from the use of "starter" fertilizer. Starter fertilizer may consist of all soil-test-recommended phosphorus and a portion of the recommended nitrogen, say 20 percent. On sandy soils, some sulfur and zinc should be included. Best results are obtained when the fertilizer is banded 2–3 inches below and 2–3 inches to the side of the seed, although topical application of dry starter fertilizer to the side of the drill may be effective when there is less residue with which to contend, such as when following soybeans and when dealing with a coarse-textured soil.

Harvest and Storage

Harvest begins before September 1 in the southern areas of the United States and may not begin in the northernmost states until late October (Figure 1.9). Most of the

TABLE 1.16   Nutrients Required for Normal Growth in Plants, Typical Concentrations, Their Major Functions, and Usual Source.

Element	Typical Concentration as % d. wt.	Usual Source[2]	Major Functions
Group I: Structural and intermediates of metabolism			
Carbon (C)	44	Atmosphere	Organic compounds.
Hydrogen (H)	6	Atmosphere	Organic compounds.
Oxygen (O)	44	Atmosphere	Organic compounds.
Nitrogen (N)	2	CF, NF, Atm.	Amino acids, proteins, co-enzymes.
Sulfur (S)	0.5	NF, CF	Amino acids, proteins.
Phosphorus (P)	0.4	CF, NF	ATP, NADP (i.e., energy system).
Group II: Enzyme activators			
Potassium (K)	2.0	CF, NF	Activates about 60 enzymes; essential for protein synthesis; responsible for turgor and stomatal movement.
Calcium (Ca)	1.5	NF, CF	Activates enzymes; essential for membrane permeability.
Magnesium (Mg)	0.4	NF, CF	Activates ATP; component of chlorophyll.
Manganese (Mn)	0.4	NF, CF	Activates enzymes; essential for photolysis of H_2O.
Group III: Redox reagents; elements that undergo reduction/oxidation (redox) by virtue of multiple valency.			
Iron (Fe)	0.015	NF	$Fe^{+3} + e^- \leftrightarrow Fe^{+2}$
Copper (Cu)	0.002	NF	$Cu^{+2} + e^- \leftrightarrow Cu^+$
Molybdenum (Mo)	0.002	NF	Reduction of nitrate (NO_3^-) by nitrate reductase and reduction of N_2 by nitrogenase of free living and nodule bacteria in legumes: $Mo^{+6} + e^- \leftrightarrow Mo^{+5}$.
Group IV: Elements of uncertain function			
Boron (B)	0.003	NF	Membrane activity ?
Chlorine (Cl)	0.01–2.0	NF	Osmosis, charge balance, photolysis of H_2O ?
Silicon (Si)[1]	?	NF	May reduce transpiration.
Zinc (Zn)	?	NF	Protein synthesis, growth hormones, may be important in reproduction.
Sodium (Na)	0.05–10.0	NF	May be essential for C_4 photosynthesis.

Source: Glass, A. D. M.: Plant Nutrition: An Introduction to Current Concepts. © 1989 Boston: Jones and Bartlett Publishers. Reprinted with permission.

1. The need and role of Si has not been universally accepted.
2. Atm. = atmosphere; CF = commercial fertilizer; NF = native fertility.

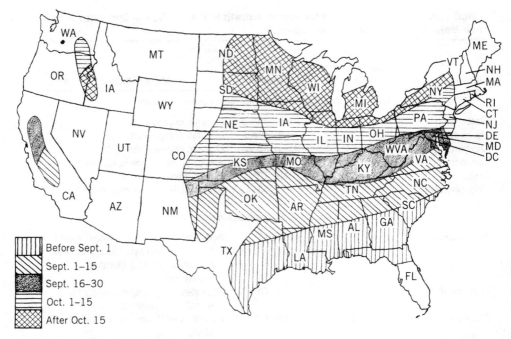

Figure 1.9 *Usual beginning harvest dates for corn for the continental United States according to the USDA-SRS. (Redrawn by Mike Hodnett)*

corn harvested for grain in the United States is field-shelled by use of a combine or a picker-sheller, with ever-decreasing acreage harvested as whole ears by mechanical pickers. Timing of beginning harvest is a compromise between waiting for seed moisture low enough for storage and maximizing yield. Delayed harvest increases the possibility of stalk lodging and increases the brittleness of ear shanks, both resulting in lower harvest efficiency. Early harvest means higher moisture and therefore increased drying and storage costs.

For maximum harvest efficiency, grain should be harvested when grain moisture is between 20 and 28 percent. Corn to be stored should be field or artificially dried to at least 15.5 percent moisture if it is to be held for fewer than six months, and dried to 13 percent or less for longer storage (Table 1.17). Southern states that experience high temperatures and high humidity recommend drying shelled corn down to 11 percent for long-term storage and drying within 48 hours to minimize aflatoxin buildup.

BIOTIC PESTS

As with most crops, corn is attacked by a large number of insects and disease pathogens, but only a small number are consistent pests requiring annual or near-annual control. While only 8 insects are considered primary pests of corn, a total of

TABLE 1.17 Number of Safe Storage Days for Shelled Corn at Different Moisture Levels.

Grain Temperature	Grain Moisture Content (%)						
	15	18	20	22	24	26	28
(°F)	---------------------- (days of safe storage) ----------------------						
40	——	195	85	54	38	28	24
50	466	102	46	28	19	16	13
60	259	63	26	16	10	8	7
70	155	37	13	8	5	4	4
80	109	27	10	6	4	3	3

Sources: Adapted from Hammond, 1983; Harner and Higgins, 1986 (Data calculated from Saul and Steele, 1966).

50 are known to damage corn in the United States (Table 1.18). Losses due to diseases vary from year to year and even from field to field (Table 1.19). Average yearly loss to all diseases of corn in the midwest has been estimated to range from 7 to 17 percent, with losses more severe in some years. The most recent, widespread epiphytotic occurrence of corn disease was the outbreak of southern corn leaf blight in 1970. This epiphytotic reached most of the U.S. Cornbelt, reducing U.S. yields by an average of 12 bushels/acre from the previous year. However, yields rebounded to record levels the following year because of lack of conditions favorable for disease incidence. Seed breeders quickly shifted from the use of susceptible Texas, or T, cytoplasm to a disease-resistant cytoplasm as a means of achieving male sterility for developing hybrids.

GRADES AND GRADING

Corn fed to animals on the farm where it was produced averaged 48 percent of the total production during 1980–1985, leaving 52 percent to move through marketing channels as corn for cash. Of this 52 percent, about 15 percent is sold directly to terminal users and shippers, 5 percent is sold directly to feedlots or other farmers, and the remaining 80 percent is sold to country elevators. There were about 5,000 country elevators in 1984 that served as the gathering point for corn sold by producers that is then sold to subterminal or terminal elevators. Corn then moves to processors or to export shippers. Country elevators also provide producers with services such as drying, storage, and also blending to provide terminal elevators with the desired quality. Specialty corn such as popcorn, waxy, high amylose, or white corn cannot be purchased through normal commodity channels, but usually is purchased through direct contract with country elevators or producers.

The first legislation to establish uniform and objective classification of grain quality was proposed in 1890. Congress rejected that proposal but passed the Grain Standards Act of 1916 that mandated that all grain sold in interstate commerce or exported to other countries be inspected "so that grain can be marketed in an orderly and timely manner and that trading in grain may be facilitated." That act, although

TABLE 1.18   Insect Pests of Corn in the United States.[1]

Insect	Description[2]	Symptom/Injury	Control
European Corn Borer *Ostrinia nubilalis* (Hbn.)	Larvae 1-inch long, dirty gray with indistinct spots.	Young larvae feed on leaves and collars before boring into stem and ear, affecting nutrient transport and causing eardrop and stem breakage.	Resistant hybrids with strong stalks; proper planting date; chemical.
Western Corn Rootworm *Diabrotica virgifera virgifera* (LeC.)	Yellowish-green beetle, 3/8-inch with or without 2 black stripes; larvae whitish with dark abdominal plate.	Eggs are laid in the soil the previous year; larvae destroy roots, causing lodging and goosenecked plants.	Crop rotation; chemical.
Northern Corn Rootworm *D. barberi* (Say)	Same as Western Corn Rootworm.	Same as Western Corn Rootworm.	Same as Western Corn Rootworm.
Southern Corn Rootworm *D. undecimpunctata howardi* (Barber)	Same as Western Corn Rootworm.	Same as Western Corn Rootworm.	Same as Western Corn Rootworm.
Lesser Corn Stalk Borer *Elasmopalpus lignosellus* (Zeller)	Bluish-green larvae with brown strips.	Feed on leaves or roots; later bore into stems.	Feild sanitation; rotation
Corn Leaf Aphid *Rhopalosiphum maidis* (Fitch)	Green with dark green head and thorax, black antennae and cornicles.	Large mid-summer populations cause wilting and chlorosis by sucking plant sap; infected plants may not produce ears.	Chemical
Black Cutworm *Agrotis ipsilon* (Hufn.)	Blackish larvae; usually found below ground.	Destroys plants to 10-inches tall by cutting plants at soil level.	No practical control; chemical.
Wireworms *Melanotus fissilis*	Slick brown worms that live several years in soil.	Destoys seed; bores into stalks causing wilting and death of plants.	Soil applied insecticides at planting.
Grub *Phyllophaga* spp.	Large, white larvae of the June Beetle.	Root pruning, causing wilting and plant death.	Broadcast soil applied insecticides at planting.
Corn Earworm *Helicoverpa zea* (Boddie)	Larvae have brown heads and body microspines; may be pink, green or brown, with white line along sides.	Eggs laid on silks predominantly; larvae feed in whorl of young leaves and in ears, consuming developing grains.	No practical control in field corn; chemical control in sweet corn.

TABLE 1.18 (*Continued*)

Insect	Description[2]	Symptom/Injury	Control
Stalk Borer *Papaipema nebris* (Guenee)	Striped larvae with purple margin behind head.	Bore into plants less than 36 inches tall; may cause lodging.	Field sanitation, especially margins; chemical.
Grasshopper *Melanoplus* spp.	Obvious.	Cut silks and destroys leaves.	Chemical if needed just before pollination through anthesis.
Seed Corn Beetle *Stenolophus lecontei* (Chaud.)	Brown beetle, 3/8-inch long; 2 black strips.	Destroys seed before germination.	Cultural-avoid planting in cold weather.
Seed Corn Maggot *Delia platura* (Meig.)	Larvae of fly; no head or legs.	Same as Seed Corn Beetle.	Same as Seed Corn Beetle.
Maize Billbug *Sphenophorus maidis* (Chittenden)	Black-snouted beetle; legless larvae.	Larvae bore into base of stalks, causing lodging.	Cultural, especially tillage.
Southwestern Corn Borer *Diatraea grandiosella* (Dyar)	White larvae with distinct black spots.	Larvae tunnel into stalks through mid season and girdle stalks late season, resulting in lodging.	Cultural; early planting and winter tillage.
Armyworm *Pseudaletia unipuncta* (Haw.)	Greenish-gray larvae with pinkish lines.	Feed on leaves and in whorl creating ragged appearance; may defoliate plants.	Chemical.
Fall Armyworm *Spodoptera frugiperda* (Smith)	Larvae with inverted "Y" on head and 4 black spots.	Same as Armyworm, plus larvae may eat into tip or shank of ears.	Chemical.
Corn Flea Beetle *Chaetocnema pulicaria* (Melsh.)	Tiny black beetles.	Strip surfaces of seedling leaves.	Chemical; cultural-avoid planting in cold weather.
Thrips *Thysanoptera* spp.	Tiny, rasping insects.	Leaves blasted.	Chemical; cultural-maintain good growing conditions.
Western Bean Cutworm *Loxagrotis albicosta* (Smith)	Light brown larvae.	Feed in tassel before entering ear at tip or base causing severe ear damage.	Chemical.
Chinch Bug *Blissus leucopterus*	Adults to 1/4-inch with red and black colors in distinct order.	Plants die from nutrient loss to these sucking insects.	Chemical; cultural.

1. Listed generally in order of importance.
2. Stage most damaging or most obvious to observer.

TABLE 1.19 Common Diseases of Corn in the United States.

Disease	Cause/Source	Symptom/Injury	Control
Per-emergence Damping off	*Pythium* spp.	Seeds rot before emerging.	Seed fungicide; plant good quality seed into warm soil.
Post-emergence Damping off	*Pythium* spp. *Fusarium moniliforme* *Aspergillus* spp. *Rhizoctonia* spp. other fungi	Seedlings becoming chlorotic; seedling death.	Seed fungicide; plant good quality seed into warm soil; other cultural control.
Early Stalk Rots	*Pythium aphanidermatum*	Water soaked first internode.	Avoid excessive N fertilization; crop rotation; field sanitation; proper surface drainage.
	Erwinia carotovora	Twisted stalk, soft rot at break point.	
Late Stalk Rots	*Cephalosporium maydis* (late wilt)	Shrunken and hollow stalk.	Same as for early stalk rots.
	Diplodia maydis	Subepidermal pycnidia near nodes.	
	Fusarium moniliforme	Pink pith.	
	Gibberella zeae	Pink pith; superficial perithecia on stalks.	
	Macrophomina phaseolina (charcoal rot)	Black sclerotia on vascular bundles.	
Leaf spot and blight diseases	*Helminthosporium maydis* (Southern Leaf Blight)	Tan colored, elongated lesions 0.4–1.2 inches; lesions may coalesce causing leaf shredding.	Resistant cultivars; crop rotation; field sanitation.
	H. turcicum (Northern Leaf Blight)	Lesions 1–6 inches long, elliptical, gray-greenish becoming tan-brown.	
	H. carbonum (Leaf spot)	Lesions 0.4–1.2 inches; grayish tan center with distinct border.	
	Erwinia stewartii (Stewart's Wilt)	Pale green-gray linear, water-soaked lesions; stunting; death.	
	Colletotrichum graminicola (Anthracnose)	Oval to broad spindle-shaped, tan to brown with yellow to reddish bordered lesions.	

TABLE 1.19 (*Continued*)

Disease	Cause/Source	Symptom/Injury	Control
Ear and Kernel Rots	*Diploidia maydis* *Aspergillus* spp. *Penicillium* spp. *Helminthosporium* spp. Others	Ranges from rotted or destroyed kernels and ears to apparently symptomless kernels.	Resistant hybrids; cultural especially field sanitation.
	Aspergillas flavus	Ranges from obvious rot to symptomless; important because of toxin produced.	None.
Rusts	*Puccinia sorghi* (Common Rust)	Cinnamon-brown rust pustulas appear about tasseling stage.	Resistant hybrids.
	P. polysora (Southern Rust)	Small, circular, orange to red rust pustulas usually appearing later in season.	
Smuts	*Ustilago maydis* (Commom Smut)	Galls up to 6 inches in diameter on any part of shoot.	Resistant hybrids; soil applied fungicides may control.
	Sphacelotheca reiliana (Head Smut)	Initial symptoms appear on floral organs.	
Downy Mildew	*Sclerospora* spp. *Sclerophthora* spp.	Chlorosis of leaves; stunting; white-striping of leaves; replacement of ears with leafy structures.	Resistant hybrids; cultural practices promoting early growth; field sanitation.
Maize Dwarf Mosaic Virus	Polyvirus group RNA virus.	Mosaic or mottle on younger leaves; progressing to narrow, light green to yellowish streaks.	Resistant hybrids; early planting; aphid control (aphid vectored).
Maize Chlorotic Dwarf Virus	Isometric virion RNA virus.	Chlorosis of whorl leaves; chlorotic banding of secondary veins; general chlorosis or reddening of leaves; stunting.	Resistant hybrids; control of johnsongrass and flea-hopper.

amended several times, remains unchanged in principle, although legislative language in the Grain Quality Improvement Act of 1986 makes it easier for the Federal Grain Inspection Service (FGIS) to evaluate and make changes in standards and criteria for identifying the quality of corn.

Corn received at the elevator is sampled by probing the lot with a device ap-

TABLE 1.20 Grades of White, Yellow, or Mixed Classes of Corn

| | | Maximum Limits of- | | |
| | | Damaged Kernels | | |
GRADE	MINIMUM TEST WEIGHT PER BUSHEL	HEAT DAMAGED	TOTAL	BROKEN CORN AND FOREIGN MATERIAL
	(lbs.)	(%)	(%)	(%)
U.S. No. 1	56.0	0.1	3.0	2.0
U.S. No. 2	54.0	0.2	5.0	3.0
U.S. No. 3	52.0	0.5	7.0	4.0
U.S. No. 4	49.0	1.0	10.0	5.0
U.S. No. 5	46.0	3.0	15.0	7.0

U.S. sample grade is corn that
  a. Does not meet the requirements for the grades U.S. Nos. 1, 2, 3, 4, or 5; or
  b. Contains 8 or more stones which have an aggregate weight in excess of 0.20 percent of the sample weight, 2 or more pieces of glass, 3 or more crotalaria seeds (*Crotalaria* spp.), 2 or more castor beans (*Ricinus communis* L.), 4 or more particles of an unknown foreign substance(s) or a commonly recognized harmful or toxic substance(s), 8 or more cockleburs (*Xanthium* spp.) or similar seeds singly or in combination, or animal filth in excess of 0.20 percent in 1,000 grams; or
  c. Has a musty, sour, or commercially objectionable foreign odor; or
  d. Is heating or otherwise of distinctly low quality.

Source: U.S. Department of Agriculture, Federal Grain Inspection Service, 1988.

proved by the FGIS that allows for extracting samples from various points in the lot. This sample is at least 5.5 pounds (2,500 g) and is randomly subdivided to a 2.2 pounds (1,000 g) working sample. Corn is classified as yellow corn, white corn, or mixed corn. Yellow corn must not contain more than 5 percent white kernels; white corn can contain not more than 2 percent yellow kernels; mixed corn does not meet the criteria for white or yellow. Corn is graded on the basis of density—i.e., test weight per bushel, damaged kernels, broken kernels, and foreign material (Table 1.20). Special grades are (1) Flint: any class of corn having at least 95 percent flint kernels; (2) Flint and Dent: any class of corn having more than 5 percent but less than 95 percent flint corn; (3) Weevily: any lot having live weevils or other live insects injurious to stored corn; and (4) Waxy: corn of any class containing at least 95 percent waxy corn.

Corn also is routinely inspected for aflatoxin, produced by the fungus *Aspergillus flavus*. Aflatoxin is toxic to humans and animals, and corn infested with this fungus is considered unfit and may be rejected by the elevator. Only about 2 percent of U.S. corn is affected annually and mostly in the southeast where conditions favorable for infestation and growth of the fungus are reoccurring—i.e., high temperatures and drought conditions. Also, immediate and rapid drying of high moisture corn can prevent the invasion and growth of *A. flavus*.

SPECIALTY CORN

Popcorn

Flint corn that has been selected for maximum popping expansion is referred to as popcorn. The consumption of popped corn dates to antiquity in the New World, and the Amerindians had selected varieties by the time of Columbus. High-quality types of popcorn, often marketed as "gourmet" or "premium," have pericarps that fragment into small, more acceptable pieces upon popping. The United States grows in excess of 200,000 acres annually.

Sweet Corn

This specialty corn is a popular vegetable and ranks third behind potato and tomato in farm value for processing among vegetable crops in the United States. The conversion of sugar to starch is slowed in sweet corn compared with field corn. A recessive mutant allele, *shrunken-2*, was discovered in the early 1950s that causes the accumulation of almost twice as much sugar in the early dough stage of kernel development as was previously possible, further enhancing the popularity of this vegetable.

Waxy Corn

This type of corn is so named because of the dull, waxy appearance of the kernels relative to the shiny, vitreous kernels of flint and dent. Waxy corn's endosperm is 100 percent amylopectin, or branched chain starch, compared to 72 percent for regular corn. This allows certain processors to bypass the separation step in the production of purified starch. Waxy corn starch is desirable in the production of certain processed foods and may cause better feed efficiency in farm animals. The United States produces about one million acres of waxy corn for domestic use and export.

QPM

Quality protein maize (QPM) contains the recessive mutant allele, *opaque-2*, that results in an improvement of protein quality. Both lysine and tryptophan amino acids are found at higher levels in QPM, both considered deficient in normal endosperm corn. Several hybrids have been developed that are competitive in yield, but other problems such as high moisture at harvest remain to be solved. No commercial cultivars of QPM are produced in the United States, although a few are grown in other countries seeking to improve the protein consumption among their population. QPM holds much promise for the future, especially in developing countries where both calories and protein of diets are deficient.

GLOSSARY

Aflatoxins: any of a number of mycotoxins that are produced by the fungus *Aspergillus flavus.*

Amylase: any of a number of enzymes that accelerate the hydrolysis of starch.

Amylopectin: a heavy molecular weight component of starch having a branched structure of glucose subunits. Corn starch is composed of amylopectin and amylose.

Amylose: a component of starch having straight chains of glucose subunits.

Blending: the mixing of two or more grain lots to establish an overall quality that may or may not be different from any one individual lot.

Country elevator: grain elevator that is the first collection point in the marketing of corn. These elevators accept, dry, and store corn from producers to be delivered to terminal elevators.

Cultivar (*cultivated variety*): an international term denoting certain cultivated plants that are clearly distinguishable from others by one or more characteristics and that will retain those characteristics when reproduced.

Dockage: material such as stems, weeds, dirt, or stones that is readily removed by ordinary grain cleaning equipment. May be referred to as foreign material.

Double-cross hybrids: hybrids developed by crossing the F_1's of two single cross hybrids.

Dry milling: a process of cleaning, conditioning, grinding, and sifting. Used to separate corn into its three physical components, germ, bran, and endosperm.

Embryo: an undeveloped plant within a seed.

Endosperm: the dead, yet nutritive tissue in seeds found in the inner bulk of the kernel, consisting primarily of carbohydrates, but also containing protein, riboflavin, and B vitamins. The endosperm provides nutrients to the seedling plant during germination, emergence, and very early seedling growth.

Enzyme: a group of catalytic proteins produced by living cells that mediate the chemical processes of life, without being destroyed or altered.

Federal Grain Inspection Service (FGIS): USDA branch responsible for setting grain standards and developing the technology to measure factors affecting grain quality. FGIS also develops sampling and inspection procedures, evaluates and approves equipment, monitors inspection accuracy, and oversees mandatory export inspection of grain.

Feed grain: grains characterized as high-energy because of their high levels of carbohydrate and low levels of crude fiber.

Germ: see embryo.

Genotype: the hereditary makeup of an individual plant or animal.

Germplasm: genotypes of crop species and related species used or of potential use in the development of cultivars. May be referred to as race stocks, breeding lines, inbred lines, or germplasm lines.

Grain: seeds or fruits of various food or feed plants including the cereal grains (i.e., wheat, corn, barley, oats, and rye) and other plants in commercial and statutory use, such as soybeans.

Hybrid: the cultivar resulting from crossing two genetically unlike inbred lines of corn.

Inbred line: a parent in the development of a hybrid cultivar developed by repeated generations of self pollination.

Intrinsic quality: characteristics important to the end use of grain that are non-visual and can only be determined by analytical tests—for example, protein, ash, oil content, or starch.

Millfeed: the material remaining after all of the food-grade flour and other components have been extracted from grain—used in animal feed and feed supplements.

Modified single-cross: a scheme of hybrid production where the female parent is the F_1 from the cross of two genetically similar parents. This F_1 is crossed with a genetically dissimilar pollen parent to produce the hybrid seed sold to the farmer.

Open pollinated: refers to seed produced on plants or to cultivars maintained with no control over which individual plants pollinate other plants within the seed increase blocks.

Pericarp: the covering of seed that is derived from the ovary wall.

Phenotype: the plant as seen; the outward expression of the genotype of a plant within the limits of its environment.

Protein: complex organic compounds composed of nitrogen, carbon, hydrogen, and oxygen that are essential to the functioning and structure of all organic cells.

Shrinkage: the loss of weight in grain due to the removal of water.

Single-cross: a scheme to produce hybrid cultivars by crossing two genetically dissimilar inbred lines of corn.

Starch: the primary fraction of the endosperm of corn composed of straight and branched chains of glucose molecules. Starch is an important component of animal diets, reacting with enzymes to form dextrose, maltose, and other sugars utilized for energy and growth.

Steepwater: water used in soaking corn during wet milling.

Tempering: the addition of water to corn and wheat during dry milling that aids in the removal of bran from the endosperm.

Terminal elevator: an elevator that takes delivery of corn from country elevators and is owned by or will sell to the manufacturer of corn products. These elevators may sell corn to the export trade or to other terminal elevators.

Three-way cross: a hybrid production scheme where the F_1 of two genetically dissimilar inbred lines of corn will be crossed with a genetically dissimilar inbred to produce hybrid corn seed.

Variety: see cultivar.

Wet milling: the process of using water in which corn is tempered, steeped, and milled to separate the grain into its four components: germ, hull, gluten, and starch. Oil is extracted from the germ; the hulls are dried and may be added to gluten to become corn gluten feed; gluten may be purified and used in several industrial products; and starch is converted into corn syrups and corn sugars.

BIBLIOGRAPHY

Ahlschwede, W. T., M. C. Brumm, D. M. Danielson, A. J. Lewis, E. R. Peo, Jr., and D. E. Reese. 1984. *University of Nebraska Swine Diet Suggestions.* Univ. of Nebraska Coop. Ext. Ser. EC 84-210.

Barber, S. A., and R. A. Olson. 1968. Fertilizer Use on Corn. pp. 163–188. *In* L. B. Nelson (ed.) *Changing Patterns in Fertilizer Use.* Soil Sci. Soc. Am., Madison, WI.

Bark, L. D. 1986. Use of Growing Degree Units. pp. 5–7. *Corn Production Handbook.* Kansas St. Univ. Coop. Ext. Ser. C-560.

Beadle, G. W. 1980. "The ancestry of corn." *Sci. Am.* 242:112–119.

Benson, G. O. 1984. *Replanting or Late Planting Decisions with Corn and Soybeans.* Iowa St. Univ. Coop. Ext. Ser. PM-1155.

Benson, G. O. 1986. *Profitable Corn Production.* Iowa St. Univ. Coop Ext. Ser. Pm-409.

Benson, G. O., and R. B. Pearce. 1987. "Corn Perspective and Culture." *In* S. A. Watson and P. E. Ramstad (eds.) *Corn: Chemistry and Technology.* Am. Assoc. Cer. Chem., St. Paul, MN.

Benz, Bruce F. 1986. "Taxonomy and Evolution of Mexican Maize." Ph.D. Dissertation. Univ. of Wisconsin-Madison.

Bird, R. McK. 1980. "Maize evolution from 500 B.C. to the present." *Biotropica* 12:30–41.

Bitzer, M. J., J. H. Herbek, G. Lacefield, and J. K. Evans. 1979. *Producing Corn for Grain and Silage.* Univ. of Kentucky Dep. of Agron. AGR-79.

Boyer, C. D., and J. C. Shannon. 1982. "The Use of Endosperm Genes in Sweet Corn Improvement." pp. 139–161. *In* J. Janick (ed.) *Plant Breeding Reviews* I. AVI Pub. Co., Westport, CT.

Brown, W. L., and E. Anderson. 1947. "The Northern Flint Corns." *Ann. Missouri Bot. Garden* 34:1–29.

Brown, W. L., and E. Anderson. 1948. "The Southern Dent Corns." *Ann. Missouri Bot. Garden* 35:255–274.

Caldwell, D. M., and T. W. Perry. 1971. "Relationships between stage of maturity of the corn plant at time of harvest for corn silage and chemical composition." *J. Dairy Sci.* 54:533–536.

Carter, G. F., and E. Anderson. 1945. "A preliminary survey of maize in the Southwestern United States." *Ann. Missouri Bot. Garden* 32:297–317.

Chicago Board of Trade. 1977. *Grains: Production, Processing, Marketing.* L. Besant, D. Kellerman, and G. Monroe (eds.) Board of Trade, Chicago, IL.

Ciba-Geigy Agrochemicals. 1979. *Maize.* Ciba-Geigy Ltd., Basle, Switzerland.

Congleton, W. F. 1983. "Growth and Development." pp. 3–5. *In* W. F. Congleton (ed.) *Irrigated Corn Production in Georgia.* Univ. of Georgia Coop. Ext. Ser. Bul. 891.

Copeland, L. O. 1988. *Grain Grading.* Michigan St. Univ. Press, East Lansing, MI.

Dicke, F. F., and W. D. Guthrie. 1988. "The Most Important Corn Insects." pp. 767–868. *In* G. F. Sprague and J. W. Dudley (eds.) *Corn and Corn Improvement.* Am. Soc. Agron., Madison, WI.

Doebley, J., J. D. Wendel, J. S. C. Smith, C. W. Stuber, and M. J. Goodman. 1988. "The origin of Cornbelt maize: The isozyme evidence." *Ec. Bot.* 42:120–131.

Earle, F. R., J. J. Curtis, and J. E. Hubbard. 1946. "Composition of component parts of the corn kernel." *Cer. Chem.* 23:504.

Fast, R. B. 1987. "Breakfast cereals: Processed grains for human consumption." *Cer. Foods World* 32:241–244.

Faw, W. F., and L. M. Curtis. 1980. *Irrigated Corn Production.* Auburn Univ. Coop. Ext. Ser. Cir. ANR-165.

Fussell, Betty. 1992. *The Story of Corn.* Alford A. Knopf, New York, NY.

Galinat, W. C. 1965. "The evolution of corn and culture in North America." *Ec. Bot.* 19:350–357.

Galinat, W. C. 1971. *The Evolution of Sweet Corn.* Univ. of Massachusetts Agri. Exp. Sta. Res. Bul. 591.

Galinat, W. C. 1973. "Intergenomic mapping of maize, teosinte, and *Tripsacum*." *Evol.* 27:644–655.

Galinat, W. C. 1979. "Botany and Origin of Maize." *In* E. Hafliger (ed.) Maize. *Ciba-Geigy Agrochemicals Tech. Monograph.* Ciba-Geigy Ltd., Basle, Switzerland.

Galinat, W. C. 1984. "The Origin of Corn." *In* C. F. Sprague and J. W. Dudley (eds.) *Corn and Corn Improvement.* Am. Soc. Agron., Madison, WI.

Glass, A. D. M. 1989. *Plant Nutrition: An Introduction to Current Concepts.* Jones and Bartlett Pub., Boston, MA.

Gomez, M. H., L. W. Rooney, R. D. Waniska, and R. L. Pflugfelder. 1987. "Dry corn masa flours for tortilla and snack food productions." *Cer. Foods World* 32:372–377.

Goodman, M. M., and R. McK. Bird. 1977. "The races of maize. IV. Tentative grouping of 219 Latin American races." *Ec. Bot.* 31:204–221.

Goodman, M. M., and W. L. Brown. 1988. "Races of Corn." pp. 33–80. *In* C. F. Sprague and J. W. Dudley (eds.) *Corn and Corn Improvement.* Am. Soc. Agron., Madison, WI.

Griffith, D. R., J. V. Mannering, D. B. Mengel, S. D. Parsons, T. T. Bauman, D. H. Scott, C. R. Edwards, F. T. Turpin, and D. H. Doster. 1987. *A Guide to No-Till Planting after Corn or Soybeans.* Purdue Univ. Coop. Ext. Ser. (Tillage) ID-154.

Hallauer, A. R., W. A. Russell, and K. R. Lamkey. 1988. "Corn Breeding." pp. 463–564. *In* C. F. Sprague and J. W. Dudley (eds.) *Corn and Corn Improvement.* Am. Soc. Agron., Madison, WI.

Hammond, C. 1983. "Harvesting, Drying and Storing." pp. 30–33. *In* W. F. Congleton (ed.) *Irrigated Corn Production in Georgia.* Univ. of Georgia Coop. Ext. Ser. Bul. 891.

Hanway, J. J. 1966. *How a Corn Plant Develops.* Iowa St. Univ. Coop. Ext. Ser. Spec., Rep. No. 48.

Hardeman, N. P. 1981. *Shucks, Shocks, and Hominy Blocks: Corn as a Way of Life in Pioneer America.* Louisiana St. Univ. Press, Baton Rouge, LA.

Harner, J. P., III, and R. A. Higgins. 1986. "Drying and Storing." pp. 33–36. *Corn Production Handbook.* Kansas St. Univ. Coop. Ext. Ser. C-560.

Hickman, J. S., and J. P. Shroyer. 1986. "Optimum Planting Practices." pp. 8–10. *Corn Production Handbook.* Kansas St. Univ. Coop. Ext. Ser. C-560.

Higgins, R. A. 1986. "Insect Management." pp. 17–20. *Corn Production Handbook.* Kansas St. Univ. Coop. Ext. Ser. C-560.

Hobbs, L., F. W. Schenck, and J. E. Long. 1986. "Corn Syrups." *Cer. Foods World* 31:852–865.

Holden, P., L. Frubish, and J. Pettigrew. 1984. "Energy for Swine." *Pork Industry Handbook*. Purdue Univ. Coop. Ext. Ser. PIH-3.

Illinois Agronomy Handbook, 1991–1992. 1990. Univ. of Illinois Coop. Ext. Ser. Cir. 1311.

Iltis, H. H., and J. F. Doebley. 1984. "Zea-A Biosystematical Odyssey." *In* W. F. Grant (ed.) *Plant Biosystematics.* Academic Press, Toronto, Canada.

Iltis, H. H. 1983. "From teosinte to maize: The catastrophic sexual transmutation." *Sci.* 222:886–894.

Jardine, D. J. 1986. "Disease Control." pp. 21–24. *Corn Production Handbook.* Kansas St. Univ. Coop. Ext. Ser. C-560.

Johnson, J. T. 1983. "Cultural Practices." pp. 6–8. *In* W. F. Congleton (ed.) *Irrigated Corn Production in Georgia.* Univ. of Georgia Coop. Ext. Ser. Bul. 891.

Jugenheimer, R. W. 1976. *Corn Improvement, Seed Production, and Uses.* John Wiley & Sons, New York, NY.

Katz, F. R. 1986. "Maltodextrins." *Cer. Foods World* 31:866–867.

Mangelsdorf, P. C. 1958. "Ancestor of corn." *Sci.* 128(3335):1313–1320.

Mangelsdorf, P. C. 1986. "The origin of corn." *Sci. Am.* 255:80–86.

Mangelsdorf, P. C., and R. G. Reeves. 1939. *The Origin of Indian Corn and its Relatives.* Texas A&M Univ. Agri. Exp. Sta. Bul. 574.

Mangelsdorf, P. C., R. S. MacNeish, and W. C. Galinat. 1964. "Domestication of corn." *Sci.* 143:538–545.

Martin, J. H., W. H. Leonard, and D. L. Stamp. 1976. *Principles of Field Crop Production.* MacMillan Pub. Co., New York, NY.

Mask, P. L., and C. C. Mitchell, Jr. (not dated). *Alabama Production Guide for Non-Irrigated Corn.* Auburn Univ. Coop. Ext. Ser. C. ANR-503.

Mask, P., J. Everest, C. C. Mitchell, and J. T. Touchton. 1989. *Conservation Tillage for Corn.* Auburn Univ. Coop. Ext. Ser. Cir. ANR-40.

McClung de Tapia, E. 1992. "The Origins of Agriculture in Mesoamerica and Central America." *In* C. W. Cowan and P. J. Watson (eds.) *The Origin of Agriculture: An International Perspective.* Smithsonian Inst. Press, Washington, DC.

Mengel, D. B. 1990. *Fertilizing Corn Grown Using Conservation Tillage.* Purdue Univ. Coop. Ext. Ser. AY-268.

Olsen, R. A., and D. H. Sander. 1988. "Corn Production." pp. 639–686. *In* G. F. Sprague and J. W. Dudley (eds.) *Corn and Corn Improvement.* Am. Soc. Agron., Madison, WI.

Pacey, D. A. 1986. "Harvesting Suggestions." pp. 30–32. *Corn Production Handbook.* Kansas St. Univ. Coop. Ext. Ser. C-560.

Perry, T. W. 1988. "Corn as a Livestock Feed." pp. 941–964. *In* G. F. Sprague and J. W. Dudley (eds.) *Corn and Corn Improvement.* Am. Soc. Agron., Madison, WI.

Reese, D. E., M. C. Brumm, A. T. Lewis, P. S. Miller, and W. T. Ahlschwede. 1992. *University of Nebraska Swine Diet Suggestions.* Univ. of Nebraska Coop. Ext. Ser. EC 92-210.

Roe, D. A. 1974. "The sharecropper's plague." *Natl. History* 838:52–63.

Sarkar, K. R., B. K. Mukherjee, D. Gupta, and H. K. Jain. 1974. "Maize." *In* Sir J. Hutchinson (ed.) *Evolutionary Studies in World Crops.* Cambridge Univ. Press, London, England.

Sauer, J. D. 1993. *Historical Geography of Crop Plants: A Selected Roster.* CRC Press, Boca Raton, FL.

Saul, R. A., and J. L. Steele. 1966. Why damaged corn costs more to dry. Agri. Eng. 47:326–329.

Segars, W. I. 1983. "Fertilization." pp. 9–12. *In* W. F. Congleton (ed.) *Irrigated Corn Production in Georgia*. Univ. of Georgia Coop. Ext. Ser. Bul. 891.

Shaw, R. H. 1988. Climate Requirement. pp. 609–638. *In* G. F. Spague and J. W. Dudley (eds.) *Corn and Corn Improvement*. Am. Soc. Agron., Madison, WI.

Shroyer, J., and J. Hickman. 1988. *Soybean Production Handbook*. Kansas St. Univ. Coop. Ext. Ser. C-449.

Smith, M. S., K. L. Wells, and G. W. Thomas. 1983. *Fertilization and Liming for Corn*. Univ. of Kentucky Dep. of Agron. AGR-105.

Smith, D. R., and D. G. White. 1988. "Diseases of Corn." pp. 687–766. *In* G. F. Sprague and J. W. Dudley (eds.) *Corn and Corn Improvement*. Am. Soc. Agron., Madison, WI.

Sturtevant, E. L. 1899. *Varieties of Corn*. USDA Off. Exp. Sta. Bul. 57.

United States Office of Technology Assessment. 1989. *Enhancing the Quality of U.S. Grain for International Trade*. OTA-F-399. U.S. Government Printing Office. Washington, DC.

United States Department of Agricultural. 1936, 1942, 1952, 1962, 1972, 1982 and 1991. *Agricultural Statistics*. U.S. Government Printing Office, Washington, DC.

United States Department of Agriculture-Economic Reporting Service. 1982. *U.S. Corn Industry*. USDA Natl. Tech. Info. Ser. PB82-173964.

United States Department of Agriculture-Economic Reporting Service. 1985. *Feed Outlook and Situation Report*. USDA Fds-296.

United States Department of Agriculture-Federal Grain Inspection Service. 1988. *U.S. Standards for Grains*. U.S. Government Printing Office, Washington, DC.

United States Department of Agriculture-Statistical Reporting Service. 1984. *Usual Planting and Harvest Dates for U.S. Field Crops*. USDA Agri. Handbook No 628.

Villegas, E., S. K. Vasal, and M. Bjarnason. 1990. "Quality Protein Maize—What Is It and How Was It Developed." pp. 27–48. *In* E. T. Mertz (ed.) *Quality Protein Maize*. Am. Assoc. Cer. Chem., St. Paul, MN.

Wallace, H. A., and W. L. Brown. 1988. *Corn and Its Early Fathers. The Henry A. Wallace Series on Agricultural History and Rural Studies*. Iowa St. Univ. Press, Ames, IA.

Walter, T. L. 1986. "Select Hybrids Carefully." pp. 2–4. *Corn Production Handbook*. Kansas St. Univ. Coop. Ext. Ser. C-560.

Watson, S. A. 1987. "Structure and Composition." pp. 53–79. *In* S. A. Watson and P. E. Ramstead (eds.) *Corn: Chemistry and Technology*. Am. Assoc. Cer. Chem., St. Paul, MN.

Watson, S. A. 1987. "Measurement and Maintenance of Quality." pp. 125–183. *In* S. A. Watson and P. E. Ramstad (eds.) *Corn: Chemistry and Technology*. Am. Assoc. Cer. Chem., St. Paul, MN.

Watson, S. A. 1988. "Corn Marketing, Processing, and Utilization." pp. 881–940. *In* G. F. Sprague and J. W. Dudley (eds.) *Corn and Corn Improvement*. Am. Soc. Agron., Madison, WI.

Wych, R. D. 1988. "Production of Hybrid Seed Corn." pp. 565–607. *In* G. F. Sprague and J. W. Dudley (eds.) *Corn and Corn Improvement*. Am. Soc. Agron., Madison, WI.

Ziegler, K. E., W. H. Vinson, and D. E. Caroll. 1990. *The 1990 Iowa Corn Yield Test Report: District 3*. Iowa St. Univ. Coop. Ext. Ser. Pm-660-3-90.

2

Wheat
*(*Triticum spp.*)*

INTRODUCTION

The exact place and date of the origin of the wheat plant that we recognize today is unknown. Unlike many of the crop species dealt with in this text, paleontologists, botanists, archaeologists, and geneticists have rather precisely identified the area of evolution, cultivation, domestication, and early movement of wheat. Evidence indicates that diploid and tetraploid wheats first appeared before 8000 B.C. in the area known as the Fertile Crescent, an area generally synonymous with the drainage basins of the Euphrates and Tigris Rivers in present day Syria and Iraq (Figure 2.1). Some authorities place wheat domestication before 16000 B.C. Hexaploid wheats, by far the most valuable and widely grown today, are thought to have evolved before 7000 B.C. in an area from just south of the Caspian Sea in northern Iran eastward into northern Afghanistan.

DOMESTICATED SPECIES

There are three basic types of wheat, based on their chromosome number, collectively forming an allopolyploid series (Table 2.1). The wheat genome is made up of a basic set of 7 (i.e., x = 7, 14, or 21 chromosomes in its reproductive cells) resulting in somatic cell nuclei, 2n, with 14 (diploid), 28 (tetraploid), or 42 (hexaploid) chromosomes. Since each set of 14 somatic chromosomes is from different diploid parents, wheats having additional sets are referred to as allopolyploids (allo = other(s), poly = many, ploid = genome). However, tetraploid and hexaploid wheats behave genetically as diploids.

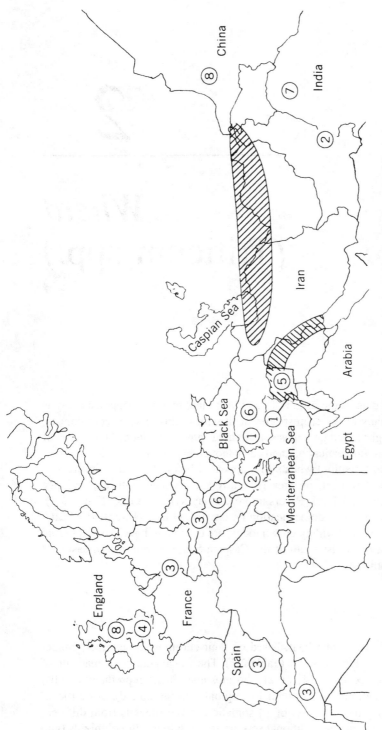

Figure 2.1 Einkorn (AA) and emmer (AABB) wheats evolved/selected in Fertile Crescent (shaded area) apparently by about 8000 B.C. and were farmed by about 7000 B.C. Evidence indicates that einkorn (AA) and emmer (AABB) wheats had reached (1) Jericho, and central and western Turkey by 7000 B.C., (2) the Indus River Valley of Pakistan and the Balkans and Greece by 6000 B.C., (3) Spain, Yugoslavia, northern Europe, and the western Mediterranean by 5000 B.C., and (4) England by 3500 B.C. Hexaploid wheats (AABBDD), by far the most important today, are believed to have evolved within a region across northern Iran, Afghanistan, and into central Asia (shaded) after the introduction of emmer (AABB) wheat from the Fertile Crescent. Club wheat (AABBDD) remains or archaeological artifacts indicate its use in (5) present-day Syria by 7000 B.C., (6) Turkey and Yugoslavia by 5000 B.C., (7) India by 4000 B.C., and (8) England and China by 3500 B.C. Source: Cook and Veseth, 1991.

TABLE 2.1 Classification of Wheats, *Triticum* spp.

Species	Subspecies[2]	Status[3]	Thrashing Habit[4]	Common Name	Chromosome Number (2n)	Genome Designation[5]
monococcum	*boeoticum*	W	—	—	14	AA
"	*monococcum*	W, PC	H	einkorn	14	AA
tauschii[1]	—	W	—	—	14	DD
turgidum	*dicoccoides*	W	—	—	28	AABB
"	*dicoccum*	PC	H	emmer	28	AABB
"	*durum*	PC, MC	F	durum	28	AABB
timopheevii	—	W	—	—	28	AAGG
aestivum	*spelta*	PC, MC	H	spelt	42	AABBDD
"	*compactum*	PC, MC	F	club	42	AABBDD
"	*aestivum*	PC, MC	F	common or bread	42	AABBDD

Sources: Adapted from Kimber and Sears, 1983; Cook and Veseth, 1991. With permission from The American Phytopathological Society. *Source:* Roger J. Veseth WSU/UI Extension Cens Tillage Spec.

1. Older classification may have as *Aegilops squarrosa* L., goat grass.
2. Classification as "subspecies" not universally accepted. Number of species of *Triticum* varies from 5 to 30, depending upon authority.
3. W = wild; PC = primitive cultivated; MC = modern cultivated (according to Cook and Veseth, 1991).
4. H = grain thrashes with glumes (hull) remaining attached to grain; F = grain thrashes free of glumes.
5. The wild grass species contributing the BB genome to wheat has not been established.

Einkorn (AA) is grown only in isolated areas of the Middle East and southern Europe today for use in making a local dark bread and as cattle or horse feed. Emmer (AABB) is grown as livestock feed in parts of Europe and the Dakotas of the United States. Durum (AABB) is used today for pastas, while the hexaploid wheats, especially *T. aestivum* subsp. *aestivum,* are used for leavened breads and pastries.

MOVEMENT FROM ORIGIN

The movement of wheat, as well as other near-eastern crop species, is inextricably tied to the development of farming in the Old World. Apparently, farming began in the Near East before 8000 B.C. with the planting of cereal grasses. Evidence indicates that a culture of cave dwellers, the Natufians, occupied the area in and around present-day Israel by 10000 B.C. These people used crude sickles made of rows of sharp stones as blades presumably for harvesting grains and other plants. The Natufians also left indications that they stored grain for future use.

It will be noted elsewhere that gathering grasses from the wild and transporting the seed heads to a common thrashing/winnowing site would have identified non-shattering mutations. Early farmers surely would have recognized these types as being more suitable for collection because they arrived at the site with more grain. They also would have recognized that they got a greater return for their labor. If they did not recognize this as a reason for "planting" crops, then there must have been some other, elusive incentive, because wild wheat and other wild grasses grow in abundance in many areas of the Near East today, especially within the Fertile Crescent, and presumably did so 12,000 years ago. The legitimate question has been posed by others, "Why go to the trouble to plant when the crop that you're planting grows naturally and in abundance?" Obviously, there was some useful purpose in planting seeds, and it may have been to increase the proportion of harvestable plants having non-shattering seed heads.

Other results from planting and harvesting cereals have been suggested. They include increases in seed production by increasing the percent seed set, inflorescence size and/or the number of inflorescences, increased seed size, and a loss of germination inhibitors—e.g., hard seededness. Some or all of these forces contributed to the acceptance of wheat and its husbandry.

Humankind as gatherers surely were attracted to the large seeds of *Triticum,* obviously providing more calories of food per unit of input labor. This characteristic of *Triticum* predisposed the grains to domestication, for when humans experimented with the new idea of farming, it meant that *Triticum* would germinate rapidly and have stronger seedlings than competing small-seeded grasses. Over the centuries, or perhaps millennia, from evolution to domestication, wheat evolved into a plant that could (1) tolerate cold, (2) survive and reproduce in rocky, shallow soil, yet be genetically capable of greater yields in better habitats, and (3) mature under conditions of limited moisture. Wheat evolved in an area characterized by relatively mild winters and hot, dry summers. Plant types genetically incapable of maturing before the onset of high temperatures and severely limited moisture would have been eliminated naturally, resulting in a crop species with a cool-season growth habit.

From their domestication somewhere within the Fertile Crescent, Einkorn and Emmer wheat spread southward along the eastern side of the Mediterranean, being farmed at Jericho by 7000 B.C., and then in northwest Africa by 5000 B.C. (Figure 2.1). It spread eastward to the Indus River Valley of Pakistan by 6000 B.C. and northwest through Turkey, the Balkans, Greece, and Yugoslavia to Spain by 5000 B.C. Movement into northern Europe probably occurred along the Rhine and Danube River basins, reaching England by 3500 B.C.

Hexaploid wheat evolved northeast of the Fertile Crescent, supposedly when tetraploid Emmer (AABB) wheat was produced in proximity of *T. tauschii*, also referred to as *Aegilops squarrosa*. Thus, a tetraploid carrying the A and B genomes hybridized with a diploid having the D genome, followed by chromosome doubling of the resulting zygote to produce a hexaploid with the genomic make up of AABBDD. It is believed that the D genome significantly expanded the climactic range of *Triticum*, adapting it to areas of severe winters and warm summers.

Hexaploid wheat probably was first carried as a contaminate in Emmer wheat. Many archaeological sites yield evidence that ancient cereal crops were not grown as pure or one-species fields, but rather that mixtures were common, either on purpose or because of a lack of removing unwanted plants. Regardless of how it was transported, hexaploid wheat, probably club wheat, spread to Syria by 7000 B.C., the Fertile Crescent and Turkey by 5000 B.C., and northwest to England by 3500 B.C. Its production had moved east to India by 4000 B.C. and to China by 3500 B.C.

TYPES OF WHEAT

The great diversity of the genus *Triticum* already has been eluded to, an allopolyploid series, predisposition to domestication, and its cool-season growth habit. Wheat is classified also on agronomic criteria, kernel color, and endosperm quality. Agronomic classifications are based on kernel hardness—i.e., soft or hard (Table 2.2). The types of wheat that compose the preponderance of worldwide and U.S. production are (1) hard red winter (HRW), (2) hard red spring (HRS), (3) hard white winter (HWW), (4) hard white spring (HWS), (5) soft red winter (SRW), (6) soft white winter (SWW), (7) soft white spring (SWS), and (8) durum, a spring growth-habit tetraploid.

Growth Habit

A unique character among agriculturally important crops evolved in some wheats whereby the germinating embryo or the seedling plant must undergo a hormonal-controlled conversion from juvenility, or vegetative growth only, to reproductive growth. This process is called vernalization and ensures that the growing point, or apical meristem, remains underground until the advent of warmer temperatures and

TABLE 2.2   Classes (Hard or Soft, Red or White, Winter or Spring, and Durum) of Wheat and a Few of the Many Products from Each.

Classes of Wheat			
T. aestivum subsp. *aestivum* (common or bread wheat, hard endosperm)	*T. aestivum* subsp. *aestivum* (soft endosperm)	*T. aestivum* subsp. *compactum* (soft endosperm)	*T. turgidum* subsp. *durum*
Hard red winter	Soft red winter	Club wheat	durum
Hard red spring	Soft white winter		
Hard white winter	Soft white spring		
Hard white spring			
Products			
Bread, rolls, bagels, English muffins, wheat germ, pizza, hot breakfast cereals	Crackers, cookies, cakes, biscuits, waffles, muffins, wheat germ	Crackers, cookies, cakes, etc.	Macaroni, spaghetti, egg noodles, wheat germ

is therefore protected from freezing and death. Winter wheats have a prostrate growth habit whereby only leaves are produced aboveground during the fall, provided that they are not planted too early, while the growing point and buds remain underground. This generalized growth habit is typical of grasses. However, winter wheats require a period of exposure to temperatures between 32 and 50°F in order for the flowering process to be initiated.

The vernalization process can occur in imbibed seeds as well as in vegetative plants. Seeds of Norstar HRW wheat for example require at least six weeks' exposure to 43°F for flowering to proceed normally (Table 2.3). Vernalization longer than eight weeks was considered unnecessary, but the data do point out that earliness of crop maturity can be influenced by the length of exposure to vernalization temperatures beyond just the minimum requirement.

TABLE 2.3   Response of Norstar HRW Wheat to Length of Vernalization at 43°F.

	Days to Heading	
Weeks	Average	Range ±
2	NH[2]	——
4	105	24
6	63	14
8	42	5
10	37	3
12	33	6

Source: Adapted from Whelan and Schaalje, 1992. Reproduced from Crop Science, Whelan E. D. P. and Schaalje p 78, 1992.

1. Following vernalization treatments, seeds were planted and pots placed in a growth chamber at 68°F and 18 hours of light for 120 days.
2. NH = Not headed at 120 days when experiment terminated.

TABLE 2.4 Days to Heading of Seven Wheat Cultivars Exposed to Varying Hours of Winter Temperatures at Gainsville, Florida, in 1985 and 1986.

Cultivar	Type	Adapted to	Weeks of Exposure[1]				
			0 (0 cu)[2]	2 (55 cu)	4 (117 cu)	6 (245 cu)	8 (376 cu)
Owens	SWS	Idaho	28	33	42	49	55
Florida 301	SRW	Florida	55	50	58	75	82
Florida 302	SRW	Florida	110	112	90	87	102
Phoenix	HWW	California	115	114	95	70	85
Hunter	SRW	——	117	111	104	90	101
Coker 983	SRW	South Carolina	NH[3]	NH	130	95	102
Caldwell	SRW	Indiana	NH	NH	NH	125	130

Source: Reproduced from *Crop Science*, Gardner, F. P. and Barnett, R. D., p 167, 1990.

1. Potted seedling plants exposed to outside temperatures during December and January in 1985 and 1986.
2. cu = cold unit, i.e. hours below 50°F.
3. NH = Not heading or less than 50% of plants heading.

The optimum vernalization temperature, or critical threshold, and its duration are cultivar-dependent. Researchers in Florida evaluated the effect of the number of hours of temperature below 50°F on the heading date of seven wheat cultivars (Table 2.4). Days to heading of Owens SWS, adapted to Idaho, and Florida 301 (reported to be a SRW type) were not reduced by exposure of seedlings to naturally occurring low temperatures, verifying in the case of Owens, and suggesting in the case of Florida 301, a spring growth habit. Response among the other cultivars varied although all responded as winter-type wheats. The delay of heading after eight weeks, or 376 hours below 50°F, was due to seedling damage from excessive low temperatures.

Kernel Color

Wheat is classified also on the basis of seedcoat color, those classifications being red or white. Red is conditioned by three dominant genes, the true whites having recessive alleles at all three loci. White kernel color could improve flour yield by allowing millers to be less stringent in removing seedcoat fragments from flour. Other than this, there appears to be no advantage of kernel color, either red or white. However, red predominates and the vast majority of crop improvement research in the United States has concentrated on red-kernel wheats.

Kernel Hardness

Wheats in the United States are classified as "hard" or "soft," depending on endosperm granularity. The endosperm of hard wheat fractionates along cell walls because of the presence of proteinaceous material on the surface of starch granules within these cells that cause the starch granules to adhere to cell walls. Soft wheats, on the other hand, do not have this strong starch granule-protein matrix and will fractionate across cells, thereby releasing cell contents. Therefore, soft wheat will, when ground, yield relatively large quantities of finely granulated flour, while hard wheats yield a more coarse product when milled under similar conditions. Soft wheat's fine-granulated flour is more difficult to control and will actually clog sifters in industrial bread-making equipment designed to process hard wheat flour.

Hard wheats normally have flinty, translucent grains (sometimes referred to as vitreous), while soft wheats have starchy or opaque-looking grains. Hard wheats also may be referred to as strong, while soft wheats are called weak, these terms having to do with gluten proteins and bread-making qualities.

Bread is made from hard wheat flour because of the cohesive and elastic properties of its gluten protein. Gluten is composed of a number of nonenzymatic proteins. Gluten is a very intricate protein complex that can be partitioned into two major components, gliadin, soluble in alcohol solutions, and glutenin that is insoluble in alcohol but soluble in dilute acids or alkali, denaturants, detergents, or disulfide-reducing agents. Gluten can be separated from flour dough by repeated washing with water to dissolve and remove starch, the gluten adhering to itself. Gluten is added to baby foods, breakfast cereals, and such, as a protein supplement, and may be added to such foods as macaroni and pretzels when a firm final texture is desired.

A flour with strong gluten, when properly prepared and baked, will yield a bread with good volume and texture. Soft wheat flour is used when the tenderness of the product is important, such as in cakes, and cookies. Soft wheats generally have lower protein than hard wheats (Table 2.5).

BREAD

Bread is the most widely consumed food product worldwide, providing a larger share of people's protein and calories than any other single product. Bread is generally made from wheat, especially in the developed nations, although a flat bread may be made from other grains such as corn, barley, oats, rice or rye.

Bread is classified as (1) yeast bread, (2) quick bread, or (3) flat bread. Yeast bread is made by mixing hard wheat flour, water or milk, and yeast, and allowing the mixture to ferment, or rise, before baking. The yeast, single-cell plants, break down sugars into alcohol and carbon dioxide (CO_2). The elastic properties of the gluten protein allow for the entrapment of CO_2, resulting in air bubbles, causing the dough to rise. The alcohol evaporates and the yeast cells are killed during the baking process.

Quick bread requires less preparation time than yeast bread, but is also "raised," or leavened, by adding baking powder to the flour. Baking powder contains baking soda—that is, sodium bicarbonate—and other leavening ingredients that react with water to form CO_2. This CO_2 is trapped by the gluten protein just as it is in yeast bread. Most home bread makers in the United States will use "self-rising" flour that has baking powder already added.

Flat breads are the staple bread products in many parts of the world. Corn is used in many Central American countries to produce tortillas. Flour tortillas are very

TABLE 2.5 Relationship of Wheat Grain Protein and Kernel Hardness, and Examples of Products in Each Class.

| | Wheat Classes | | | High Protein | |
	Soft	Mixed	Hard	Hard	Durum
Protein	8.5–9.5	9.5–12.5	12.5–13.5	13.5–15.0	14.0–15.0
Typical Products	Cakes, biscuits, pastries	Puddings, thickeners, noodles	Bread, high-protein flour	Mixed with other flours	Pasta

popular in the southwest United States. Rice flour is used in the Far East to produce flat bread; "chapatty," a flat bread made from coarse wheat flour, is eaten in the Middle East and India.

PASTA

Durum wheat is used in the production of pasta products such as macaroni and spaghetti in the Americas and Europe, but only about 15 percent of durum wheat goes to this purpose in the Middle East and North Africa. In these areas, the major products from durum flour are local breads, couscous (a North African dish of steamed coarse-ground wheat—usually durum—served with meat and broth), and bulgar (a cracked wheat cereal).

Durum wheat is harder and more vitreous than bread wheats and is normally ground into a coarse flour called semolina. Durum wheat will average about 1 percent higher protein than bread wheat, production practices and other considerations being comparable. Durum cultivars grown in the United States have spring growth habits, although winter types do exist, and the grains are yellowish to yellowish-brown (amber) in color. Technically, most durum is white, but its translucent endosperm gives it an amber appearance. Red durum wheats exist but they are used primarily as feed and are not used for pasta in the United States. Durum wheat is high in carotenoid pigments, which are responsible for the yellowish color of pasta, an important criteria for pasta conoisseurs that is unrelated to other quality parameters.

The gluten of durum is usually weaker than in bread wheats, although, as noted, durum is significant in making local bread in some countries. Types of durum have been developed that have somewhat stronger gluten, but even so they still result in poor loaf volume relative to bread wheat. Also, durum kernels are physically much harder than bread wheat grains, thereby making grinding to flour more difficult. Some flour is produced in milling as a by-product of semolina production.

The higher protein and weaker gluten are key elements that make durum especially suited for the production of pasta. Most of the proteins in durum wheat are found inside cells as enveloping structures around starch granules. These proteins appear to be hydrophobic and account for the tendency of pasta made from durum semolina to maintain shape, with little swelling, surface disintegration, or stickiness even if overcooked. Bread wheat farina, a coarsely ground flour, can be used to produce pasta but it is lacking in yellow color and is not resistant to overcooking— i.e., it absorbs more water and tends to disintegrate.

MOVEMENT TO THE UNITED STATES AND CULTIVAR DEVELOPMENT

Wheat was introduced into the New World by Christopher Columbus on his second voyage of 1493–1496. The next recorded introduction was by the Spanish soon after the capture of Tenochtitlan, present-day Mexico City, from the Aztec Amerindians. By 1735, wheat was being exported from the Mexico City area to the West Indies. Wheat, as well as seed of other grain crops, was taken northward from central Mexico by the Spanish as they searched for gold and set up missions along their routes in efforts to "Christianize" the Amerindians. By this route, wheat probably was first introduced into present-day United States through Texas. A mission was established in 1682 at Presidio, Texas, along the Rio Grande River, and wheat was reportedly growing prior to that date. The Spanish had explored the Presidio area in 1535 and again in 1582.

Wheat was introduced into New England by the Pilgrims landing at present-day Plymouth, Massachusetts, in 1620. There is no doubt that wheat seeds were introduced into the New England and Atlantic Coast colonies, but the first planting of European grain occurring in 1621 may have been barley and not wheat. However, the list of supplies that were brought from England in 1628 with the John Endecott expedition that settled Massachusetts Bay included "Wheat, Rye, Barley, Oats. . . ."

Early efforts at communal living at Plymouth were abandoned by 1623, and food shortages experienced in 1621 and 1622 were replaced with excess by 1624, to the point of trade among the colonists. By 1645, the English colonists were efficient in producing wheat, barley, oats, and rye, as well as maize and beans.

Prior to 1644, spring wheats predominated in Massachusetts because of the severe winterkill experienced with winter wheats. A disease epidemic in 1664 induced producers to return to winter wheat cultivars that matured before the onset of weather conducive to wheat "blast and mildew." Carrier (1923) reports that after the disease epidemic of 1664, wheat was never again a major crop in Massachusetts.

Neighboring colonists of Connecticut and Rhode Island had similar problems with wheat production as were experienced by Massachusetts farmers and farmers today. The government of Connecticut, in order to promote the production of English grain (i.e., wheat) gave 100 acres of land to a farmer for each team of horses that he possessed. A provision of this law was that 20 acres were to be seeded to wheat the first year, 80 acres the second, and all 100 acres by the third year. Total wheat production soared, prices plummeted, and farmers protested. The solution was characteristically "government policy." Growers were ordered to sell their wheat only to two buyers who were forced to pay twice the going rate and were required to export all of the 1644 crop. Apparently, the two buyers took the wheat but did not meet their financial obligations to all producers. As in Massachusetts, diseases and insects became so prevalent that by 1680 there was no danger of overproduction during the remaining colonial years. Early colonists in New England turned to a white flint maize as their bread flour, and wheat bread was unknown in many New England colonial households.

Wheats from around the world were introduced into the present-day continental U.S. by immigrants looking for better and more fulfilling lives. Early on, seeds were brought in by individuals for their own use, and later, even until today, specimens were collected from foreign lands by seedsmen, breeders, and organizations such as the United States Department of Agriculture (USDA) and state agricultural experiment stations (SAES). As noted, the first documented introductions were brought in by English colonists. Wheats from Sweden and the Netherlands were introduced into New York, New Jersey, and Delaware by 1638; and Spanish wheat was growing in California by 1770. By 1844, nearly 30 cultivars of wheat were grown in only one county in New York State, and a number of cultivars were reportedly grown in Ohio by 1858, and in Missouri by 1881. Four hundred and seventy-eight cultivars were evaluated by the New Mexico Agricultural Experiment Station in 1892. Many cultivars were known by a number of different names. A farmer might select or establish a cultivar by name in his community, and sell seed to a farmer who migrates to another community where he might produce and sell seed under another name. Obviously, there are many other scenarios in which a land race cultivar, perhaps better named a cultivar complex, could be known by a number of synonyms.

Prior to 1860, soft red winter and soft white winter wheats, both *T. aestivum* subsp. *aestivum* and subsp. *compactum,* were the primary types grown in the United States, and that mostly in the eastern states. Table 2.6 shows a few of the many early introductions, their growth habit, and their dates of entry into the United States. Note that of the five earliest documented introductions, three are SRW types and two are SWS types. The SRWs were introduced into the eastern states while the two whites were introduced into and grown in the Western states, from Arizona to Washington. As will be discussed subsequently, the soft wheats are unacceptable today for breads, but were used for that purpose during the early days of the United States. The reasoning was simply economics. Farmers could produce more per acre of soft wheats than they could with hard wheats. Acceptance of hard wheats by industry and farmers would await the development and acceptance of the roller mill, and the introduction of adequate germplasm.

Many events of history and science have to come together to catapult a plant species into major crop status. Looking back, these events do not seem isolated, but simply part of the industry picture today. Examples include the cotton gin, a simple idea responsible for making the U.S. a major world power, and the discovery of F_1 hybrid vigor resulting from the hybridization of Northern Flint with Southern Dent corn. Development of the hard wheat industry also demonstrates this meeting of events, germplasm and people.

Introduction of Hard Red Spring Wheat

The tremendous acreage of HRS wheat in Minnesota, North Dakota, South Dakota and Montana (i.e., the Northern Great Plains of the United States) began with a few seeds of one plant. David Fife of Otonabee, Ontario, Canada, received a sample of wheat seed in 1842 from a friend in Scotland, who had obtained it from a shipload of wheat at the port in Danzig, Germany. It was thought at the time that the seed had originated in Russia, but later evidence indicated that its origin was Galicia, Poland. Believing the seed to be of a spring growth habit, Fife planted the seeds accordingly. However, the seed turned out to be winter wheat, thereby not receiving sufficient vernalization for flowering, save for one plant. Fife harvested, planted, and increased the seed from this single plant and developed a cultivar called Red Fife.

TABLE 2.6   Early Wheat Introductions into the United States that Became Progenitors of Many, More-Modern Cultivars.

Year of Entry	Name	Synonyms	Origin	Class[1]
<1700	Michigan Amber	Red May	unknown	SRW
1819	Mediterranean	Numerous	Italy	SRW
1820	Sonora	Red Chaff, White Sonora, Ninety-Day	Mexico	SWS
1822	Purplestraw	Alabama Bluestem	unknown	SRW
1852	Pacific	Bluestem, White	Australia	SWS
1860	Red Fife	Fife	Poland	HRS
1864	Arnautka	——	N.E. Asia	D
<1870	Little Club	——	Chile	W
1870	Big Club	Big Four, Chile Club	Chile	W
1874	Turkey	Kharkof, Crimean (+27 more)	N.E. Asia	HRW
<1880	Poole	Harvest King	unknown	SRW
<1883	Valley	Indiana Swamp	unknown	SRW
<1883	Rice	Zimmermann, Red May	unknown	SRW
1890	Goldcoin	Fortyfold	unknown	W
1895	Harvest Queen	Black Sea, Red Cross	unknown	SRW
1896	Pelissies	——	Algeria	D
1898	Kubanka	——	N.E. Asia	D
1903	Monad	——	N.E. Asia	D
1903	Pentad	——	N.E. Asia	D
1903	Kota	RBR3	N.E. Asia	HRS
1912	Marquis	——	Canada	HRS
1914	Baart	——	S. Africa	W
1914	Bunyip	——	Australia	W
1918	Onas	——	Australia	W
1918	Golden Ball	——	S. Africa	D
1920	Federation	——	Australia	W

Source: Adapted from Cox, 1991; Clark, 1936; Clark and Bayles, 1942.

1. SRW = Soft Red Winter; HRW = Hard Red Winter; HRS = Hard Red Spring; W = White; SWS = Soft White Spring; D = Durum.

This cultivar was introduced into the United States in 1860 by J. W. Clarke of Wisconsin who reportedly produced an "excellent crop" that year. Red Fife was used as one of the parents of Marquis, a cultivar considered the "king" of wheats for about 20 years, from the 1920s until the 1940s. Marquis was grown in 24 states in 1939, accounting for 3.2 million acres. Marquis also was used as a parent in many subsequent cultivars.

Introduction of Hard Red Winter Wheat

The success of HRW wheat production, predominantly in Texas, Oklahoma, Kansas, Colorado, and Nebraska (also known as the Southern and Middle Great Plains of the United States), traces a similar history. The introduction of Turkey HRW wheat into the United States by Russian Mennonite immigrants in 1872–1874 provided the basis for the HRW wheat industry in the United States. These immigrants settled in Marion County, Kansas, in 1874 and produced their first crop of wheat in 1874–75. It is believed that the winter growth habit along with a hard endosperm evolved in an area north of the Caucasus Mountains in Northeast Asia (part of the former U.S.S.R.), primarily the area in and around the Crimea, a peninsula extending into the Black Sea from the north. A number of other introductions of Turkey-type wheats by 1902 were made by other Mennonite immigrants and as the result of expeditions to the Crimea area by plant breeders and other scientists.

The fame and success of Turkey wheat spread rather slowly from its meager beginnings. Rumor has it that the farmers around the Mennonite community realized its worth and immediately took up the new type. This is not substantiated by history. In some wheat growing areas of Kansas, it was at least a decade before Turkey wheat became popular, and it was not considered the standard hard wheat until 1898. It was 1900 before almost 40 percent of Nebraska's wheat acreage was planted to HRW wheat.

The reasons for this slow acceptance may be many, but two probably predominate: first, simply the lack of seed and lack of distribution facilities; second, and perhaps most important, the hard grain texture of Turkey wheat slowed its acceptance. Commercial millers were equipped to grind soft wheat and not hard types. In fact, Turkey was discounted as much as 15 cents per bushel by millers until the value of its hard flour came to be appreciated. Also, except for the Mennonite women, housewives were not accustomed to using hard wheat flour.

Once hard wheat flour was determined to be better for hard breads and such breads were accepted by the baking trade and their customers, Turkey quickly became the standard against which all other hard wheats were compared. By 1919, Turkey was grown on 83 percent of Nebraska's wheat acreage, 82 percent of Kansas, 73 percent of Oklahoma, 67 percent of Colorado, 52 percent of Iowa, 31 percent of Utah, and on 27 percent of Illinois wheat acreage. Nationwide in 1919, Turkey was grown in 32 states on over 21 million acres. Turkey lead all other cultivars in acreage in Kansas until 1939, and in Nebraska until 1949. Incredibly, 100,000 acres of Turkey wheat were grown in the United States in 1969, 95 years after its introduction by a few immigrants seeking a better way of life in the Great Plains of the United States.

Introduction of Durum Wheat

Another group of Russian immigrants settled in North Dakota in 1864, bringing with them wheat seed, called Arnautka, from their native country. This seed was *T. turgidum* subsp. *durum*, a tetraploid (AABB) referred to today as durum wheat. M. A. Carleton, cerealist with the USDA from 1894 until 1918, collected wheat specimens in Northeast Asia (formally southern Russia), introducing Kubanka, a high-yielding durum wheat that was subsequently grown in North Dakota.

Introduction of White Wheat

Several white wheats were introduced into the United States during the 1800s. These include Sonora, Pacific, Little Club, Club, and Goldcoin. As will be noted later, millers preferred and the country predominantly produced white wheat by the mid-1800s. Baart white wheat was introduced into Arizona in the early 1900s and its production was well established in that state by 1914. Baart originated in South Africa and arrived in the United States via Australia. Production spread to Oregon, Idaho, Washington, and California by 1929, occupying over 750,000 acres. Other early introductions of white wheat from Australia include Bunyip in 1914, Onas in 1918, and Federation in 1920. These introductions became the foundation germplasm from which modern white wheat cultivars were derived.

MODERN CULTIVAR DEVELOPMENT

As with other crops, production of introduced cultivars (probably "cultivar complex" more adequately describes these introductions as they were most likely mixtures of many genotypes) gave way to cultivars "selected" from those introductions. Red Fife HRS, as noted earlier, was selected from an introduction of wheat by David Fife of Canada. Earl G. Clark of Sedgwick, Kansas, selected a very popular Blackhull HRW in 1912 when he found three plants having glumes with black strips growing in a field of Turkey. By 1939, over 8,000,000 acres were grown in Texas, Oklahoma, and Kansas. From 1819 until 1936, 132 cultivars of all classes of wheat were sold or distributed by private individuals or companies and by USDA/SAES. Of these 132, 4 were of unknown origin, 24 were introductions from other countries, 74 were direct selections, and 32 resulted from hybridizations. Clark and Bayles described 190 cultivars of all classes of wheat grown in the United States in 1942. Today there are hundreds of wheat cultivars available to U.S. producers. Most, if not all, resulted from planned hybridizations and selection. Many of these have ancestral ties to the earliest introductions into the United States—e.g., Mediterranean SRW is ancestral to many HRS and white wheats, as well as SRW cultivars. Likewise, the original, successful HRW wheat, Turkey, has been used as a parent in HRS, HRW, W, and SRW cultivars. All of the early white introductions are ancestral to only white wheat cultivars, except for Goldcoin, which was used in the development of SRW wheats. Durum wheat introductions were the basis of improved durum cultivars only.

Cultivar development in the United States by intentional hybridization began in 1886 with the release of Fulcaster, the result of crossing Fultz SRW with Lancaster SRW. Fultz was an awnless selection from Lancaster, which apparently was a selection from Mediterranean SRW. Fulcaster was developed by S. M. Schindel of Hagerstown, Maryland. This cultivar became extremely popular, being grown on over a million acres in the eastern United States as late as 1939. Fulcaster was distributed under an amazing 60 plus synonymous names.

A. N. Jones of Newark, New York, released 15 cultivars resulting from intentional hybridizations from 1886 until 1906. Cyrus G. Pringle of Charlotte, Vermont, released Chambers and Surprise, both SWW, and Defiance, a SWS cultivar. E. S. Carman, former editor of the *Rural New Yorker,* produced Rural New Yorker No. 6 in 1883 by crossing wheat and rye. After the establishment of the USDA in 1862 and especially SAESs in 1887, development of wheat cultivars slowly shifted from private seedsmen to public scientists. From about 1890, A. E. Blount, working at the Colorado AES, developed over a dozen cultivars by planned hybridizations. Some of his wheats were evaluated in New South Wales and were used by Farrer to produce white wheats such as Bunyip that were then introduced into the United States. W. J. Spillman at the Washington AES developed and released four club wheat cultivars in the early 1900s. Ceres HRS, developed by L. T. Waldron at the North Dakota AES, was one of the most popular cultivars of the 1920s and 1930s in the Northern Great Plains. Released in 1926, three million acres were grown by 1932 and five million by 1933. Other noted scientists of the early 1900s responsible for the development of improved wheat cultivars include H. K. Hayes with the Minnesota AES, E. F. Gaines at the Washington AES, J. H. Parker with the Kansas AES, and E. S. McFadden with the USDA in South Dakota.

Scientists and breeders making meaningful contributions to cultivar development or germplasm improvement since the early pioneers in the field are too numerous to note in such a text as this. However, the student of wheat must be apprised of a Japanese wheat cultivar, Norin 10, its impact on world wheat production, and a few of the men involved with its incredible contribution. Norin 10 was developed by G. Inazuka and S. Asanuma in Japan in 1935 from a cross of Fultz Daruma and Turkey. During the U.S. occupation of Japan after World War II, Dr. S. S. Salmon collected seed of Norin 10 and forwarded the sample to the United Sates in 1946. This Japanese cultivar was subsequently used as a parent in 97 cultivars of U.S. wheats: 19 HRW, 34 SRW, 21 HRS, and 23 white wheat cultivars. Norin 10, along with two other Japanese and two Korean introductions, contributed genes for reduced height, commonly called semi-dwarf genes, which made possible the development of short-statured wheat cultivars that could be fertilized heavily for maximum yields without lodging. These oriental cultivars, or their derivatives, were used by Dr. Norman Borlaug at the International Maize and Wheat Improvement Center (CIMMYT) to develop germplasm/cultivars adapted to the tropics. For his work in developing such wheat cultivars adapted to Mexico, India, Pakistan, Iran, and the Mediterranean basin, Dr. Borlaug was awarded the Nobel Peace Prize in 1970.

HISTORICAL EVENTS

Much of the recorded and presumed history of humankind is intertwined with wheat, its production, its harvest, and its consumption. It would take volumes to record all of the interactions of humans and wheat, but four deserve special mention: (1) wars, (2) sources of power, (3) the Deere plow, and (4) the McCormick Reaper.

Wars

Former President of the United States Herbert Hoover is quoted as saying, "The first word in war is spoken by guns, but the last word has always been spoken by bread." Examples of this truism abound. Napoleon I pursued a retreating Russian army in

1812 so vigorously that they outraced their food supply (part of which surely was bread from wheat). The retreating Russians consumed and/or destroyed all food supplies (as well as other items that could provide aid and comfort to the enemy) and when Napoleon's troops reached Moscow, they found the city virtually empty and destroyed by fire. By this time, the Russian winter was approaching and during the return to France, Napoleon lost 500,000 men to hit-and-run attacks of Russian Cossacks, harsh winter conditions, and hunger. Hitler would make a similar mistake 130 years later.

The U.S. Civil War has been called a war of bread versus cotton. The northern states produced wheat, the south cotton; one capable of sustaining a nation and feeding an army, the other could only afford men to produce and sell to foreign markets for money with which to buy bread.

During World War II, the United States exempted farmers from military service. Wheat from the United States and Canada provided the bread that energized the armies of many nations that banded together to defeat Hitler. Other examples are beyond the scope of this text, but one need only to consider that ancient wars began after harvest because no nation can afford to contemplate war without food for its army.

Sources of Power

Transitions of power sources have played major roles in the evolution of humankind from an almost totally agrarian society to industrialized, complex societies with bustling cities of millions. Societies reach a critical mass in which people are almost forced to look for better ways to perform tasks, or remain slaves to burdensome, time-consuming technology. To appreciate the importance of power sources, one must gain some insight into the machinery of converting wheat to flour.

Archaeologists suggest that wheat (grain) milling predates our recorded history. It seems obvious that humans would have first consumed grains whole but at some point realized that crushed grains mixed with water (and later cooked) were more easily consumed, and perhaps in an empirical way, they learned that crushed grains provided more energy. The most primitive source of power is human power. From before the beginning of agriculture, humans gathered wheat, placed the grains on one stone, and hit them with another—a mortar and pestle system. The pounding of grains gave way to rubbing stones together, the bottom stone having a depression or "saddle" and a smaller stone used by the miller in a back-and-forth motion. Still using human labor, querns became the accepted method of milling. The simplest quern consists of two round stones, one atop the other. Grains are placed in between and ground when the top stone is rotated. This was the basic principle of milling into the 16th century A.D.

The quern was improved by adding grooves in the bottom stone to discharge the flour and a hole in the center of the top stone where grain could be added. Later, large "hoppers" would complete the transition from home industry to commercial industry. The acceptance of bread in the Hebrew, Greek, and Roman cultures, combined with the aggregation into cities, mandated mass production. This necessitated large mills, which required energy, either from slaves or beasts of burden. Water power was harnessed first by the Romans and first appeared in England in 762 A.D. Wind-powered grinding mills came on the scene in the 1100s in Europe and England and for the next seven centuries these two sources of power, wind and water, drove the wheat mills of the world. James Watt provided the next source of power with his invention of the steam engine in 1769. The first mill powered by steam was built in England in 1780. The first such mill in the United States was built in Pittsburgh around 1800 and by 1870, 23 percent of the U.S. mills had steam power. Steam power, an external combustion system, was replaced by the internal combustion engine, first described in principle by W. Cecil to the Cambridge Philosophical Society in 1820.

Transition of power sources had played a major role in the production and milling of wheat, as well as in its transportation. But in the early days of the United States, it was a lack of power, human power, that spurred inventiveness. The opening up of the Great American Plains, from Texas to the Dakotas and then west to the Rocky Mountains meant that land, cheap land, often free land, and lots of it was available. But it also meant farming in semi-arid environments, with hot summers and extremely cold winters. It meant finding a crop that supplied the farmers with food and a living. The genetic diversity of wheat solved the crop question, but tilling the earth under such extreme conditions could be near impossible. The Deere Plow was a giant step forward in the ability of the American farmer to till the earth in preparation for planting, and the McCormick Reaper made it possible to harvest the large acreage necessary for profitable wheat production on the American Great Plains.

The Deere Plow

The early settlers of the New World had only wooden plows with which to turn the earth in preparation for cropping. Obviously, such tools were fragile and were dwarfed by the difficulty of tilling virgin soil, with tree stumps and roots, or with prairie grasses in the Great Plains. The first improvement of this situation came in 1819 when Jethro Wood patented a cast iron plow that soon became standard. However, many soils of the Midwest and Great Plains would stick to the plow, making it difficult for man and beast.

John Deere, a blacksmith in Grand Detour, Illinois, observing this plight of local farmers in 1836, fashioned a circular steel saw blade into a moldboard. Deere's moldboard worked as he had hoped, the furrow slice fell away cleanly from the steel surface, and the moldboard was actually polished as it sliced through the soil. By 1857, Deere was producing 10,000 plows per year, and Deere and Company went on to become one of the largest industrial corporations in the United States.

The McCormick Reaper

To appreciate the significance of a mechanical harvester, one must be cognizant of its predecessors. From ancient times, humans cut the heads from wheat by any sharp instrument, beginning with crude, sharpened rocks. By the time the United States was settled, the sickle, a knife with a curved blade somewhat perpendicular to the handle, and the scythe, essentially a larger sickle attached at a right angle to long wooden handle, were used. The advantage of the scythe was that a person could stand upright while working, but the disadvantage was that the wheat had to be cut close to the ground, meaning the handling of a lot of straw. The cut stalks then were gathered and tied into bunches called "sheaves" that were stacked into groups called "shocks." In wetter climates, the dried wheat was then placed in a barn, awaiting the winter work of thrashing the grain free from the heads. Thrashing was accomplished by flail, a wooden handle with a short, heavy free-swinging stick attached to one end, or by trampling the heads with animals, usually cows. The trampled or flailed wheat was then cleaned of chaff, or glumes, and straw by winnowing—i.e., tossing into the air and allowing the wind to blow chaff and straw particles away while the heavier grain fell back in place. The machine age of harvesting wheat began with Cyrus McCormick of Virginia and Obed Hussey of Maryland.

In 1831, both men, unknown to each other, succeeded in building a horse-drawn reaper. Both reapers utilized a toothed sickle bar that moved back and forth horizontally. Hussey received the first patent, but McCormick also was granted a patent the following year, 1834. There was a long and often bitter struggle between Hussey and McCormick, with the brash, business-minded McCormick finally winning, although Hussey may have had originally the better machine. Hussey, a quiet, unassuming Quaker, died in 1860, the result of a freak train accident. This "reaper war" was the beginning of the industrialization of the American wheat farm.

There is an interesting footnote of American political history that was reported in "Invention and Technology" in 1990. In 1851, John H. Manny was sued by McCormick for infringement, although McCormick no longer held an exclusive patent. Manny won the case, represented by Edwin M. Stanton, later Secretary of War under Abraham Lincoln. McCormick was represented by Abraham Lincoln and William H. Seward, who would later become Lincoln's Secretary of State. The $1,000 fee received by Lincoln may have helped finance his senatorial campaign, which featured head-on debates with Stephen Douglas that established Lincoln as a viable presidential candidate.

DEVELOPMENT OF U.S. INDUSTRY

The successful establishment of the New England colonies and the success of spring wheat production after 1641 made mills of some type a necessity. Surely, women ground wheat to produce flour for home use, but the commercial operation of a larger mill would let these pioneers grind enough flour for a long-term supply, freeing the household for other work. The importance of mills accessible to the public was recognized very early, and settlements such as Boston offered exclusive rights (i.e., monopolies) to anyone willing to build and operate such a mill. The miller in turn had to set aside certain millstones or days for custom work. Gradually,

TABLE 2.7   Wheat Flour Milling Production in the United States and Specified States, 1992.

State or Geographical Area	Wheat Ground for Flour	Daily Capacity	Total Flour Production
	(×1000 bu.)	(cwt)	(×1000 cwt)
California and Hawaii	52,557	87,433	23,736
Florida	17,886	27,865	7,993
Georgia	10,552	—1	4,444
Illinois	44,081	69,038	19,219
Indiana	—1	—1	—1
Iowa	16,019	20,960	7,112
Kansas	89,301	135,580	39,590
Michigan	17,585	29,800	7,878
Minnesota	75,487	123,350	34,360
Missouri	54,883	84,900	24,335
New York	70,512	110,224	31,667
North Carolina	23,219	33,760	10,227
Ohio	52,000	81,950	23,115
Oklahoma	—1	31,000	—1
Oregon and Washington	29,979	59,800	13,365
Pennsylvania	33,057	50,152	14,560
Tennessee	33,511	48,364	14,702
Texas	32,365	44,060	14,174
All other states	137,609	222,780	61,789
Total U.S.	833,339	1,293,844	370,829

Source: USDC, Current Industrial Reports. Flour Milling Products. Summary for 1992.

1. Data withheld to avoid disclosing figures for individual companies.

as commercial business expanded, the miller was regulated by the state to ensure honest weights and measures, quality, and to prevent adulteration.

New York City became the first commercial center for wheat milling. Exporting wheat flour appears to have begun in the 1640s and the need for some form of inspection to ensure quality and maintain a good reputation was recognized by 1674. In 1686, the flour barrel was placed on the city's coat of arms, indicating the importance of the milling industry in colonial New York. The flour barrel to this day remains a part of New York City's seal. By 1725, wheat and flour were being regularly exported to the West Indies and then on to southern Europe. As the country grew westward, new centers of milling emerged: Lancaster and Philadelphia, Pennsylvania; St. Louis, Missouri; Chicago, Illinois; Minneapolis, Minnesota; and Kansas City, Kansas. Railroad expansion from 1850 to 1875 made wheat grown almost anywhere in the United States accessible to established milling centers. Today the leading states in milling capacity are Kansas and Minnesota and, although it produces very little wheat, New York remains a major state in flour production, a fact made possible by modern transportation (Table 2.7). Nine other states or geographical areas produced in excess of one billion pounds of flour in 1992.

PROCESSING WHEAT

The soft wheats predominated U.S. production before the introduction of Red Fife HRS in 1860 and Turkey HRW in 1874. Acreage of Red Fife and its derivatives expanded quickly in the Northern Great Plains as did acreage of Turkey types in Texas, Oklahoma, and Kansas. Millers preferred winter wheat because of its reputation for better milling qualities, but farmers in the upper Midwest and Great Plains could not consistently produce winter wheats because of their susceptibility to winter kill. However, farmers in this northern tier of states were enthusiastic about Red Fife and its derivative cultivars because of its spring growth habit and yield potential. A somewhat similar scenario was played out in the southern Great Plains. Only in this case, the farmers wanted a winter wheat to take advantage of limited winter and spring moisture and to avoid having spring wheat heading during early and mid-summer when extreme droughts and high temperatures can occur. Turkey HRW wheat met these criteria and became the pattern for present day wheat cultivars for northern Texas, Oklahoma, Kansas, and Nebraska.

In both cases, wheat millers opposed the new wheats, preferring instead soft wheats that were easier to mill. The traditional mill consisted of low, or flat, grinding where the stones are set close together and turned at high speed. This system crushed all of the components of the wheat kernel into an indiscriminate flour. This flour then contained seedcoat fragments that "speckled" the flour, thus accounting for the fact that millers preferred, and the country in the mid 1800s predominantly produced, white wheats. This flour also would absorb moisture readily, a characteristic that hastened the deterioration of the flour and shortened its shelf life. This indiscriminate mix contained the oil-rich embryo, or germ, that was not easily removed by sifting. Without refrigerated or sealed storage, the oil would become rancid, again shortening the shelf life of the flour.

Sifting, or bolting as it was called in trade parlance, would separate some of these fractions, especially in soft winter wheats having a thicker seedcoat that would crack into relatively large fragments. These could be sifted out, resulting in a whiter, much preferred flour that was less likely to spoil. But forward-thinking millers in Minnesota were among the first to realize that the HRS wheats would come to dominate acreage in the northern Great Plains. These millers began in 1865 to experiment with "high grinding," a process whereby the mill stones were set farther apart and turned at slower speeds. The effect was to first crack the wheat kernel rather than to instantly pulverize it, and then to extract the flour by a succession of grindings. One problem remained: low flour yield because of the inability to sift out medium-size endosperm-bran particles from endosperm free bran particles. The problem was solved by Edmund LeCroix by adding a device that sent a blast of air into the mixture of middlings and bran as they were being sifted, an idea that dates to 1814 in Austria. The bran particles, being lighter than the medium-size endosperm particles, are lifted upward and into a separate holding area. Thus was born the middlings purifier. This process meant consistent, high-quality flour that for a while was called "patent" flour (because a patent was issued to George Christian for the process). Because of the better bread-making quality of the Red Fife type HRS wheats, the new process established hard, high-gluten wheat as superior to soft wheats.

One last great improvement in milling took place in the 1800s that helped establish the United States as a major producer of wheat and wheat products. This was the introduction of rollers to replace the grindstones. Millers first began to experiment with rollers in 1873 at Winona, Minnesota. The first rollers were smooth and were used only for regrinding of the purified middlings—i.e., medium-size particles of endosperm and bran. Corrugated rollers that proved superior to grindstones for the initial phase of wheat grinding were perfected and patented by John Stevens, a Neenah, Wisconsin, miller. By 1881, the process of gradual size reduction by use of rollers was well established in Minneapolis and spread rapidly to other milling centers. No all-millstone mill of any significance was built in the United States after 1880, and the grindstone virtually disappeared from the American scene by 1900.

MODERN MILLING

Because the primary product of wheat processing is flour, essentially all wheat is dry milled, a process designed to separate the anatomical parts of the wheat kernel, those being the embryo (germ), seedcoat (bran), and endosperm (flour). (Some low-grade flour, called clears, is processed by wet milling procedures to separate the wheat endosperm, or flour, into its chemical constituents, starch and gluten.) The milling process consists of four basic steps: cleaning, tempering, fractionating, and separating (Figure 2.2).

Only normal, relatively clean grain is used for the milling of flour and milling by-products. Grain that is moldy, exceptionally dirty, or insect-infested is rejected. Grain entering a mill facility is inspected and classified. Mills may blend wheats of different chemical quality before the bulk enters the milling process. Bulk grain is sized by means of sieves to ensure proper breakage of kernels as they pass through the break rollers. The wheat is cleaned of all foreign objects and washed to remove any impurities adhering to the kernels.

Grains are conditioned for breakage by adding moisture, a process called tempering. The purpose of this process is to toughen the bran so that it will break into larger pieces that are more easily removed and also to soften the endosperm, thereby making it easier to mill. The soft wheats are moisturized to 15 to 15.5 percent, while hard wheats are wetted to 16.5 percent. Durum wheat is tempered to 17 to 17.5 percent moisture; however, white durum may be tempered to only 16 to 16.5 percent. Heat is often applied to speed the uptake of moisture, but care must be taken not to overheat the grain as denaturation of the gluten can occur.

Figure 2.2 *A generalized schematic of wheat dry milling. Source: Matz, Samuel A.; The Chemistry and Technology of Cereals as Food and Feed, 2nd ed.; 1991; reprinted by permission of Pan-Tech International, Inc., P.O. Box 4548, McAllen, TX 78502.*

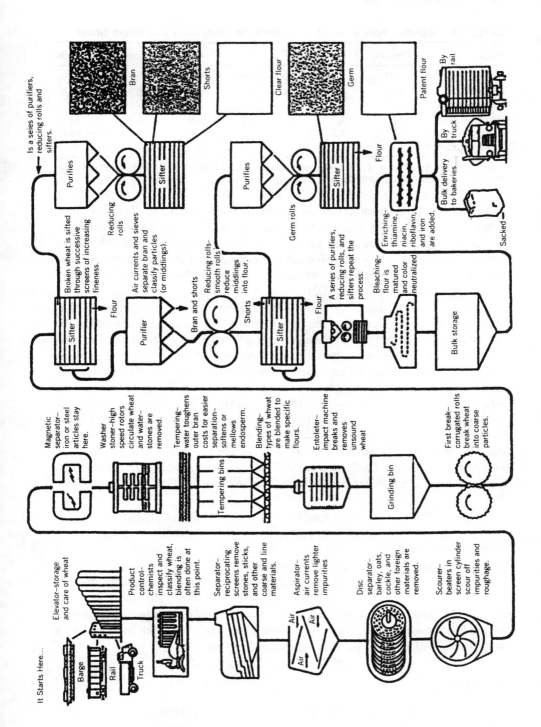

It Starts Here...

Barge · Rail · Truck

Elevator—storage and care of wheat

Product control—chemists inspect and classify wheat, blending is often done at this point.

Separator—reciprocating screens remove stones, sticks, and other coarse and line materials.

Aspirator—air currents remove lighter impurities

Air

Disc separator—barley, oats, cockle, and other foreign materials are removed.

Scourer—beaters in screen cylinder scour off impurities and roughage.

Magnetic separator—iron or steel articles stay here.

Washer stoner—high speed rotors circulate wheat and water—stones are removed.

Tempering—water toughens outer bran costs for easier separation—softens or mellows endosperm.

Tempering bins

Blending—types of wheat are blended to make specific flours.

Entoleter—impact machine breaks and removes unsound wheat

Grinding bin

First break—corrugated rolls break wheat into coarse particles.

Is a series of purifiers, reducing rolls and sifters.

Purifies

Reducing rolls

Sifter

Broken wheat is sifted through successive screens of increasing fineness.

Flour

Purifier

Air currents and sieves separate bran and classify particles (or middlings).

Bran and shorts

Reducing rolls—smooth rolls reduce middlings into flour.

Shorts

Sifter

Flour

A series of purifiers, reducing rolls, and sifters repeat the process.

Bleaching—flour is matured and color neutralized

Bulk storage

Bran

Shorts

Clear flour

Purifies

Germ rolls

Sifter

Flour

Germ

Enriching—thiamine, niacin, riboflavin, and iron are added.

Patent flour

Bulk delivery to bakeries....

By rail

By truck

Sacked

79

TABLE 2.8 Nutritive Value of White Flour, Enriched White Flour, and Whole Wheat.

Flour	Protein	Fat	Ca	Fe	Thiamine	Riboflavin	Niacin
	(%)	(g./lb.)	------------------------- (mg./lb.) -------------------------				
Unenriched (all purpose)	10.5	4.5	73	3.6	0.28	0.21	4.1
Whole wheat	13.3	9.1	186	15.0	2.49	0.54	19.7
Enriched (all purpose)	10.5	4.5	73	13.0	2.00	1.20	16.0

Source: Adapted from Watt et al., 1975. With permission from B. K. Watt and A. L. Merrill, Handbook of the Nutritional Contents of Food, 1975. Dover Publications.

In simple terms, wheat kernels are passed through a series of "break" rollers. After each break, usually four or five, the ground stock is separated by a combination of sieves and purifiers—i.e., aspiration by air. The first objective is to break the kernels into large pieces with additional breaks separating the endosperm from the bran and germ, finally reducing the endosperm to flour fineness. At each break, some flour is produced (Figure 2.2).

The break rollers consist of two rollers revolving in opposite directions, thus moving in the same direction at the closest point along a line perpendicular to the axis of the rollers, referred to as the nip. Break rollers are always corrugated and one revolves faster than the other, thus causing a shearing action, as well as a crushing action, that scrapes pieces of endosperm from bran in later breaks. The final rollers are smooth and referred to as reducing rolls, because they are designed to reduce the size of endosperm pieces resulting from the break rollers. Products from flour milling are bran, shorts, clear flour (may be referred to as clears), germ, and white flour. White flour enrichment with thiamine, riboflavin, niacin, and iron was advocated as early as 1940, and by mid-1941, 40 percent of household flour was being enriched, and by early 1942, 80 percent. This cereal enrichment with vitamins and iron is credited as a major step in the improvement in the health of Americans (Table 2.8).

By-products of the dry milling industry include the bran, germ, and a number of products that are mixtures of flour and particles of bran and/or embryo with some endosperm attached. These mixtures are too difficult and uneconomical to separate and so are used in feeds for cattle, poultry, and small animals. A small amount is used in breakfast foods and baked products such as bran muffins. These by-products are higher in protein, vitamins, and minerals than white flour. The wheat germ and germ oil are used in breakfast foods and specialty breads. Glutamic acid, another by-product, is used to produce monosodium glutamate, a salt-like substance with almost no flavor that enhances the flavor of other foods.

Wet Milling

As noted previously, some wheat clears is wet-milled into wheat starch and gluten. The gluten is used predominantly to increase the protein content of such products as baby foods and breakfast cereals. Wheat starch has properties that make it unique relative to corn starch, but wheat starch is clearly a by-product of the production of gluten. No significant amount of whole grain wheat is used for wet milling in the United States today because of economics. However, starch and gluten manufactures utilized 24 percent of Australia's flour production in 1988, and worldwide consumption increased from 97 tons in 1980 to 264 tons in 1988.

U.S. AND WORLD PRODUCTION

The United States averaged 11 bushels/acre on 15,408,000 acres for all wheats produced in 1866, thus totaling 169,703,000 bushels (Table 2.9). Of this production, 12,647,000 bushels, 7.5 percent, were exported. The producer received $2.06/bushel, the great war between the states having just ended the previous year, holding the price received much higher than during peace time. Acreage devoted to wheat increased to 38,000,000 by 1880 and reached an all-time high of 73,700,000 in 1919, in response, no doubt, to the increased price received as a result of World War I, that being $2.16/bushel, a price received not matched again until 1973. In more modern times, the United States grows SRW, HRW, HRS, durum, and white wheat on about 70,000,000 acres (Figure 2.3).

The first million bushel crop was produced in 1915, and the 2-million bushel mark was reached in 1975 (Table 2.9). Average yield ranged from 10 to 15 bushels/ acre from 1866 through 1925, reached 20 bushels/acre for the first time in 1956, 30 bushels/acre by 1969, and was 39.5 bushels/acre in 1990. The lowest average price received by producers in the United States was $0.38/bushel during the depths of the Great Depression in 1932, and the all-time high was $4.09/bushel in 1974. The United States exports a large amount of wheat, averaging 58.6 percent of production from 1986 through 1990. The United States averaged giving 42,752,000 bushels/ year to world food programs from 1986 through 1989.

Production in most states has increased dramatically since the 1930s (Table 2.10). States with the greatest production in 1990 were Kansas, producing 472,000,000 bushels of HRW, and North Dakota that produced 385,220,000 bushels of HRS and durum. In 1990, the northeast region, including Michigan, produced over 60,000,000 bushels of predominately white wheat. The region including the Mississippi River Valley and the eastern states produced over 500,000,000 bushels of SRW, and the Western states produced 373,000,000 bushels of white and HRW, including about 10,000,000 bushels of durum. The southern Great Plains produced 990,463,000 bushels of mostly HRW and the northern Great Plains produced 688,395,000 bushels of HRS and 112,689,000 bushes of durum. Over 90 percent, 103,700,000 bushels, of the durum wheat produced in the United States in 1990 was produced in North Dakota. Harvested acres, average yield, and total production of winter, durum, and other spring wheats (durum is a spring wheat) are shown in Table 2.11 for 1990.

Wheat is a worldwide crop, one that is harvested somewhere every month of the year. Although the United States is a major producer of wheat, 2,736,428,000 bushels, and ranked second only to the People's Republic of China in world produc-

TABLE 2.9 Production Statistics for All Wheats Grown in the United States from 1866–1990.

Year	Harvested Acres (×1000)	Average Yield (bu./ac.)	Average Yield (lb.¹/ac.)	Total Production (×1000 bu.)	Total Production (×1000 lb.)	Average Price Received (cents/bu.)	Exports as % of Production²
1866	15,408	11.0	660	169,703	10,182,180	206.2	7.5
1870	20,945	12.1	726	254,429	15,265,740	104.2	20.7
1875	28,382	11.1	666	313,728	18,823,680	101.0	23.8
1880	38,096	13.2	792	502,257	30,135,420	95.2	37.5
1885	35,095	11.4	684	399,931	23,995,860	77.2	24.2
1890	36,686	12.2	732	449,042	26,942,520	83.7	24.3
1895	38,998	13.9	834	542,119	32,527,140	50.5	24.0
1900	49,203	12.2	732	599,315	35,958,900	62.1	36.8
1905	46,306	15.2	912	706,026	42,361,560	74.7	14.3
1910	45,793	13.7	822	625,476	37,528,560	90.8	11.4
1915	60,303	16.7	1002	1,008,637	60,518,220	96.1	24.4
1919	73,700	12.9	774	952,097	57,125,820	216.3	23.3
1920	62,358	13.5	810	843,277	50,596,620	182.6	43.8
1925	52,443	12.8	768	668,700	40,122,000	143.7	16.2
1930	62,614	14.2	852	886,470	53,188,200	67.1	14.8
1932	57,839	13.1	786	756,927	45,415,620	38.2	5.4

Year						
1935	51,229	12.2	732	626,344	37,580,640	83.2	2.5
1940	53,273	15.3	918	814,646	48,878,760	68.2	5.0
1945	65,167	17.0	1020	1,107,623	66,457,380	149.0	28.8
1950	61,607	16.5	990	1,019,344	61,160,640	200.0	36.7
1955	47,290	19.8	1188	935,094	56,105,640	198.0	36.7
1956	49,768	20.2	1212	1,005,397	60,323,820	197.0	53.5
1960	51,879	26.1	1566	1,354,709	81,282,540	174.0	46.6
1965	49,560	26.5	1590	1,315,603	78,936,180	135.0	64.8
1969	47,146	30.6	1836	1,442,679	86,560,740	125.0	41.8
1970	43,564	31.0	1860	1,351,558	81,093,480	133.0	54.8
1974	65,368	27.3	1638	1,781,918	106,915,080	409.0	57.1
1975	69,499	30.6	1836	2,126,927	127,615,620	355.0	55.1
1980	70,984	33.4	2004	2,374,306	142,458,360	391.0	63.1
1985	64,704	37.5	2250	2,424,115	145,446,900	308.0	37.5
1990	69,283	39.5	2370	2,736,428	164,185,680	261.0	39.0

Source: Adapted from USDA Agricultural Statistics, 1936 and following years.

1. One bu. of U.S. No. 1 grade wheat = 60 lbs.

2. Percent exports calculated as gross exports, including in later years flour and other products expressed as wheat equivalent (i.e., exports divided by current year production ×
100). Does not consider imports of wheat or wheat products.

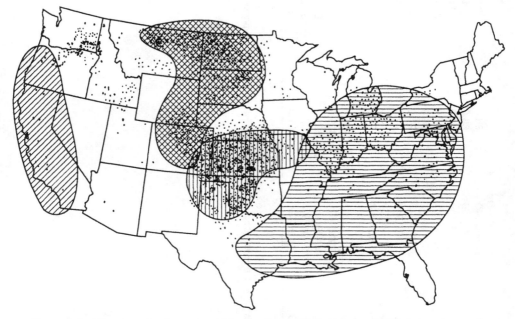

Figure 2.3 *Wheat production areas of the continental United States. The southeastern and eastern states primarily produce soft red winter; hard red winter wheat is produced primarily on the southern Great Plains; hard red spring is produced primarily on the northern Great Plains; white wheat production predominates in the western states; durum is produced in the northern Great Plains. Some production of all wheats occur outside of the shaded areas. (Drawing by Mike Hodnett)*

tion, it produced only 12.7 percent of the world's supply in 1990 (Table 2.12). World acreage devoted to wheat was 572,195,000 in 1990, averaging 37.8 bushels/ acre and producing a total of 21,606,000,000 bushels.

PLANT MORPHOLOGY

Wheat is typical of many annual grasses in its generalized growth and structure, even with the distinct differences in growth habits within and across several species. Owing to its economic importance worldwide, morphology and physiology of the wheat plant have been studied in great detail. Plants may reach seven feet in height, although less than four feet is common of normal-height commercial cultivars with the modern semi-dwarf cultivars being usually less than three feet, depending on moisture, fertility, day length, and, of course, genetic makeup. The mature wheat plant is composed of roots, stems (culms), leaves, and heads (also called ears). The wheat inflorescence is botanically a spike.

The first visible roots are the seminal, or seedling roots, that appear shortly after germination (Figure 2.4a). These roots provide the young seedling with moisture and nutrient uptake during early growth stages. The seminal roots originate in the

TABLE 2.10 Production of All Wheats in the United States by State and Region.

State	1934	1940	1950	1960	1970	1980	1990
				$(\times 1000\ bu)^1$			
Michigan	10,976	18,290	29,666	33,926	22,035	35,200	41,250
New Jersey	1,127	1,265	1,667	1,617	1,216	1,849	1,247
New York	4,284	8,082	11,172	7,380	6,579	6,000	7,105
Pennsylvania	14,654	17,675	18,986	15,782	9,834	9,250	10,500
Northeast[2]	31,041	45,312	61,501	58,705	39,664	52,299	60,102
Alabama	66	75	170	1,300	2,324	6,630	6,650
Arkansas	297	341	252	4,420	10,725	32,870	49,000
Delaware	1,539	1,273	1,020	775	798	1,080	3,060
Florida	—[3]	—[3]	—[3]	—[3]	1,102	—[3]	1,815
Georgia	756	1,892	1,350	2,070	3,600	19,800	20,650
Illinois	29,248	39,285	27,632	46,226	35,748	75,360	88,800
Indiana	32,040	27,934	32,193	41,844	29,799	53,900	50,440
Kentucky	4,250	5,625	3,720	5,191	6,120	13,825	20,000
Louisiana	—[3]	—[3]	—[3]	1,218	957	1,876	12,870
Maryland	7,934	6,897	5,162	4,588	4,181	3,686	9,880
Mississippi	—[3]	—[3]	120	1,110	4,930	9,300	15,600
Missouri	21,266	32,547	23,782	37,648	31,222	89,010	76,000
North Carolina	4,340	6,645	5,340	7,966	8,514	10,500	22,550
Ohio	33,350	42,121	46,596	52,500	35,927	67,130	79,650
South Carolina	765	2,725	1,947	3,358	2,835	6,912	14,440
Tennessee	3,392	4,968	3,050	3,408	7,378	17,100	17,640
Virginia	8,092	8,168	6,768	6,656	7,260	10,582	12,220
West Virginia	1,974	1,711	1,221	756	462	342	552
Wisconsin	207	1,682	540	1,680	1,422	4,365	10,085
East	149,516	183,889	160,890	222,714	195,304	424,268	511,902
Colorado	3,760	12,354	35,184	66,121	66,684	110,300	86,950
Kansas	79,663	126,553	178,060	290,640	299,013	420,000	472,000

(continued)

TABLE 2.10 (Continued)

State	1934	1940	1950	1960	1970	1980	1990
				(×1000 bu)[1]			
Nebraska	15,008	34,634	87,714	85,712	97,204	108,300	85,500
New Mexico	561	1,680	760	4,546	5,152	10,500	8,100
Oklahoma	37,348	58,290	42,363	121,278	98,202	195,000	201,600
Texas	25,749	29,911	18,992	84,645	54,408	130,000	130,200
Wyoming	481	2,256	5,130	5,541	6,259	8,620	6,113
Southern Great Plains	162,570	265,678	368,203	658,483	626,922	982,720	990,463
Iowa	3,288	7,603	4,994	2,985	1,400	3,496	3,375
Minnesota	790	32,069	1,220	26,543	22,882	102,556	138,620
Montana	8,820	51,676	22,800	79,517	85,167	119,800	145,865
North Dakota	—[3]	93,930	—[3]	127,500	152,826	179,650	385,220
South Dakota	168	26,261	3,965	46,156	39,282	62,425	128,004
Northern Great Plains	13,066	211,539	32,979	282,701	301,557	467,927	801,084
Arizona	1,000	780	700	858	10,350	17,200	9,266
California	8,384	11,745	13,671	7,744	22,175	85,500	48,165
Idaho	8,208	26,292	19,992	35,031	42,734	96,030	99,600
Nevada	40	491	120	457	810	1,800	980
Oregon	8,874	16,960	18,450	26,542	27,658	77,400	57,616
Utah	1,606	5,466	5,216	5,106	5,976	8,942	7,170
Washington	21,247	44,180	58,960	65,102	97,075	160,220	150,080
West	49,359	105,914	117,109	140,840	206,778	447,092	372,877
Total U.S.	405,552	812,374[4]	740,682	1,363,443	1,370,225	2,374,306	2,736,428

Source: Adapted from USDA Agricultural Statistics, 1936 and following years.

1. One bushel of U.S. No. 1 grade wheat = 60 lbs.
2. Regions roughly indicative of type of wheat produced in 1990: Northeast and East—soft red winter; Southern Great Plains—hard red winter; Northern Great Plains—hard red spring and durum; West—white and hard red winter.
3. No production reported.
4. Includes 42,000 bushels produced in Maine.

TABLE 2.11 Production Statistics for the United States by Class, 1990.

Class	Harvested Acres	Yield	Total Production
	($\times 1000$)	(bu.[1]/ac.)	($\times 1000$ bu.)
Winter Wheat	49,901	40.7	2,030,874
Durum	3,507	34.9	122,430
Other Spring	15,875	36.7	583,124

Source: Adapted from USDA Agricultural Statistics, 1991.
1. One bu. of U.S. No. 1 grade wheat = 60 lbs.

scutellar, the part of the embryo that is in contact with the endosperm, and epiblast regions of the wheat seed (Figures 2.4b and 2.5). These roots may be the primary roots of wheat in the upper rhizosphere, functioning throughout the life of the plant unless destroyed by mechanical means or biotic pests.

Additional roots, called nodal, coronal, or adventitious, arise from nodes within the crown, the crown being the meristematic region where the shoot and root meet (Figure 2.4c). The first crown roots grow laterally to some degree before becoming strongly tropic. This growth pattern means that the wheat plant is well adapted to arid climates where maximum soil moisture extraction is important. These roots may penetrate over three feet deep, with much deeper penetration possible.

The crown is composed of differentiated nodes, unexpanded internodes, leaves, etc., plus the developing head. The meristic region of cell division and differentiation (that is, the crown) is located as deep as two inches below the soil surface until the developing head is thrust upward by the expanding internodes.

The development of the wheat plant is keyed to the development of stem nodes and the expansion of internodes. All leaves, adventitious roots, and tillers arise from buds associated with crown or stem nodes. The embryo has five nodes already differentiated before germination: the scutellar, or cotyledonary, the epiblast, coleoptilar, and the first and second true leaf nodes. During germination and early seedling growth, the coleoptile elongates upward until it encounters sufficient light to cause growth to stop or it reaches its maximum length. The first true leaf then emerges through the coleptile.

The wheat stem (after vernalization in winter wheat and/or the critical photoperiod is reached in photosensitive wheats) elongates by a process called jointing, a joint being synonymous with node. The early leaves arise from lower nodes with unexpanded internodes, giving the juvenile plant a typical short-grass appearance. Once the plant shifts from vegetative to reproductive growth, up to 25 reproductive, or rachis, nodes are differentiated and begin expanding in the meristic region. The maximum number of spikelets, subdivisions of the spike containing one to nine florets, is set by the late tillering stage of growth, well before stem elongation, and the maximum number of florets is set by the time the stem begins to elongate.

The inflorescence, or head, is pushed upward as the stem internodes elongate below the meristematic region, a process called jointing. The spike is enclosed in the uppermost leaf sheath—i.e., the flag leaf sheath—as it moves upward (Figure

TABLE 2.12   Production of All Wheats in Specified Countries for 1990/1991[1] Producing at Least 50 Million Bushels.

Country/Continent	Harvested Acres	Yield	Total Production
	(×1000)	(bu.[2]/ac.)	(×1,000,000 bu.)
Canada	35,551	33.8	1,202
Mexico	2,347	61.1	143
United States	69,254	39.6	2,736
N. and C. America	107,188	38.1	4,081
Argentina	14,079	28.4	400
Brazil	8,151	14.0	114
Chile	1,151	50.7	58
South America	25,161	24.2	610
Austria	687	75.1	52
Belgium-Luxembourg	553	92.5	51
Bulgaria	2,873	65.1	187
Czechoslovakia	3,065	80.4	247
Denmark	1,319	110.0	145
France	12,795	96.5	1,234
Germany	6,002	93.2	560
Greece	2,174	28.4	62
Hungary	2,769	81.8	226
Italy	6,849	43.4	298
Poland	5,634	58.9	332
Spain	4,955	35.2	175
Romania	5,582	46.4	259
United Kingdom	5,064	101.6	514
Yugoslavia	3,639	63.2	234
Europe	67,140	71.5	4,803
Former Soviet Union	119,047	31.4	3,742
Egypt	1,828	86.1	157
Morocco	6,718	19.8	133
South Africa	3,829	16.4	66
Africa	22,371	22.6	504
Afghanistan	3,952	15.3	61
People's Republic of China	75,960	47.4	3,608
India	58,050	31.5	1,831
Iran	16,055	18.7	301
Pakistan	19,377	27.4	530
Saudi Arabia	1,754	75.4	132
Syria	2,717	23.3	63
Turkey	21,613	27.2	588
Asia	208,848	35.1	7,335
Australia	22,813	24.2	553
Oceania	22,914	23.8	560
World Total	572,195	37.8	21,606

Source: Adapted from USDA Agricultural Statistics, 1991.

1. Refers to years of harvest. Harvests in the Northern Hemisphere were combined with those immediately following in the Southern Hemisphere.
2. Based on 60 lbs./bu.

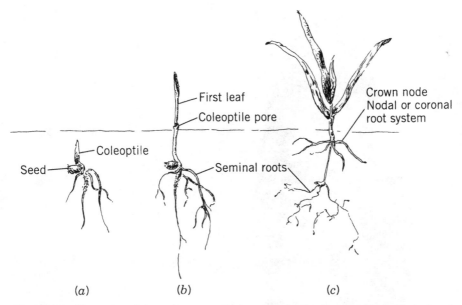

First leaf
Coleoptile pore
Crown node
Nodal or coronal
root system
Coleoptile
Seed
Seminal roots

(a)   (b)   (c)

Figure 2.4 *Illustration of the early stages of wheat germination and seedling growth showing the establishment of the seminal roots and early stage of coleoptile elongation (a), the emergence of the first leaf through the coleoptile pore (b), and the development of the coronal or adventitious roots from the plant crown (c). (Drawing by Mike Hodnett)*

2.6). The inflorescence usually has fully emerged before flowering begins. The inflorescence of wheat is a spike, with spikelets borne singularly at nodes, or joints, on alternate sides of a zig-zag, flattened central axis called the rachis (Figure 2.7). Each spikelet can be composed of one to nine florets, or flowers, but two to five is normal. Florets are borne on an axis termed a rachilla or secondary rachis. Each flower is composed of two empty glumes (may be termed a sterile flower) subtending a set of fertile glumes termed the lemma and palea, which enclose the sexual organs: three stamens and a single ovary with a feather-like stigma borne on the connecting style. Each flower is potentially one wheat kernel. On awned plants, it is the lemma that terminates in an awn.

STAGES OF GROWTH

There are at least three accepted systems of identifying the stages of plant and kernel development in wheat. The Haun scale and Zadoks scale are much more precise than the Feekes-Large scale, which is more popular. The Zadoks scale (Table 2.13) divides the growth stages of wheat into ten primary stages (10, 20, 30, etc.) and each of these is divided into ten secondary stages (01, 02, 11, 12, 21, 22, etc.) for a total of 100 stages. The problem with the Zadoks scale is that a plant could require more than one code to be adequately described. For example, consider a plant with four leaves, two tillers, and with the main stem elongated to the point of one visible

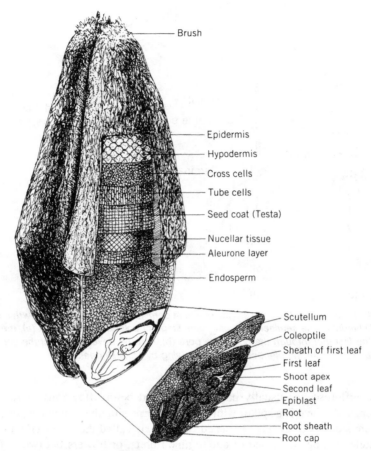

Brush

Epidermis

Hypodermis

Cross cells

Tube cells

Seed coat (Testa)

Nucellar tissue

Aleurone layer

Endosperm

Scutellum

Coleoptile

Sheath of first leaf

First leaf

Shoot apex

Second leaf

Epiblast

Root

Root sheath

Root cap

Figure 2.5 *Cross-section of the wheat kernel and embryo.* Source: *Fahn, A.; Plant Anatomy, 2nd ed.; Elsevier Sciences, Inc., 1974; reprinted by permission of Butterworth-Heinemann Ltd. (Embryo redrawn by Mike Hodnett.)*

node. This plant would be described as attaining growth stages 14, 22, and 31, all three being necessary to precisely describe the plant. The Feekes-Large scale, on the other hand, would put this plant simply at stage 6.

PRODUCTION PRACTICES

Cultivar Choice

Information on cultivars adapted to particular locales should be obtained from state agriculture experiment stations through local or county extension personnel. Other sources include seed company advertisements and experience of neighboring producers. Wheat producers should always look for multiple-year data or experience and plant only a part of their acreage to any new cultivar in order to evaluate it under their management strategies.

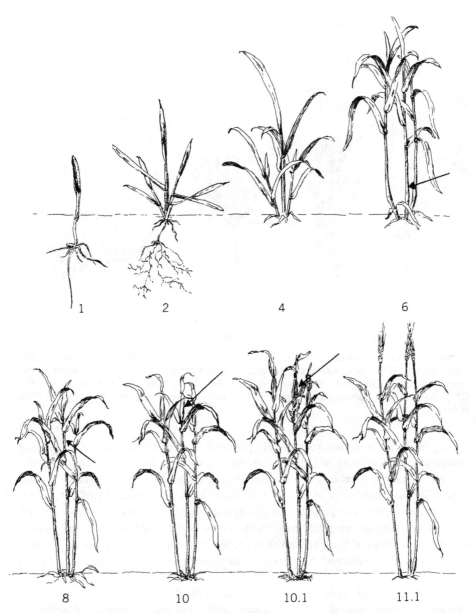

Figure 2.6 *Selected growth stages of the wheat plant according to the Feekes-Large scale. Position of the developing ear is identified by arrows. Stages are described in Table 2.13. (Drawing by Mike Hodnett)*

The cultivar development industry in the United States, both public and private, is a mature and sophisticated industry supported by an extensive, public evaluation process. Scientists involved in cultivar development and evaluation consider an array of details in such a manner that the local producer need only to consider a few agronomic aspects of new cultivars. Producers look for such characters as grain

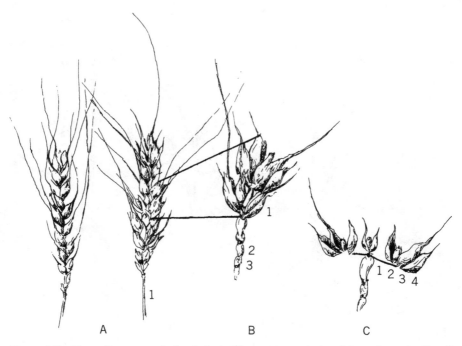

Figure 2.7 *The spike or ear of wheat plant: A1 = uppermost stem internode supporting the inflorescence; B1 = two individual spikelets with multiple florets; B2 = a flattened rachis internode; B3 = rachis node; C = an individual spikelet at maturity with three wheat kernels; C1 = palea; C2 = kernel; C3 = lemma; C4 = outer glume of spikelet. (Drawing by Mike Hodnett)*

yield potential (sometimes forage yield potential if planting for grazing, or grazing and grain), susceptibility to lodging, earliness, and sometimes resistance to specific and reoccurring insects or diseases. Breeders may consider also resistance or susceptibility to a number of insects and diseases, bushel test weight (indicates flour yield and therefore is an indicator of quality), day length sensitivity (many cultivars in the United States are photoperiod sensitive although emphasis is usually given simply to earliness of heading or maturity), winter hardiness, vernalization requirement in winter wheats, and even baking characteristics such as loaf volume, mixing time, etc.

Yield potential of wheat has been improved over the millennia since its domestication, but especially during this century following the application of scientific principles of plant breeding. Application of biotechnology promises even more dramatic improvements. Research conducted in Kansas and reported in 1988 indicated that yield potential increased at a rate of 14.5 pounds of grain/acre/year from the early 1900s through 1987 (Figure 2.8). Days to heading decreased by .1 day/year, or about 7 days since 1919, and plant height decreased by .2 inches/year, or about 14 inches.

Other agronomic or quality traits included in cultivar test reports will vary by locale and cultivar-dependent production hazards, including reaction to biotic pests.

TABLE 2.13   Stages of Growth in Wheat According to Feekes-Large and Zadoks Scales

Zadoks Stage	Event	Feekes-Large Stage
	Germination	
01	Imbibition begins	
03	Imbibition complete	
05	Radical emerges	
07	Coleoptile emerges	
09	Leaf at coleoptile tip	
	Seedling through Main Stem Growth	
10	1st leaf through coleoptile	1
11	1st leaf unfolded	
12	2 leaves unfolded	
13–18	3–8 leaves unfolded, respectively	
19	9 or more leaves unfolded	
	Tillering	
20	Main stem only	
21	Main stem and 1 tiller	2
22–28	Main stem and 2–8 tillers, respectively	
29	Main stem and 9 or more tillers	3
	Winter dormancy	3
	Stem Elongation	
30	Pseudostem[1] erection	4–5
31	1st node detectable	6
32–36	2nd through 6th nodes detectable	7
37	Flag leaf visible	8
39	Flag leaf ligule/collar visible	9
	Booting-Anthesis	
40	——	
41	Flag leaf sheath extending	
45	Boots slightly swollen	10
47	Flag leaf sheath opening	
49	First awns visible	
50	First spikelet visible	10.1
53	25% of inflorescence emerged from boot	10.2
55	50% of inflorescence emerged from boot	10.3
57	75% of inflorescence emerged from boot	10.4
59	Inflorescence emergence complete	10.5
60	Beginning of anthesis	10.51
65	Anthesis 50% complete	
69	Anthesis complete	

(*continued*)

TABLE 2.13 *(Continued)*

Zadoks Stage	Event	Feekes-Large Stage
	Kernel Development	
70	——	
71	Kernel watery	10.54
73	Early milk	
75	Medium milk	11.1
77	Late milk	
80	——	
83	Early dough	
85	Soft dough	11.2
87	Hard dough	
90	——	
91	Kernel hard (cannot be divided by thumbnail)	11.3
92	Kernel hard (cannot be dented by thumbnail)	11.4
93	Kernel loosening in glumes during daytime-ready for harvest	
94	Overripe, straw dead and collapsing	

Source: Adapted from Cook and Veseth, 1991; Large, 1954; Zadoks, et al. 1974.
1. Pseudostem: the early vestiges of an erect stem in wheat caused be the expanding whorl of leaves before internodes are detectable.

In cases where the producer contracts to produce wheat of a given quality, such as high-protein or wheats for specific end uses, the producer may need to seek additional information on adapted cultivars from their SAES or State Extension Specialists.

Plant Height Many of today's wheat cultivars carry genes for the semi-dwarf characteristic that reduces final plant height, thereby reducing lodging potential under high nitrogen fertility. Wheats carrying the semi-dwarf trait generally exhibit a shorter coleoptile. The risk of planting these seed too deep for coleoptile emergence is greater than with standard-height cultivars. Producers in the Great Plains may consider growing standard-height cultivars when planting into soil where sufficient moisture for germination is too deep for emergence of semi-dwarf cultivars.

Winter Hardiness The ability of a cultivar to survive extremely cold temperatures is another cultivar-dependent trait of interest to wheat producers. Winter wheat is most sensitive to low temperatures in the fall before plants have undergone a "hardening" process and in the spring after growth resumes (Table 2.14). Leaves of hardened spring wheats can tolerate temperatures as low as 14°F, while leaves of hardened winter wheats tolerate temperatures as low as −30°F. Of course, cold injury or heat injury increases with the length of exposure. Hardened crowns of winter wheat will die after exposure to 0°F for 24 hours but can tolerate 5°F for about 5 days.

While winter hardiness is desirable as far south as northern Georgia and Alabama, it is a must for winter cultivars grown in the central Great Plains, Great Lakes

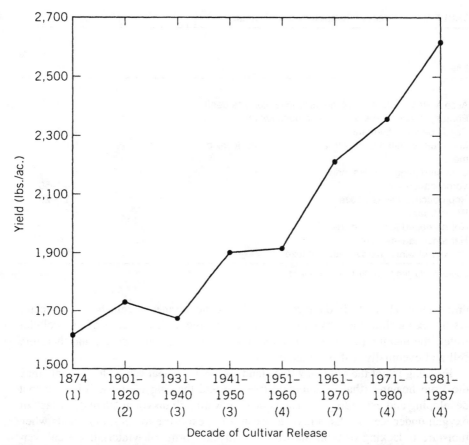

Figure 2.8 *Yields of HRW wheat cultivars by decade of release, averaged across three locations in Kansas in 1986 and 1987. Numbers in parenthesis are the number of cultivars within decades of release that were evaluated. The 1874 cultivar is Turkey and is for historical reference. Source: adapted from Cox, et al. 1988.*

region and the inland Pacific Northwest. However, cultivars with excellent winter hardiness will stay dormant longer and therefore will be, generally, lower yielding than a less-winter-hardy cultivar that begins to grow earlier in the spring. Planting a less-winter-hardy cultivar in those areas will result in higher yields, unless temperatures drop abruptly following an early period of warmer temperatures, in which case severe frost damage could greatly decrease yields.

The winter-hardening process is not well understood. It is apparent that winter-hardy types will suffer some root death, tillers may die, and new tillers will be initiated from adventitious buds and not from intercalary meristem tissue. Wheat leaf cells shift during hardening from having one large vacuole with a thin cytoplasm to having thickened cytoplasm and numerous small vacuoles. The internal structure of wheat leaves is composed of loosely packed mesophyll cells such that extracellular ice can be tolerated without injury. However, as the leaf temperature

TABLE 2.14 Effect of Temperature on Wheat.

Effect	Approximate Temperature Range
	(°F)
Accelerated respiration, plants may lose biomass-death	88–104
Photosynthesis decreases to 0 carbon fixation	75–86
Seeds become dormant	70–75
Maximum vegetative growth and reproductive development	50–75
Cold hardening process induced	≤50
Vernalization	<40–50
Reproductive tissue freeze	≤27–29
Roots freeze	≤26
Not hardened leaves freeze	≤20
Hardened leaves freeze	≤10
Crowns of winter hardy cultivar freeze	≤−11

Source: Adapted from Cook and Veseth, 1991.

drops below about 26°F, damage to the plasma membranes of cells begins to occur. As the extracellular ice crystals continue to enlarge at the expense of intercellular water, the membrane breaks down and intracellular freezing occurs, and therefore cell and eventually leaf or tissue death.

Freezing is not the only way that stands of winter wheat can be lost. Plants may die from heaving (that is, alternate freezing and thawing of moist soil such that developing and expanding ice results in an upward pull on young plants), by lack of oxygen under ice, or plants may desiccate from exposure to cold, dry winds when moisture is lacking or is in the frozen form. All forms of winterkill are cultivar-dependent.

Winter wheat cultivars that break dormancy very early in the spring are susceptible to frost damage. Frost damage occurs to developing heads after the onset of stem elongation, or jointing. With stem elongation, the apical meristem emerges from its natural protection below ground, and the rapidly dividing meristematic cells are susceptible to freezing temperatures. Spring wheats usually are planted after the threat of freezing temperatures.

Wheat cultivars developed for one region of the United States should not be planted to large acreage in other regions without extensive testing by SAES, extension personnel, or individual farmers. Winter wheats developed for Kentucky or central Texas would winterkill in more northern states, while cultivars developed for Wisconsin might never receive sufficient vernalization in Florida to initiate heading.

Early Maturity Earliness is the overall expression of (1) vernalization requirement, (2) sensitivity to photoperiod, and (3) rapid plant development. Cultivars that have little vernalization requirement, are non-photoperiodic, and have rapid development will be earlier maturing. Early maturity of winter wheat cultivars allows for seed set and grain fill before the onset of damaging summer heat and drought.

Spring wheats that mature early also take greater advantage of spring rains, and production can occur before late summer drought and heat.

Photoperiod Many wheat cultivars are photoperiod-sensitive. This is not a consideration of most U.S. producers because they are provided information on overall earliness (i.e., heading dates) through cultivar performance trials of SAES, extension agents, and cultivar development companies. However, in mild, wet climates, such as the United Kingdom, northern Europe, and the U.S. Pacific Northwest, the growing season can be extended to almost 12 months by incorporation of photoperiod-sensitive genes, thereby increasing yields. In more arid and/or hotter areas of production such as southern Europe, India, Australia, and the southeast and southwest United States, cultivars may be photoperiod-insensitive in order to initiate reproduction solely on temperature. Such non-photoperiodic cultivars would likely initiate reproductive growth too early in more northern areas of the United States and therefore be more susceptible to frost damage.

Date of Planting

Obviously, recommended planting dates for wheat vary by production area and the intended purpose of the crop—i.e., grain, silage, or grazing and grain (Table 2.15). However, the underlying principles behind the choice of planting date are the same, regardless of location.

Spring wheat should be planted as early as possible without undue risk of damage from cold temperatures and frost. Winter wheats should be planted early enough to allow for vernalization. Planting winter wheat too early, however, can result in excessive fall growth and soil moisture depletion in some dryland areas such that winter survival is reduced. Higher temperatures also can be detrimental to germination and early seedling development (Table 2.14). Recommended planting dates in the southeast United States also represent a trade-off between planting early enough to allow for maximum tillering before the onset of cold weather and planting late enough to avoid late summer heat, moisture stress, insects, and disease pathogens that can be greater in late summer.

Spring wheats should be planted when the soil temperature is at least 40°F, although maximum germination and emergence occur at 55 to 75°F. Planting when soil temperatures are below 40°F will result in delayed emergence and reduced vigor. On the other hand, late planting can result in reduced yields because of heat stress during tillering and during initiation of reproductive growth. Temperatures above 64°F at these stages of growth (stages 21 to 29 in Table 2.13) can result in fewer spikelets per head. Data from Mandan, North Dakota, indicated an average decrease in yield of 1 percent per day delay in planting past the earliest suitable date.

Recommended planting dates may be determined by factors other than agronomic ones. Pink and speckled snow molds offer a unique example. Unlike most fungi, snow mold fungi have the unique ability to grow at temperatures from about 30 to 35°F. These molds can be controlled in the Pacific Northwest by

**TABLE 2.15 Recommended Planting Dates
of Wheat for Grain in Several States and Regions
in the United States**

State or Region	Recommended Planting Date Range
Winter Wheats	
Michigan	September 1–October 5
North Dakota	September 5–October 5
Montana	September 1–October 20
Texas	September 1–November 20
Kansas	September 10–October 20
Arkansas	October 1–November 1
Kentucky	October 15–October 30
North Central Texas	October 15–November 15
Gulf Coast Texas	October 10–November 20
Piedmont area of Georgia	October 25–November 15
Alabama	November 1–December 1
Central Arizona	November 15–December 1
North Florida	December 1–December 15
Spring Wheats	
Rio Grande Plain, Coastal Plains and East Texas	December 20–January 15
Arizona	December 15–January 15
Southern Idaho	April 15–May 25
Minnesota	April 15–June 1
Montana	April 10–May 30
North Dakota	April 15–June 5
South Dakota	April 1–May 20
Washington	March 1–May 15

planting very early or very late, neither of which maximizes winter hardiness. Researchers in Idaho recommend planting winter wheat, predominately white winter wheat, very late in the planting season in areas likely to have snow mold. The late planting is timed such that emergence occurs just before snow cover. The oversummered inoculum of the fungus responsible for snow mold attacks young wheat leaves pressed against the soil by the weight of the snow. Just emerged or emerging seedlings, therefore, offer a much smaller target.

On the other hand, scientists in Washington recommend very early planting as a means, not of control, but of surviving or offsetting the effects of snow molds. By planting in late August, plants can reach maximum size before the onset of low temperatures and snow, and can therefore better tolerate the loss of leaf area to these fungi. This strategy also takes advantage of a characteristic of many monocots, that being the ability to store carbohydrates in the crown for vigorous vegetative growth when conditions allow. At least one wheat cultivar, Sprague, has been released by Washington State University that carries this trait.

Hessian fly is probably the most widespread pest of wheat that influences planting dates in the United States. This insect can cause (1) tillers to stop growing, (2) stem lodging, (3) poor grain fill, (4) stunting of seedlings, and/or (5) plant death.

This insect is distributed throughout the East, Southeast, Great Plains, and the Pacific Northwest. It attacks both spring and winter wheats and is especially damaging in the Southeast United States where conditions are favorable for many generations per year.

Adults emerge from pupae oversummered on wheat stubble or grasses. Females oviposit eggs, 200+ per female, on the upper leaf blade surface of wheat (or grass hosts in the absence of wheat). Larvae migrate down the leaf's central groove to just above the attachment of the leaf sheath to the stem. Here they feed on the stem until pupation, followed by adult emergence and the cycle repeats.

The Hessian fly is particularly important in the Southeast where up to six generations per year can occur, a generation requiring only 35 days at 70°F. In the coastal plains of the Southeast, two or three fall generations are possible, and early-planted winter wheat can be destroyed. Another two to three spring generations of Hessian fly then attack wheat during stem elongation and/or heading. Losses in wheat production to the Hessian fly are estimated at $15 to 20,000,000 annually in Georgia alone.

There are some cultivars that are resistant to certain Hessian fly biotypes, a biotype being a grouping of individuals that are phenotypically identical with all other members of its species, yet having physiological or biochemical similarities different from other members of the species. However, along with field sanitation and crop rotation, adjustment of planting dates is the most common way to control this pest in much of the United States. Fall-seeded winter wheat is planted after the emergence and death of the oversummer generation of Hessian fly, assuming clean or herbicide tillage. This is normally referred to as the Hessian fly-free date, FFD, and works very well in more northern areas. In the more southern areas, planting after the FFD will not eliminate the Hessian fly, but will prevent its building to large populations on wheat before cooler temperatures slow or stop its population growth. In the hotter southern areas, as noted above, other control methods must be employed.

Seeding Rate and Depth

The growth and development of the wheat plant are limited by genetics and environment. With reduced interplant competition, and given adequate moisture and nutrients, wheat cultivars developed for North America will initiate and produce additional stems and therefore heads, called tillers. Given higher-density stands, the wheat plant will tiller less. This trait allows for a wide range in seeding rates, although producers should strive to obtain the stand density that research has shown to optimize yield.

Seeding rates vary by locale, expected moisture outlook in the case of dryland production, fertility, date of planting, method of planting, purpose of the crop (i.e., grain, or grain and grazing), and whether planting a spring or winter wheat. Spring wheats tend to tiller less and therefore are seeded at rates 25 to 50 percent greater than winter wheats grown in the same area.

In terms of pounds of seed to plant, a poor guide but a guide nonetheless,

TABLE 2.16   Seeding Rates in Pounds per Acre for Winter Wheat in Selected Areas of the United States.

State or Region	Planting Rate Ranges		
	Dryland	Irrigated	+Grazing
	---------------------------- (lbs./ac.) ----------------------------		
Alabama	60–90	—	90–120
Texas Gulf Coast	60–75	—	90–120
Rio Grande Valley, Texas	60	75	—
Central Texas	60	75	75
High Plains, Texas	30	60–75	75
East Kansas	60–90	60	120
West Kansas	20–45	—	—
Arizona	—	100–150	—
Michigan	90–120	—	—

recommended rates for dryland production of winter wheat vary from 20 pounds in western Kansas to over 100 pounds in wetter climates (Table 2.16). Higher seeding rates should be used when planting later than recommended, planting into a dry seedbed, aerial seeding, broadcast seeding, planting no-till, or when planting for grazing and grain. Rates under irrigation range from 60 to 150 pounds/acre, and rates for grazing and grain range from 75 to 120 pounds/acre. Broadcasting seed and then disking or harrowing results in about 20 percent of the seed being either too deep or too shallow. Those two shallow are subject to desiccation after germination, and those planted too deep will leaf-out under the soil and die.

Planting wheat with a drill into a prepared seedbed will result in the optimum placement of seeds and number of seeds per unit area. Producers should strive for uniform stands to decrease lodging potential, to more uniformly extract soil moisture and nutrients from the rhizosphere, and to more efficiently intercept incoming light energy, all of which will optimize yield. Planting in rows as narrow as possible for expected moisture conditions helps accomplish these goals and will help control weeds.

Planting equipment should be calibrated to place the appropriate number of seeds per linear foot of row to maximize yield. The number of seeds per pound varies from 12,000 to 20,000, depending on cultivar and the environment in which the planting seeds were produced. Planting 60 pounds/acre at a 6-inch inter-row spacing could result in planting 8 seeds per foot of row if the cultivar averaged 12,000 seeds/pound or 14 seeds per linear foot if the cultivar averaged 20,000 seeds/pound (Table 2.17). Planting too few seeds per acre could result in excessive tillering that may delay maturity and/or increase weed competition; too many plants per acre is an unnecessary expense and may result in increased lodging and decreased yields.

Seeds should be planted into moisture if possible. Cultivars that have long coleoptiles can be planted up to three inches deep, while semi-dwarf cultivars should not be seeded more than two inches deep. In either case, deep placement of seeds can result in delayed emergence and erratic stands.

TABLE 2.17 Number of Seeds per Linear Foot of Row at Varying Pounds Planting Seed per Acre, Row Widths, and Seeds per Pound.

Planting Rate	Seeds per Pound	Inches Between Rows				
		4	6	8	10	12
(lbs./ac.)						
30	12,000	2.8	4.1	5.5	6.9	8.3
	16,000	3.7	5.5	7.3	9.2	11.0
	20,000	4.6	6.9	9.2	11.5	13.8
60	12,000	5.5	8.3	11.0	13.8	16.5
	16,000	7.3	11.0	14.7	18.4	22.0
	20,000	9.2	13.8	18.4	23.0	27.5
90	12,000	8.3	12.4	16.5	20.7	24.8
	16,000	11.0	16.5	22.0	27.5	33.1
	20,000	13.8	20.7	27.5	34.4	41.3
120	12,000	11.0	16.5	22.0	27.5	33.1
	16,000	14.7	22.0	29.4	36.8	44.1
	20,000	18.4	27.5	36.7	45.9	55.1

Tillage

Appropriate soil tillage for wheat production spans the entire range of possibilities across the United States, from full tillage using moldboard plowing to no-till. Producers of row crops have traditionally looked at tillage as a means for controlling weeds, management of previous crop residue, and seedbed preparation. But tillage also is used to help control diseases, and sometimes insects, that overwinter or oversummer on crop residue, and certain tillage operations allow for percolation of limited precipitation and/or help control wind erosion.

In areas such as the southeast and east where rainfall is plentiful, moldboard plowing is often recommended. Deep plowing and full or near full tillage allows producers to create a seedbed conducive to (1) maximum seed-soil contact for rapid moisture imbibition and germination, (2) optimum seed placement, (3) maximum fertilizer distribution, (4) maximum mechanical weed control, and (5) in-furrow insecticide or fungicide treatments if needed. Higher yields are produced on the sandy loam and loamy sands of Alabama, Georgia, and South Carolina with mold-board tillage than with disk tillage alone or no-till (Table 2.18). A three-year study from 1982 through 1984 at Quincy, Florida, indicated that moldboard and chisel plowing resulted in higher wheat yields than no-till. Moldboard plowing plus other tillage operations are used in parts of the Great Plains (e.g., eastern Kansas) as well, although moisture conservation is a major concern.

In areas of the southeast United States, SRW wheat is often doublecropped with soybean, corn, or sorghum. Producers usually disk only to partially destroy the summer crop residue and to prepare the seedbed as quickly as possible. Moldboard plowing would cost these producers time, deep moisture, and the seedbed would require time to firm such that seed could be placed at the proper depth and have

TABLE 2.18   Effects of Tillage on Wheat Grain Yields in the Southeast.

| State | Soil | Tillage System | | |
		Deep-till	Disk	No-till
		---------------------- (bu./ac.) ----------------------		
Alabama	Varied	48	43	36
Alabama	Sandy Loams	46	39	33
Georgia	Sandy Loams	50	45	32
South Carolina	Loamy Sand	40	36	37

Source: Adapted from Martin and Touchton, 1982; Smith and Hudson, 1990.

good seed-soil contact. Producers in this section of the United States may aerial seed wheat into what will become the previous crop, either by fixed-wing aircraft, helicopter, or by use of a cyclone seeder fixed to a high-clearance tractor. Seed sown this way are scattered onto a non-prepared seedbed within standing corn or soybean. Aerial seeding must be done before leaf drop in soybean and as early as possible in September into standing corn. The current crop shades the ground, preventing topsoil moisture loss and permitting the wheat seed to germinate and the seedling to become established. The summer crop is harvested in the usual manner with some wheat plant losses in the wheel tracks. Very little wheat is aerial seeded.

Moldboard, and perhaps deep disking, buries residue harboring Hessian fly, wheat stem sawfly, and other destructive insects. Control of volunteer wheat plants, either by mechanical or chemical means, can reduce the incidence of wheat streak mosaic virus by controlling the wheat leaf curl mite, particularly in the Plains states. Destruction of previous crop residue also helps control diseases such as *Rhizoctonia* root rot in higher rainfall areas of the East and parts of the Pacific Northwest, or where wheat is grown with supplemental irrigation.

Much of the HRW and HRS wheat production in the United States occurs in areas receiving very little rainfall. In much of the Great Plains and Northern Prairies, snowfall can make up the preponderance of winter moisture, with the bulk of the moisture in these areas coming during the winter and spring. This situation mandates production practices that maximize moisture conservation and minimize soil erosion. Thus, residue management becomes a major issue.

In much of the Great Plains that receives less than about 20 inches of annual precipitation, rotations involving a prolonged fallow period are used to store adequate moisture for winter wheat production. The fallow period may be 9 to 14 months in duration, during which time all vegetation must be controlled to achieve maximum soil moisture conservation. Conventional methods of vegetation control include tillage with a V-shaped sweep-type implement after winter wheat harvest when volunteer wheat and weed pressure warrant. This operation leaves upwards of 90 percent of the surface residue for erosion control (Table 2.19). The land will stay fallow during the subsequent winter, and tillage will resume in the spring after volunteer wheat and weeds emerge. A disk or V-blade will be used during the spring and summer, plus other tillage as necessary to prepare the seedbed for planting in the fall.

TABLE 2.19 Crop Surface Residue Remaining after Various Tillage Operations.

Implement	Percent Residue Remaining after One Pass	
	Idaho	Indiana
Moldboard plow	20	3–5
One way disk plow	40	——
Chisel (7–9 inches deep)	75	50–80
Blades or sweeps >24 inches wide	90	——
Rod weeder-plain rod	90	——
Rod Weeder-semi chisels	80	——
Heavy duty cultivator with 16–22 in. sweeps, 2 to 3 inches deep	85	——
Field cultivator	75–80	50–80
Offset disc or tandem disc	50	30–80
Spring tooth harrow	50	——
Drag type spike tooth harrow	40	——
Deep furrow drill with shovel or disk opener	80	——
Double-disc or single-disc opener	90	90–95
Winter weathering	——	75–95

Source: Adapted from McDole and Vira, 1980; Hill et al., 1989.

Minimum tillage, including no-till, has received a lot of attention from producers and research scientists in the American Great Plains in recent years. Advantages of minimum tillage are (1) reduced fuel costs, (2) reduced equipment costs, (3) decreased operator time, (4) less soil compaction, (5) less soil erosion, (6) improved soil moisture retention in many instances, and (7) maintenance of soil organic matter. No-till production is favored by some, because herbicides can now be used to control weeds, thereby preventing loss of soil moisture through tillage operations, and maintaining maximum residue cover for prevention of soil erosion. Soil disturbance is limited to opening a narrow slot in the soil for fertilizer and seed placement. Usually, less than 10 percent of the soil surface is disturbed during no-till production, and it appears to be an appropriate production system for some of the Great Plains winter wheat area (Table 2.20).

Factors limiting acceptance of no-till seem to be a lack of labeled herbicides for vegetation control, the cost of herbicides being perceived as more expensive than utilization of already owned tillage equipment, suitable equipment for planting through previous crop residue, and, in some instances, the discarding of field sanitation as a control measure for biotic pests. In some areas, particularly the east and Pacific Northwest that receive substantial rainfall, wheat seed germination and early seedling growth are adversely affected by no-till or minimum-till. A phytotoxic substance in wheat straw or an environment conducive to the survival of and infection by seedling disease organisms such as *Rhizoctonia*, *Pythium* and *Gaeumannomyces* spp. is believed to be the major cause of this phenomenon. Some evidence suggest that minimum tillage in the northern Great Plains will allow wheat

TABLE 2.20   Effects of Tillage Treatments on Yields of Winter Wheat.

Tillage Treatment	Sydney, NE[1] 1970–1977	Akron, CO[2] 1975–1987	Fargo, ND[3]	
			1982/85	1983–84
		------------------------------------- (bu./ac.) -------------------------------------		
Conventional	36 a[4]	42 b	7 b	49 a
Reduced-till	37 a	44 b	33 ab	53 a
No-till	38 a	48 a	61 a	47 a

Source: Adapted from Cox et al., 1986; Fenster and Peterson, 1979; Smika, 1990.

1. Conventional = moldboard plowed, 2–3 passes with field cultivator and 1–2 passes with a rotary rod weeder; Reduced-till = stubble mulch fallow with 2–4 passes with 90–150 V-blades and 1–2 passes with rotary rod weeder; No-till = vegetation controlled by herbicides.
2. Conventional = stubble mulch fallow with 2 passes with blade tillage and rod weeder as needed before planting; Reduced-till = one blad tillage operation plus rod weeder as needed before planting; No-till = vegetation controlled by herbicides.
3. Conventional = moldboard plow plus disk and harrow operations; Reduced-till = 2 diskings; No-till = vegetation controlled by herbicides, stubble left 8 inches high to trap and hold snow. Winter injury occurred in 1982 and 1985 but not in 1983 and 1984.
4. Yields within columns followed by the same letter are not statistically different ($P \leq 0.05$).

rust pathogens to overwinter in areas where susceptible spring wheats are presently grown.

Fertility

Production of 100 bushels of wheat will require 120 pounds nitrogen (some sources put nitrogen removal as high as 2 pounds/bushel), 56 pounds of phosphorus, as P_2O_5, 31 pounds of potassium, as K_2O, 12 pounds of sulfur, 13 pounds of magnesium, and 5 pounds of calcium (Table 2.21). Wheat grown for grazing and grain will require additional nutrients as will wheat grown for grazing alone. The need for applied fertilizer can be determined by soil tests, plant tissue tests, producer experience, personal field trials, and/or knowledge of nutrient removal, along with expected yields. Soil tests prior to planting are the best investment that producers can make to ensure top yields and quality. Suspected in-season nutrient deficiency problems can be evaluated quickly by plant tissue analysis, and appropriate measures can be taken to correct any problems. However, such analyses and corrections must be made early because wheat, as with other crop plants, requires the majority of its nutrients during rapid growth stages that occur from jointing until the boot stage. The exception to this generalization is that later-applied nitrogen, especially approaching and around the boot stage of growth, will increase grain protein. This is especially important where producers of HRW and HRS wheats have contracted for production of high-protein wheat.

It is important to supply all nutrients in such a fashion that ensures their availability throughout the life of the wheat plant, but especially at certain critical growth stages. Wheat will, as will all non-legume row crops, respond more dramatically to added nitrogen than to any other element. As nitrogen is susceptible to leaching, at least two applications are recommended, one preplant, or soon thereafter, and one

TABLE 2.21 Nutrients Removed by a Bushel of Wheat.

Nutrient	Amount Removed	
	Grain	Grain + Straw
	------------- (lbs./bu.) -------------	
Nitrogen (N)	1.20	1.52
Phosphorus (P_2O_5)	0.56	0.70
Potassium (K_2O)	0.31	1.03
Sulfur (S)	0.12	0.22
Calcium (CaO)	0.05	0.25
Magnesium (MgO)	0.13	0.35
	-------------- (oz./bu.) --------------	
Copper (Cu)	0.012	——
Manganese (Mn)	0.036	——
Zinc (Zn)	0.055	——

Source: Adapted from Laloux et al., 1980; Whitney, 1986; Smith and Hudson, 1990.

topdress or spring application. Even in the more arid production areas, this ensures adequate nitrogen throughout the season. Also, in more arid winter wheat production areas, splitting the recommended nitrogen allows producers to evaluate winter survival before committing their investment capital. It also allows these producers to make such investments closer to the sale of their wheat and recoupment of their capital. Splitting the recommended nitrogen into three or four applications may be advantageous on sandier soils. Producers should be aware that excessive preplant or fall-applied nitrogen, intentional or unintentional because of no soil test, on winter wheat may lead to excessive vegetative growth, winter kill, increased incidence of certain diseases, lodging, and decreased milling and flour quality, plus excess nitrates leach into and pollute groundwater. Dryland wheat producers on the Great Plains also must be concerned that too much fall-applied nitrogen will lead to excess vegetation that will deplete the soil of moisture needed for grain development the following spring.

Dryland wheat is produced with modest inputs in the semiarid Great Plains and parts of the U.S. North Prairies (i.e., eastern North and South Dakota, and Western Minnesota), where less than about 20 inches of precipitation occurs per year. Until recently, expected yields were not sufficient to justify application of inorganic nitrogen to wheat in a fallow rotation in these areas. The average yields during 1940 to 1949 were about 15 bushels/acre in the Dakotas, and fallowing for 9 to 14 months allowed enough time for some mineralization of soil mineral nitrogen and for recycling the organic nitrogen in crop residue. However, improved yield potential, about 34 bushels/acre for the Dakotas in 1990, resistance to lodging, and a decrease in soil organic matter have lead to nitrogen fertilizer being recommended every year in many of these states and situations. Producers continue to be cautioned about absolute amounts for the reasons noted above and should sample their soils in incremental units to at least two feet deep, since nitrates are mobile and will move up and down in the soil with the wetting front. There also is a trend to continuous cropping rather than using a fallow rotation, which requires the use of at least some inorganic nitrogen. When the previous year's crop produced a large amount of residue, producers should increase their fall nitrogen application by about 15 to 20 pounds/acre per ton of excess straw, but not exceeding 50 pounds/acre. This extra nitrogen is required because microbial breakdown of plant material temporarily ties up organic nitrogen, making it unavailable to the current crop.

Wheat responds well to phosphorus and potassium in soils deficient in these macronutrients. Phosphorus is particularly important in early seedling growth, especially in developing a vigorous root system that will aid in winter survival. Phosphorus and potassium may be found in sufficient amounts in many areas and soil types, but maintenance applications should be part of most fertility strategies. Again, soil tests are the best guide in determining appropriate amounts of phosphorus and potassium.

Where full or near-full tillage is practiced, preplant fertilizer should be incorporated into the upper few inches of soil. Additional topdress nitrogen can be surface-applied.

No-till production poses some unique opportunities for fertilizer management, because applied fertilizer cannot be broadcast and incorporated. Starter fertilizer, usually nitrogen only, can be added into the drill with seed, but should not total more than 20 pounds/acre of nitrogen or nitrogen and potassium combined because it creates salt problems. Nitrogen applied to the soil surface also may be lost to volatilization. Surface-applied nitrogen will move into the soil with precipitation or irrigation and eventually be available to wheat plants. The same is true for sulphur. Potassium is rarely deficient, and so maintenance amounts applied topically will keep potassium at acceptable levels in most situations. Phosphorus, on the other hand, poses some real problems. It is so immobile that surface-applied phosphorus probably will not help much with the current crop, especially in more arid environments and on soils particularly deficient in phosphorus. Strip tillage with fertilizer placed below the seed at planting or other fertilizer banding systems should be considered for soils deficient in phosphorus when using no or minimum tillage.

One other point relative to nitrogen fertility is that the grain protein of previous wheat crops can indicate the adequacy of your nitrogen fertility program. Research from Colorado suggests that fields consistently producing grain with less than 11.1 percent protein likely have nitrogen deficiencies sufficient to limit yield and protein content. Protein content consistently between 11.1 and 12 percent may indicate the need for additional nitrogen, while fields producing grain with greater than 12 percent protein may have adequate nitrogen for maximum yields. These associations may not hold in other states or under other production conditions.

Harvest and Storage

Harvest of winter wheat may begin before June 1 in the southeast and as late as late summer in Montana (Figure 2.9). Spring wheat harvest may begin in mid-July in Minnesota, South Dakota, Colorado, Wyoming, and the Pacific Northwest, while usually beginning after August 1 in the more northern areas of Minnesota, North

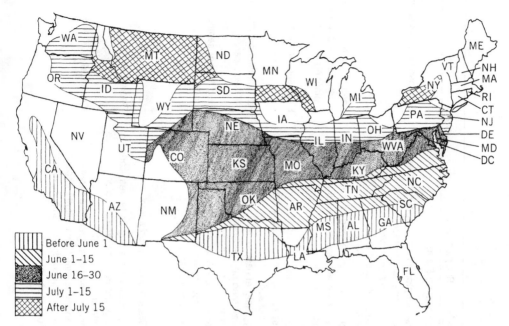

Figure 2.9 *Usual beginning harvest of winter wheat in the continental United States.* Source: *USDA-SRS, 1984. (Redrawn by Mike Hodnett)*

Dakota, Montana, and Washington. Grain should be harvested in a timely fashion and before shattering or sprouting of seeds in the head occurs. Wheat harvested at less than 12.5 percent moisture is dry enough for storage. However, producers may harvest wheat at moisture levels as high as 20 percent if they can dry the grain quickly after it is placed in storage bins. High-moisture wheat delivered immediately to commercial buyers will, of course, be devalued according to the amount of water in the grain.

High-moisture grain must be dried to less than 12.5 percent moisture, and preferably cooled, within four days if possible, to ensure against the growth of molds and deprivation by insects. Heat must be added in areas of high humidity, and air flow rate must be appropriately adjusted. At 85 percent relative humidity (RH), wheat will equilibrate at about 18.5 percent moisture, but air alone at a RH of 60 percent will dry wheat to 12 to 12.5 percent moisture over time. The higher the moisture of stored grain, the faster heated air should be moved through the storage bin, up to 4 cubic feet per minute (cfm) at a grain moisture of 20 percent but only 1.5 cfm for 14 percent moisture grain.

BIOTIC PESTS

Wheat is attacked by a number of insects and disease pathogens (Tables 2.22 and 2.23). Producers should become familiar with life cycles, appearance, and damage caused by insects and disease pathogens that are reoccurring pests in their area.

TABLE 2.22 Common Insect Pests of Wheat in the United States.

Insect	Description[1]	Symptom/Injury[2]	Control
		Pests of Planted Seeds and Seedling Plants	
Fall Armyworm *Spodoptera frugiperda* (Smith)	Light tan to green to black larvae; 3 yellowish-white lines on back, bordered by wider dark strip and a wavy, yellow strip; prominent inverted "Y" on head.	First symptom is small "windowpanes" in leaves; entire seedling can be consumed by larger larvae; partial to complete stand loss.	Field sanitation; chemical.
False Wireworm *Eleodes* spp.	Brown to yellowish brown larvae, some may be black; prominently jointed segments.	Soil dwelling larvae feed on germinating seed and young seedling; partial stand failure.	Seed treatment if larvae detected, especially if planting in dry seedbed.
Grubs *Phyllophaga* spp.	White, C-shaped larvae of May and June beetles.	Feed on roots of seedlings; young seedlings may appear water stressed.	Field sanitation; chemical.
Lesser Cornstalk Borer *Elasmopalpus lignosellus* (Zeller)	Bluish green caterpillar with brown strips.	Eggs laid at base of pseudostem; larvae feed on seedlings; plant death may occur.	Field sanitation; chemical if necessary.
Grasshopper *Melanoplus* spp. and *Camnula pellucida* (Scudder)	Obvious.	Consume plants at any stage but can destroy stands of seedling wheat during dry years.	Field border sanitation; chemical.
		Pests of Older Seedlings Affecting Tillering and Reproduction	
Army Cutworm *Euxoa auxiliaris* (Grote)	Tan-colored larvae to 1.25 inches with dark stripes; night feeders of foliage.	Typical larvae feeding on leaves, causing apparent thin and off-color stands.	Chemical.
Pale Western Cutworm *Agrotis orthogonia* (Morrison)	White larvae to 1.25 inches with brown head.	Larvae feed below ground on roots and emerging tillers; plants appear water stressed; especially severe in dry areas or years.	Chemical.

Pest	Description	Damage	Control
Chinch Bug *Blissus leucopterus* (Say)	Similar in appearance to aphids; nymphs extract plant sap.	Nymphs hatch just below soil surface; feed on roots and tiller bases; plants may be stunted with small heads.	Chemical.
Greenbug *Schizaphis graminum* (Rondani)	Light green, soft, teardrop shaped with dark green stripe on back; antennae about 2/3's body length.	Aphid injects a phytotoxin as it feeds on plant sap, causing necrosis of surrounding leaf tissue; entire leaves and plant can be affected with heavy infestation.	Chemical; resistant cultivars.
Russian Wheat Aphid *Diuraphis noxia* (Mordvilko)	Light green, elongated aphid to 1$\frac{1}{16}$ inches with short antennae.	Aphid injects a phytotoxin causing white or purple longitudinal streaks; plants may appear flattened with tightly rolled leaves.	Chemical; resistant cultivars.
English Grain Aphid *Sitobion avenae* (Fabricius)	Light green, soft-bodied nymphs and adults.	Extract plant sap without injecting phytotoxin; may carry barley yellow dwarf virus.	Not usually necessary unless vectoring barley yellow dwarf virus.

Pests Occurring Primarily during Stem Elongation and Heading

Pest	Description	Damage	Control
Armyworm *Pseudaletia unipuncta* (Haworth)	Larvae to 1.5 inches with 5 stripes lengthwise of body.	Night feeders; cut leaves, awns and heads may be found on ground.	Chemical, rarely needed.
Billbugs *Sphenophorus* spp.	Larvae white, short, fat and legless to about 1/4 inch.	Feed on first internode above crown; may cut stems that may break over and/or turn white.	Rotation; usually not needed.
Hessian Fly *Mayetiola destructor* (Say)	Adults are small black flies.	Larvae feed on stem underneath leaf sheath; seedlings and young tillers may die or be stunted with dark green, broadened leaves; plants infested at later stages may lodge; small, brown "flaxseed" puparia present behind leaf sheaths.	Resistant cultivars; planting date; field sanitation.

(continued)

109

TABLE 2.22 (Continued)

Insect	Description[1]	Symptom/Injury[2]	Control
Wheat Stem Maggot Meromyza americana (Fitch)	Pale green, very slender maggot to 1/4 inch.	White heads while flag leaf remains green; heads can be pulled from sheath without resistance; end of stem appears chewed.	Usually not required; field sanitation.
Wheat Stem Sawfly Cephus cinctus (Morton)	Larvae are pale yellow with prominent wrinkles; to 1/2 inch.	Larvae hatch from eggs laid inside of stems and feed along its length; notches base of stem causing stems to break; heads produce shriveled grain.	Resistant cultivars and field sanitation.
European Corn Borer Ostrinia nubilalis (Hubner)	Flesh colored caterpillar, inconspicuously marked with small, round, brown spots; to 1 inch long.	Larvae bore into stem base causing tillers to lodge or die.	Chemical before larvae bore into stem.

Pests of Stored Grain

Insect	Description[1]	Symptom/Injury[2]	Control
Granary Weevil Sitophilus granarius (Linne)	Dark brown snouted beetle to 1/16 inch; oval, shallow pits on prothorax; individual grains may contain small, fatbodied, legless white grubs.	Kernels with holes or completely eaten out leaving empty shells; damage occurs above 50°F and reproduction of weevil requires a minimum of 60°F; can feed at any depth of grain bin.	Storage bin and area sanitation; chemical.
Rice Weevil Sitophilus oryzae (Linne)	Same as Granary Weevil except pits on prothorax are round.	Same as granary weevil.	Same as granary weevil.

Insect	Description	Feeding/Damage	Stage most damaging[1]
Lesser Grain Borer *Rhizopertha dominica* (Fabricius)	Very small, brown to black beetle to 1/8 inch; larvae are whitish, curved-bodied, grub-like, found inside kernel.	Same as granary weevil.	Same as granary weevil.
Red Flour Beetle *Tribolium castaneum* (Herbst)	Small, dark red, narrow-bodied beetles to 1/6 inch; larvae are brownish white, six-legged, to 1/6 inch.	Feed on broken and damaged kernels at all depths of bin; requires 60°F for reproduction.	Same as granary weevil.
Confused Flour Beetle *Tribolium confusum* (Duval)	Essentially the same as the red flour beetle to the nonentomologist.	Same as red flour beetle.	Same as red flour beetle.
Saw-Toothed Grain Beetle *Oryzaephilus surinamensis* (Linne)	Similar to red flour beetle but more slender and darker; six, fine saw-tooth-like projections on each side of thorax.	Same as red flour beetle.	Same as red flour beetle.
Cadella Beetle *Tenebroides mauritanicus* (Linne)	Black beetle to 1/2 inch; head and thorax distinctly separated from body; whitish to gray-white larvae with black heads.	Same red flour beetle.	Same as red flour beetle.
Angoumois Grain Moth *Sitotroga cerealella* (Olivier)	Brownish gray moths less than 1/2 inch with long fringe hairs on wings; larvae whitish to 1/2 inch.	Attacks whole or cracked kernels, or dockage; feeds in top 6 inches of grain pile.	Same as granary weevil.
Indian Meal Moth *Ploida interpunctella* (Hubner)	Small moths with basal half of front wings much lighter in color than tip; whitish to greenish larvae to 1/2 inch.	Feeds only on cracked kernels and dockage in top 6 inches; produces silky webbing on surface of stored grain; usually produces a musty, sweet odor.	Same as granary weevil.

1. Stage most damaging or apparent to observer.
2. Plant stage(s) most likely for infestation and damage of winter wheat.

TABLE 2.23   Common Diseases of Wheat and Their Pathogens in the United States.

Disease	Cause/Source	Symptom/Injury	Control
Speckled, Pink, or Sclerotinia Snow Mold	*Typhula* spp., *Fusarium nivale, Sclerotinia borealis*	Fungi attack leaves of young plants of winter wheat in northwestern and northeastern states while plants are dormant and covered with snow; occurs when snow falls on unfrozen ground and the temperature at the soil-snow interface is 30–34°F for 3–4 months; stands can be lost.	Rotation, seed treatment.
Snow Rot	*Pythium* spp.	Rotted leaves may be only fragments held together by veins; most damaging when wheat is trapped under snow or ice with liquid water at soil interface; localized as snow molds; stand may be lost.	Rotation, seed treatment.
Seeds Rots and Seedling Blights	*Fusarium* spp., *Cochliobolus sativus*	Seed borne from infection of spikelets during kernel development; soil borne inoculum rarely infects germinating seed or seedlings; poor, uneven stands.	Plant seed produced in dry climates; rotation, seed treatment.
Common Root Rot or Dryland Root Rot	*Cochliobolus sativus, Fusarium* spp.	Dark brown to black subcrown internode or brown to reddish brown lesions on roots, especially crown roots; poor, uneven stands; disease favored by dry soils.	Deep placement of seeds as fungi live near soil surface.
Take-all	*Gaeumannomyces graminis* var. *tritici*	Soil borne on old crop residue; infects seminal roots at stages 12–13; subsequent lesions randomly on roots in upper 10 inches of soil; plant to plant progress by root contact and can eventually kill stems up to 2 inches above soil line leaving area shiny black; patches of dead stalks with shriveled grain; whiteheads; favored by wet soil.	Rotation; good fertility; summer fallow, avoid following brome or weedy grasses.

TABLE 2.23   (*Continued*)

Disease	Cause/Source	Symptom/Injury	Control
Root Rot	*Pythium* spp., *Rhizoctonia* spp.	Soil borne fungi; plants stunted with small leaves, otherwise normal in appearance; plants rarely killed	Rotation; seed treatment.
Barley Yellow Dwarf	Barley Yellow Dwarf Virus vectored by aphids.	Plants stunted; yellowing of leaves from tip to upper third (may be red or purple); lab test necessary for an accurate identification.	Eliminate aphids; cultivars vary in level of tolerance but no effective resistance.
Wheat Streak Mosaic	Virus vectored by wheat curl mite.	Plants stunted; leaves mottled-turning yellow; prostrate tillers.	Eliminate mites by destroying volunteer wheat; plant 2–3 weeks after volunteers destroyed; late planting of winter wheat; mite resistant cultivars.
Wheat Yellow Mosaic—also Wheat Spindle Streak Mosaic	Virus vectored by soilborne fungus, *Polymyxa graminis*.	Intermittent streaks of light green and yellow mottling and mosaic parallel to the veins; plants can be stunted with few tillers or small heads.	Resistant cultivars; not controlled by rotation.
Soilborne Wheat Mosaic	Virus vectored by soilborne fungus, *Polymyxa graminis*	Occurs exclusively on winter wheat; yellowing of leaves of plants in wet areas of field in spring; light green and yellow mosaic patterns on leaves; stunting; plants appear to recover but have fewer and smaller heads.	Resistant cultivars; not controlled by rotation.
Stripe Rust	*Puccinia striiformis*	Occurs in Pacific Northwest where temperatures during tillering and heading less than 60°F; reddish pustules generally in longitudinal strips.	Resistant cultivars; early-maturing cultivars may provide escape
Leaf Rust	*Puccinia recondita*	Small, round, red pustules on blades and sheaths; grain shriveled; leaves may die.	Same as stripe rust.
Stem Rust	*Puccinia graminis*	Oblong dark red, rough pustules on leaves, stems and heads; grain shriveled.	Same as stripe rust.

(*continued*)

TABLE 2.23   (Continued)

Disease	Cause/Source	Symptom/Injury	Control
Powdery Mildew	*Erysiphe graminis*	White-gray powdery tufts on foliage, stems and heads.	Resistant cultivars; chemical; planting date.
Septoria Leaf Blotch	*Septoria* spp.	Tan-colored, rectangular lesions with black pimples.	Foliar fungicides; crop rotation; resistant cultivars; field sanitation.
Septoria Glume Blotch	*Septoria* spp.	Oblong lesions on leaves; heads purple.	Same as Septoria Leaf Blotch.
Cephalosporium stripe	*Cephalosporium gramineum*	Yellow stripes $1/8$ inch wide, full length of leaves; premature death; whiteheads.	Crop rotation.
Pseudocerosporella Foot Rot; also called Eyespot or Strawbreaker Foot Rot	*Pseudocerosporella herpotrichoides*	Eyespot lesions at base of stems after jointing; lodging.	Resistant cultivars; foliar fungicides.
Common Bunt; also called Stinking Bunt	*Tilletia caries, T. foetida*	Grain replaced by black spores; fishy odor.	Seed treatment with systemic fungicide.
Loose Smut	*Ustilago tritici*	Black smut spores on head that easily blow away.	Seed treatment with systemic fungicide or seed from field certified disease free.
Scab	*Fusarium* spp.	Some florets are white; shriveled grain.	Crop rotation but not following corn.
Storage Molds	*Penicillium* spp., *Aspergillus* spp.	Seed are infected in field so bin sanitation not effective control; fungi prefers 75°F but will grow at 40°F and requires at least 12.5% grain moisture.	Keep grain moisture below 12.5%; temperature below 40°F if possible.

GRADES AND STANDARDS

The search for ways to identify the quality of wheat produced by the farmer probably dates to antiquity. About 750 B.C., the prophet Amos in the Bible refers to merchants selling screenings and sweepings with wheat. The Chicago Board of Trade established the first descriptive grades for wheat in 1856. Individual states and private industry establish a myriad of grades and descriptions, totaling 338 by

1906 when the U.S. Congress established the Division of Grain Standardization within the USDA. The first U.S. congressional bill providing for federal grain standards and inspections was introduced by Senator Paddock in 1890, and the USDA published the first official grades for wheat in 1917. Current standards and grades as published by the USDA Federal Grain Inspection Service are delineated below.

Terms Defined

Definition of Wheat Grain that, before the removal of dockage, consists of 50 percent or more common wheat (*Triticum aestivum* subsp. *aestivum* L.), club wheat [(*T. aestivum* subsp. *compactum* Host.)], and durum wheat (*T. turgidum* subsp. *durum* Desf.) and not more than 10 percent of other grains for which standards have been established under the U.S. Grain Standards Act and that, after the removal of the dockage, contains 50 percent or more of whole kernels of one or more of these wheats.

Definitions of Other Terms

(a) **Classes.** There are seven classes for wheat: Durum wheat, Hard Red Spring wheat, Hard Red Winter wheat, Soft Red Winter wheat, White wheat, Unclassed wheat, and Mixed wheat.

    (1) **Durum wheat.** All varieties of white (amber) durum wheat. This class is divided into the following three subclasses:

        (i) **Hard amber durum wheat.** Durum wheat with 75 percent or more of hard and vitreous kernels of amber color.

        (ii) **Amber durum wheat.** Durum wheat with 60 percent or more but less than 75 percent of hard and vitreous kernels of amber color.

        (iii) **Durum wheat.** Durum wheat with less than 60 percent of hard and vitreous kernels of amber color.

    (2) **Hard red spring wheat.** All varieties of Hard Red Spring wheat. This class shall be divided into the following three subclasses:

        (i) **Dark northern spring wheat.** Hard Red Spring wheat with 75 percent or more of dark, hard, and vitreous kernels.

        (ii) **Northern spring wheat.** Hard Red Spring wheat with 25 percent or more but less than 75 percent dark, hard, and vitreous kernels.

        (iii) **Red spring wheat.** Hard Red Spring wheat with less than 25 percent of dark, hard, and vitreous kernels.

    (3) **Hard red winter wheat.** All varieties of Hard Red Winter wheat. There are no subclasses in this class.

    (4) **Soft red winter wheat.** All varieties of Soft Red Winter wheat. There are no subclasses in this class.

(5) **White wheat.** All varieties of white wheat. This class is divided into the following four subclasses:

   (i) **Hard white wheat.** White wheat with 75 percent or more of hard kernels. It may contain not more than 10 percent of white club wheat.

   (ii) **Soft white wheat.** White wheat with less than 75 percent of hard kernels. It may contain not more than 10 percent of white club wheat.

   (iii) **White club wheat.** White club wheat containing not more than 10 percent of other white wheat.

   (iv) **Western white wheat.** White wheat containing more than 10 percent of white club wheat and more than 10 percent of other white wheat.

(6) **Unclassed wheat.** Any variety of wheat that is not classifiable under other criteria provided in the wheat standards. There are no subclasses in this class. This class includes

   (i) Red durum wheat.

   (ii) Any wheat that is other than red or white in color.

(7) **Mixed wheat.** Any mixture of wheat that consists of less than 90 percent of one class and more than 10 percent of one other class, or a combination of classes that meet the definition of wheat.

(b) **Contrasting classes.** Contrasting classes are

   (1) Durum wheat, White wheat, and Unclassed wheat in the classes Hard Red Spring wheat and Hard Red Winter wheat.

   (2) Hard Red Spring wheat, Hard Red Winter wheat, Soft Red Winter wheat, White wheat, and Unclassed wheat in the class Durum wheat.

   (3) Durum wheat and Unclassed wheat in the class Soft Red Winter wheat.

   (4) Hard Red Spring wheat, Durum wheat, Hard Red Winter wheat, and Unclassed wheat in the class White wheat.

(c) **Damaged kernels.** Kernels, pieces of wheat kernels, and other grains that are badly ground-damaged, badly weather-damaged, diseased, frost-damaged, germ-damaged, heat-damaged, insect-bored, mold-damaged, sprout-damaged, or otherwise materially damaged.

(d) **Defects.** Damaged kernels, foreign material, and shrunken and broken kernels. The sum of these three factors may not exceed the limit for the factor defects for each numerical grade.

(e) **Dockage.** All matter other than wheat that can be removed from the original sample by use of an approved device according to procedures prescribed in FGIS instructions. Also, underdeveloped, shriveled, and small pieces of wheat kernels removed in properly separating the material other than wheat and that cannot by recovered by properly rescreening or recleaning.

(f) **Foreign material.** All matter other than wheat that remains in the sample after the removal of dockage and shrunken and broken kernels.

(g) **Heat-damaged kernels.** Kernels, pieces of wheat kernels, and other grains that are materially discolored and damaged by heat that remain in the sample after the removal of dockage and shrunken and broken kernels.

(h) **Other grains.** Barley, corn, cultivated buckwheat, einkorn, emmer, flaxseed, guar, hull-less barley, nongrain sorghum, oats, Polish wheat, popcorn, rice, rye, safflower, sorghum, soybeans, spelt, sunflower seed, sweet corn, triticale, and wild oats.

(i) **Shrunken and broken kernels.** All matter that passes through a $0.064 \times \frac{3}{8}$ oblong-hole sieve (a metal sieve 0.032 inch thick with oblong perforations 0.064 inch by 0.375 ($\frac{3}{8}$) inch).

Principles Governing the Application of Standards

Basis of Determination Each determination of heat-damaged kernels, damaged kernels, foreign material, wheat of other classes, contrasting classes, and subclasses is made on the basis of the grain when free from dockage, and shrunken and broken kernels. Other determinations not specifically provided for under the general provisions are made on the basis of the grain when free from dockage, except the determination of odor is made on either the basis of the grain as a whole or the grain when free from dockage (Table 2.24).

Grades and Grade Requirements for Mixed Wheat

Mixed wheat is graded according to the U.S. numerical and U.S. sample grade requirements of the class of wheat that predominates in the mixture, except that the factor wheat of other classes is disregarded.

Special Grades and Special Grade Requirements

(a) **Ergoty wheat.** Wheat that contains more than 0.30 percent of ergot.

(b) **Garlicky wheat.** Wheat that contains in a 1,000 gram portion more than two green garlic bulblets or an equivalent quantity of dry or partly dry bulblets.

(c) **Light smutty wheat.** Wheat that has an unmistakable odor of smut, or which contains in a 250-gram portion, smut balls, portions of smut balls, or spores of smut in excess of a quantity equal to 14 smut balls, but not in excess of quantity equal to 30 smut balls of average size.

(d) **Smutty wheat.** Wheat that contains, in a 250-gram portion, smut balls, portions of smut balls, or spores of smut in excess of a quantity equal to 30 smut balls of average size.

(e) **Treated wheat.** Wheat that has been scoured, limed, washed, sulfured, or treated in such a manner that the true quality is not reflected by either the numerical grades or the U.S. sample grade designation alone.

TABLE 2.24 Grades and Grade Requirements for All Classes of Wheat Except Mixed Wheat.

	Minimum Limits of —— Test Weight per Bushel		Maximum Limits of ——						
			Damaged Kernels					Wheat of Other Classes[4]	
Grade	Hard Red Spring Wheat or White Club Wheat	All Other Classes and Subclasses	Heat Damaged Kernels	Total[2]	Foreign Material	Shrunken and Broken Kernels	Defects[3]	Contrasting Classes	Total[5]
	(lbs.)	(lbs.)	(%)	(%)	(%)	(%)	(%)	(%)	(%)
U.S. #1	58.0	60.0	0.2	2.0	0.5	3.0	3.0	1.0	3.0
U.S. #2	57.0	58.0	0.2	4.0	1.0	5.0	5.0	2.0	5.0
U.S. #3	55.0	56.0	0.5	7.0	2.0	8.0	8.0	3.0	10.0
U.S. #4	53.0	54.0	1.0	10.0	3.0	12.0	12.0	10.0	10.0
U.S. #5	50.0	51.0	3.0	15.0	5.0	20.0	20.0	10.0	10.0

U.S. Sample Grade is wheat that

(a) Does not meet the requirements for the grades U.S. Nos. 1, 2, 3, 4, or 5; or

(b) Contains 32 or more insect-damaged kernels per 100 grams of wheat; or

(c) Contains 8 or more stones or any number of stones which have an aggregate weight in excess of 0.2 percent of the sample weight, 2 or more pieces of glass, 3 or more crotalaria seeds (*Crotalaria* spp.), 2 or more castor beans (*Ricinus communis* L.), 4 or more particles of an unknown foreign substance(s) or a commonly recognized harmful or toxic substance(s), 2 or more rodent pellets, bird droppings, or equivalent quantity of other animal filth per 1,000 grams of wheat; or

(d) Has a musty, sour, or commercially objectionable foreign odor (except smut or garlic odor); or

(e) Is heating or otherwise of distinctly low quality.

Source: USDA-FGIS, 1988.

1. These requirements also apply when Hard Red Spring wheat or White Club wheat predominate in a sample of Mixed wheat.

2. Includes heat-damaged kernels.

3. Defects include damaged kernels (total), foreign material, and shrunken and broken kernels. The sum of these three factors may not exceed the limit for defects for each numerical grade.

4. Unclassed wheat of any grade may contain not more than 10 percent of wheat of other classes.

5. Includes contrasting classes.

118

GLOSSARY

Blending: the process of combining measured amounts of different lots from bins and mixing them into a uniform blend by grain assemblers or millers.

Bolt: to sift through a cloth or sieve.

Bran: the outer covering of the wheat kernel composed of seedcoat, nucellar tissue, tube and cross cells, hypodermis and endodermis, which are separated from the endosperm and embryo, or germ, during the usual process of commercial milling.

Break flour: produced by the break rolls during commercial milling. Particles of endosperm are to be further reduced to flour.

Break system: the stage in the milling process in which kernels are broken through a series of successively closer-set pairs of rollers to separate the endosperm from the bran coat.

Broken kernel: kernels separated into two or more pieces, exclusive of insect boring or surface consumption.

Clear flour: flour remaining after a patent flour cut has been removed; normally higher in ash and protein than patent but of lower market value because of color.

Club wheat: *Triticum aestivum* subsp. *compactum;* usually white wheat cultivars; may be winter or spring; heads are usually awnless, elliptical, oblong, or clavate in shape, and short, compact, or laterally compressed.

Coarse breaks: the break rolls that grind the larger particles in a classified break system where break stock is classified as coarse and fine by size and ground on separate rolls.

Crop year: the U.S. officially designated production-marketing year for a commodity. For wheat, the crop year is June 1 to May 31.

Durum wheat: *Triticum turgidum* subsp. *durum;* 14 pairs of chromosomes; spring growth habit; very hard; high protein; used primarily for pasta.

Endosperm: the starchy portion of the wheat kernel that is ground into flour. The endosperm is the seed-stored nutrient supply for the embryo during germination and early seedling growth.

Ethanol: grain alcohol made from almost any kind of grain containing a reasonable amount of starch.

Family flour: commonly called all-purpose flour; used for baking bread, cakes, biscuits, etc.

Fancy patent flour: the most finely ground flour.

Farina: very pure endosperm of non-durum wheat ground to about medium screen size; may be used in pasta, but will overcook more easily than pasta made from durum wheat.

Fine breaks: break rolls used to further reduce the smaller particles in a classified break system of milling.

First clear: a portion of a straight run flour remaining after a patent flour has been removed; higher in protein than the patent flour produced but poorer in color and therefore with a lower commercial value; about 20 to 25 percent of the flour produced in hard wheat mills may be first clear.

Flour extraction rate: percent of flour produced from milling 100 pounds of wheat kernels.

Gluten: the rubber-like proteinaceous material in wheat responsible for its superior bread-making quality; a measure of flour quality; one-third of the weight of wet gluten approximates the protein content of the flour.

Grain reserve: wheat stored by the U.S. government.

Hard red spring (HRS): *Triticum aestivum* subsp. *aestivum;* also called common or bread wheat; 21 chromosome pairs—A, B, and D genomes; spring seeded; not winter hardy; does not require vernalization for reproduction; may be referred to as red, dark northern, or northern; high protein; hard endosperm; primarily used to produce bread flour.

Hard red winter (HRW): *Triticum aestivum* subsp. *aestivum;* also called common or break wheat; 21 chromosome pairs—A, B, and D genomes; fall seeded; winter hardiness varies; requires vernalization for reproduction; may be referred to as dark hard, hard, or yellow hard; medium- to high-protein wheat; hard endosperm; primarily used to produce bread flour.

Hard wheat: generic for wheat having a vitreous endosperm suitable for bread flour or semolina; yields coarse, gritty flour that is free-flowing and easily sifted; flour consists of regular-shaped particles that are mostly whole endosperm cells.

Middlings: pieces of endosperm not ground into flour; commercially, a byproduct of milling composed of coarse material during break reduction; usually used for animal feed.

Middlings rolls: a pair of smooth rolls used to reduce middlings to flour particle size; may be called reduction rolls.

Millfeed: any of the byproducts of the milling industry used as livestock feed.

Pasta: products made principally from durum wheat flour; includes such food products as macaroni, spaghetti, and noodles.

Patent flour: highest-value grade flour with good dress and color.

Product stream: any one of 125 to 150 mill streams in the flour manufacturing process; different grades of flour are produced by blending individual streams.

Semolina: a coarse endosperm extracted from durum wheat; used for pasta.

Shorts: an inseparable mixture of bran, endosperm, and wheat germ remaining after flour extraction in milling; used for animal feed.

Soft red winter (SRW): *Triticum aestivum* subsp. *aestivum;* a common wheat; 21 chromosome pairs—A, B, and D genomes; fall seeded; winter hardiness varies; requires vernalization for reproduction; low- to medium-protein content; soft or floury endosperm; used primarily for cakes, pastries, etc., where product tenderness is important.

Soft wheat: generic for wheat having a chalky, nonvitreous endosperm suitable for pastry flour; yields a very fine flour consisting of irregular-shaped fragments of endosperm cells that cling together and are difficult to sift.

Spring wheat: wheat seeded in the spring and harvested the following summer or fall; does not require vernalization to reproduce.

Stocks: wheat in storage or transit; sometimes includes processed products inventoried.

Straight flour: flour extracted from a blend or mill mix of wheat without division or addition of flour from other runs.

Elevator: a point of accumulation and distribution in the movement of grain; terminal elevators usually receive grain by railroad carload; country elevators receive grain by truck, usually from farmers.

Test weight: a quality test used to determine weight per bushel; for wheat, bushel weight standard is 60 pounds.

Vernalization: the low-temperature triggered, hormonal-controlled conversion from vegetative growth to reproductive growth in winter wheat; requires exposure to temperatures below 50°F; length of exposure required is cultivar-dependent.

Source: Heid, 1979.

BIBLIOGRAPHY

Anonymous. 1985. *High-yield Management for Small Grains*. Union Carbide.

Atkins, I. M. 1980. *A History of Small Grain Crops in Texas: Wheat, Oats, Barley, Rye 1582–1976*. Texas A&M Univ. Agri. Exp. Sta. B-1301.

Barnett, R. D., and H. H. Luke. 1980. *Florida 301: A New Wheat for Multiple Cropping Systems in North Florida*. Univ. of Florida Agri. Exp. Sta. Cir. S-273.

Besant, L., D. Kellerman, and G. Monroe. 1977. *Grains Production Processing Marketing*. Chicago Board of Trade, Chicago, IL.

Bitzer, M. J., D. F. Miles, G. D. Lacefield, and J. H. Herbek. 1987. *Grain and Forage Crop Guide for Kentucky*. Univ. of Kentucky Agri. Exp. Sta. Agr-18.

Bitzer, M. J., J. H. Herbek, and H. C. Vaught. 1980. *Producing Small Grains for Grain and Silage*. Univ. of Kentucky Coop. Ext. Ser. Agr-32.

Bizzarri, O., and A. Morelli. 1988. "Milling Durum Wheat." pp. 161–190. *In* G. Fabriani and C. Lintas (eds.) *Durum Wheat: Chemistry and Technology*. Am. Assoc. Cer. Chem., St. Paul, MN.

Briggle, L. W., and B. C. Curtis. 1987. "Wheat Worldwide." *In* E. G. Heyne (ed.) *Wheat and Wheat Improvement*. Crop Sci. Soc. Am., Madison, WI.

Briggle, L. W. 1980. "Origin and Botany of Wheat." pp. 6–13. *In* E. Hafliger (ed.) *Wheat*, Ciba-Geigy, Switzerland.

Brooks, H. L. 1986. *Insect Control*. Kansas St. Univ. Coop. Ext. Ser. C-529.

Brown, B. D. 1991. *Idaho Fertilizer Guide: Irrigated Wheat*. Univ. of Idaho Coop. Ext. Ser. Curr. Info. Ser. No. 373.

Brumfield, K. 1968. *This Was Wheat Farming*. Superior Pub. Co., Seattle, WA.

Carrier, L. 1923. *The Beginnings of Agriculture in America*. McGraw-Hill Book Co., New York, NY.

Clark, J. A. 1936. *Improvement in Wheat. Yearbook of Agriculture, 1936*. USDA, Washington, DC.

Clark, J. A., and B. B. Bayles. 1942. *Classification of Wheat Varieties Grown in the United States in 1939*. USDA Tech. Bul. 795.

Colburn, E., and D. Pennington. 1978. *Keys to Profitable Small Grain Production in the Rio Grande Plain*. Texas A&M Univ. Coop. Ext. Ser. MP-1380.

Cole, J. E., G. D. Alston, and C. O. Spence. 1977. *Keys to Profitable Small Grain Production in North Central Texas*. Texas A&M Univ. Coop. Ext. Ser. L-870.

Cook, R. J., and R. J. Veseth. 1991. *Wheat Health Management*. Am. Phytopath. Soc. (APS) Press, St. Paul, MN.

Cox, D. J., and D. R. Shelton. 1992. "Genotype-by-tillage interactions in hard red winter wheat quality evaluation." *Agron. J.* 84:627–630.

Cox, D. J., J. K. Larsen, and L. J. Brun. 1986. "Winter survival response of winter wheat: Tillage and cultivar selection." *Agron. J.* 78:795–801.

Cox, T. S. 1991. "The Contribution of Introduced Germplasm to the Development of U.S. Wheat Cultivars." pp. 25–47. *In* H. L. Shands and L. E. Wiesner (eds.) *Use of Plant Introductions in Cultivar Development Part 1*. Crop Sci. Soc. Am., Madison, WI.

Cox, T. S., and J. P. Shroyer. 1984. *Pedigree of Hard Red Winter Wheat Varieties*. Kansas St. Univ. Coop. Ext. Ser. AF 124.

Cox, T. S., J. P. Shroyer, L. Ben-Hui, R. G. Sears, and T. J. Martin. 1988. "Genetic improvement in agronomic traits of hard red winter wheat cultivars from 1919 to 1987." *Crop Sci.* 28:756–760.

Dennis, R. E., R. K. Thompson, A. D. Day, and E. B. Jackson. (not dated). *Growing Wheat in Arizona.* Univ. of Arizona Coop. Ext. Ser. Bul. A-32 (revised).

Dick, J. W., and V. L. Youngs. 1988. "Evaluation of Durum Wheat, Semolina, and Pasta in the United States." pp. 237–248. *In* G. Fabriani and C. Lintas (eds.) *Durum Wheat: Chemistry and Technology,* Am. Assoc. Cer. Chem., St. Paul, MN.

Fahn, A. 1982. *Plant Anatomy.* Pergamon Press, Elmsford, NY.

Feillet, P. 1988. "Protein and Enzyme Composition of Durum Wheat." pp. 93–120. *In* G. Fabriani and C. Lintas (eds.) *Durum Wheat: Chemistry and Technology.* Am. Assoc. Cer. Chem., St. Paul, MN.

Feldman, M. 1976. "Wheats." pp. 120–128. *In* N. W. Simmonds (ed.) *Evolution of Crop Plants.* Longman Press, New York, NY.

Fenster, C. R., and G. A. Peterson. 1979. *Effects of No-tillage Fallow as Compared to Conventional Tillage in a Wheat-Fallow System.* Univ. of Nebraska Agri. Exp. Sta. Res. Bul. 289.

Feyerherm, A. M., G. M. Paulsen, and J. L. Sebaugh. 1984. "Contribution of genetic improvement to recent wheat yield increases in the U.S.A." *Agron. J.* 76:985–990.

Finney, K. F., W. T. Yamazaki, V. L. Youngs, and G. L. Rubenthaler. 1987. "Quality of Hard, Soft, and Durum Wheats." pp. 677–748. *In* E. G. Heyne (ed.) *Wheat and Wheat Improvement.* Crop Sci. Soc. Am., Madison, WI.

Food and Agriculture Organization of the United Nations (FAO). 1991. *Production Yearbook.* Vol 45. FAO, Rome, Italy.

Freed, R. D., E. H. Everson, L. O. Copeland, D. W. Fulbright, and J. L. Clayton. 1983. *Wheat Variety Performance in Michigan.* Michigan St. Univ. Coop. Ext. Ser. Bul. E-1352.

Gardner, F. P., and R. D. Barnett. 1990. "Vernalization of wheat cultivars and a triticale." *Crop Sci.* 30:166–169.

Gies, J. 1990. "The great reaper war." pp. 20–28. *Invention and Tech.*

Girbach, D. S. 1977. *Winter Wheat Production.* Michigan St. Univ. Coop. Ext. Ser. Bull. E-1049 SF-1.

Goos, R. J., D. G. Westfall, and A. E. Ludwick. 1984. *Grain Protein Content as an Indicator of Nitrogen Fertilizer Needs in Winter Wheat in Eastern Colorado.* Colorado St. Univ. Coop. Ext. Ser. Service in Action .555.

Goos, R. J., D. G. Westfall, and A. E. Ludwick. 1982. *Nitrogen Fertilization of Dryland Winter Wheat in Eastern Colorado.* Colorado St. Univ. Coop. Ext. Ser. Service in Action 554.

Gusta, L. V., and T. H. H. Chen. 1987. "The Physiology of Water and Temperature Stress." pp. 115–150. *In* E. G. Heyne (ed.) *Wheat and Wheat Improvement.* Crop Sci. Soc. Am. Madison, WI.

Halvorson, A. D., and J. L. Havlin. 1992. "No-till winter wheat response to phosphorus placement and rate." *Soil Sci. Soc. Am. J.* 56:1635–1639.

Handcock, J. F. 1992. *Plant Evolution and the Origin of Crop Species.* Prentice Hall, Englewood Cliffs, CA.

Harlan, J. R. 1981. "The Early History of Wheat: Earliest Traces to the Sack of Rome." p. 1–20. *In* L. T. Evens and W. J. Peacock (eds.) *Wheat Science—Today and Tomorrow.* Cambridge Univ. Press, Cambridge, MA.

Harlan, J. R., and D. Zohary. 1966. "Distribution of wild wheats and barley." *Sci.* 153:1074–1080.

Hart, L. P., D. W. Fulbright, A. Ravenscroft, and K. Z. Haufler. 1981. *Wheat Spindle Streak Mosaic Virus.* Michigan St. Univ. Coop. Ext. Ser. Bul. E-808.

Heid, W. G., Jr. 1979. *U.S. Wheat Industry.* U.S. Department of Commerce. Natl. Tech. Info. Ser. PB-299930.

Hill, L. D. 1990. *Grain Grades and Standards.* Univ. of Illinois Press, Urbana and Chicago, IL.

Hill, P. R., J. V. Mannering, and J. R. Wilcox. 1989. *Estimating Corn and Soybean Residue Cover.* Purdue Univ. Coop. Ext. Ser. AY-269.

Hodges, R. J., Jr., E. C. Gilmore, and M. E. McDaniel. 1979. *Keys to Profitable Small Grain Production in East Texas and Coast Prairie.* Texas A&M Univ. Coop. Ext. Ser. B-1198.

Horder, T. J. H., C. Dodds, and T. Moran. 1954. *Bread.* Constable and Co., Ltd., London, England.

Hoseney, R. C. 1986. *Principles of Cereal Science and Technology.* Am. Assoc. Cer. Chem., St. Paul, MN.

Inglett, G. E., and L. Munck. 1980. *Cereals for Food and Beverages.* Academic Press, New York, NY.

Kephart, K. D., and J. C. Stark. 1989. *Irrigated Spring Wheat Production Guide for Southern Idaho.* Univ. of Idaho Ext. Ser. Bul. 697.

Kimber, G., and E. R. Sears. 1987. "Evolution in the Genus *Triticum* and the Origin of Cultivated Wheat." pp. 154–164. *In* E. G. Heyne (ed.) *Wheat and Wheat Improvement.* Crop Sci. Soc. Am., Madison, WI.

Klepper, B., R. W. Rickman, and C. M. Peterson. 1982. "Quantitative characterization of vegetative development in small cereal grains." *Agron. J.* 74:789–792.

Klepper, B., R. W. Rickman, and R. K. Belford. 1983. "Leaf and tiller identification in wheat plants." *Crop Sci.* 23:1002–1004.

Knutson, A., E. P. Boring III, G. J. Michels Jr., and F. Gilstrap. 1993. *Biological Control of Insect Pests in Wheat.* Texas A&M Univ. Coop. Ext. Ser. B-5044.

Kozmin, P. A. 1920. *Flour Milling.* D. Van Nostrand Co., New York, NY.

Kuhlmann, C. B. 1929. *The Development of the Flour-Milling Industry in the United States.* Riverside Press, Boston, MA.

Laloux, R., A. Falisse, and J. Poelaert. 1980. "Nutrition and Fertilization of Wheat." pp. 19–24. *In Wheat.* Ciba-Geigy Ltd. Basle, Switzerland.

Large, E. C. 1954. "Growth stages in cereals: Illustration of the Feekes scale." *Plt. Path.* 3:128–129.

Lovelace, D. A., and W. B. Gass. 1977. *Keys to Profitable Small Grain Production in the Rolling Plains and Edwards Plateau.* Texas A&M Univ. Coop. Ext. Ser. MP-1324.

Marcellos, H., and M. J. Burke. 1979. "Frost injury in wheat: Ice formation and injury in leaves." *Aust. J. of Plt. Physiol.* 6:513–521.

Martin, G. W., and J. T. Touchton. 1982. *Tillage requirements for optimum wheat yield.* Auburn Univ. Agr. Exp. Sta. Highlights of Agri. Res.

Martin, J. H., W. H. Leonard, and D. L. Stamp. 1967. *Principles of Field Crop Production.* MacMillan Pub. Co., New York, NY.

Mask, P., A. K. Hagan, and C. C. Mitchell Jr. 1987. *Planting Small Grains.* Auburn Univ. Coop. Ext. Ser. Cir. ANR-497.

Masle, J., G. Doussinault, and B. Sun. 1989. "Response of wheat genotypes to temperature and photoperiod in natural conditions." *Crop Sci.* 29:712–721.

Matz, S. A. 1991. *The Chemistry and Technology of Cereals as Food and Feed.* Van Nostrand Reinhold/AVI Pub., New York, NY.

McDole, R. E., and S. Vira. 1980. *Minimum Tillage for Soil Erosion Control under Dryland Crop Production.* Univ. of Idaho Coop. Ext. Ser. Curr. Info. Ser. No. 523.

McKinney, H. H. 1934. *Field experiments with vernalized wheat.* USDA Cir. 325.

Metcalf, R. L., and R. A. Metcalf. 1993. *Destructive and Useful Insects: Their Habits and Control.* McGraw-Hill Inc., New York, NY.

Miller, T. D., and S. Livingston. 1987. *Profitable Small Grain Production in the Texas Gulf Coast.* Texas A&M Univ. Coop. Ext. Ser. B-1587.

Montandon, J. L., S. McClure, F. Laws, and S. G. Perry. 1994. *Delta Agricultural Digest.* Argus Agronomics, Clarksdale, MS.

Noggle, G. R., and G. J. Fritz. 1983. *Introductory Plant Physiology.* Prentice-Hall, Inc., Englewood Cliffs, NJ.

Office of Technology Assessment. 1989. *Enhancing the Quality of U.S. Grain for International Trade United States Congress.* OTA-F-399. Washington, DC.

Patrick, C. D., and E. P. Boring III. 1990. *Managing Insect and Mite Pests of Texas Small Grains.* Texas A&M Univ. Coop. Ext. Ser. B-1251.

Paulsen, G. M. 1986. "Growth and Development." pp. 3–6. *In Wheat Production Handbook.* Kansas St. Univ. Coop. Ext. Ser. C-529 Revised.

Petr, F. C., and Z. W. Daughtrey. 1978. *Keys to Profitable Small Grain Production in the High Plains.* Texas A&M Univ. Coop. Ext. Ser. MP-1390.

Purvis, O. N., and F. G. Gregory. 1937. "Studies in vernalization of cereals I: A comparative study of vernalization of winter rye by low temperature and by short days." *Ann. Bot.* (NS) 1:569–591.

Quick, J. S., and R. D. Crawford. 1983. "Bread baking potential of new durum wheat cultivars." *Proc. Sixth Internatl. Wheat Genet. Sym.* 1983:851–856.

Quisenberry, K. S. 1938. "Survival of wheat varieties in Great Plains winterhardiness nursery, 1930–1937." *Agron. J.* 30:399–405.

Quisenberry, K. S. 1974. "Let's Talk Turkey." *Transactions Kansas Acad. of Sci.* 77:135–144.

Renfrew, J. M. 1973. *Palaeoethnobotany: The Prehistoric Food Plants of the Near East and Europe.* Columbia Univ. Press, New York, NY.

Roelfs, A. P., and D. L. Long. 1987. "*Puccinia graminis* development in North America in 1986." *Plant Dis.* 71:1089–1093.

Salmon, S. C., O. R. Mathews, and R. W. Lenkel. 1953. "A Half Century of Wheat Improvement in the United States." pp. 3–152. *In* A. G. Norman (ed.) *Advances in Agronomy.* Vol. 5. Academic Press, New York, NY.

Sauer, J. D. 1993. *Historical Geography of Crop Plants: A Selected Roster.* CRC Press, Boca Raton, FL.

Schmidt, J. W. 1974. "The role of Turkey wheat germplasm in wheat improvement." *Transactions Kansas Acad. of Sci.* 77:159–172.

Shroyer, J. P., and J. S. Hickman. 1986. *Planting Practices.* Kansas St. Univ. Coop. Ext. Ser. C-529.

Shroyer, J. P., J. Martin, and R. Sears. 1982. *Wheat Emergence Problems.* Kansas St. Univ. Coop. Ext. Ser. AF 94.

Smika, D. E. 1990. "Fallow management practices for wheat production in the Central Great Plains." *Agron. J.* 82:319–323.

Smith, R. W., and R. D. Hudson. 1990. *Wheat Production Guide.* Univ. of Georgia Coop. Ext. Ser. MP 431.

Stecker, J. A., D. H. Sander, F. N. Anderson, and G. A. Peterson. 1988. "Phosphorus fertilizer placement and tillage in a wheat-fallow cropping system." *Soil Sci. Soc. Am. J.* 52:1063–1068.

Steen, H. 1963. *Flour Milling in America.* Greenwood Press, Westport, CT.

Stuckey, R., J. Nelson, G. Cuperus, E. Oelke, and H. M. Bahn. (not dated). *Wheat Pest Management.* Ext. Ser./USDA, Wheat Ind. Res. Com., and Natl. Assoc. of Wheat Growers Found.

Thom, W. O., M. J. Bitzer, and K. L. Wells. 1981. *Nitrogen Fertilization of Wheat.* Univ. of Kentucky Agron. Dept. AGR-87.

Tisdale, S. L., and W. L. Nelson. 1956. *Soil Fertility and Fertilizers.* MacMillan Pub. Co., New York, NY.

United States Department of Agriculture. 1936, 1942, 1952, 1961, 1972, 1982 and 1992. *Agricultural Statistics.* U.S. Government Printing Office, Washington, DC.

United States Department of Commerce. 1993. *Flour Milling Products: Summary for 1992.* USDC Ec. and St. Adm. Bur. of the Cen. M20A(92)-13.

Unger, P. W. 1991. "Organic matter, nutrient, and pH distribution in no- and conventional-tillage semiarid soils." *Agron. J.* 83:186–189.

Walter, T. L., and R. G. Sears. 1986. "Varieties." pp. 7–8. *In Wheat Production Handbook.* Kansas St. Univ. Coop. Ext. Ser. C-529 Revised.

Westfall, D. G., and R. H. Follett. 1988. *A New Technique for Phosphorus Fertilization of Winter Wheat.* Colorado St. Univ. Coop. Ext. Ser. Service in Action .557.

Watt, B. K., and A. L. Merrill. 1975. *Handbook of the Nutritional Contents of Foods.* Dover Pub., Inc., New York, NY.

Wheat Flour Institute. 1966. *From Wheat to Flour: The Story of Man . . . in a Grain of Wheat.* Wheat Flour Institute, Chicago, IL.

Whelan, E. D. P., and G. B. Schaalje. 1992. "Vernalization of embryogenic callus from immature embryos of winter wheat." *Crop. Sci.* 32:78–80.

Whitney, D. A. 1986. *Wheat Fertilization.* Kansas St. Univ. Coop. Ext. Ser. C-529.

Wiese, M. V., R. Loria, K. Dimoff, and N. Kilmer. 1979. *A Derivation of Optimal Planting Dates for Winter Wheat in Michigan.* Michigan St. Univ. Agri. Exp. Sta. R.R. 387.

Wiese, M. V., and A. V. Ravenscroft. 1979. *Control of Common Bunt.* Michigan St. Univ. Agri. Exp. Sta. Res. Rep. 385.

Wilhelm, W. W., L. N. Mielke, and C. R. Fenster. 1982. "Root development of winter wheat as related to tillage practice in Western Nebraska." *Agron. J.* 74:85–88.

Willis, W. C. 1986. *Disease Control.* Kansas St. Univ. Coop. Ext. Ser. C-529.

Worland, A. J., M. D. Gale, and C. N. Law. 1987. "Wheat Genetics." pp. 129–172. *In* F. G. H. Lupton (ed.) *Wheat Breeding.* Chapman and Hall, London, England.

World Book Encyclopedia. 1987. World Book, Inc., Chicago, IL.

Wright, D. L., R. D. Barnett, E. C. French, and M. Swisher. 1983. *Wheat for Grazing and Grain.* Univ. of Florida Coop. Ext. Ser. AF 150.

Wrigley, C. W., and G. J. McMaster. 1988. "Wheat: Right for Bread, but for Much More Too." pp. 21–40. *In* Y. Pomeranz (ed.) *Wheat is Unique.* Am. Assoc. of Cer. Chem., St. Paul, MN.

Yamazaki, W. T., M. Ford, K. W. Kingswood, and C. T. Greenwood. 1981. "Soft Wheat Production." pp. 1–32. *In* W. T. Yamazaki and C. T. Greenwood (eds.) *Soft Wheat: Production, Breeding, Milling, and Uses.* Am. Assoc. Cer. Chem., St. Paul, MN.

Zadoks, J. C., T. T. Chang, and C. F. Konzak. 1974. A decimal code for the growth stages of cereals. *Weed Res.* 14:415–421.

3

Sorghum (Sorghum bicolor (L.) Moench)

INTRODUCTION

Sorghum is a major cereal crop both worldwide (Table 3.1) and in the United States (Table 3.2). It ranked a distant fifth in total grain produced worldwide in 1991 at 64 million tons, while 607 million tons of wheat and 528 million tons of corn were produced worldwide. In the United States, production was just over 16 million tons, well behind the 59 million tons of wheat and 209 million tons of corn produced that year. Sorghum is characterized by a tremendous amount of morphological variability, so much so that as late as 1978 taxonomists were struggling with ways to classify species and/or subspecies. Even the area of greatest diversity is in dispute, with Vavilov placing the evolution of sorghum in the northeast quadrant of Africa, while Doggett (1980) references Hartley (1958) who placed the origin of the tribe to which sorghum belongs in the Indo-Malaysian area. Doggett suggested that the center of variability for the grass tribe *Andropogoneae*, from which grain sorghum developed, is in the Congo or central region of Africa.

The domestication of sorghum is inextricably tied to the origins and evolution of African agriculture. Mann et al. (1983) provided a detailed review of the then-current views on the subject, reviewing five scenarios for the origins of African agriculture and therefore the domestication of sorghum.

1. Agriculture, that being the planting and harvesting of wheat (*Triticum* spp.) and barley (*Hordeum* spp.), and confining of sheep and goats, was introduced into the Nile Valley from Asia Minor approximately 5000 B.C. This technology then spread southward along the Nile and westward into the Sahara during the second Saharan wet period occurring about 5000 to 2500 B.C.

TABLE 3.1   World Production of Sorghum, Wheat, Rice, Barley, and Maize (Corn) for 1991.

Grain	Acres	Yield	Total Production	Increase*
	(×1000)	*(lb./ac.)*	*(×1000 t)*	*(%)*
Wheat	552,801	2,197	607,194	3.4
Rice	366,464	3,127	572,896	3.1
Maize (corn)	319,001	3,308	527,610	3.6
Barley	188,150	1,985	186,662	3.2
Sorghum	110,414	1,153	63,655	2.4

Source: Adapted from FAO, 1991; Maunder, 1990.
* Annualized rate of increase in production from 1961–1965 to 1981–1985.

This theory runs counter to the long-accepted view that agriculture—i.e., planting selected seeds and harvesting entire inflorescences of grain—had its origins in Africa. One school of thought is that the idea or technology originated along the Nile, where the Wadi Kubbaniya (present-day Sudan) people planted crops in the damp pond beds following seasonal flooding of the Nile and returned only when the crops were ready to harvest (similar to the "decrue" system employed in parts of West Africa today). The maturation of this technology (i.e., planting, husbandry and harvest) occurred in Asia Minor and returned to Africa to be adopted as new technology.

2. Primitive cooking implements dating to 5000 B.C. have been found along the Niger River in western Africa, which prompted a theory that the Mande-speaking people of this area independently developed an agrarian society. Although this theory of independent development seems to attract few followers, Mann et al. provide a scenario that combines the finding of these artifacts with the reintroduction of agriculture from Asia Minor—i.e., scenario 3 below.

3. If agricultural technology was introduced from Asia Minor, after its origins in a decrue system in Africa as early as 7000 to 12000 B.C., then it could have moved from the Nile Valley westward during the Sahara and Sahel wet period of 10000 to

TABLE 3.2   Production Statistics for Grain Crops Produced in the United States, 1991.

	Acres	Average Yield	Total Production
	(×1000)	*(lb./ac.)*	*(×1000 t)*
Maize (corn)	68,812	6,082	209,233
Wheat	57,667	2,061	59,414
Sorghum	9,816	3,306	16,221
Barley	8,410	2,651	11,145
Rice	2,749	5,618	7,721
Oats	4,794	1,615	3,879
Rye	395	1,383	273

Source: FAO, 1991; USDA Agricultural Statistics, 1992.

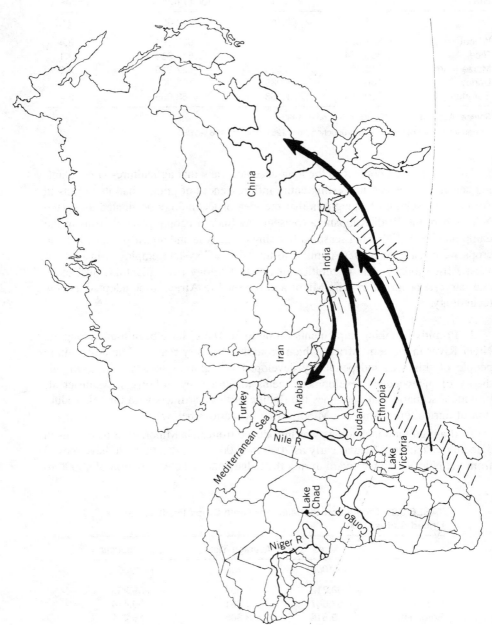

Figure 3.1 Direction and spread of Sorghum bicolor from its origin in northeast Africa. (Drawing by Mike Hodnett)

6000 B.C. Assuming this to be the case, then the expansion of the Sahara desert after this interpluvial period could have forced agrarian people southward where they encountered an established hunting and gathering Mande population along the Niger River in present-day Cameroon and Nigeria. The continued drought, which would last until about 5000 B.C., could have forced the Mande to adopt agriculture by that date and would account for the cooking artifacts.

4. A chain of highland plateaus and mountains stretches across Africa from southwestern Egypt to the central Sahara and then south to within a few hundred miles of the Niger River (Saharan Fertile Crescent). Some evidence indicate that cereals were cultivated in the Sahara between 6100 and 4800 B.C. Wheat, barley, sorghum, and bulrush millet are grown in these highland areas even today, so movement from the Nile Valley to the Niger Valley could have occurred conceivably even during dry periods of history.

5. The final scenario is that agriculture developed in the area of Ethiopia and then spread westward to Lake Chad and on to the Niger River, northward into Asia Minor, etc. (Figure 3.1).

MOVEMENT FROM ORIGIN

Conventional wisdom is that the Hamitic people, also called Afro-Mediterranean, migrated into northeastern Africa from the Mediterranean region during a time when the desert conditions were much less harsh than today. These people of caucasoid origin are tall, with narrow noses and brown skin. Today they occupy parts of Ethiopia, the Sahara, and the Sudan. Many remain today as herders of sheep, cattle, goats, and camels. Present-day Hamitic, or Cushite, groups occupying the area of northeastern Africa are the Beja, living predominantly between the Nile and the Red Sea, the Berber, Fulani, and Somali peoples.

These early Hamites probably migrated through the Middle East, through the Iraqi-Kurdistan area, where agriculture was practiced, according to some authorities, by 7000 B.C. It seems reasonable to accept that these Hamites would have brought primitive wheat and barley into northeast Africa. As sorghum grows wild in that area today, it is also reasonable to assume that these people encountered it as well, and adopted it as a part of their diet. While geographical impediments restricted, or slowed, the spread of sorghum westward, no such barriers limited its spread southward. Apparently, these Hamites moved as far south as Kenya, perhaps Tanzania. Archeological excavations along the Njoro River revealed tools and utensils that carbon-dated to 1000 B.C. However, the oldest carbonized sorghum grains date only to the 12th century A.D. It seems illogical to assume that sorghum was not grown and consumed much earlier than 1000 B.C.

Although the time and means of the spread of sorghum to western Africa are unclear, the adoption and movement into southern Africa appear to be straightforward. Figure 3.1 displays a combination of these events. Doggett suggested that about the time of Christ, Malaysian crops such as bananas, *Musa* spp., and yam, *Dioscorea alata,* were introduced into east Africa, spread into the Congo, and were carried overland as far as the Benue River. Harris (1907) and Shaw (1976), on the other hand, proposed that African yam, *D. cayenenis,* and oil palm, *Elaeis guineensis,* that are native to west Africa could have been distributed throughout the Congo before the introduction of bananas and the Malaysian yam. These and other high-calorie food crops provided the energy necessary for the expansion of the Bantu people, who occupied the area around the Benue River in present-day Cameroon and southeastern Nigeria. The Bantu population enlarged and spread into the Congo forest and then south and east throughout the savannah country lying south of the Congo. This relatively rapid expansion occurred about 2,000 years ago with the Bantu reaching east Africa about 500 A.D. Current wisdom credits the Bantu with spreading sorghum into central and southern Africa, after their initial encounter with sorghum in eastern Africa. The Bantu people reached Botswana by the tenth century A.D. and Zimbabwe by the 1300s. The Bantu today comprise a large group of black-skinned people, 60,000,000 to 80,000,000 in size, that speak the Bantu languages. The Bantu make up a major part of the population of the African continent that lies south of 5°N latitude. There are 300 autonomous Bantu groups today that vary in size from a few hundred to several million. The best known groups are the Zulu, once feared warriors in South Africa, and the Swahili people.

Sorghum probably moved out of Africa to India as stores on board Arabian, lateen-rigged sailing ships. By the time of Christ, sea-trade routes had been established between the east coast of Africa, Arabia, and the west coast of India. These routes were known as "dhow" routes, and the sailors who sailed them were called dhow sailors, so named for the type of ship used. These trade routes predate the time of Christ and depended on very reliable and alternating winds, the northeast and southwest monsoons. With these winds, seafarers based in Arabia could travel down the coast of Africa to Mozambique or to the southern tip of India on winds of the northeast monsoons that occur from November to March. They could then return on southwesterly winds during the southwest monsoon seasons of April to October (Figure 3.1). In near recent times, dhow traders would take on grain sorghum as ships stores from the Lindi-Kilwa area of southern Tanzania. They preferred sorghum called "Msumbija," having palatable grains that were hard enough to store well. It appears reasonable, therefore, that sorghum growing along the east coast of Africa would be used as ship's stores, with intentional excess or remanent taken to India on the prevailing winds the next year. Evidence suggests that sorghum was established in India before the time of Christ, but not earlier than 1500 B.C.

Records documenting sea trade between India and China date to about 600 A.D., and Chinese coins and pottery of the eighth century A.D. have been found in Kilwa and Tanzania. Grain sorghum could have been introduced into China through sea traders as it was in India. However, the Chinese grain sorghum, called kaoliang, probably was developed from introductions of sorghum from Arabia or India via the so-called silk road.

Evidence points to sorghum being introduced into the Middle East and Mediterranean areas through Arabia rather than over land routes from Ethiopia and Egypt. The Persian name for sorghum, juar-i-hindi, indicates an Indian origin. The dhow traffic between Africa, India, and Arabia could have introduced and reintroduced sorghum into the area over a period of several hundred years, with only the better and more adapted types being accepted.

Grain sorghum, called guinea and chicken corn, was first introduced into the Americas with the slave trade. Later introductions included durra types in 1874, kafir in 1876, milo about 1880, shallu in 1890, feterita in 1906, and hegari in 1908. Sorgos, sorghum types for forage and/or syrup, were introduced in 1857, and sudangrass was introduced as a forage in 1908.

EVOLUTION OF RACES

As with all crops, indeed with all plant species, distinct and true breeding biotypes of sorghum have evolved over the many thousands of years since speciation began. The general process of a genetically diverse population becoming isolated from the base population, followed by the intermating and survival of only the "fittest" individuals, had ample opportunity to operate relative to grain sorghum. As noted above, sorghum was moved by humans several thousands of miles in all directions from the generally accepted Ethiopian center. Added to this natural speciation process was and is the influence of human selections.

The basic difference between a wild, or native, and a domesticated cereal is the shattering character—i.e., the presence of an abscission zone at spikelet or rachis nodes that causes seed to fall from the panicle. As long as humans existed as hunter-gatherers, we collected cereal seeds by hand stripping or by beating the panicles against some sort of collection vessel where the cereals were growing wild. There was no need to transport whole plants or whole inflorescences to the camp site or to a common thrashing site. The on-site thrashing of shattering heads would disperse seeds some distance from the parent plant, and humans most likely dropped grains on the way back to camp and/or as they moved on. This probably gave a selective advantage of the shattering type over the non-shattering type, as wide dispersion of non-shattering types could occur only through animals, wind, or water. However, when humans became cognizant of the need to farm if they were to remain in one place for any length of time, they needed non-shattering types of plants. These types could weather until all plants in the field were mature and would allow people to harvest and transport whole inflorescences to a common thrashing site. These non-shattering types have been termed "domesticated," and in grain sorghum came about by mutation of only two genes. People simply had to recognize the advantage of such types.

As the very early domesticated sorghum type plants were selected and dispersed, genetic adaptation and intercrossing followed by selection and continued intercrossing in isolated ecosystems gave rise to new and stable sorghum biotypes. Reintroduction of native biotypes and introduction of biotypes evolved in other locations offered additional opportunity for intercrossing, and development and selection of additional biotypes. This movement and evolution of biotypes gave rise to five basic races of sorghum, and these can combine to produce ten rather distinct intermediate races (Table 3.3).

TABLE 3.3   Races and Intermediate Races of Cultivated Sorghum.

Basic Races	Intermediate Races
SBR-1: Bicolor	SIR-1: guinea-bicolor
SBR-2: Guinea	SIR-2: caudatum-bicolor
SBR-3: Caudatum	SIR-3: kafir-bicolor
SBR-4: Kafir	SIR-4: durra-bicolor
SBR-5: Durra	SIR-5: guinea-caudatum
	SIR-6: guinea-kafir
	SIR-7: guinea-durra
	SIR-8: kafir-caudatum
	SIR-9: durra-caudatum
	SIR-10: kafir-durra

Source: Adapted from Harlan and de Wet, 1972.

The most diverse and primitive biotype, or race, is *Sorghum bicolor* subsp. *bicolor* race bicolor. The primitive bicolor race probably gave rise to all other races of *S. bicolor.* The early progenitor of bicolor probably was *S. bicolor* race *aethiopicum* or race *verticilliflorum,* and may have been domesticated as early as 4000 B.C. Some authorities place the evolution of *S. bicolor* subsp. *bicolor* in west Africa, in which case the progenitor was more likely *S. bicolor* race *arundinaceum.* The race bicolor is distributed everywhere in Africa that sorghum is grown and is ancient, yet not native, in India, Indonesia, and China.

The current wisdom then is that the early *S. bicolor* subsp. *bicolor* race bicolor was moved out of its center of origin in northeast Africa, and the forces noted above resulted in the evolution of other races, with the possible exception of race kafir. Race bicolor was moved from east Africa to the savannah area west of Lake Chad to near the Atlantic coast of Africa before the time of Christ. Race guinea was selected from race bicolor growing in the wet tropical habit of that area and probably was the first evolved "specialized" grain sorghum. Guineas are today the dominant sorghum of western Africa. Race bicolor was then moved eastward by the Bantu, reaching east Africa about the time of Christ.

Other races of *S. bicolor* are caudatum, kafir, and durra. The caudatum race of grain sorghum was also a direct selection from early bicolor, selected apparently by people of the Chari-Nile language group. These people predominately occupy East and Central Sudan today. The evolution of race kafir is uncertain. It may have been derived from race bicolor after being carried south of the Congo from the north African savannah country. However, electrophoretic data of sorghum proteins suggest that the kafir type is more closely related to the wild grass *Sorghum bicolor* race *verticilliflorum* than any of the other four cultivated sorghum races. This, plus the fact that none of the kafirs moved into India through the same channels as the guinea types, suggests a more recent evolution and may indicate independent domestication relative to other cultivated races.

The cultivated race durra is a classic example of an evolving species being taken out of its center of origin, adapted biotypes selected, and those returned to its center of origin or diversity. Race bicolor was apparently taken to India before 1000 B.C., where the durra race evolved, was selected, and cultivated. About 550 A.D., Arabs acquired durra seed and established it in Arabia and parts of Africa. Durra was then introduced into Ethiopia around 615 A.D. by Arabic Muslims and became the economic base of the Muslim states in that area. The predominance of white grain within durra restricts its adaptation to areas of low rainfall and low grain mold risk. However, white grain sorghum is desirable as human food because of its low tannin content, and plant improvement programs in developed countries continue to expand its area of adaptability.

The cultivated races are distinctive in panicle morphology, grain size, and yield potential as they exist in the feral state. Below is a brief comparison:

Bicolor: the most primitive of the five cultivated races; open panicles of medium size; very small, dark grains covered by glumes; and a slightly stiff rachis (Figure 3.2a).

Guinea: Long, loose, glabrous, hanging panicles; sessile spikelets that open at maturity; small to medium grains; and low yielding (Figure 3.2b).

Caudatum: Medium to large panicles that are dense to slightly open; sessile and pedicellate spikelets; grains are "turtle-backed" with one side flat and the other side bulging; high yielding (Figure 3.2c).

Kafir: Erect panicles, mostly semicompact and cylindrical; hairy spikelets; glumes much shorter than grain; high yielding (Figure 3.2d).

Durra: Stiff and compact panicles covered with dense pubescence; peduncles often recurved, but occasionally erect (Figure 3.2e).

BOTANICAL CLASSIFICATION

Although the above delineates the classification of cultivated sorghums as we know them in the United States, there are numerous "types" or "groups" within each of these found in the feral state in Africa and Asia. A detailed treatment of the botanical classification of the genus *Sorghum* is beyond the scope of this text, but a cursory review is necessary for the understanding of the evolution and improvement of grain sorghum. The following table is adapted from de Wet and Huckabay (1976), whose resources were Snowden (1936, 1955) (Table 3.4).

One should note that cultivated race caudatum is classified as a group within race kafir by the Snowden classification. Harlan and de Wet (1972), in their simplified version, list caudatum as a separate cultivated race because of their characteristic "turtle-backed" grains. Other than the grain shape, taxonomic studies reveal little difference in caudatum and kafir. The simplified classification attempts to place each of Snowden's species, groups in Table 3.4, and other working group classifications into one of the five races or one of the ten intermediate races shown in Table 3.2. This simplified system attempts to have enough flexibility to account for all of the morphological variability found in *Sorghum bicolor*. However, supporters of the

Figure 3.2. *Illustration of the five basic races of grain sorghum: (a) bicolor, (b) guinea, (c) caudatum, (d) kafir, (e) durra. (Drawing by Mike Hodnett)*

TABLE 3.4   Classification and Distribution of *Sorghum Bicolor*.

Subspecies	Variety	Race	Group	Feral Distribution
halapense	nc[1]	nc	controversum	E. and S. India
halapense	nc	nc	halapense	Med.-N. India
halapense	nc	nc	miliaceum	N.W. Pakistan-India
halapense	nc	nc	propinquum	S.E. Asia
bicolor	aethiopicum	nc	aethiopicum	N. Nigeria-Ethiopia
bicolor	aethiopicum	nc	lanceolatum	Senegal-Sudan
bicolor	aethiopicum	nc	virgatum	Egypt and Sudan
bicolor	arundinaceum	nc	arundinaceum	W. Africa
bicolor	arundinaceum	nc	vogelianum	Tropical W. Africa
bicolor	verticilliforum	nc	brevicarinatum	Kenya and Tanzania
bicolor	verticilliforum	nc	castaneum	N.E. Congo
bicolor	verticilliforum	nc	machrochaeta	Congo to Sudan
bicolor	verticilliforum	nc	panicoides	Ethiopia
bicolor	verticilliforum	nc	pugionifolium	Punjab of India
bicolor	verticilliforum	nc	somaliense	Somaliland
bicolor	verticilliforum	nc	usambarense	Tanzania
bicolor	verticilliforum	nc	verticilliflorum	South of Ethiopia
bicolor	bicolor	Guinea	conspicum	Tanzania and Mozambique
bicolor	bicolor	Guinea	exsertum	W. tropical Africa
bicolor	bicolor	Guinea	gambicum	W. tropical Africa
bicolor	bicolor	Guinea	guineense	W. tropical Africa-Uganda
bicolor	bicolor	Guinea	margaretiferum	Sierra Leone to N. Nigeria
bicolor	bicolor	Guinea	mellitum	W. Africa-S. Africa
bicolor	bicolor	Guinea	roxburgii	E. Africa-India-Burma
bicolor	bicolor	kafir	caffrorum	Africa
bicolor	bicolor	kafir	caudatum	Equatorial Africa
bicolor	bicolor	kafir	cariaceum	Congo-Tanzania
bicolor	bicolor	kafir	dulcicaule	Congo
bicolor	bicolor	kafir	nigricans	Tropical Africa
bicolor	bicolor	durra	cernuum	Asia Minor-India
bicolor	bicolor	durra	durra	E. Africa-Arabia-India
bicolor	bicolor	durra	rigidum	Sudan
bicolor	bicolor	durra	subglabrescens	E. Africa-Arabia
bicolor	bicolor	bicolor	ankolib	N.E. Africa
bicolor	bicolor	bicolor	basutorum	S. Africa
bicolor	bicolor	bicolor	bicolor	Arabia-Burma
bicolor	bicolor	bicolor	dochna	India-Burma
bicolor	bicolor	bicolor	elegans	W. Africa-E. Africa
bicolor	bicolor	bicolor	melaleucum	N.E. Africa
bicolor	bicolor	bicolor	membranaceum	E. Africa, India and China
bicolor	bicolor	bicolor	miliiforme	E. Africa and N.E. India
bicolor	bicolor	bicolor	nervosum	E. Asia
bicolor	bicolor	bicolor	notabile	N. Nigeria-Sudan
bicolor	bicolor	bicolor	simulans	Malawi
bicolor	bicolor	bicolor	splendidum	S.E. Asia

Source: Adapted from de Wet and Huckabay, 1967, J. M. J. de Wet and J. F. Huckabay, Evolution 21: 787–802 © 1967, Society for the Study of Evolution.

1. nc = not classified.

Snowden classification accurately point out that the vast diversity of the *Sorghum* genus cannot be represented by so few categories. Table 3.5 compares Snowden's species and groups as identified by others with the simplified system of Harlan and de Wet.

PLANT MORPHOLOGY

Grain sorghum has a growth habit typical of any grass species. The roots are fibrous, arising from the lower nodes of the stem, except for the radicle of the germinating seed (see chapter on rice for a more complete discussion of grass morphology). Roots may extend as far as five feet and penetrate to six feet, with the majority in the top three feet of soil. Sorghum stems vary in thickness because of feral or cultivated status, competition from surrounding plants, radiant energy, moisture, etc. Basal diameters may reach over one inch, decreasing toward the inflorescence. Stems of all cultivated sorghums are erect and solid. Root primordia buds are found at each node and will develop if in contact with soil. Prop roots may grow from lower nodes above the soil, especially in tall cultivars or biotypes of grain sorghum and in the sweet or forage types. A single bud is found at each node that may give rise to branches, called tillers or suckers, if they are from nodes at or near the soil surface. Some types also branch at upper nodes, but branching of this type is more often the result of stem damage in cultivated sorghum. Feral sorghum can reach over 12 feet, but cultivated grain sorghum in the United States is usually less than four feet from soil level to the top of the panicle.

The inflorescence is a branched panicle containing both stalked and sessile racemes. The flower stalk, or peduncle, is the final "internode" above the flag leaf. The peduncle gives rise to the rachis or main branch of the panicle, which branches again to rachilla that support the racemes. Within each raceme are two to seven pairs of spikelets. One of each pair is sessile and fertile, while the other is pedicled and sterile. The sessile, fertile spikelet will contain two florets, the upper perfect and therefore fertile, and the lower floret normally will be sterile, consisting of a lemma only. Occasionally the lower floret will be fertile and produce a grain. The pedicled spikelet will vary from being very small with only two glumes to having staminate florets producing viable pollen. Very rarely will the pedicled spikelet have florets with functional ovaries that produce grain. The terminal sessile spikelet of each raceme usually will have only pedicled, staminate florets.

MOVEMENT TO UNITED STATES AND CULTIVATED DEVELOPMENT

The first grain sorghums introduced into the New World came by way of slaves imported from West Africa. These introductions were probably used as ship's stores for the voyage across the Atlantic. These were probably race guinea and became known as guinea corn and chicken corn. Although grain sorghum is a major food crop in much of the world, and indeed was domesticated as such, the fact that its

TABLE 3.5   Snowden's Species, Groups of Table 3.4, and Groups as Identified by Others of *Sorghum Bicolor* Classified According to Harlan and de Wet Simplified System.

Snowden's and Other Working Groups	Harlan and de Wet Simplified System
Aterrimum	Shattercane
Drummondii	Shattercane
Nitens	Shattercane
Margaritiferum	Guinea
Guineesnse	Guinea
Mellitum	Guinea-bicolor
Conspicum	Guinea
Roxburgii	Guinea-kafir
Gambicum	Guinea
Exsertum	Bicolor
Basutorum	Kafir-bicolor
Nervosum	in part bicolor
	in part Caudatum-bicolor
	in part Kafir-bicolor
Melalcucum	Guinea-bicolor
Ankolib	Durra-bicolor
Splendidum	Bicolor
Dochna	Bicolor
Bicolor	Bicolor
Miliiforme	Kafir-bicolor
Simulans	Kafir-bicolor
Elegans	in part Guinea-caudatum
	in part Kafir-bicolor
Notabile	in part Guinea-caudatum
	in part Caudatum-bicolor
Cariaceum	Kafir
Caffrorum	Kafir
Nigricans	in part Kafir-caudatum
	in part Caudatum
Caudatum	in part Caudatum
	in part Guinea-caudatum
	in part Durra-caudatum
Dulcicaule	Guinea-caudatum
Rigidum	Durra-bicolor
Durra	Durra
Cernuum	Durra
Subglabrescens	Durra-bicolor
Roxburgii	Guinea
Roxburgii/Shallu	Guinea-kafir
Membranaceum	Durra (by glumes)
Kaoliang (Chinese sorghums)	Bicolor
Nevosum-kaoliang	Kafir-bicolor
Bicolor-broomcorn	Bicolor
Bicolor-sorgos & others	Kafir
Bicolor/Kafir	Bicolor
Dochna	Bicolor

(continued)

TABLE 3.5 (*Continued*)

Snowden's and Other Working Groups	Harlan and de Wet Simplified System
Dochna/Leoti	Bicolor
Dochna/Amber	Bicolor
Dochna/Collier	Caudatum-bicolor
Dochna/Honey	Bicolor
Dochna/Roxburgii	Guinea-caudatum
Dochna/Kafir	Kafir
Dochna/Nigricans	Caudatum
Dochna/Durra	Durra
Elegans	No specimen
Caffrorum	Kafir-caudatum
Caffrorum/Darso	Kafir-caudatum
Caffrorum/Birdproff	Kafir-caudatum
Caffrorum/Roxburghii	Guinea-kafir
Caffrorum/Bicolor	Caudatum-bicolor
Caffrorum/Feterita	Durra-caudatum
Caffrorum/Durra	Kafir
Nigricans	Caudatum
Nigicans/Bicolor	Kafir
(1) Dobbs	Guinea-caudatum
(2) Nigricans/Guinea	Guinea-caudatum
Nigricans/Feterita	Caudatum
(1) Dobbs	Caudatum
Nigricans/Durra	Durra
Caudatum	Durra
Caudatum/Kaura	Durra-caudatum
Caudatum/Guinea	Bicolor
Caudatum/Bicolor	Caudatum-bicolor
Caudatum/Dochna	Bicolor
Caudatum/Kafir (Hegari)	Caudatum
Caudatum/Nigricans	Caudatum
Zera-Zera	Caudatum
Caudatum/Durra	Caudatum
Durra	Durra
Durra/Roxburgii	Guinea-bicolor
Durra/Membranaceum	Guinea-durra
Durra/Bicolor	Durra-bicolor
Durra/Dochna	Durra-bicolor
Durra/Kafir	Kafir-caudatum
Nandyal	Durra
Durra/Nigricans	Caudatum
Durra/Kaura and others	No specimen
Cernuum	Guinea-bicolor
Subglabrescens	Durra-bicolor
Subglabrescens/Milo	Durra
Sudanese	Bicolor

Source: Reproduced from Harlan, J. R. and de Wet, J. M. J., A Simplified Classification of Cultivated Sorghum, Crop Science 12: 172–176.

first introduction into present-day United States was as a food for slaves probably destined it to become established as a food for the poor only and consequently found use predominately as a feed grain in the United States. The term chicken corn may denote that it was quickly established as a corn (feed grain) of appropriate size for chickens. At any rate, grain sorghum is today a feed grain in this country, while being a major food grain in most other countries where it is produced. It also is reasonable to assume that, since the industrialization of the food industry took place in the northern tier of states, wheat and corn would predominate research and equipment development to the exclusion of grain sorghum, a regional crop of the South only used for feed. This same phenomenon occurs today with the major food companies ignoring gossypol-free cottonseed, which has protein quantity and quality for human consumption superior to many other seed crops.

Deliberate introduction of sorghum began in 1857 with the importation of a "sorgo" type used for the production of syrup. These are referred to today as cane sorghum, sweet sorghum or cane, not to be confused with sugar cane. In that year, seed of Chinese Sugarcane, a sorgo or syrup sorghum, was sent to Texas by the U.S. Indian Service for production in the Brazos and Comanche Amerindian reservations. Additional early introductions were:

- 1874: Brown and white seeded durras called "gyp" corn, as they were falsely believed to have originated in Egypt;
- 1876: Kafir type from South Africa;
- 1880: Milo, milo maize, or giant milo; race and origin unknown;
- 1890: Shallu, a guinea-kafir intermediate from India;
- 1906–8: Feterita, race caudatum, or intermediate race durra-caudatum, from Sudan;
- 1906–8: Hegari type, race caudatum, from Sudan; and
- 1906–8: Pink kafir from South Africa.

All of the early sorghums were for forage or were dual-purpose—i.e., they were grazed, baled, and/or seed harvested as feed. Early farmers in the more arid plains of Oklahoma and Texas quickly realized the difficulty and uncertainty of producing corn as a feed for work animals and turned to the more drought-tolerant grain sorghum. The Texas Agricultural Experiment Station Bulletin No. 13 published in 1890 reported 23 varieties (i.e., cultivars) of sorghum available to Texas producers. That publication noted that farmers valued both the stalk and grain as feed stuffs. The need for more drought-tolerant crops in the arid plain states resulted in the majority of early plant improvement in grain sorghum being accomplished in Texas, Oklahoma, and Kansas.

The early sorghums in the United States were tall-growing and late-maturing, facts that we will return to subsequently. But the milo maize introduction of 1880, later referred to as Giant Milo, spread to Texas by 1890 and was particularly tall-growing. In fact, the common Texas folk tale of this sorghum was that it took a ladder to reach the head, an axe to cut the stalk, and a grubbing hoe to uproot the stubble. Seed of tall sorghums of that time were harvested by a person standing on a mule-pulled wagon to cut off heads and then of course accumulate the heads in the wagon for transport.

Shortly after the turn of the century, farmers found early-maturing heads of Giant Milo, which gave rise to a cultivar called Standard Milo. (Note that much of the literature will use the term "milo" to refer to grain sorghum of the durra race. Much of this race is characterized by a recurved peduncle such that the head is upside down at maturity. Throughout the ensuing discussion of cultivar development, this text also will use the term milo to be consistent with the literature on this subject.) Also about 1900, a farmer in Oklahoma selected Dwarf Yellow Milo from Standard Milo, seed of which were purchased by A. B. Conner of the U.S. Department of Agriculture (USDA) and distributed to farmers in north Texas near the town of Chillicothe. This distribution probably occurred between 1905 and 1910, as the Texas Agricultural Experiment Station and the USDA began research efforts at Chillicothe in 1905. An earlier-maturing cultivar, Early White Milo, was selected from the yellow milos about 1910.

A short-statured guinea-kafir was selected from an unidentified introduction of guinea-kafir about 1880. This material gave rise to Blackhull Kafir, a farmer-selected, short-statured cultivar. Sunrise and Dawn Kafir cultivars that were released in west Texas in 1905 may have been selected from Blackhull Kafir. Shorter-growing feterita and hegari-type cultivars were selected and released in Texas by 1920.

Although sorghum is predominantly self-pollinated, out-crossing does occur. A natural hybridization of a kafir and milo gave rise to sorghum types suitable for mechanical harvest—i.e., short-statured, lodging resistant, and more uniform maturity. Such hybridizations were selected by a Kansas farmer, H. W. Smith, and gave rise to such cultivars as Buff Kafir, Header Milo, and Dwarf Duallo. Another natural hybridization produced Darso, a drought- and chinch-bug-resistant cultivar.

Recognizing that natural hybridizations gave rise to usable and improved types, H. N. Vinall and A. B. Cron began the second phase of grain sorghum improvement in 1914 with the deliberate hybridization of a feterita biotype from Sudan with Blackhull Kafir. Chiltex and Premo cultivars were released from that cross in 1923. J. C. Stephens and J. R. Quinby released Bonita from a cross of Chiltex with a hegari type, and Quadroon from a milo and kafir combination. J. B. Sieglinger also was working with milo and kafir crosses in Oklahoma and released Beaver in 1928. Beaver had the desirable characteristic of erect heads, lacking the curvature of the peduncle that is characteristic of many of the milo sorghum. Beaver was also short-statured and suited to mechanical harvest.

Perhaps Sieglinger's greatest contribution came with the release of Wheatland in 1931, also from a milo and kafir cross. The use of the wheat combine for harvesting sorghum was just becoming popular and Wheatland proved well adapted for such harvest. In 1937, W. P. Martin found a single plant of Wheatland that was resistant to root and stalk rot (*Periconia circinata*). Martin saved and increased seed of this plant and released it as Martin's Milo in 1941. This cultivar, identified in 1937, proved so popular that it was the number-one cultivar in the United States until the development of hybrids in the 1950s. Several improved cultivars were released by scientists in Texas, Oklahoma, and Kansas during this phase of grain sorghum improvement.

Scientists in the 1940s began to believe that they had reached a yield plateau, with any future advances promising to be slow at best. Heterosis (i.e., the improved performance of a hybrid over that of either parent) had been observed during the deliberate hybridization phase, and of course everyone had taken note of the success of the hybrid corn industry in the United States by this time. The discovery of a cytoplasmic male sterile system reported by Stephens and Holland in 1954 quickly lead to the development of commercial sorghum hybrids. The greatest yield increases came early-on, with another, albeit higher level, yield plateau soon realized. Later improvements would be smaller yield increments, and advances in disease and insect resistance and in quality. Yields in the United States increased over 300 percent from 1950 to 1990.

Tall, late-maturing cultivars were no longer acceptable, yet scientists knew that genes for resistance to biotic pests and abiotic stresses could be found in accessions of the Indian World Sorghum Collection. Many of these were photoperiodic, resulted in hybrids too tall and too long-season to be of immediate value. These types were not adapted to the United States and required a considerable investment of time (years) to extract the desirable genes from the tropical parent and place them into non-photoperiodic, short-statured, early-maturing breeding lines. Scientists also recognized that U.S. sorghum hybrids were kafir and milo crosses, and that this rather narrow genetic base could cause problems should new disease or insect biotypes evolve.

CONVERSION PROGRAM

J. C. Stephens established a program to "convert" many of the accessions in the world sorghum collection to day neutrality and reduced height. The scheme is designed to move four dwarf genes and four maturity genes from BTx 406 into the tall, late-maturing, exotic types in the world collection. This is accomplished by

TABLE 3.6 Average Heights of Cultivars Having Varying Maturity Requirements at Chillicothe, Texas.

Cultivar	Genotype	Days to Bloom	Plant Height (in.)
Plainsman	$dw_1Dw_2dw_3dw_4$	64	20
Double Dwarf Yellow Milo	$dw_1dw_2Dw_3dw_4$	83	24
Dwarf White Sooner Milo	$dw_1Dw_2Dw_3dw_4$	61	37
Texas Blackhull Kafir	$Dw_1Dw_2dw_3dw_4$	74	39
Dwarf Yellow Milo	$dw_1Dw_2Dw_3dw_4$	83	42
Tall White Sooner Milo	$Dw_1Dw_2Dw_3dw_4$	62	50
Durra	$Dw_1Dw_2Dw_3dw_4$	62	63
Sumac	$Dw_1Dw_2Dw_3dw_4$	75	65
Standard Yellow Milo	$Dw_1Dw_2Dw_3dw_4$	89	68
Standard Broomcorn	$Dw_1Dw_2dw_3Dw_4$	74	81

Source: Adapted from Quinby, J. R. and Karper, R. E. Inheritance of Height in Sorghum, Agron. J. 46:211. 1954.

TABLE 3.7 Days to Floral Initiation of Eight Sorghum Accessions When Exposed to Varying Photoperiod and Temperature Strategies.

Accession	Photoperiod	Day/Night Temperature (F)					Genotype
		90/84	90/73	90/63	73/68	63/52	
	(hrs.)	------------- Days to floral initiation--------------					
100M	10	30	20	26	24	147	$Ma_1Ma_2Ma_3Ma_4$
	17	80	66	58	66	NFB[1]	
SM 100	10	29	23	26	29	147	$ma_1Ma_2Ma_3Ma_4$
	17	38	40	49	56	NFB[1]	
Kalo	10	29	23	34	26	87	$ma_1ma_2Ma_3Ma_4$
	17	29	34	34	37	87	
Combine	10	30	19	22	21	109	$Ma_1Ma_2ma_3Ma_4$
Hegari	17	68	49	36	40	125	
Hegari	10	33	18	21	19	84	$Ma_1Ma_2Ma_3ma_4$
	17	49	38	37	31	94	
Early	10	29	18	21	20	77	$Ma_1Ma_2ma_3ma_4$
Hegari	17	44	35	34	29	87	

Source: Reproduced from Quinby, J. R., Hesketh, J. D., and Voight, R. D., Influence of Temperature . . . Crop Science 13:243–246, 1973.

1. Had not formed a flower bud when the experiment was terminated at 200 days.

making crosses of each exotic accession to BTx 406 in Puerto Rico. Plants of the second generation after the cross, the F_2 generation, are grown in Texas and a short, early-maturing plant is selected. Seeds of this plant are returned to Puerto Rico to be backcrossed to the exotic parent and the cycle begins again. Five backcrosses are made. By this means, the final product will have the cytoplasm and 98.4 percent of the genes of the exotic parent plus the four dwarf and four maturity genes from BTx 406. This program will make genes, and therefore plant characters, in over 20,000 exotic grain sorghum lines more available to scientists worldwide. Over 1,500 such converted lines have been released since 1963. The effects of the four plant height and four maturity genes are shown in Tables 3.6 and 3.7, respectively.

STAGES OF GROWTH

The use of binomial nomenclature developed by Karl vol Linne, Latin = Carolus Linnaeus, revolutionized the language of scientists by providing a means of referring to plants and animals in a way as to avoid mistakes in communication. A plant may have many common names, but once properly classified, it will have only one Latin designation. With the advent of herbicides, and the need for better management to maximize profits, producers and scientists also must be able to communicate relative to the physiological or morphological state of agricultural plants. Identification of growth stages improves this communication.

The life of the grain sorghum plant, and therefore field, is divided into ten stages, zero through nine (Figure 3.3). Stages one through three depends on recognition of the leaf collar. A leaf is considered developed whenever the leaf collar (i.e., the

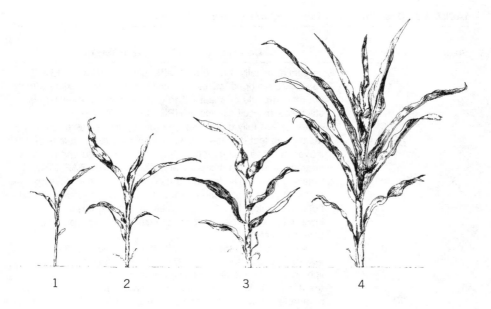

Figure 3.3 *Illustration of seven of the ten stages of growth of grain sorghum. See Table 3.8 for stage descriptors. (Drawing by Mike Hodnett)*

TABLE 3.8   Description of Grain Sorghum Stages of Growth.

Stage	Days after Emergence[1]	Description (DESCRIPTIVE NAME)
0	0	Emergence, usually 3 to 10 days after planting. (EMERGENCE)
1	10	Three fully developed leaves, i.e., collar of each leaf visible without tearing blade or sheath. The growing point or meristem is still below the soil surface. (THREE-LEAF STAGE)
2	20	Five leaves are fully developed; the root system is developing rapidly and leaf disease at lower nodes may result in loss of leaves at those nodes; rate of growth is increased and remains near constant until physiological maturity. (FIVE-LEAF STAGE)
3	30	Normally, 7 to 10 fully developed leaves; reproductive development initiated; final number of leaves has been determined and potential panicle size will be established during this stage; 3 lower leaves may be lost; culm or stalk growth is rapid. (no descriptive name)
4	40	Flag leaf is visible in the whorl with all except the final 3 or 4 leaves fully expanded; with 80% of the total leaf area present, light interception is approaching maximum; any reference to the number of leaves or leaf number should be from the top as a variable number of leaves have been lost from the base of the culm, the flag leaf is leaf number 1; about 20% of the final dry weight has occurred. (no descriptive name)
5	50	All leaves are fully expanded; panicle is near full size and enclosed in the flag leaf sheath; culm elongation is essentially complete; peduncle elongation begins; boot stage of inflorescence development occurs during this stage of growth. (BOOT)
6	60	Peduncle has grown rapidly since stage 5 and the panicle is extended out of flag leaf sheath by stage 6; half of the plants in a field are in some stage of bloom, or for an individual plant, flowering has progressed half-way down the panicle; time required to reach half bloom depends upon maturity of the hybrid and environmental conditions. (HALF-BLOOM)
7	70	Between the previous stage and the point in time when grains are at soft-dough, stage 7, approximately half of the total dry weight of the grain is accumulated; culm weight is decreased by about 10% during grain filling; lower leaves continue to be lost with only 8 to 12 functional leaves remaining at this time. (SOFT-DOUGH)
8	85	Approximately three-fourths of the grain dry weight has accumulated; grain contents have become more solid; culm weight has declined to it's lowest dry matter weight. (HARD-DOUGH)
9	95	Maximum total dry weight of the plant and panicle has occurred; the time from flowering to physiological maturity varies by hybrid but represents about one-third of the total time from planting; grain moisture is usually between 25 and 35%; remaining functional leaves may remain green; branches may develop from upper nodes if temperature and moisture are adequate. (PHYSIOLOGICAL MATURITY)

Source: Reproduced from Vanderlip, R. L. and Reeves, H. E. Growth Stages of Sorghum . . . Agron. J. 64: 13–16. 1972.

1. Approximate number of days for hybrids of RS 610 maturity grown at Manhattan, Kansas.

juncture of the leaf blade and leaf sheath) is visible without tearing the leaf blade or sheath. Because early stages of development depend on counting leaves, we should note that the first leaf (i.e., the lowest) on a sorghum plant has a round tip while all others have pointed tips. Therefore, if the lowest leaf has a pointed tip rather than a rounded tip, then one knows that at least one leaf has been lost from the plant. Stages of growth are outlined in Table 3.8.

USES OF GRAIN SORGHUM

Grain sorghum is used almost exclusively as a feed for cattle, swine, and poultry in the United States, while it is a major food crop in most other areas of the world, especially the developing nations of North Africa and Asia. As noted earlier, grain sorghum was introduced into the United States through slaves and therefore received very little attention as a food grain by the non-slave community. It also was introduced into a society that had a major native grain crop—i.e., corn. From a nutritional viewpoint, grain sorghum suffers from similar deficiencies as many other grains (Table 3.9). These are low palatability, low protein content on a whole-grain basis, deficiency of at least one essential amino acid, lack of several vitamins, and lower calcium and phosphorus content. Sorghum does compare well with other feed grains in total carbohydrates indicating its suitability as a feed grain from that perspective.

Grain sorghum has been a stable food source in many parts of Africa and Asia for hundreds of years. People of these regions have developed preparation techniques and means to supplement grain sorghum in a manner as to utilize the relatively poor

TABLE 3.9 Comparison of the Nutritive Value of Selected Grain and Oilseed Crops.

Crop	Calories	Protein	Fat	Ca	P	K	Total Carbohydrates
	(per 100 g.)	(%)	(%)	------- (mg./lb.)-------			(%)
Barley (pearled)	349	8.2	1.0	16	189	160	78.8
Corn (whole, ground)	355	9.2	3.9	20	256	284	73.7
Rice (milled)	363	6.7	0.2	24	94	92	80.4
Rye (whole grain)	334	12.1	1.7	38	376	467	73.4
Sorghum (whole grain)	332	11.0	3.3	28	287	350	73.0
Peanut (whole seed)	564	26.0	47.5	69	401	674	18.6
Soybean (whole seed)	403	34.1	17.7	226	554	1,677	33.5

Source: Adapted from Watt and Merrill, Handbook of the Nutritional Contents of Foods, © 1975, Dover Publications, Inc.

protein quality of sorghum grains. Although it is not a source of human nutrition in the United States to date, no text would be complete without documentation of the ways that sorghum grains are prepared in other cultures.

Traditionally, whole grains may be ground into flour or pearled before grinding by hand removal of the pericarp. This is accomplished by employing some type or variation of a mortar and pestle after seeds are soaked to about 20 percent moisture. Sorghum food products around the world include the following:

Immature grain: Dimpled, sugary sorghums are used widely in India, and a high-lysine grain is consumed this way in Ethiopia. Grain is gathered while in the hard dough stage and parched or roasted while still green. These products are called vani or hurda in India.

Boiled: This is the simplest way that sorghum grains are consumed. Mature seeds are harvested and boiled whole, sometimes mixed with blood from cattle. Grains may be pearled before boiling to produce a product resembling rice. This method of preparation is found in India, Bangladesh, China, Ethiopia, Kenya, Botswana, Nigeria, and other more isolated areas of Africa.

Popped: Pop sorghums, similar to popcorn, are popular in India.

Fermented or sour foods: In most, if not all, of the places where grain sorghum is consumed as fermented or sour porridges, precise control of bacteria and yeast is not practiced and indeed often is not possible. Souring occurs from lactobacilli and later from yeasts. The acidity of the grain-water mixture helps prevent or at least slows the growth of undesirable pathogens. In some cultures, each new batch is mixed with a bit of the previous batch to start the fermentation process. These products are called ambali in India, motozo wa Ting in Botswana, obushera in Western Uganda, and ogi in Nigeria. These foods may be consumed as porridges or, by allowing the fermentation process to continue, they may be consumed as beer.

Leavened breads: Much of the grain sorghum crop in Ethiopia, Sudan, and Nigeria is consumed as bread. The leavened bread dough is prepared by mixing flour, whole grain or dehulled, water, and yeast-containing starter mix from a previous batch. Fresh or packaged yeast can be used. Leavened bread is called injera in Ethiopia, risra in Yemen, and "masa" in Nigeria.

Unleavened breads: Breads made without yeast are popular from Nigeria to India. Roti is unleavened bread made from sorghum and consumed extensively in India. Chapatty is also popular in India and is usually made from a mixture of cereal flours. Ouitta is an unleavened bread used to a small extent in Ethiopia. Sorghum Bread, waina, containing hot peppers, spices, and peanut oil is consumed in Nigeria. Tortillas are often made of grain sorghum in Latin America.

Stiff porridges: Sorghum is consumed as porridge in practically every sorghum-producing country. Grain, usually dehulled, is ground into flour, course or fine, and is added to boiling water. The product is a porridge stiff enough to mould without being sticky or crumbly. This product, with some local variation in color or additives, is called ugali in East Africa, mafo in Somalia, nsima in Malawi and Zambia, sadza in Zimbabwe, genfo in Ethiopia, guma or asida in Sudan and tuwo in Nigeria. Sankati is made in India from a mixture of fine and coarsely ground sorghum flour.

Grain sorghum can be divided into two grain types, those containing tannins and those without tannins. Grain sorghum that contains tannins possesses a pigmented testa layer just under the pericarp and outside the aleurone layer (Figure 3.4). Tannin-containing sorghum grains are brown as a result of the tannin molecules in the testa, whereas other sorghum grains vary from yellow to red to white. Color is confined to the seed coat, or testa in the case of tannin, except for yellow, which may be found in the endosperm.

Apparently, tannins protect grain sorghum kernels against insects, birds, and microorganisms, and inhibit preharvest germination. These are all desirable agronomic characteristics, but these compounds also affect the nutritional quality of grain sorghum. Tannin levels vary from 0.016 to 0.037 percent in yellow sorghums grown in the United States, while tannin levels of the brown or high-tannin sorghum cultivars vary from 1.57 to 4.80 percent. Almost all, about 98 percent, of the

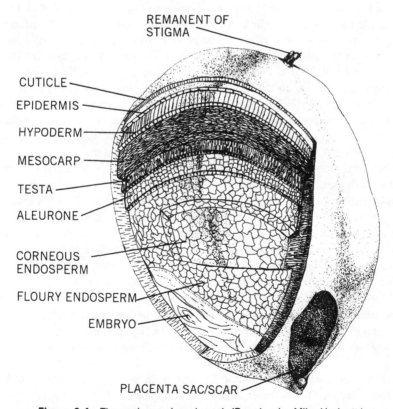

Figure 3.4   *The grain sorghum kernel. (Drawing by Mike Hodnett.)*

cultivars grown in the United States are of the yellow or lower-tannin type. However, there are areas of the world where birds and/or fungal attack make high-tannin sorghums desirable, even though the digestibility and nutrient value are reduced.

Condensed tannins also have been indicted as harmful to people and animals consuming large quantities. The first evidence that ingesting large quantities of tannins could be harmful came with the discovery that the Xhosa people, Bantu origin, of Transkei, Africa suffered a high rate of esophageal cancer. A ten-year study, from 1953 to 1962, of the food and folk medicine utilized by the people of Transkei revealed a diet of plants high in condensed tannins, especially catechin tannin. Further study of areas such as Honan Province, China; Turkmen and Uzbek regions of the former USSR; northern Iran; Curacao Island, West Indies; the Normandy peninsula of France; Bombay, India; Kenya; and northern Chile having exceptionally high levels of esophageal cancer revealed that catechin tannin was a common thread connecting diets and consumption habits. In some of these locations, such as Curacao Island, dark grain sorghum is a staple in their diet.

The tannins impart an astringent taste that is repugnant to most animals, thereby making these types bird-resistant. The antifungal properties protect grains from fungal attack when grown in warm, moist climates and also act as an inhibitor to moisture uptake, thus preventing preharvest germination and improving storability. Tannins apparently bind to proteins and digestive enzymes. These tannin-protein molecules are then too large to be absorbed during the digestive process, effectively reducing the availability of feed (or food) protein. In fact, Bruce Boren, associate director of the National Grain Sorghum Producers Association, suggests that there is enough tannin in high-tannin sorghum to precipitate—i.e., combine with—more protein than is contained in the grain. Pancreatic amylase also can combine with tannin molecules, thus affecting starch digestion, both actions resulting in abut a 20 percent reduction in metabolizable energy relative to non-tannin sorghum. Although this may be a worst-case scenario, there seems little doubt that high-tannin grain sorghum is nutritionally inferior to low- or no-tannin sorghums.

The outward manifestations of the consumption of high-tannin sorghum are reduced growth rates and diminished feed efficiency. Some scientists report reduced feed intake although the actual amount of food may be affected very little under no-choice situations. Animals fed high-tannin grain sorghum grow at a reduced rate, thereby resulting in smaller animals requiring less feed bulk. These animals grow more slowly because of poor feed utilization rather than failure to eat.

The major animal enterprises in the United States are swine, cattle, and poultry. The effects of high-tannin sorghum as the sole feed source for swine are typical: reduced feed efficiency and reduced growth rates. The effects in ruminant nutrition are more complex. Increased levels of tannin consumption may actually protect the animal against bloat and rumen proteolysis of dietary proteins. Cattle fed high-tannin sorghum will, however, gain weight more slowly than animals that are fed corn.

The effects of dietary tannin in poultry have been extensively studied. Reduced growth rates, reduced feed efficiency, and lower egg production have been reported. Supplementation of poultry diets high in tannins with protein and/or methionine or choline, but not lysine, overcame the deleterious effects of tannins. Supplementary protein usually is added to poultry rations for other dietary reasons, so producers may rarely see the kinds of problems associated with higher-tannin sorghums that can be encountered by swine and cattle producers.

A number of ways to reduce the tannin levels of feed sorghum have been evaluated. Dehulling is the most obvious way, as the testa layer containing the tannin would be removed with the seed coat. A major problem with this approach is that the endosperm of sorghum is soft, and an unacceptable amount of endosperm is lost in the process. Efforts are underway to breed for high-tannin sorghum with improved dehulling characteristics and to improve dehulling equipment.

Treatment of high-tannin grain with a water extract of wood ash for 12 hours reduced tannin by 96 percent, while the same treatment followed by germination for four days reduced tannin content by more than 99 percent (Table 3.10). Soaking grains in wood ash and water is a common practice in some areas of Africa where high-tannin grain sorghum is grown and consumed as food. Other traditional ways of preparing high-tannin sorghum such as germination or fermentation alone are not effective in reducing assayable tannin. To date, these processes are cumbersome and unnecessary in the United States for producing an acceptable sorghum feed because of the availability of other grains and because of the accepted practice of adding protein supplements. As also noted above, the United States produces primarily non-tannin sorghum.

Livestock feed rations can and do utilize grain sorghum, although low-tannin sorghums are the norm in the United States. Sorghum is often fed as the only grain in rations with protein supplementation, usually with something such as soybean meal, plus salt, vitamins, trace minerals, etc. The driving force for using sorghum versus corn is economics. Corn has about 1600 calories of energy/pound, while grain sorghum is slightly lower with approximately 1500 calories/pound, thereby

TABLE 3.10 Effects of Four Treatments of Grain of a Low-Tannin and a High-Tannin Sorghum Hybrid.

Cultivar	Treatments[1]	Protein	Tannin[2]	In Vitro Protein Digestibility	Rate Weight Gain
		(%)			(%)
Rs 610	1	8.8	0.00	84.2	45.5
	2	9.0	0.14	56.6	27.9
	3	8.9	0.07	81.2	36.5
	4	9.3	0.02	52.3	14.3
BR 64	1	8.0	10.04	13.3	7.2
	2	8.4	6.38	27.4	1.5
	3	8.3	0.39	75.4	52.5
	4	8.7	0.03	53.7	30.4

Source: Adapted from Mukuru, 1990.

1. Treatments: 1 = control; 2 = grains soaked in water for 12 hours, then germinated for four days; 3 = grains mixed with wood ash slurry and soaked in water 12 hours; 4 = same as 3 plus germination for four days.
2. Catechin equivalent.

TABLE 3.11   Components of Recommended Pig Starter Diets.

Diet Component	Pounds of Diet Component When Pig Weights (lbs.) are		
	< 10	10–15	25–40
Sorghum (or corn)	722	1013	1102
Soybean meal (44%)	620	630	540
Dried whole whey	250	300	300
Oats groats	200	—	—
High fat product	150	—	—
Deflourinated Phosphate (32% Ca 18% P)	30	30	31
Ground Limestone	6	7	7
Trace minerals	3	3	3
Vitamins	10	10	10
Copper sulfate	2	2	2
Lysine HCl, 98%	2	—	—
Antibiotic supplement	5	5	5
Total	2000	2000	2000

Source: Data supplied by D. A. Knabe, Texas A&M University.

having about 95 percent of the feeding values of corn (high-tannin sorghum has only 80 to 90 percent of the feeding value of corn) (Table 3.9). Grinding or pelleting of any grain will improve its feed efficiency by making the grains more digestible as a result of exposing the endosperm—i.e., the starch or energy source. A typical swine starter ration is shown in Table 3.11. Note that either corn or sorghum is recommended.

U.S. AND WORLD PRODUCTION

Kansas was the leading producer of grain sorghum among U.S. states in 1990 with 184,800,000 bushels (9,979,200,000 pounds), followed by Texas with 135,200,000 bushels (7,300,800,000 pounds) and Nebraska with 107,800,000 bushels (5,821,200,000 pounds) in 1990 (Table 3.12). These three states produced 75 percent of the grain sorghum produced in the United States that year. They were followed by Missouri, 42,460,000 bushels, Arkansas, 18,150,000, Oklahoma, 15,400,000 bushels, Illinois, 14,625,000 bushels, South Dakota, 14,300,000 bushels, and Colorado with 10,340,000 bushels. Nine other states produced less than 10,000,000 bushels each.

Acreage in the United States has remained fairly steady at 13,000,000 to 15,000,000 from 1960 through the middle 1980s, but acreage numbers for 1990 indicate a downturn in land area devoted to sorghum in the United States (Table 3.13). Texas, Kansas, and Nebraska grew fewer acres in 1990 than in 1960, 1970, or 1980. Production statistics for grain before 1920 are not available in normally

TABLE 3.12   Production of Grain Sorghum by State, 1935–1990.

State	\[Total Grain Production (×1000 bu.)[1]\] 1935[2]	1940[2]	1950	1960	1970	1980	1990
Alabama	—	—	595	672	748	1,122	990
Arizona	1,134	842	3,520	6,060	12,670	1,950	—
Arkansas	876	1,139	493	780	9,500	5,887	18,150
California	4,736	4,680	4,674	17,420	29,304	11,096	—
Colorado	994	3,978	1,236	7,344	10,270	12,250	10,340
Georgia	—	—	—	840	684	1,968	1,200
Illinois	—	—	—	676	840	3,658	14,625
Indiana	—	—	54	1,272	845	416	—
Iowa	—	—	—	2,365	2,002	1,330	—
Kansas	9,680	27,638	44,689	166,960	145,960	149,640	184,800
Kentucky	—	—	—	1,012	486	1,300	2,604
Louisiana	—	—	32	204	2,960	476	8,320
Mississippi	—	—	—	510	6,032	1,330	5,525
Missouri	2,346	4,960	615	23,868	12,826	41,280	42,640
Nebraska	2,678	6,954	4,850	86,102	77,520	121,800	107,800
New Mexico	2,816	3,024	8,417	9,080	17,499	10,280	3,250
North Carolina	—	—	690	3,760	2,597	2,232	1,840
Oklahoma	13,160	15,664	17,520	23,168	23,306	16,320	15,400
South Carolina	—	—	152	240	288	330	264
South Dakota	—	3,349	1,175	6,300	9,010	10,725	14,300
Tennessee	—	—	—	1,056	816	1,470	4,235
Texas	60,075	55,666	144,566	277,680	329,616	181,700	135,200
Virginia	—	—	—	304	675	387	—
Total[3]	98,495	127,894	233,278	637,673	696,454	579,197	571,483

Source: USDA Agricultural Statistics, 1936 and following years.

1. One bu. of U.S. No. 1 grade grain sorghum = 57 lbs.
2. Includes grain and forage sorghum harvested for seed.
3. Total includes minor production in other states.

circulated documents or perhaps because most production was used to feed live-stock on the farms where it was produced. Average yields ranged in the teens, except for 1920 and the dust bowl year of 1934, up until the 1950s when the contribution of hybrid cultivars, commercial fertilizers, and power machinery pushed yields higher. By the 1980s, average U.S. yields were above the 60 bushels/acre mark. Peak production occurred in 1985 with over one billion bushels produced. Exports vary with world supplies but have ranged from 16 to 50 percent of production since 1980.

The United States produced 25 percent of the world's total production of grain sorghum in 1991 on about 9 percent of the total world acreage (Table 3.14). Twenty other countries produced at least 10 million bushels in 1991. Average yields ranged from a low of 3.6 bushels/acre in Niger to 89.4 bushels/acre in France. The United States is the leading country in production of grain sorghum, followed by India, the People's Republic of China, Nigeria, Mexico, and Sudan.

TABLE 3.13   Grain Sorghum Harvested for Grain in the United States, 1920–1990.

Year	Harvested Acres	Average Yields	Total Production[1]	Average Price Received	Exports[2]
	(×1,000 ac.)	(bu./ac.)	(×1,000 bu.)	(dollars/bu.)	(%)
1920	4,027	21.8	87,734	0.94	——
1925	3,887	14.2	55,244	0.75	——
1930	3,449	10.8	37,203	0.56	——
1931	4,667	16.3	76,047	0.26	——
1934	2,647	7.9	18,521	1.00	——
1935	4,222	12.9	54,634	0.56	——
1940	5,943	13.5	80,363	0.48	<1
1943	6,889	15.9	109,536	1.14	<1
1945	6,324	15.2	96,063	1.19	1
1950	10,346	22.6	233,536	1.05	32
1955	12,866	18.8	242,526	0.98	27
1960	15,601	39.7	619,954	0.84	11
1965	13,029	51.6	672,698	1.00	40
1970	13,568	50.4	683,179	1.14	21
1975	15,403	49.0	754,354	2.36	30
1980	12,513	46.3	579,343	2.91	51
1985	16,782	66.8	1,120,271	1.93	16
1990	9,089	63.1	573,303	2.12	40

Source: Adapted from USDA Agricultural Statistics, 1936 and following years.

1. One bu. of U.S. No. 1 grade grain sorghum = 57 lbs.
2. Exports calculated as a percent of annual domestic production.

PRODUCTION PRACTICES

Cultivar Choice

Considerations in choosing a hybrid cultivar of grain sorghum include maturity requirement, insect and disease resistance, lodging susceptibility, and purpose or utilization of the grain. Producers should carefully consider their particular production constraints and determine if their choice of hybrid could eliminate or reduce those constraints.

There are four basic color classes of grain sorghum: yellow, white, brown, and mixed. Most hybrids grown in the United States are yellow sorghums having grains that are yellow, salmon pink, red, or white-spotted. These are often called red-seeded, bronze, hetero-yellow, and full-yellow endosperm hybrids. To be classified as yellow, a given lot (rail box car, elevator, truck load, etc.) can have no more than 10 percent brown-colored grains. The feeding value of these grains ranges from 90 to 95 percent of the value of corn except for the full-yellow endosperm grains which may equal corn. Hybrids of this class are not bird-resistant as they have none or very low levels of condensed tannins.

White sorghums are increasing in popularity because they can enter the food

TABLE 3.14 Countries Producing at Least Ten Million Bushels, Continental Totals and World Production Statistics for 1991.

Country/Continent	Harvested Acres	Average Yield[1]	Total Production[2]
	(×1,000)	*(bu./ac.)*	*(×1,000 bu.)*
Burkina FASO	3,199	13.7	43,852
Cameroon	1,235	12.7	15,760
Chad	1,225	11.7	14,381
Egypt	343	74.9	25,807
Mali	1,853	15.5	28,723
Niger	5,105	3.6	18,597
Nigeria	11,362	16.6	189,120
Sudan	11,594	10.0	115,875
Tanzania	1,408	13.4	18,912
Uganda	605	24.7	14,972
Africa	47,007	12.6	596,122
Mexico	3,409	50.3	172,060
United States	9,816	58.9	579,968
N. & C. America	14,254	54.0	772,595
Argentina	1,670	52.9	88,689
Brazil	447	23.9	10,717
Colombia	635	45.8	29,077
Venezuela	679	35.6	24,270
S. America	3,623	44.2	130,791
China	3,952	55.8	221,231
India	37,050	11.4	425,520
Pakistan	1,087	9.4	10,244
Asia	43,633	15.4	676,143
France	170	89.4	15,287
Europe	408	70.7	28,920
Australia	993	35.3	35,184
Oceania	993	35.3	35,224
World totals	110,414	20.5	2,275,862

Source: Food and Agriculture Organization, 1991.

1. One bu. of U.S. No. 1 grade grain sorghum = 57 lbs.
2. Values may not multiply nor add to totals because of rounding.

market as well as the feed market in the United States. The standards are therefore a bit more strict with only 2 percent non-white grains allowed. White sorghums have about 100 percent of the feeding value of corn and contain no tannins, thereby making them susceptible to bird deprivation.

The testa of brown sorghums, as noted earlier, contain condensed tannins and are therefore bird-resistant. These types also withstand field weathering and some fungi attack much better than white and other low-tannin sorghums. Brown sorghums containing not more than 10 percent other colors cannot be sold through commercial channels in the United States and are used on farms as livestock feed, most often for swine. The feeding value of high-tannin sorghums ranges from 75 to 90 percent of the feeding value of corn.

Lodging of grain sorghum plants usually is caused by factors other than cultivar. However, early-maturing cultivars often have a smaller diameter culm than mid- or late-maturing hybrids and are therefore more likely to go down. Even so, susceptibility to lodging is cultivar-dependent, and producers with fields having a history of lodging problems should pay careful attention to research data relative to lodging susceptibility.

Some hybrids offer resistance to leaf and grain diseases and low-to-moderate resistance to midge and greenbug. High numbers of midge or greenbug will overwhelm the resistance available to date, which means that the producer must continue to monitor insect levels regardless of cultivar and be prepared to treat fields with insecticides should the need arise. Open or less compact heads or panicles can reduce the incidence of mold in humid areas and reduce losses to larvae of *Heliothis viresence* and *Helicoverpa zea*, tobacco budworm and corn ear worm, respectively.

The final consideration in choosing a grain sorghum hybrid is the maturity requirement. Hybrids are on the market that require as few as 60 days or as many as 150 days from planting to maturity. Producers and seedsmen alike use the terms early, medium, and late to designate maturity requirements. The number of days from planting to maturity within each of these three classes varies across the United States, because the length of growing season varies widely. An early-maturing hybrid in Georgia requires 95 to 110 days to mature while early-maturing hybrids in Arizona require 95 days or less (Table 3.15). Hybrids having the longest season requirement within the limitations of expected moisture availability may outperform earlier-maturing hybrids grown within the same moisture and temperature constraints. If irrigation is available, then producers should plant the latest-maturing hybrids that will mature before first frost or freeze. In the southeast where rainfall patterns permit, full-season hybrids that require 110 to 120 days to reach physiological maturity are recommended. This recommendation contrasts with northeast Texas where early-maturing hybrids are suggested for early plantings and medium-

TABLE 3.15   Classification of Sorghum Cultivars by Maturity Requirements.

	Days from Emergence to Maturity	
Maturity Group	Central Arizona[1]	Georgia
Very short season	< 80	–
Short season	85–95	95–110
Med season	96–120	111–120
Full season	121+	121+

Source: Adapted from Dennis, 1981; Duncan, 1985.

1. Days required in central Arizona. The time required will usually be more at higher and less at lower elevations.

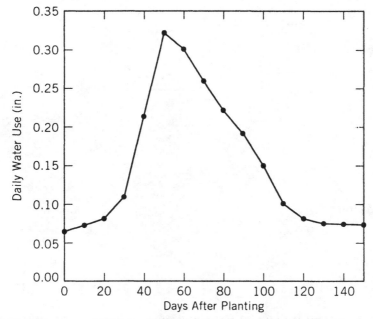

Figure 3.5 *Daily water use from planting to maturity for grain sorghum.* Source: *Adapted from Alexander et al., 1987*

maturing hybrids are recommended for late plantings. Planting early hybrids early and later-maturing hybrids later allows for grain development, or "head fill," before or after the expected heat and drought conditions of July. Maximum daily water use of grain sorghum can be as much as 0.3 inches/day around blooming and can remain above 0.2 inches/day for the following 30 days (Figure 3.5). If irrigation is available, then medium- to late-maturing hybrids are recommended unless planting is delayed.

Date of Planting

Planting is recommended in northeast Texas any time after seed depth soil temperatures reach 55°F for three consecutive days at about 7 A.M. This early planting date for early-maturing hybrids is desirable in some areas so that blooming and grain fill will occur before the onset of high temperatures and mid-summer drought conditions. In most other scenarios, planting should not occur until soil temperatures at four inches deep reach 65°F. This ranges from before mid-April in south Texas to after May 1 in more northern areas of production (Figure 3.6). This temperature allows for quick germination, emergence, and good seedling growth. Planting dates should be timed such that physiological maturity will be reached before the first autumn freeze. This planting data could be August 1 for doublecropped or crop salvage operations in the south or no later than early June in more northern states.

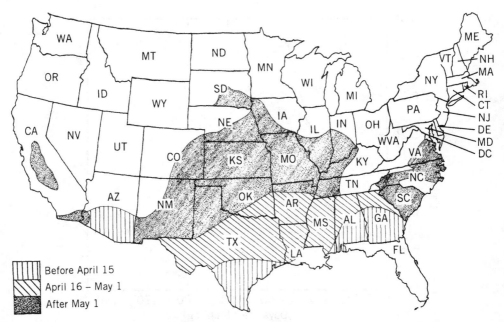

Figure 3.6 *Usual start of planting dates for grain sorghum.* Source: *USDA-SRS, 1984. (Re-drawn by Mike Hodnett)*

Tillage

Tillage operations associated with grain sorghum production span the range of tillage, from maximum, or conventional, to no-till. With conventional tillage in the drier plains of the midwest, stalks and crop residue are destroyed soon after harvest and fields may be disked, subsoiled, and/or bedded with furrow dikes to impound precipitation, thus allowing percolation and soil storage of winter precipitation. Beds may be reworked in late winter or early spring to adjust final bed height for planting and to reestablish furrow dikes in non-wheel furrows. A similar scenario of pre-plant tillage with the exception of furrow dikes may be carried out in the humid southeast where fall subsoiling or chiseling will shatter hard or traffic pans and allow winter and spring rain to percolate into the subsoil, as well as allowing for root penetration the following summer. Where bedding is practiced, the beds are "knocked down" to their final height just ahead of planting so that seed can be placed into moist soil. Any preplant tillage system must provide for good seed-soil contact to encourage and speed germination and emergence.

At the other end of the pre-plant tillage spectrum is no-till, where seeds are planted in standing or existing stubble (may be mowed) with no soil disturbance of any kind other than opening a drill for placement of seed and a starter fertilizer where recommended. Conventional or no-till are not rigid systems but rather concepts, with a number of possible systems of tillage between the two. Some tillage, such as ridge tilling, strip tilling, or reduced till, is often utilized in grain sorghum production.

No matter where one is on the spectrum from full tillage to no-tillage, the goals of tillage remain the same: (1) provide for good seed-soil contact; (2) control weeds; (3) conserve moisture; (4) preserve or improve tilth; and (5) control wind and water erosion. Reduced tillage can result in (1) fuel conservation, (2) improved trafficability in wet weather, (3) use of marginal and sloping lands unusable with conventional tillage, (4) as much as a 50 percent reduction in labor requirements, (5) improved water infiltration, (6) reduced water runoff, (7) reduced soil moisture evaporation, and (8) reduced equipment requirements. Disadvantages of no-tillage include (1) reduced weed control options, (2) potential for increased insect and disease problems, and (3) increased problems from traffic and natural hardpans in some southeastern soils. Federal legislation passed in 1985 places the United States on a path to mandate the reduction of top soil lost to wind and water erosion. Grain sorghum as well as other grains fit well into conservation tillage. Soil loss from a silty clay loam soil in Kansas with a 5 percent slope was 10.1 tons/acre when the tillage included moldboard plowing, disking twice, and then planting. Disking one time and then planting reduced soil erosion by 35 percent while planting without any pre-plant tillage reduced soil erosion to only 3.1 tons/acre (69 percent reduction). Percent ground cover from the previous corn crop was 4 percent following full tillage, 15 percent when planting was preceded by only one disking, and 39 percent when the new crop was planted with no pre-plant tillage.

Top soil loss continues to be a severe problem in the southeastern United States where losses are in the 5 tons/acre/year range. In one study, 23 tons/acre/year were being lost from a 6.7-acre watershed in northeast Georgia when continuously cropped with conventional-tilled soybeans. When double cropped with barley and grain sorghum grown without tillage, soil erosion was reduced to less than one ton/acre/year.

Long-term studies in Kansas suggest that grain sorghum can be successfully produced with reduced till or no-till strategies (Table 3.16). Averaged over 45 test years and five soil types, grain sorghum produced no-till, and reduced till yielded 58 bushels/acre, while conventional till yields averaged 61 bushels/acre. It has been projected that 65 percent of the major grain crops in the United States will be grown without tillage by the year 2000.

TABLE 3.16 Long-Term Yields of Grain Sorghum When Grown under Different Tillage Strategies at Five Sites in Kansas.

Site	Soil Type	Number of Years Tested	No-till	Reduced-till	Conventional
			----------------- (bu./ac.) -----------------		
1	Satanta loam	8	63	61	58
2	Harney silt	20	45	51	51
3	Smolan silt	5	84	——	85
4	Woodson silty clay	6	59	64	66
5	Woodson silty clay	6	70	72	77
	Average	——	58	58	61

Source: Adapted from Hickman and Shroyer, 1988.

Double-cropping/Rotations

Grain sorghum lends itself well to a number of rotations as well as to producing two crops per year on the same land or three crops in two years (multicropping). There has been limited effort and success in ratooning early-maturing grain sorghum in parts of the southeast. Yields of the ratooned sorghum are usually less than 50 percent of the first cutting.

Common cropping sequences vary across the United States. In the rolling plains of Texas, multicropping sequences are (1) grain sorghum-small grain-grain sorghum, and (2) small grain-cotton-winter fallow-grain sorghum, while single-crop rotation most likely will be grain sorghum-cotton-grain sorghum. Further north, the most common multicropping scenario is winter wheat-grain sorghum-fallow-winter wheat. In the Southeast, the growing season is such that producers can continuously double-crop a winter grain and summer grain sorghum. Common summer rotations in the Southeast will include soybean, cotton, or peanut while in the Midwest grain sorghum will be rotated with soybean. Rotating with a dicot or broadleaf crop is preferred over rotating with a monocot or grass crop such as corn or rice, because different chemical families of herbicides can be used to reduce grassy weed pressures on grain sorghum, plus life cycles of common disease pathogens and insects attacking each crop are disrupted.

The most interesting crop sequence that the author is aware of is double-cropping sorghum with crawfish. The crawfish (*Procambarus* spp.) will feed on components of a vegetative detrital, or disintegration, system in ponds with vegetation—in this case, sorghum plant residue. Crawfish have been harvested as an incidental crop in Louisiana rice fields for many years, with intentional double-cropping since the 1960s. Following the establishment of a permanent flood on the summer rice crop, about 50 pounds/acre of adult brood crawfish are stocked. The crawfish burrow into the soil and the rice is managed for grain production. In late summer the water is removed from the rice paddy, and the grain matures and is combined. Regrowth is encouraged and the field is reflooded in October to a depth of about 14 inches. The adult crawfish and their young emerge from their summer burrows and feed on the deteriorating rice stubble and associated organisms. The crawfish are then harvested the following spring. Crawfish can be managed such that restocking is not necessary. Following spring harvest, fields (or crawfish ponds) are drained, forcing the crawfish to burrow into the soil, after which the field is prepared for planting and the cycle repeats.

Research reported in 1989 suggested that grain sorghum can be produced using standard production practices and double-cropped with crawfish in a rice-crawfish-grain sorghum-crawfish rotation or a grain sorghum-crawfish system where crawfish ponds are previously established and restocking is not necessary. Grain sorghum would be favored over soybean in this system because the biomass of sorghum far exceeds soybean. Another advantage of sorghum over soybean and rice in this double-crop rotation is that the regrowth of sorghum exceeds that of rice and certainly exceeds soybean since the soybean dies at maturity. Table 3.17 reflects these considerations and the yield of crawfish in this system.

TABLE 3.17   Grain, Vegetative Biomass, and Crawfish Yields in Rice-Crawfish and Sorghum-Crawfish Double-cropping Systems in Louisiana during 1985 and 1986.

System[1]	Grain Yield	Biomass[3] Original	Biomass[3] Ratoon	Crawfish Yield
	----------------------------(lbs./ac.) ----------------------------			
Sorghum	3375a[2]	1985b	2297a	1981a
Rice	2845a	2935a	511b	1950a

Source: Adapted from Brunson, 1989.

1. Crawfish ponds had been previously established.
2. Values within columns are not statistically different if they are followed by the same letter.
3. Dry weight basis.

Grain sorghum is particularly well-suited to double-cropping because of its short growing season requirement. In the more humid southern United States, irrigated sorghum can be planted as late as August 1, or where rainfall amounts/patterns will allow. This versatility in planting means that winter crops such as wheat can be produced without on-farm or otherwise available drying facilities necessary for harvesting high-moisture wheat in order to get the summer crop planted early enough for full season production. And since grain sorghum competes well under no-till culture, valuable soil moisture and time can be saved.

Occasionally, one finds reports that suggest depressed yields of the winter crop following grain sorghum in rotations. Various explanations have been postulated to explain this phenomenon, including (1) depletion of soil moisture by the sorghum plants, (2) the "tying-up" of soil nitrogen by microbes during the decomposition of the relatively large amount of sorghum residue, and (3) the production of a water-soluble extract that is toxic to emerging wheat seedlings.

It is well established that the roots of grain sorghum are particularly efficient at extracting soil moisture, up to 90 percent of available moisture to a depth of 35 inches. This is due to the massive number of secondary roots, up to twice as many as corn even though the primary roots of both crops are equally extensive.

Although the sorghum plant has been shown to contain a water-soluble chemical that is detrimental to wheat germination, there seem to be few practical consequences in the United States. Many rotations or cropping sequences on U.S. farms involve only summer crops, allowing for the decay and leaching of such compounds. Georgia reported slightly depressed winter wheat yields following grain sorghum from 1980 through 1983 relative to wheat following soybean, 51 versus 54 bushels/acre, respectively. However, double-cropping of grain sorghum and wheat or other winter small grains and winter green manure crops is recommended in states with sufficient growing season length to allow such systems. If toxic compounds are produced by grain sorghum, they must not occur in sufficient quantities to cause consistent and reoccurring problems.

TABLE 3.18 Recommended Plant Populations for Grain Sorghum for Grain.[8]

State	Annual Rainfall (in.)				
	20	21–26	27–32	>32	Irrigated
	------------- (plants/ac. ×1000)-------------				
Kansas	24	35	45	70	100
Georgia[1]				65	100
Georgia[2]				32	
Alabama[3]				60	100
Alabama[4]				40	100
Texas[5]					78
Texas[6]					70
Texas[7]	32	32			105

Source: Adapted from Alexander et al., 1987; Duncan, 1985; Hickman and Shroyer, 1988; Mulkey et al., 1985.

1. Loam and sandy clay soils of good fertility and gentle slopes.
2. Loamy sands, sandy loams, sloped, eroded hillsides.
3. Soils of good fertility and water holding capacity with gentle slopes.
4. Extremely sandy, drought prone soils.
5. Winter garden area of Southwest Texas.
6. Blackland Prairie of Central Texas.
7. Rolling Plains of Texas.
8. Final stand can usually vary by a ±25% without effecting yield, assuming adequate distribution of plants.

Seeding Rates/Plant Populations

The appropriate number of sorghum plants per acre (ppa) varies with irrigation, total rainfall, rainfall patterns, and other growing conditions. Plant populations may be as high as 100,000 ppa or more under full irrigation, or as low as 24,000 ppa under very limited rainfall on the American Great Plains (Table 3.18). Pound basis seeding rates vary from 2 pounds/acre under severely limited water to about eight pounds/acre with full irrigation coupled with good management.

Pound basis is not the most desirable way to determine planting rates. Grain sorghum cultivars usually vary from about 12,000 to about 20,000 seeds/pound, and will vary within a cultivar because of variation in environmental condition during the year of production. Many seed companies in the United States aid producers in planting an appropriate number of seeds by identifying the number of seeds/pound on the certification tag attached to each bag of seed. If such information is not provided, the producer can make the determination, contact the seed company representative, or ask their County Extension Service.

The following formula determines seeding rate per acre when one knows the number of plants per linear foot of row desired:

[43,560/(row width (ft)] (desired no. of plants per foot of row)

The derived value is divided by the expected field emergence, usually 65 to 75 percent, to allow for losses to seedling diseases, bird predation, or non-viable seed. For example, assume a row width of 40 inches, a desired stand of three plants/foot of row and 75 percent emergence to viable plants. Substituting in the formula:

$$[(43,560 \text{ ft.}^2/3.3 \text{ ft}) \times 3]/0.75 = 52,272 \text{ seeds/acre}$$

If the producer is planting a hybrid having 15,000 seeds/pound, then he or she would plant 52,272/15,000 = 3.48 pounds/acre.

Grain sorghum is grown in a variety of planting patterns. Under expected moisture stress of north and northwest Texas, sorghum may be planted in a skip-row pattern of 2 planted rows and 1 row not planted, commonly called a 2-in-and-1-out pattern. Georgia, on the other hand, recommends that rows be spaced 30 inches for moisture conservation, via reduced evaporation, and for weed control, as canopy closure occurs much quicker than with 40-inch rows. In fact, canopy closure may not occur with 40-inch rows. Grain sorghum is planted in rows 20 to 30 inches wide in most areas of the United States, because most farms produce other grain crops that also perform well at 30-inch inter-row spacing, and therefore they can set all of their equipment for a common row spacing. Grain sorghum is grown in rows much more narrow than 30 inches, 15 inches in Kansas or 26 inches in areas of Texas. In most cases, unless intensely managed and fully irrigated, the recommended number of plants per acre as a final stand does not change as a result of between-row distance. Producers simply plant seeds further apart in the row as the distance between rows narrows. The scientific principle behind this truism is that narrowing the distance between rows and increasing the distance between plants within the row maximizes photosynthesis per unit land area by maximizing energy entrapment.

Producers should strive for optimum plant densities to optimize yields. Although some hybrids do tiller—that is, produce fruiting culms in addition to the main culm—sorghum has little morphological variability that will compensate for severely reduced strands, say 50 percent below recommended. On the other hand, excessively high plant densities can result in reduced individual plant performance because of interplant competition for nutrients, moisture, and energy. High plant densities often result in stalks with smaller diameters that lodge easily. Smaller panicles resulting from interplant competition may lower yields more than low plant densities if both are produced under an environmental stress such as drought.

Fertility Requirements

Grain sorghum is an efficient user of soil nutrients because of its extensive fibrous root system, having about twice the mass of secondary roots as corn. As with all crops, fertilizer and lime requirements are best determined by soil test; however, primary and secondary nutrient needs can be estimated by knowing the approximate yield potential of each field. Table 3.19 shows the amount of these nutrients required to produce 5,600 pounds of grain. Since the amount of plant biomass is not linearly related to grain produced, the amount removed in the stover (i.e., above-

TABLE 3.19   Approximate Amount of Nutrients Removed by 5,600 Pounds Grain Sorghum.

Element	Grain	Stover[1]
	-------------- (lbs.) --------------	
Nitrogen (N)	90	76
Phosphorus (P_2O_5)	35	20
Potassium (K_2O)	22	110
Sulfur (S)	9	10
Magnesium (Mg)	7	10
Calcium (Ca)	1.4	18.9
Copper (Cu)	0.014	0.02
Manganese (Mn)	0.056	0.11
Zinc (Zn)	0.07	0.14

Source: Adapted from Tucker and Bennett, 1968, Changing Patterns in Fertilizer Use, Soil Science Society of America.

1. The amount of stover is not linearly related to grain yield.

ground vegetative mass) cannot be estimated as easily as that in the grain. For example, an unchecked infestation of midge could reduce a potential yield of 5,000 pounds of grain to 2,500 pounds, but the amount of plant vegetative biomass would not change.

A soil nitrogen profile can be and should be made when previous management and cropping history suggest accumulation of nitrogen, either through heavy application of commercial nitrogen, use of legumes, or heavy applications of manure. Fields should be sampled to a depth of two feet, and the sample should be air-dried as soon as possible after collection to ensure an adequate nitrogen profile. Unless a chemical analysis is requested by the producer, soil test nitrogen recommendations may be based on soil type, cropping sequence, and historical yields or yield potential. A general rule of thumb is that grain sorghum requires about 2 pounds of applied nitrogen to produce 100 pounds of grain. This amount allows for some loss of nitrogen through leaching or volatilization. If grain sorghum follows grain sorghum or another high-residue crop, and the residue is not decomposed at planting, then nitrogen rates should be increased.

Nitrogen may be applied preplant or as a split application. Split applications (i.e., some (usually half) applied before and some after planting—often called sidedressing) are recommended in sandier soils that are subject to leaching. When nitrogen is applied as a split application, the crop should be sidedressed about 10 to 25 days after planting, but before the five-leaf stage—i.e., growth stage 2. Nitrogen utilization by the sorghum plant increases very rapidly after this stage of growth with about 70 percent of the required nitrogen taken up by growth stage 5 (Table 3.20). Nitrogen applied 35 days or later after emergence will not affect seeds/head but may influence seed size. Potassium, phosphorus, and other elemental requirements of grain sorghum should be applied preplant, broadcast and incorporated throughout the rooting zone, or applied banded at planting. Potassium is particularly important, as inadequate amounts will result in weaker stalks and increased lodging.

In some instances, especially in reduced tillage systems, a starter fertilizer application is made at planting. Dry formulations having 18 to 26 percent nitrogen and 23

TABLE 3.20   Cumulative Estimated Amounts of Primary Nutrients Taken Up by Grain Sorghum during the Growing Season in Georgia.

Nutrient	Stage of Growth				
	0–2	3	5	8	9
	------------------------ (%) ------------------------				
N	5	38	70	85	100
P_2O_5	3	27	60	87	100
K_2O	8	48	80	95	100
	5600 lbs./ac.				
	------------------------ (lbs.) ------------------------				
N	5	38	70	85	100
P_2O_5	2	16	36	52	60
K_2O	6	38	64	76	80

Source: Adapted from Duncan, 1985.

to 46 percent P_2O_5 are applied at 70 to 100 pounds/acre, and liquid formulations having 10 to 21 percent nitrogen and 18 to 34 percent P_2O_5 are applied at 10 to 15 gallons/acre. Starter fertilizer, dry or liquid, is banded 2 inches to the side and 2 to 4 inches below the seed to avoid direct seed-fertilizer contact that could significantly reduce emergence and survival.

In no-till production systems, fertilizer is surface-applied. Winter legumes are especially useful in replenishing soil nitrogen, and starter fertilizer should be considered to provide an adequate amount for early seedling growth. Urea and urea-containing fertilizer should be avoided because appreciable amounts of nitrogen can be lost through volatilization. An appreciable amount of nitrogen can be supplied through winter legumes as a cover crop or legumes such as soybean in rotation. Winter legume cover crops add 50 to 100 pounds/acre while soybean will add 30 to 60 pounds/acre.

Soils testing below 5.5 pH should be limed to bring the pH to 6.5. If sorghum is grown in rotation with a legume or other crop requiring a higher pH, soils should be limed to the higher level.

Biotic Pests

Grain sorghum is attacked by a number of insect and disease pests, but only a few consistently cause economic damage requiring pesticidal control (Tables 3.21, 3.22 and 3.23). Losses to disease in the humid rainbelt can reach 50 percent under conditions of cool wet weather favoring a number of organisms such as anthracnose and fusarium in combination with susceptible hybrids. Losses usually are well below 10 percent when good management is practiced. Across the nation, midge, chinch bug, and greenbug are the most troublesome insects. Midge is probable the most serious insect pest in the rainbelt, while greenbug and chinch bug are relatively more serious in the Plains states. Most of the insects attacking sorghum also attack corn, and so the reader is referred to Table 1.18 of Chapter 1. Other insects attacking sorghum are identified in Table 3.22.

**TABLE 3.21 Estimated Annual Losses and Cost of Control for
Several Insect Pests of Sorghum in Georgia, 1978–1982.**

Insect	Percent Yield Loss	Cost per Acre (dollars)
Sorghum midge	7.8	1.52
Lesser cornstalk borer	2.6	0.13
Sorghum webworm	2.3	0.40
Fall armyworm	0.9	0.46
Corn earworm	0.8	0.20
Stinkbugs	0.3	0.03
Chinchbug	0.1	0.08
Greenbug	0.04	0.08
Total	14.86	2.90

Source: Adapted from Gardner et al., 1985.

Birds can be a local problem in many sorghum production areas of the United States, especially with small acreage near wildlife refuges. As larger acreage is planted, the effect of birds is diluted. The most troublesome birds are redwing blackbird, grackle, brown-headed cowbird, and starling. Bird deprivation is more serious in the eastern United States, as there are many more roosting areas than in the more arid and more open areas of the Plains and Midwest.

In areas where sufficient wild feed and roosting sites prevail, flocks of 10,000 or more blackbirds can be found. Flocks of this size can do considerable damage, especially in the milk-to-soft-dough stages of sorghum grain development. Producers should pay close attention to the number of birds in their area. Estimates are

TABLE 3.22 Insects Attacking Grain Sorghum.[1]

Insects	Description	Symptom/Injury	Control
Greenbug *Schizaphis graminum* (Rondani)	Tiny, light green, soft-bodied; numerous, whitish cast off skins may be present.	Small seedlings showing reddening with some dying.	Chemical.
Sorghum webworm *Nola sorghiella* (Riley)	Greenish larvae thickly covered with hairs or spines; 4 red or brown stripes; to ¹/₂-inch long.	Destroys developing grains.	Open panicle hybrids; chemical.
Sorghum midge *Contarinia sorghicola* (Coquillet)	Tiny, orange to reddish larvae generally not visible to unaided eye; white pupa cases prevalent under severe infestation.	Larvae feed in developing grain; heads appear "blasted."	Early planting; field sanitation; chemical.

1. Most of the insects attacking sorghum also attack corn, therefore the reader is referred to Table 1.18 in this text.

TABLE 3.23 Common Diseases of Grain Sorghum.

Disease	Cause/Source	Symptom/Injury	Control
Seeds and seedling			
Seed Rots	*Fusarium* spp., *Aspergillus* spp., *Pythium* spp.	Thin stands; shoots yellowish; seeds and roots rotted.	High-quality planting seed; seed treatment.
Damping-off	Same as seed rots	Death at or prior to emergence.	Same as seed rots.
Seedling Blight	*Fusarium* spp.	Stunting; reddish leaves; red to black areas of roots.	Same as seed rots.
Root and stalk rots			
Root Rots	*Fusarium* spp.	Plants easily uprooted; root epidermis strips off easily; shoot growth stunted.	Resistant cultivars; rotations; field sanitation.
Fusarium Stalk Rot	*Fusarium* spp.	Premature plant death; stalk tissue salmon to dark red.	Same as root rot.
Charcoal Rot	*Macrophomina phaseolina*	Yellowing of main culm near base; stalk disintegration with small black sclerotia scattered throughout.	Resistant cultivars; avoid excessive plant populations; maintain soil moisture $\geq$80% during grain maturation if possible.
Foliar fungal diseases			
Leaf Blights	*Exserohilum turcicum*	Large, elliptical spots with grey centers and tan to reddish borders.	High-quality planting seed; rotation.
Sooty Stripe	*Ramulispora sorghi*	Elongated lesions with broad, yellow margins; soot-like growth on lesions.	High-quality planting seed; rotation.
Rust	*Puccinia purpurea*	Small brown to rust colored pustules on upper and lower leaf surfaces.	Resistant cultivars; proper irrigation.
Crazy Top Downy Mildew	*Sclerophthora mascrospora*	Leaves are stiff, leathery, upright with roughened, blistery appearance; if heads are present the glumes are often proliferated to give "crazy top" appearance.	Resistant cultivars; avoid areas susceptible to flooding or poor drainage.
Foliar bacterial diseases			
Bacterial Stripe	*Pseudomonas andropogunis*	Long, narrow, reddish stripes; lesions usually confined between veins.	Rotations; good crop management.

(continued)

TABLE 3.23 *(Continued)*

Disease	Cause/Source	Symptom/Injury	Control
Bacterial Streak	*Xanthomonas holcicola*	Narrow, water-soaked, translucent streaks ⅛-inch wide and 1 to 6 inches long.	Rotations; good crop management.
Other fungal or bacterial leaf diseases			
Grey Leaf Spot	*Cercospora sorghi*	Dark purple spots with grey cast during sporulation; elongate to round, ¼-inch or larger.	Rotations; high-quality planting seed.
Anthracnose	*Colletotrichum graminicola*	Small, ⅛ to ¼-inch circular to elliptical spots; tan, orange, red, or blackish depending on hybrid.	Rotations; high-quality planting seed.
Zonate Leaf Spot	*Gleocercospora sorghi*	Circular, reddish-purple bands alternating with tan zones giving bull's eye appearance; may extend several inches.	Rotations; high-quality planting seed.
Bacterial Spot	*Pseudomonas syringae*	Water-soaked spots about ⅓-inch in diameter occurring first on lower leaves; later turning tan with red borders.	Rotations, good crop management.
Panicle diseases			
Head Smut	*Sporisorium reilianum*	Part or all of head replaced by smut galls.	Resistance cultivars.
Covered Kernel Smut	*Sphacelotheca sorghi*	Kernels, usually all, replaced by smut galls.	Chemical seed treatments; high-quality planting seed.
Head Molds	*Fusarium* spp. *Colletotrichum* spp. other fungi	See foliar symptoms for these fungi as foliar and head diseases usually occur together.	Avoid hybrids maturing during periods of high rainfall probability; irrigate at boot stage if needed, control midge.
Viral diseases			
Maize Dwarf Mosaic Virus (MDMV)	Polyvirus group RNA virus	Mosaic patterned light and dark green on unfolding leaves; cool nights may cause red and necrotic areas resembling blight; flowering may be delayed and seed underdeveloped.	Resistance cultivars; aphid control (aphid vectored).

TABLE 3.23   (*Continued*)

Disease	Cause/Source	Symptom/Injury	Control
Small Seed	May be same as MDMV	Red to black lesions on panicle branches; often limited to point of seed attachment, appearing as a black dot inside florets.	No practical control but if MDMV then controlling aphid should help.

that one redwing blackbird will destroy 3 ounces per day. This means that 1,000 birds can destroy 185 pounds per day if sorghum grains are in the milk stage. By the dough stage of development, birds may destroy only 0.1 as much grain per day. Fright techniques will decrease bird damage, especially early on. Gas-operated "guns," balloons, and perhaps synthetic webbing may scare birds away or keep them agitated, which will decrease feeding.

GRADES AND GRADE REQUIREMENTS

Grain sorghum that is sold into commercial channels in the United States is inspected by the USDA Federal Grain Inspection Service. Grades and grade requirements are shown in Table 3.24.

TABLE 3.24   Grade and Grade Requirements for All Classes of Sorghum.

Grade[1]	Minimum Test Weight per Bushel	Moisture	Maximum Limits of—		Broken Kernels, Foreign Material and other Grains
			Damaged Kernels		
			Total	Heat-Damaged Kernels	
	(lbs.)	------------------------------- (%) -------------------------------			
U.S. No. 1	57.0	13.0	2.0	0.2	4.0
U.S. No. 2	55.0	14.0	5.0	0.5	8.0
U.S. No. 3[2]	53.0	15.0	10.0	1.0	12.0
U.S. No. 4	51.0	18.0	15.0	3.0	15.0

U.S. sample grade shall be sorghum that:
  (a) does not meet the requirements for grades U.S. Nos. 1, 2, 3, or 4,
  (b) contains more then 7 stones that have an aggregate weight in excess of 0.2 percent of the sample weight or more than 2 crotalaria seeds (*crotalaria* spp.) per 1,000 grams of sorghum,
  (c) has a musty, sour, or commercially objectionable foreign odor (except smut odor), or
  (d) is badly weathered, is heating, or is of distinctly low quality.

Source: USDA-FGIS, 1988.

1. Special grades supplemental to the designated grades are (1) smutty: sorghum that is covered with smut spores or that contains 20 or more smut masses in 100 grams of sorghum; and (2) weevily: sorghum that is infested with live weevils or other live insects injurious to stored grain.

2. Sorghum that is distinctly discolored shall not be graded higher than U.S. No. 3.

BIBLIOGRAPHY

Adams, J. E., G. F. Arkin, and E. Burnett. 1976. *Narrow rows increase dryland grain sorghum yields.* Texas A&M Univ. Agri. Exp. Sta. MP-1248.

Adams, D., and M. L. May. 1989. "General Culture." pp. 1–5. *In Growing Grain Sorghum in Arkansas.* Univ. of Ark. Coop. Ext. Ser. MP 297.

Alexander, U. U., C. Coffman, E. P. Boring III, N. McCoy, and D. Weaver. 1987. *Profitable grain sorghum production in the Rolling Plains.* Texas A&M Univ. Coop. Ext. Ser. Bul. 1577.

Almond, M., W. C. Smith, and G. P. Savage. 1979. "A comparison of two contrasting types of grain sorghum in the diet of the growing pig." *Anim. Prod.* 29:143–150.

Arkin, G. F., E. Burnett, and R. Monk. 1978. *Wide-bed, narrow-row sorghum yields in the Blackland Prairie.* Texas A&M Univ. Agri. Exp. Sta. MP-1377.

Baker, H. G. 1962. Comments on the thesis that there was a major center of plant domestication near the headwaters of the river Niger. *J. African His.* 3:229–233.

Black, R. 1988. "Irrigation." pp. 11–13. *In Grain Sorghum Production Handbook.* Kansas St. Univ. Coop. Ext. Ser. C-687.

Brunson, M. W. 1989. *Double cropping crawfish with sorghum in Louisiana.* Louisiana St. Univ. Agri. Exp. Sta. Bul. 808.

Butler, L. G. 1989. "Effects of condensed tannin on animal nutrition." pp. 391–402. *In* R. W. Hemingway and J. J. Karchesy (eds.) *The Chemistry and Significance of Condensed Tannins.* Plenum Press, New York, N.Y.

Butler, L. G. 1990. "The nature and amelioration of the antinutritional effects of tannins in sorghum grain." pp. 191–205. *In* G. Ejeta, E. T. Mertz, L. Rooney, R. Schaffert, and J. Yohe (eds.) *Sorghum Nutritional Quality.* Proc. Internatl. Conf. on Sorghum Nutritional Quality. Purdue Univ., West Lafayette, IN.

Chang, S. I., and H. L. Fuller. 1964. "Effect of tannin content of grain sorghums on their feeding value for growing chicks." *Poultry Sci.* 62:2420–2428.

Dennis, R. E. 1981. *Sorghum for grain in Arizona.* Univ. of Arizona Coop. Ext. Ser. T 8104.

de Wet, J. M. J., and J. P. Huckabay. 1967. "The origin of *Sorghum bicolor:* II. Distribution and domestication." *Evol.* 21:787–802.

Doggett, H. 1988. *Sorghum.* Tropical Agricultural Series. John Wiley & Sons, Inc., New York.

Duke, J. A., and A. A. Atchley. 1986. *CRC Handbook of Proximate Analysis Tables of Higher Plants.* CRC Press, Boca Raton, FL.

Duncan, R. R. 1985. "Agronomic principles and management tactics." pp. 5–17. *In* R. R. Duncan (ed.) *Proceedings Grain Sorghum Short Course.* Univ. of Georgia Agri. Exp. Sta. Spec. Pub. No. 29.

Food and Agriculture Organization of the United Nations. 1991. *FAO Production Yearbook.* Vol. 45. FAO, Rome, Italy.

Gardner, W. A., B. R. Wiseman, and R. D. Hudson. 1985. "Management strategies for sorghum midge and other key pests of grain sorghum in Georgia." pp. 25–32. *In* R. R. Duncan (ed.) *Proceedings Grain Sorghum Short Course.* Univ. of Georgia Agri. Exp. Sta. Spec. Pub. No. 29.

"Grain Sorghum Production." 1992. pp. 157–166. *In* G. Rutz (ed.) *Delta Agricultural Digest.* Farm Press Pub., Clarksdale, MS.

Guenzi, W. D., T. M. McCalla, and F. A. Norstadt. 1967. "Presence and persistence of phytotoxic substances in wheat, oat, corn, and sorghum residues." *Agron. J.* 59:163–165.

Haney, R. L. 1989. *Milestones Marking Ten Decades of Research.* Texas Agri. Exp. Sta., College Station, TX.

Hargrove, W. L., and G. W. Langdale. 1985. "Sorghum in no-tillage production systems." pp. 46–53. *In* R. R. Duncan (ed.) *Proceedings Grain Sorghum Short Course.* Univ. of Georgia Agri. Exp. Sta. Spec. Pub. No. 29.

Harlan, J. R., and A. Stemler. 1976. "The Races of Sorghum in Africa." pp. 465–478. *In* J. R. Harlan, J. M. J. de Wet, and A. B. L. Stemler (eds.) *Origins of African Plant Domestication.* Moulton Pub., The Hague, Netherlands.

Harlan, J. R., and J. M. J. de Wet. 1972. "A simplified classification of cultivated sorghum." *Crop Sci.* 12:172–176.

Harris, P. R. 1967. "New lights on plant domestication and the origins of agriculture: A review." *The Geographical Review* 57:90–107.

Hartley, W. 1958. "Studies on the origin, evolution, and distribution of the Gramineae (1). The tribe Andropogoneae." *Aust. J. Bot.* 6:116–125.

Hickman, J., and J. Shroyer. 1988. "Seedbed preparation and planting practices." pp. 3–6. *In Grain Sorghum Production Handbook.* Kansas St. Univ. Coop. Ext. Ser. C-687.

Hirrel, M. C. 1989. "Diseases and Their Control." pp. 15–18. *In Growing Grain Sorghum in Arkansas.* Univ. of Arkansas Coop. Ext. Ser. MP 297.

Holden, P., L. Frobish, and J. Pettigrew. 1984. "Energy for swine." pp. 1–4. *In Pork Industry Handbook.* Purdue Univ. Coop. Ext. Ser. PIH-3.

Kumar, R., and M. Singh. 1984. "Tannins: their adverse role in ruminant nutrition." *J. Agri. Food Chem.* 32:447–453.

Mann, J. A., C. T. Kimber, and F. R. Miller. 1983. *The origin and early cultivation of sorghums in Africa.* Texas Agri. Exp. Sta. Bul. 1454.

Mask, P. L., A. Hagan, and C. C. Mitchell Jr. 1988. *Production guide for grain sorghum.* Auburn Univ. Coop. Ext. Ser. Cir. ANR-502.

Maunder, A. B. 1990. "Importance of sorghum on a global scale." pp. 8–16. *In* G. Ejeta, E. T. Mertz, L. Rooney, R. Schaffert and J. Yohe (eds.) *Sorghum Nutritional Quality.* Proc. Internatl. Conf. on Sorghum Nutritional Quality. Purdue Univ., West Lafayette, IN.

McCullough, M. E., E. E. Worley, and L. R. Sisk. 1981. *Evaluation of sorghum silage as a feedstuff for growing cattle.* Univ. of Georgia Agri. Exp. Sta. Rep. No. 366.

Meyer, R. O., and D. W. Gorbert. 1985. "Waxy and normal grain sorghums with varying tannin contents in diets for young pigs." *Anim. Feed Sci. Tech.* 12:179–186.

Morton, J. F. 1978. "Economic botany in epidemiology." *Econ. Bot.* 32:111–116.

Mukuru, S. Z. 1990. "Traditional food grain processing methods in Africa." pp. 216–221. *In* G. Ejeta, E. T. Mertz, L. Rooney, R. Schaffert and J. Yohe (eds.) *Sorghum Nutritional Quality.* Proc. Internatl. Conf. on Sorghum Nutritional Quality. Purdue Univ., West Lafayette, IN.

Mulkey, J. R., J. Drawee, and E. L. Albach. 1985. *Irrigated sorghum response to plant population and row spacing in southwest Texas.* Texas A&M Univ. Agri. Exp. Sta. PR-4293.

Quinby, J. R., and R. E. Karper. 1954. "Inheritance of height in sorghum." *Agron. J.* 46:211–216.

Quinby, J. R., J. D. Hesketh, and R. L. Voigt. 1973. "Influence of temperature and photoperiod on floral initiation and leaf number in sorghum." *Crop Sci.* 13:243–246.

Reichert, R. D., M. A. Mwasaru, and S. Z. Mukuru. 1988. "Characterization of colored grain sorghum lines and identification of high tannin lines with good dehulling characteristics." *Cer. Chem.* 65:165–170.

Rooney, L. W., and S. O. Serna-Saldivar. 1990. "Sorghum." pp. 233–271. *In* K. J. Lorenz and K. Kullp (eds.) *Handbook of Cereal Science & Technology.* Marcel Dekker, Inc., New York, NY.

Sell, D. R., J. C. Rugler, and W. R. Featherston. 1983. "The effects of sorghum tannin and protein level on the performance of laying hens maintained in two temperature environments." *Poultry Sci.* 62:2420–2428.

Shaw, T. 1976. "Early crops in Africa: A review of the evidence." pp. 107–153. *In* J. R. Harlan, J. M. J. de Wet, and A. B. L. Stemler (eds.) *Origins of African Plant Domestication.* Moulton Pub., The Hague, Netherlands.

Shechter, Y., and J. M. J. de Wet. 1975. "Comparative electrophoresis and isozyme analysis of seed proteins from cultivated races of sorghum." *Am. J. Bot.* 62(3):254–261.

Shroyer, J. P., R. L. Vanderlip, D. L. Bark, J. A. Schaffer, and T. L. Walter. 1987. *Probability of sorghum maturing before freeze.* Kansas St. Univ. Coop. Ext. Ser. AF-162.

Snowden, J. D. 1955. "The wild fodder *Sorghums* of the section *Eusorghum.*" *J. Linn. Soc. London* 55:191–260.

Snowden, J. S. 1936. *The cultivated races of Sorghum.* Allard and Son Pub., London, England.

Stephens, J. C., and R. F. Holland. 1954. "Cytoplasmic male-sterility for hybrid sorghum seed production." *Agron. J.* 46:20–24.

Thomas, C. H. 1963. *A preliminary report on the agricultural production of the red swamp crawfish (Procambarus clarkii (Girard)) in Louisiana rice fields.* Proc. Ann. Conf. S.E. Assoc. of Game and Fish Commissioners 17:180–186.

Touchton, J. T., and W. L. Hargrove. 1981. "Fertility practices." pp. 7–20. *In* R. R. Duncan (ed.) *Proceedings Grain Sorghum Short Course.* Univ. of Georgia Agri. Exp. Sta. Spec. Pub. No. 29.

Tucker, B. B., and W. F. Bennett. 1968. "Fertilizer Use on Grain Sorghum." pp. 189–220. *In* R. C. Dinauer (ed.) *Changing Patterns in Fertilizer Use.* Soil Sci. Soc. Am., Madison, WI.

United States Department of Agriculture. 1936, 1942, 1952, 1961, 1972, 1982 and 1992. *Agricultural Statistics.* U.S. Government Printing Office, Washington, D.C.

United States Department of Agriculture-Federal Grain Inspection Service. 1988. *U.S. Standards for Grains.* U.S. Government Printing Office, Washington, D.C.

United States Department of Agriculture-Statistical Reporting Service. 1984. *Usual Planting and Harvest Dates for U.S. Field Crops.* USDA Agri. Handbook No. 628.

Vanderlip, R. L. 1979. *How a Sorghum Plant Develops.* Kansas St. Univ. Coop. Ext. Ser. Bul. 5-3.

Vanderlip, R. L., and H. E. Reeves. 1972. "Growth stages of sorghum [*Sorghum bicolor* (L.) Moench]." *Agron. J.* 64:13–16.

Watt, B. K., and A. L. Merrill. 1975. *Handbook of the Nutritional Contents of Foods.* Dover Pub., Inc., New York.

Walter, T. 1988. "Selection of grain sorghum hybrids." pp. 2–3. *In Grain Sorghum Production Handbook.* Kansas St. Univ. Coop. Ext. Ser. C-687.

Whitney, D. 1988. "Fertilizer Requirements." pp. 8–10. *In Grain Sorghum Production Handbook.* Kansas St. Univ. Coop Ext. Ser. C-687.

Wright, D. L., and D. L. Prichard. 1987. *Grain sorghum and silage.* Univ. of Florida Coop. Ext. Ser. Agron. Facts No. 213.

Yohe, J. 1990. "Relevance of International Sorghum and Millet Research to U.S. Agriculture." pp. 17–19. *In* G. Ejeta, E. T. Mertz, L. Rooney, R. Schaffert and J. Yohe (eds.) *Sorghum Nutritional Quality.* Proc. Internatl. Conf. on Sorghum Nutritional Quality. Purdue Univ., West Lafayette, IN.

4

Barley
*(*Hordeum vulgare *L.)*

INTRODUCTION

Barley is one of the world's oldest crop plants, having been domesticated probably before 8500 B.C. Harvesting and consumption of barley predate this by perhaps several hundred years. Artifacts of ancient people began to include tools that could have been used for "cutting" grass heads and grinding seeds by about 10000 B.C. Wooden sickles with rows of sharp stones as blades have been found at archaeological sites in the Jordan Valley region of the Middle east. Presumably, these wooden sickles were from the Natufian, a cohesive culture of cave dwellers that occupied this region by about 9000 B.C. Indeed, at least one report suggests that barley was consumed by humans as early as 15000 B.C.

Remains from the oldest archaeological sites are of two-row, hulled type barley, but the six-row, naked type (i.e., grain thrashes free of glumes or husks) appears in the record before 6000 B.C. Evidence of barley production is associated usually with emmer and/or einkorn wheat at many of the archaeological sites around the Mediterranean Sea, in Asia, and in Europe. Barley, as did emmer and einkorn wheat, evolved in or around the Fertile Crescent and was farmed in Spain by 5000 B.C.; England by 3000 B.C.; and China by about 7000 B.C. (see Figures 2.1 and 2.2 in Chapter 2). Apparently, barley was more important than wheat early on in the development of European culture, as the shift in preference from barley to wheat as the bread grain of choice for the common people of Greece and Rome is documented in written history during the first century B.C. This should coincide with the spread of yeast bread-making technology from its discovery in Egypt about 2600 B.C. (note discussions on sources of power and bread in Chapter 2 of this text). Since barley may have been the more common and plentiful cereal grain, the non-

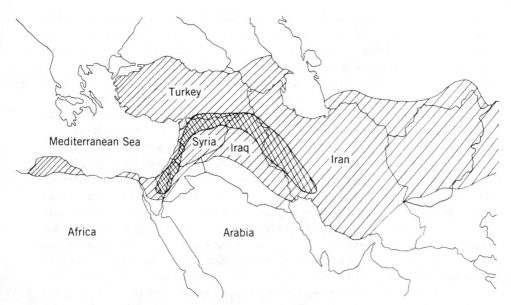

Figure 4.1 *Distribution of wild barley. Large areas of primary habitats may occur within the Fertile Crescent (cross-hatched). Large stands may occur within hatched area also but in more isolated habitats. Source: Adapted from Harlan and Zohary, 1966. (Drawing by Mike Hodnett.)*

elite, economically and socially, would have been the last parts of society to afford yeast bread that required wheat flour and not simply cereal grain meal as needed for flat breads.

The evolution of barley is rather straightforward compared to most crops dealt with in this text. The cultivated barley of today is believed to have evolved from a wild, two-row *Hordeum* that has been classified as *H. spontaneum* and is found growing wild in many areas of Southwest Asia and Northern Africa today (Figure 4.1). *Hordeum spontaneum* is the only truly "wild type" ancestor of cultivated barley. Other ancestors have been proposed, especially an extinct six-row ancestor, but there appears to be no substantial evidence that six-row domesticated barley was derived from any ancestral form other that *H. spontaneum*. Indeed, only three gene mutations can account for all substantial differences between *H. spontaneum* and *H. vulgare*. Cultivated barley is homozygous recessive at the Bt, vv, and n loci. These three mutations from the dominant to the recessive confer the non-shattering head character, six rows of fertile, single-flowered spikelets rather than two, and the free thrashing character, respectively. Wild and domesticated barley are both diploid, 2n = 14, and are interfertile, giving rise to combinations of shattering propensity, number of rows of fertile spikelets, and adherence of glumes to grains. It appears that the most accepted classification today of wild and cultivated barley is *Hordeum* section *Hordeum vulgare* subspecies *spontaneum* for the wild progenitor of domesticated/cultivated barley that is classified as *Hordeum* section *H. vulgare* subspecies *vulgare*.

A question that we have dealt with superficially when considering the domestication of crop species is "Why this one and not another?" Ease of harvest and calorie
return for imputed labor are but two of the many reasons to explain the crop choices
of gathering and farming humankind of 10 to 12 millennia ago. Ofer Bar-Yosef of
the Institute of Archaeology, Hebrew University in Jerusalem, Israel, and Mor-
dechai Kislev of the Life Sciences Department at Bar-Ilan University at Ramat-Gan,
Israel, suggest that at least seven factors determined which species in the diverse
Jordan Valley grass flora were selected for domestication. These criteria were (1)
grain size and weight; (2) local abundance of each respective species; (3) generation
length—i.e., annual or perennial; (4) seed dormancy; (5) ploidy level; (6) harvesting efficiency; and (7) ease of dehusking the grains as preparation for consumption.
The impact of large, heavy kernels and local abundance have been discussed for
wheat and appear obvious. Perennial species should evolve more slowly than annuals, thereby not providing the diversity of types from which early humans selected
candidate species for domestication. Bar-Yosef and Kislev suggest that considering
only those criteria (i.e., size and weight, abundance, and generation time) elimi-
nates all but four of the 23 relative large-seeded grasses native to the Jordan Valley.
Those four are (1) *H. spontaneum;* (2) *Aegilops peregrina* (see footnote of Table 2.1
in Chapter 2 regarding *Aegilops* genus), a common species throughout the Jordan
Valley and having large, heavy grains; (3) *Triticum dicoccoides* (emmer wheat
according to Bar-Yosef and Kislev, although emmer is classified as *T. dicoccum* by
some authorities) that is locally common with large, heavy grains; and (4) *Avena
sterilis* (genus of cultivated oats) that is very common but has smaller kernels (Table
4.1).

Early farmers would realize over time the effects of seed dormancy, although
they would have recognized it only in that fewer plants resulted from planting the
same amount of seed of one species versus another. The other advantage of dormancy is that seeds will not germinate while still on the plant during inclement weather.

**TABLE 4.1   Comparisons of a Few of the Wild Grasses Found in the Jordan Valley
That Were or Were Likely Early Domesticates.**

Species	Abundance[1]	Kernel Volume (mm³)	Weight (mg)	Ploidy Level[2]
Triticum dicoccoides	3	37	20	4x = 2n
Hordeum spontaneum	5	33	24	2x = 2n
Aegilops peregrina	5	32	20	4x = 2n
Avena sterilis	5	19	11	6x = 2n
Hordeum bulbosum	5	16	10	2x or 4x = 2n
Bromus sterilis	4	11	6	4x = 2n
Aegilops speltoides	2	10	5	2x = 2n

Source: Adapted from Bar-Yosef and Kislev, 1989.

1. 1 = rare to 5 = very common.
2. X = basic chromosome number such that 2x = diploid status; 2n = number of sets of chromosomes in
somatic cells.

This aspect of dormancy, unlike dormancy at planting, would be a very real advantage to gatherers and early farmers.

Ploidy level is important because in a true diploid, such as barley, a mutation of only one gene is required for conversion from articulating heads to disarticulating heads, a mutation away from maximum natural survival and distribution to maximum harvestability. In polyploid species, there could be a gene, or genes, regulating the shattering character in each of the genomes, thus mandating more than one mutation, but also mandating that they all occur at the same time, or that each have some fitness benefit to the plants such that one mutation is not lost in the population before the other mutation(s) occurs. This logic has intuitive appeal but would be difficult to prove experimentally and probably has not been demonstrated.

Barley is not superior to other grasses, especially the wheats, relative to the last two criteria of Bar-Yosef and Kislev. The glumes, or husks, of wild barley adhere very tightly to the kernel, and the wheats, by nature of having multiple flowers per spikelet versus single-flowered spikelets in barley, would return the gatherer/farmer more grains per head. Nonetheless, there seems to be general agreement that barley was domesticated before wheat.

TYPES OF BARLEY

While *Hordeum* does not display the diversity of *Triticum,* there are distinct types of barley, agronomic, morphologically, and industrially. Barley tolerates a wider ecological range than any of the other major grains. It has been or is currently grown from inside the Arctic Circle, where the soil thaws to only a few inches during the summer, to the tropics. It will tolerate saline and more alkaline soils better than other grains and is grown from near sea level to elevations above 15,000 feet. Barley is, however, as with any plant species, more productive under favorable conditions. Barley is particularly well suited to production in areas of relatively high rainfall, low relative humidity, and cool temperatures during maturation.

Growth Habit

Barley may have either winter or spring growth habit just as in the case of wheat. True winter types require vernalization, a process independent of winter hardiness, morphological form, and photoperiod requirement. Winter barley requires two to ten weeks of temperatures below about 50°F, with lower temperatures, to about 40°F, generally influencing reproductive initiation in fewer days. The process is cultivar-dependent, as in wheat. Winter barley is sometimes fall seeded as far north as Pennsylvania in the east and also in California and the Pacific Northwest, although not in large acreage in some areas because of the lack of sufficient winter hardiness in adapted cultivars. Of the four small-grain species, only oats are less winter-hardy than barley, while wheat, and especially rye, can tolerate much colder temperatures. Identification of greater winter hardiness could significantly expand acreage in the upper north, Northern Prairies, Northern Great Plains, and the northern Cordillera states. Today, winter barley is primarily restricted to more southern states, especially those where the crop is used for winter pasture as well as for grain. There are no official acreage surveys that distinguish between winter and spring barley in most states, nor by the USDA on a national basis. However, it is estimated that about 25 percent of the total U.S. barley acreage is planted to winter types.

End Use

Barley cultivars are classified as malting or feed barleys in the United States, although barley is a major food grain in many parts of the world, such as the semiarid regions of north Africa, southwest and southern Asia, the Himalayas, other mountain highlands of Tibet and Nepal, and the Andean countries of South America. Barley also is grown for food in northern European countries where it is the most adapted grain to their short growing season. The hull is removed and kernels are ground into grits or flour for making flat bread or cooked as a gruel or porridge. Whole kernels are often boiled or parched. There are few food products in the United States that include barley, most notable being soups and baby foods.

Most of the barley produced in the United States is fed to livestock as cracked, rolled, or ground grain, plus being used for forage or silage in some areas. The protein content ranges from 8 to 18 percent, averages about 13 percent across cultivars, and is comparable to wheat grown under similar climatic and soil conditions. Barley is usually higher in protein than corn, rice, or sorghum. Barley, like other cereal grains, is nutritionally lacking in the amino acid lysine, and is considered to have about 95 percent of the feed value of corn. However, in areas of the United States where corn cannot be grown because of an extremely short growing season, cool summer temperatures, or too little rain (85 percent of U.S. barley is grown dryland), barley is often the principal feed grain. These areas include the Northern Great Plains, especially Western Minnesota, North Dakota, and Montana, and the Pacific Northwest states of Idaho and Washington.

Barley is produced for malting in relatively limited areas of the United States because of the impact of climatic conditions on characteristics such as protein content that affect malting quality. Malting barley should have plump kernels, moderately low protein, thereby mandating lower nitrogen fertilization, and a mealy rather than a glassy, vitreous appearance. Production of such barley is cultivar- and climate-dependent. Malt barley should be produced where the growing season is relatively long with cool temperatures, uniform moisture, and proper nutrient supply. Production in the United States is concentrated today along the Red River Valley of Minnesota and North Dakota, and the irrigated intermountain valleys of Idaho, Wyoming, and Montana.

Malting is a process of allowing germination to begin so that two enzymes, α-amylase and β-amylase, will be activated and begin the conversion of endosperm starch to dextrin and fermentable sugars. Wheat and rye also produce these enzymes during germination, but barley has three advantages over these two grains for malting. First, barley has glumes, or hulls, that adhere very tightly to the kernel after thrashing. The coleoptile of the germinating barley kernel elongates inside these glumes and is therefore protected from being broken during processing. The coleoptile that is expanding inside the glumes is sometimes called the acrospire. Second, the hulls aid in filtration during the separation process. And third, the kernel texture of steeped barley is firmer than that of wheat and rye, thereby being less damaged when handled at high moisture. Barley also is used for malting to the exclusion of rye and wheat because of tradition, probably dating back several thousand years.

Two-Row versus Six-Row

The barley inflorescence is a terminal spike with sets of three spikelets attached alternately at nodes of the rachis. In two-row barley, only the central spikelet is fertile and produces a single kernel. This situation means a single column of kernels along each side of the rachis. In six-row barley types, the lateral spikelets are fertile, allowing for three columns of kernels along each side of the rachis.

Kernel shape varies between two-row and six-row types. All kernels of two-row barley are symmetrical with size varying by location within the spike; larger kernels are produced in the middle of the spike. In six-row barley, one-third of the kernels, those produced by the central spikelets, are symmetrical. The remaining two-thirds are produced in lateral spikelets and have a slightly twisted appearance.

Awns

Barley may be awned or awnless, an awn being a slender, apically tapering extension of the glumes and the lemma. These extensions are broken off when harvested with modern combines. Awns may be rough or smooth, with the former type usually having barbs or hairs that angle toward the apex. Awnless cultivars exist although outnumbered by awned cultivars. Awns may be replaced by a trifurcate (that is, three-forked) appendage called and resembling a hood, and such cultivars are called hooded barleys.

Hulled/Hull-less

In most barley cultivars, the lemma and palea adhere to the kernel, a trait desirable for malting. Cultivars where their grain thrashes free of these appendages are called "naked" or "hull-less." Naked barleys are most frequently found in areas of the world where barley is used for human food. In countries like the United States where barley is a livestock feed or a malting grain, hulled cultivars predominate.

Aleurone Color

The aleurone is a layer of cells, usually two to four cells thick, lying just inside the husks, the pericarp, testa, and nucellar tissues, and is botanically a part of the endosperm. However, unlike the remainder of the endosperm, these are living cells that are filled with dense cytoplasm that contains a wide array of organelles: rough endoplasmic reticulum, mitochondria, proplastids, numerous lipid-containing spherosomes, and aleurone grains. These aleurone grains contain hydrolytic enzymes responsible for the conversion of starch to sugars. Grain color is determined in the aleurone layer and may be colorless, white, yellow, blue, or various shades of these.

MOVEMENT TO THE UNITED STATES AND CULTIVAR DEVELOPMENT

Many authorities assume that barley, like wheat, was first introduced into the New World by Columbus on his first voyage of 1492, as he left "stores" behind at La Navidad, near present-day Limonade, Haiti, that included "seeds for sowing crops." We are certain that he returned with barley, along with wheat and "vines" in 1493.

Barley was introduced into present-day United States in 1602 by Gosnold on Martha's Vineyard and the Elizabeth Islands off the coast of Massachusetts, and by the London Company in present-day Virginia in 1611. By 1648, barley culture was widespread in the Virginia colony but soon declined in favor of tobacco production. Barley was planted in New Foundland, which at that time included Manhattan Island, as early as 1626 and was planted by the Massachusetts Bay Colony in 1629. Barley probably was introduced by the Pilgrims at Plymouth, Massachusetts, in 1620, as it was a common practice to take seeds of familiar crops when immigrating to new lands. The English colonists most likely brought two-row barleys such as the Chevalier and Thorpe types that were common to England, whereas European colonists such as the Dutch and German brought types adapted to the mainland.

As with wheat, barley was introduced into the U.S. Southwest by Spanish Missionaries during the 1500s or 1600s. Seeds of a number of plant species, both crop and weed, and including barley, have been recovered from adobe brick of buildings erected by these Spanish Missionaries and their followers throughout northern Mexico, Arizona, and California during that period of our history. The earliest bricks to contain barley seeds were used in the 1701 construction of the San Cayetano del Tumacacori Mission in southern Arizona. Obviously, the crop would have predated the construction by at least one year and possibly by decades, since we know that the first Spanish incursion into present-day United States was in 1535. The barley kernels found at San Cayetano de Tumacacori, and 11 of 13 other such structures, were identical to and classified as a Coast-type barley, a cultivar group typical of barley found in northern Africa and Spain. This type barley is well adapted to arid conditions such as those found in the Southwest United States and Mexico. This cultivar group continues to be the dominant type in these areas today.

The first barleys cultivated by English and European settlers were most likely grown for malt to be used in the production of beer. Although not well adapted to the humid areas of the Atlantic seaboard, demand increased with population and by 1796, barley was the leading agricultural commodity in Rhode Island. These early barleys were Thorpe and Chevalier types from England and Hanna types typical of Germany. All three cultivar groups have two-row inflorescences and spring growth habit. Apparently, a six-row barley also was grown early on in New England, but its origin and name have gone unrecorded in commonly circulated literature. Although grown primarily for malt, barley was used as a feed for livestock and occasionally as flour during years when wheat and/or corn flour was in short supply.

Winter barley was introduced into the Southeast United States supposedly from Switzerland and/or the Balkans. The date of its introduction and early varietal selections are unknown. The 1936 United States Department of Agriculture (USDA) Yearbook of Agriculture simply states that it was introduced into the mountain region of the Southeast and "has long been grown" as Tennessee Winter and as Union Winter.

As the population of the fledgling United States grew and the country expanded westward, ecological conditions that were suited to the production of high-quality barley were finally encountered in western New York State. By 1839, 60 percent of the barley produced in the United States was grown in New York, and reached 69 percent by 1849 (Table 4.2). With transportation limited and expensive, the expanding and westward movement of English and European citizenry, plus new immigrants, mandated that barley be grown for malt production near every emerging center of population. Barley production flourished around cities such as Detroit, Michigan; Cincinnati and Toledo, Ohio; Pittsburgh, Pennsylvania; St. Louis, Missouri; and Chicago, Illinois. Production of barley for malt increased steadily through the first half to three quarters of the 19th century in New York, Maine, Ohio, Pennsylvania, Michigan, and Illinois.

While barley was produced in the Northeast and into the upper Midwest primarily for malt, it was produced in the arid Southwest United States and California as livestock feed, other grains being less productive in these areas. Following the discovery of gold in California in 1848 and the subsequent influx of a great number of people, the demand for barley malt, as well as increased demand for livestock feed, caused production to soar from 0.2 percent of U.S. production in 1848 to 28 percent by 1859.

By the close of the 19th century, the barley growing areas in the United States had been fairly well established. Transportation by railroad and later by truck freed the American farmer to cultivate the crop best adapted to his locale and to market those products miles away. Production at the turn of the 20th century flourished along the Red River Valley of eastern North and South Dakota, western Minnesota and Iowa, southeastern Wisconsin and the Imperial Valley of California. Other centers of production would develop during this century, including areas of Idaho, Washington and Oregon of the Pacific Northwest, and Kansas and Colorado of the Southern Great Plains. However, by 1990, the Northern Great Plains states of North and South Dakota, Minnesota, and Montana would produce over 60 percent of U.S. barley.

The first immigrants to the United States, whether of European or Spanish ancestry, brought seeds of the barley cultivars that were being grown in the locale they left behind (Table 4.3). In the west, barley of the Coast cultivar group proved well adapted to the arid conditions. In the east, immigrants brought barley of the Chevalier, Thorpe, and Hanna cultivar groups that were not as well adapted as one

TABLE 4.2 Production of Barley for All Purposes for Selected States, 1839–1919.

State	1839	1849	1859	1869	1879	1889	1899	1909	1919
					(×1000 bu.)				
New York	2,520	3,585	4,187	7,435	7,792	8,220	2,943	—	—
Maine	355	152	802	659	—	—	—	—	—
Ohio	212	324	1,664	1,715	1,707	—	—	—	—
Pennsylvania	210	166	531	530	—	—	—	—	—
Massachusettes	165	112	—	—	—	—	—	—	—
Michigan	128	75	308	835	1,204	2,522	—	—	4,803
New Hamphire	122	70	—	—	—	—	—	—	—
Virginia	87	—	—	—	—	—	—	—	—
Illinois	82	111	1,036	2,480	1,230	1,197	—	—	4,227
Rhode Island	66	—	—	—	—	—	—	—	—
Wisconsin	—	210	707	1,645	5,043	15,226	18,700	22,156	12,192
Kentucky	—	95	—	—	—	—	—	—	—
California	—	—	4,415	8,783	12,464	17,548	25,149	26,442	21,897
Iowa	—	—	467	1,961	4,023	13,406	18,059	10,964	5,353
Indiana	—	—	382	—	—	—	—	—	—
Minnesota	—	—	—	1,032	2,973	9,101	24,314	34,928	14,849
Nebraska	—	—	—	—	1,745	1,822	2,035	—	4,405
Oregon	—	—	—	—	921	—	1,515	2,378	—
North Dakota	—	—	—	—	—	1,571	6,752	26,366	12,053
Washington	—	—	—	—	—	1,269	3,641	5,835	—
South Dakota	—	—	—	—	—	—	7,032	22,396	12,816
Idaho	—	—	—	—	—	—	—	4,598	—
Kansas	—	—	—	—	—	—	—	2,222	8,325
Others	213	237	1,326	2,687	4,897	6,450	9,494	15,060	21,105
U.S. totals	4,162	5,167	15,826	29,761	43,997	78,333	119,635	173,344	122,025

Source: Adapted from Harlan, 1925.

TABLE 4.3 Cultivar Groups of Barley Introduced into the United States from 1620 to the Early 1900s, Their Origins and Other Typifying Characteristics.

Cultivar Group	Origin	Year of Introduction[4]	Inflorescence Type	Hull Adherence	Awns
Thorpe	England	1602–1620	2-row	hulled	yes
Chevalier	England	1602–1620	2-row	hulled	yes
Hanna	Austria	1901	2-row	hulled	yes
Smyrne	Asia Minor	1901	2-row	hulled	yes
Bohemian	Czechoslovakia	?	2-row	hulled	yes
Baku	C. Asia	?	2-row	naked	yes
Zeocriton	Asia	?	2-row	hulled	yes
Black Hull-less	C. Asia	?	2-row	naked	yes
Hannchen	Sweden	1904	2-row	?	yes
Horn[1]	Europe	1909	2-row	hulled	?
Coast	North Africa	before 1701	6-row	hulled	yes
Tennesee Winter	Switzerland (?)	before 1936	6-row	hulled	yes
Nepal	India	1840	6-row	naked	hooded
Manchuria	PRC	1861	6-row	hulled	yes
Oderbrucker[2]	PRC	about 1890	6-row	hulled	yes
Stavropol[3]	Russia	before 1900	6-row	hulled	yes
Odessa	Russia	1902	6-row	hulled	yes
Club Mariout	Egypt	1904	6-row	hulled	yes
Trebi	Turkey	1905	6-row	hulled	yes
Artic	Russia	?	6-row	hulled	yes
Himalaya	C. Asia	?	6-row	naked	yes

1. May be considered as a Chevalier type.
2. May be considered as a Manchuria type.
3. May be considered as a Coast type.
4. Year first introduced into present-day U.S.

would hope. The introduction and selection of better cultivars during the early days of the United States are not as easily traced as in some other crops, but the general areas of production of the cultivar groups as of 1925 are shown in Figure 4.2.

The vast majority of improved cultivars of barley, as with all major U.S. row crops, have been developed this century following the principles of scientific plant breeding. However, as we have seen with other crops, introductions and farmer or seedsmen selections within introduced material were responsible for the development of outstanding cultivars prior to 1900. One such example in barley is the introduction of the Manchuria cultivar group from the People's Republic of China. A German citizen whose name has been lost from generally circulated records was traveling along the Amur River in eastern Manchuria in 1850, where he observed an unusually vigorous type of barley, long indigenous to the region. Collecting seeds, he returned to Germany where, a few years later, Dr. Herman Grunow of Mifflin, Wisconsin, observed and was equally impressed with this Manchurian barley. About 1861, Dr. Grunow brought seeds back to the United States. In 1872, he sent a sample to the Wisconsin Agricultural Experiment Station (AES). Manchuria proved

Figure 4.2 Generalized areas of adaptation of cultivar groups of barley in 1925. Source: Adapted from Harlan, 1925. (Redrawn by Mike Hodnett)

so well adapted that selections were distributed to farmers the following year. Within a few years, selections of the original Manchuria cultivar group were grown as Manshury, Mansury, Mandsheuri, Manchuria, Minnesota 6, and North Dakota 787.

A Manchuria group was again imported from Russia by the Ontario AES in 1881. Samples were forwarded to the Minnesota and North Dakota Experiment Stations in 1894 and subsequent selections were released as Minnesota 32 and Minnesota 105.

One other documented introduction that had an almost immediate and significant impact on production was Trebi, a cultivar group found growing in irrigated culture in an isolated area along the southern shore of the Black Sea. Seeds were collected in 1905 by the USDA from near Samsun, Turkey, and tested in 1909 at the Minnesota AES and in 1910 at Chico, California. Results were disappointing both years. However, when Trebi was grown under irrigation at Aberdeen, Idaho, in 1910, results were much more promising. From 1910 through 1917, Trebi averaged 80 bushels/acre. The first farmer to receive seed of Trebi in 1918 produced 90 bushels/acre on 28 acres. Irrigated acreage devoted to Trebi increased from the Western Dakotas to Oregon.

The first barley cultivar developed through controlled hybridization was developed by F. H. Horsford of Charlotte, Vermont. Horsford released a number of cultivars beginning about 1880 that combined the naked kernel and hooded awn traits from the Nepal group (see Table 4.3) with improved yield potential. These were sold under such names as Beardless, Success, and Success Beardless. Robert Withecombe of Union, Oregon, also used Nepal as a parent to produce Union Beardless that he released in 1910 for production in the Pacific Northwest. Other privately developed cultivars have made significant improvements in yield and quality of barley.

The USDA established a regular program of exploration, collection, and maintenance of barley germplasm in 1894. Since that date, thousands of accessions have been added, including over 210 cultivars developed in the United States during this century. Early public scientists of note who worked to produce improved cultivars were C. A. Moores of the Tennessee AES, R. A. Moore at the Wisconsin AES, and W. M. Hays at Minnesota. Moore and his associate L. A. Stone developed improved cultivars of the Oderbrucker cultivar group, releasing Wisconsin Pedigree No. 5 and Wisconsin Pedigree No. 6 about 1908. W. M. Hays was responsible for the wide distribution of the original, unselected Manchuria introduction, called Minnesota 105, and a selection from Manchuria named Minnesota 184. Moores of Tennessee was an early proponent of hybridization and selection, releasing Tennessee No. 5 and Tennessee No. 6 in 1915. These two cultivars were hooded and resulted from the hybridization of Tennessee Winter with a hooded, spring barley, perhaps one developed by Horsford.

HISTORICAL EVENTS

In addition to the development of mass transportation mentioned earlier, four events in U.S. history significantly impacted barley production: (1) the California gold rush of 1849; (2) the McKinley Tariff of 1890; (3) prohibition; and (4) the dust-bowl years.

Gold

John A. Sutter, a pioneer trader, settled in a large land grant in the Sacramento Valley of California in 1839. Sutter hired James W. Marshall to build a sawmill on the American River, and it was here that Marshall discovered gold in 1848. The first "forty-niners" reached San Francisco by steamer on February 28, 1849. Thousands of others followed, swelling California's population from 20,000 on January 1, 1849 to 107,000 by December 31, 1849. Even larger numbers poured into California over the next few years. Because of this influx of people, California was admitted to statehood in 1850. More people meant the need for more barley, for both beer and livestock. The California territory produced no significant acreage of barley in 1848, having only 20,000 inhabitants, but by 1859, the state of California produced 4,415,000 bushels, 28 percent of U.S. production.

McKinley Tariff

Regardless of the California situation, western New York State remained the center of U.S. barley production, supporting a large malting and brewing industry located in the Northeast. Acreage in New York increased every decade through 1889 (Table 4.2). However, even with this increased acreage, the demand was so great that maltsters were forced to import grain, and the closest grain was just across the border in Ontario, Canada. But through the efforts of William McKinley, 25th President of the United States and a member of Congress from 1876 until 1891 (except for ten months during 1884–1885), the United States passed legislation raising tariffs to protect American industries. This legislation placed a 30 cents/ bushel tariff on barley, and maltsters in New York suddenly could not compete with the malt produced in Wisconsin and Minnesota. Breweries could buy malt produced in these states more cheaply than that produced in New York because all of their barley was produced in the United States, while New York malt houses had to pay higher prices for their barley because of the tariff. In the decade following the new tariff, barley production in New York plummeted by 64 percent while production in Wisconsin, Iowa, and Minnesota increased by 23, 35, and 167 percent, respectively.

Prohibition

Barley production increased rather steadily in the United States, reaching 100,000,000 bushels in 1895 and 200,000,000 bushels by 1915. One would expect that the great American social experiment called prohibition would have dealt a devastating blow to the production of barley, and in certain states it did. Compared with 1909, production for 1919 in Wisconsin dropped 45 percent, Iowa dropped 51 percent, Minnesota production fell 57 percent, and production in North and South Dakota dropped 54 and 43 percent, respectively. Total U.S. production in 1919 was 37 percent lower than it was in 1915. But production rebounded rather quickly with 40,000,000 more bushels produced in 1920 than in 1919, a 30 percent increase, and production rose to 300,000,000 bushels by 1928. Before prohibition, much of the best quality barley went into malting channels, with the least quality being used as livestock feed. With the vanishing of most of the malting industry, the farmer was forced to feed his high-quality barley. Its desirability as a feed was recognized by many for the first time, especially in the Northern Great Plains states of the Dakotas, and the arid areas of western Minnesota where corn production was a tenuous endeavor.

Dust Bowl

Although the 1930s were a disastrous decade worldwide, the American barley producers of the Great Plains states were particularly hard-hit by the severe droughts of 1931, 1933, 1934, 1936, and 1937. In 1931, 1,844,000 acres (14 percent of planted acreage) were planted but not harvested; in 1933, 3,707,000 acres (27 percent); 1934, 4,823,000 acres (40 percent); 1936, 3,749,000 acres (31 percent); and 1937, 1,611,000 acres (14 percent). Producers in these states found no alternative crop that had the drought resistance necessary for the area. Recall that corn producers of the Midwest turned to soybeans and hybrid corn for some relief from the devastating droughts. The barley farmer endured.

PROCESSING

Food

Barley consumption as a food is overshadowed by its use as a feed grain and its use in the brewing industry. Only 5,000,000 bushels of unmalted barley were consumed as food in 1959, about 1 percent of U.S. production that year, and annual per capita consumption for 1981 through 1990 averaged about the same as for 1959 (Table 4.4). Barley used for food purposes usually is pearled, a process whereby the hulls and outer layers of the kernel are removed by abrasion. Pearling machinery is designed such that kernels are in constant motion so that the abrasive action removes the outer layers uniformly from all grains in the batch. Once all of the hulls, kernel coating, most of the embryo, and part of the outer layer of the endosperm are removed, the remainder is marketed as pearled barley that is used primarily for soups and dressings. One hundred pounds of barley grain yields about 35 pounds of pearled barley. Some flour is produced during the pearling process and the remainder of the ground material will be used as feed.

TABLE 4.4   Per Capita Consumption (Lbs./Person) of Grains as Food in the United States, 1910–1990.

Year	Wheat	Corn	Rice	Oats	Barley	Rye
1910	214.0	59.4	8.2	3.3	3.5	4.9
1915	206.1	53.1	5.3	3.6	3.2	4.6
1920	192.3	48.7	5.2	5.6	3.1	4.5
1925	179.6	44.1	5.3	4.4	3.9	3.6
1930	175.5	45.7	5.3	6.0	5.4	3.2
1935	160.6	36.3	5.6	4.0	1.5	2.5
1940	157.8	35.6	5.9	4.0	1.1	2.4
1945	164.2	37.5	4.9	3.6	1.5	2.4
1950	138.1	28.9	5.1	3.3	1.4	1.5
1955	125.9	25.1	5.5	3.3	1.0	1.4
1960	120.8	24.1	6.1	3.6	1.1	1.1
1965	115.9	28.8	7.6	3.4	1.2	1.1
1970	113.9	32.4	6.7	3.2	1.2	1.2
1975	116.9	45.9[2]	7.6	3.2	1.2	1.0
1980[1]	119.7	51.9	9.4	8.2	1.1	0.9
1985	128.7	84.3	9.0	8.5	1.0	0.8
1990	140.6	94.9	16.2	12.8	0.9	0.8

Source: Adapted from USDA Agricultural Statistics, 1936 and following years.

1. Increase in consumption of wheat, corn, rice and oats probably reflects a heightened emphasis on healthy diets.
2. Increase in corn consumption reflects industrialization of an enzymatic process perfected in 1957 to convert corn starch to fructose.

TABLE 4.5   Percent of Total Domestic Use of Barley as Feed and Residual, and Food, Alcohol, and Seed.

Year	Feed and Residual[1]	Food, Alcohol and Seed
1970	67	33
1975	56	44
1980	50	50
1985	64	36
1986	63	37
1987	59	41
1988	49	51
1989	52	48
1990	54	46
1991	56	44

Source: Adapted from USDA Agricultural Statistics, 1982 and 1992.
1. Residual includes small quantities for other uses and waste.

A good-quality flour, with the exception of gluten content, can be produced from pearled barley or from whole kernel processing once the hulls are removed. Flour is usually produced by roller milling, just as with wheat flour, and is used in baby foods and a few other specialty products. Large kernel barley with white aleurone is preferred for pearling and milling.

Feed

The largest use of barley in the United States, averaging 57 percent of domestic use from 1985 through 1991, is as livestock feed, especially for ruminants and horses (Table 4.5). Although the amino acid balance and total protein in barley is better than corn or grain sorghum, its feeding value is less than corn because it contains only 50 to 60 percent as much starch (Table 4.6). Because the palea and lemma, or hulls, remain attached to the kernel in all U.S. cultivars, fiber content is too high for barley use as more than a small percentage of poultry rations, and swine will not

TABLE 4.6   Relative Feed Values as Total Digestable Nutrients of Unprocessed Grains.

Grain	Total Digestible Nutrients	
	Cattle	Swine
Corn	100	100
Barley	91	88
Grain Sorghum	88	96
Oats	84	79
Wheat (HRW)	97	99

Source: Adapted from Church, D. C., *Livestock Feeds and Feeding,* 3/e, © 1991, p. 118. Adapted by permission of Prentice Hall, Upper Saddler River, New Jersey.

TABLE 4.7 Uses of Barley Malt and Malt Products.

Malt Type	Products
Brewer's	Beer, ale, and animal feeds
Distiller's	Alcohol, livestock and poultry feeds
Specialty	Breakfast cereals, sugar coloring, dark beers, and coffee substi-
High dried	tutes
Dextrin	
Caramel	
Black	
Malt enriched products	Malted milk for beverages and baby foods; malt flour for wheat flour supplement, and for human food and animal feed; malt syrups for medicinal, textile, baking, breakfast cereals and candies; malt sprouts for dairy feeds, vinegar, and industrial fermentation.

make maximum weight gains with barley as the only feed grain. Barley usage as a percent of all feed grains, including wheat and rye, used in the United States from 1975 through 1988 ranged from 3.0 to 5.1, averaging 3.7 percent/yr. The amount of barley sold for use as feed depends on the relative price of corn and other acceptable feed grains. Barley is usually cracked or ground to various particle sizes before feeding, with a limited amount steamed and rolled.

Malt

The second largest use of barley in the United States is for malt that is used primarily in the production of beer, but is used also in a number of other products (Table 4.7). Malting is a controlled, limited germination that is designed to activate the production of alpha and beta-amylases that hydrolyze starch to dextrin and fermentable sugars. A generalized malting scheme is outlined below.

1. Barley grain to be malted is cleaned and sorted by kernel size;
2. Selected kernels are steeped, usually by immersion, to a predetermined moisture level;
3. Water is removed and grains are held at a temperature below 64°F for the germination process to proceed;
4. Germination is allowed to proceed for a pre-set length of time to allow for the "modification" of the endosperm by enzymatic action;
5. The process is halted by drying and cooking with hot, dry air (called kilning);
6. The emerged sprouts, primarily rootlets, are separated before storage because they impart a bitter taste during subsequent processing; and
7. The dried malt is stored at 5.5 to 6.0 percent moisture; sprouts are used in animal feeds, primarily as a high protein (about 25 percent) source for dairy cows.

U.S. PRODUCTION

Barley production, as noted earlier, expanded as the population of the United States expanded, reaching 422,196,000 bushels (20,265,408,000 pounds at 48 pounds/bushel) in 1990 (Table 4.8). The United States produced one hundred million bushels in 1895, reached 200,000,000 bushels in 1915, 300,000,000 bushels 13 years later, 400,000,000 bushels in 1955, and topped 500,000,000 bushels in 1982. Average per acre yields were in the low 20 bushel range through about 1950, reached the 30 bushels/acre range in 1958, and today are about 55 bushels/acre. U.S. gross exports, as a percent of current year production, have never been especially large, reflecting the fact that throughout much of our history barley was grown for on-farm feed in areas where corn is not an enticing alternative feed crop or for local beer production. Exports did increase during the early years of Prohibition, 1919 through the late 1920s, to 16 to 26 percent of production, a level not reached in post-Prohibition years until the 1950s.

Historically, at least back to 1866, the highest price received by producers was $1.49/bushel in 1868, a price not matched until 1973 when the farm value surged to $2.14/bushel (Table 4.8). The price plummeted to 22 cents/bushel in 1932 at the height of the Great Depression, but rebounded to over $1.00/bushel during the World War II and post-war years of 1944 through 1954. Inflationary pressures of the 1970s pushed market prices to over $2.00/bushel.

The leading states in barley production are North Dakota, Montana, and Minnesota in the Northern Great Plains, and Idaho and Washington in the Pacific Northwest (Table 4.9). These five states produced 60 percent of all barley in 1980 and 78 percent in 1990. For 1981 through 1987, 93 percent of the barley acreage in Minnesota was planted to malting-type cultivars, 81 percent in North Dakota, 69 percent in South Dakota, 41 percent in Montana, and 38 percent in Idaho. During that same period of time, only 8 percent of barley acreage in Washington was planted to malting types and only about 3 percent of California acreage. In these two states, as with the rest of the United States, essentially all barley is grown for feed.

WORLD PRODUCTION

The United States was the fifth largest barley-producing country in 1990–1991, behind Germany, Canada, and France (Table 4.10). The countries with the highest average per acre yields are in Europe, where Switzerland averaged an amazing 107+ bushels/acre, and Belgium-Luxembourg, Denmark, France, Ireland, Netherlands, and Czechoslovakia all averaged over 100 bushels/acre in 1990–1991. Much of Europe has average July temperatures of 54 to 76°F and moderate precipitation throughout the year, near perfect conditions for barley. European countries produced 69 percent of the world's barley for 1990–1991 on only 41 percent of the world's acreage devoted to barley. Barley is produced as a feed grain in northern Europe where the summer season is too short and too cold for maize, and in all of Europe for malting. The near perfect climate and extraordinarily high yields in Europe are

TABLE 4.8 Production Statistics for Barley Grown in the United States, 1866–1990.

Year	Harvested Acres (×1000)	Average Yield (bu./ac.)	Average Yield (lb./ac.)[1]	Total Produced (bu. ×1000)	Total Produced (lb. ×1000)	Average Price Received (¢/bu.)	Exports as % of Production[2] (%)
1866	754	24.0	1,152	18,095	868,560	95.2	—
1868	1,064	21.8	1,046	23,200	1,113,600	149.0	<1
1870	1,331	21.8	1,046	29,047	1,394,256	85.3	1
1875	1,702	19.3	926	32,812	1,574,976	85.6	1
1880	1,990	22.7	1,090	45,261	2,172,528	66.3	2
1885	2,862	22.3	1,070	63,963	3,070,224	55.7	<1
1890	3,250	21.5	1,032	69,880	3,354,240	62.1	2
1895	4,185	25.0	1,200	104,475	5,014,800	32.8	8
1900	4,703	20.5	984	96,588	4,636,224	40.7	7
1905	6,658	25.8	1,238	171,639	8,238,672	39.4	11
1910	7,546	18.9	907	142,419	6,836,112	60.7	7
1915	7,279	28.4	1,363	206,976	9,934,848	52.0	15
1917	8,453	21.6	1,037	182,209	8,746,032	123.2	16
1919	6,579	19.9	955	131,086	6,292,128	124.4	26
1920	7,439	23.0	1,104	171,042	8,210,016	84.4	16
1925	8,186	23.5	1,128	192,779	9,253,392	61.4	16
1930	12,666	24.0	1,152	303,752	14,580,096	40.4	4
1932	13,346	22.6	1,085	302,042	14,498,016	22.0	3
1933	10,009	15.6	749	155,825	7,479,600	43.3	4

(continued)

TABLE 4.8 *(Continued)*

Year	Harvested Acres (×1000)	Average Yield (bu./ac.)	Average Yield (lb./ac.)[1]	Total Produced (bu. ×1000)	Total Produced (lb. ×1000)	Average Price Received (¢/bu.)	Exports as % of Production[2] (%)
1934	7,095	16.7	802	118,348	5,680,704	69.2	3
1935	12,436	23.2	1,114	288,667	13,856,016	37.9	3
1940	13,525	23.0	1,104	311,278	14,941,344	39.7	1
1945	10,454	25.5	1,224	266,994	12,815,712	101.0	3
1950	11,155	27.2	1,306	303,772	14,581,056	119.0	13
1955	14,523	27.8	1,334	403,065	19,347,120	92.0	26
1960	13,856	31.0	1,488	429,005	20,592,240	84.0	20
1965	9,166	42.9	2,059	393,055	18,866,640	102.0	20
1970	9,712	42.8	2,054	416,091	19,972,368	97.3	20
1973	10,295	40.5	1,944	417,434	20,036,832	214.0	22
1975	8,617	44.0	2,112	379,162	18,199,776	242.0	6
1979	7,522	50.9	2,443	382,798	18,374,304	229.0	14
1980	7,275	49.6	2,381	360,956	17,325,888	279.0	21
1985	11,603	51.0	2,448	591,383	28,386,384	198.0	3
1990	7,529	56.1	2,693	422,196	20,265,408	214.0	19

Sources: Adapted from USDA Agricultural Statistics, 1936 and following years.

1. One bu. = 48 lbs.

2. Percent exports calculated as gross exports, including flour and malt expressed as bushel equivalent grain, i.e. (exports ÷ current year production) × 100. Does not consider imports of barley or barley products.

TABLE 4.9 Production of Barley in the United States by State and Region, 1935–1990.

State/Region	1935	1940	1950	1960	1970	1980	1990
				(×1000 bu.)			
Maine	168	116	204	—	—	—	—
Vermont	135	174	26	—	—	—	—
New York	4,158	3,668	2,754	828	510	517	488
New Jersey	30	182	640	1,344	1,050	795	—
Pennsylvania	1,595	3,770	6,461	7,056	9,250	3,750	4,200
Northeast	6,086	7,910	10,085	9,228	10,810	5,062	4,688
Arkansas	—	176	84	510	32	—	—
Delaware	—	145	336	680	882	1,225	2,516
Georgia	—	—	110	319	376	258	—
Illinois	1,813	4,270	1,176	2,013	533	—	—
Indiana	572	1,160	624	1,715	430	—	—
Kentucky	299	1,500	1,679	2,584	2,322	1,595	1,210
Maryland	759	2,008	2,635	4,042	4,268	3,640	4,992
Michigan	5,326	6,720	3,910	2,590	1,127	1,113	1,419
Mississippi	—	—	—	102	—	—	—
Missouri	1,254	4,560	1,640	4,384	836	—	—
North Carolina	254	360	892	2,176	3,276	3,000	1,645
Ohio	450	840	728	2,709	850	416	—
South Carolina	—	—	340	742	882	1,012	279
Tennessee	306	1,452	1,054	1,300	720	168	—
Virginia	936	2,160	2,714	4,680	5,916	4,590	5,695
West Virginia	81	282	406	492	484	396	—
Wisconsin	25,878	24,262	8,979	1,314	1,683	1,534	3,312
East	37,928	49,895	27,307	32,352	24,617	18,947	21,068
Colorado	5,436	10,100	9,536	18,513	14,445	15,925	10,400
Kansas	4,538	18,176	3,683	18,980	7,141	2,091	759
Nebraska	15,686	21,136	4,650	6,786	1,620	950	1,215
New Mexico	264	345	594	2,021	960	1,995	—

(continued)

TABLE 4.9 (Continued)

State/Region	1935	1940	1950	1960	1970	1980	1990
				(×1000 bu.)			
Oklahoma	1,485	7,310	702	15,336	18,130	1,650	370
Texas	2,646	4,336	1,625	8,822	4,224	1,080	320
Wyoming	1,950	2,142	4,228	2,944	5,922	8,645	10,530
South Great Plains	32,005	63,545	25,018	73,402	52,442	32,336	23,594
Iowa	15,444	12,870	1,696	832	164	—	—
Minnesota	58,752	58,320	36,308	29,882	21,534	34,638	43,750
Montana	3,312	4,208	23,430	40,726	65,132	44,100	85,800
North Dakota	45,558	27,952	51,504	80,066	68,705	48,000	138,670
South Dakota	43,130	27,523	18,942	14,940	12,144	15,180	17,940
North Great Plains	166,196	130,873	131,880	166,446	167,679	141,918	286,160
Arizona	910	1,120	7,850	10,858	10,640	4,500	2,400
California	36,642	34,328	60,010	68,880	61,776	44,144	9,440
Idaho	5,236	8,250	14,840	14,964	36,300	58,960	59,250
Nevada	216	560	792	444	930	1,960	360
Oregon	3,302	5,325	10,784	16,224	17,775	10,075	12,600
Utah	1,620	4,387	6,204	6,794	7,353	10,804	7,885
Washington	2,108	3,915	8,763	23,544	19,458	32,250	37,050
West	50,034	57,885	109,243	141,708	154,232	162,693	128,985
Total U.S.	292,249	310,108	303,533	423,136	409,780	360,956	464,495

Source: Adapted from USDA Agricultural Statistics, 1936 and following years.

TABLE 4.10   Production Statistics for Countries Producing over 4,000,000 Bushels of Barley in 1990/1991.

Country/Continent	Harvested Acres	Average Yield	Total Production[1]
	(×1000)	(bu./ac.)	(×1000 bu.)
Canada	11,614	55.1	639,436
Mexico	593	35.0	20,664
United States	7,526	56.2	422,097
North America	19,733	54.9	1,082,197
Argentina	329	42.4	13,914
Brazil	259	37.2	9,643
Chile	79	62.1	4,913
Colombia	124	37.9	4,684
Uruguay	173	35.3	6,107
South America	1,519	31.8	48,400
Former USSR	64,694	37.4	2,412,178
Algeria	2,964	12.5	36,736
Egypt	136	57.5	7,806
Ethiopia	2,347	23.3	54,553
Morocco	5,965	16.6	98,177
South Africa	319	37.8	12,031
Tunisia	1,359	16.2	21,950
Africa	14,002	17.1	240,437
Afghanistan	605	19.0	11,480
PRC	2,984	60.5	180,466
Cyprus	161	30.1	4,822
India	2,448	27.9	68,237
Iran	6,546	23.4	153,832
Iraq	3,458	14.0	48,216
Japan	262	60.6	15,888
Korea	395	67.0	26,450
Pakistan	383	15.8	6,016
Saudi Arabia	185	93.0	17,220
Syria	2,470	9.3	22,960
Turkey	8,398	36.1	303,072
Asia	28,659	30.1	865,179
Belgium-Luxembourg	264	102.9	27,185
Denmark	2,248	101.9	229,049
France	4,347	106.8	463,792
Germany	6,454	99.5	642,513
Greece	605	36.5	22,042
Ireland	585	104.2	60,982
Italy	1,153	67.7	78,156
Netherlands	99	101.9	10,056
Spain	10,767	40.2	432,291
United Kingdom	3,777	96.2	362,768
European Community	30,435	76.6	2,331,680
Austria	721	96.9	69,798
Bulgaria	889	69.6	61,762
Czechoslovakia	1,835	101.4	186,022

(*continued*)

TABLE 4.10   *(Continued)*

Country/Continent	Harvested Acres	Average Yield	Total Production[1]
	(×1000)	(bu./ac.)	(×1000 bu.)
Finland	1,200	65.8	78,982
Hungary	734	85.0	62,359
Norway	427	78.7	33,568
Poland	2,900	66.8	193,645
Romania	1,850	65.8	121,734
Sweden	1,139	85.6	97,442
Switzerland	148	107.3	15,888
Yugoslavia	605	52.5	31,777
Europe	42,919	76.6	3,286,678
Australia	6,200	30.1	186,206
New Zealand	240	83.3	19,975
Oceania	6,439	32.0	206,181
World Total	177,966	45.8	8,141,249

Source: Adapted from USDA Agriculture Statistics, 1992.

1. Values do not sum to totals because of rounding.

sharp contrasts with production in the subtropical, dry climate of northern Africa. Tunisia averaged 16.2 bushels/acre in 1990/1991, while Morocco averaged 16.6, and Algeria averaged only 12.5 bushels/acre.

PLANT MORPHOLOGY

The barley plant is very similar to the wheat plant in growth and development, and therefore will not be discussed in great detail here. As germination proceeds, the coleorhiza, or root sheath, emerges from the micropylar end of the seed, producing seminal roots. The coleoptile emerges from the micropyle also and elongates underneath the husks, emerging from the apex of the husk and continuing to grow until it reaches the soil surface. The first leaves appear, emerging through the pore at the tip of the coleoptile. Each leaf is attached to a node, sometimes called a joint, that is yet to be visible above ground. Permanent roots, also called adventitious roots, form and grow from the crown, or area where stem and root tissue meet. Several tillers may form.

The mature barley plant consists of (1) roots; (2) stems or culms (both main stem and tillers) that are made up of hollow internodes and usually five to seven solid nodes; (3) leaves arising singularly and alternately at nodes on opposite sides of the stem and consisting of a sheath, ligule, auricles, and blade; (4) a single ear or head at the top of each stem—a spike consisting of spikelets, alternately borne in sets of three at each node of the rachis; (5) spikelets consist of two glumes and a floret; (6) self-pollinating florets that are comprised of a lemma and palea that enclose the reproductive organs; and (7) kernels that consist of the caryopsis only in naked

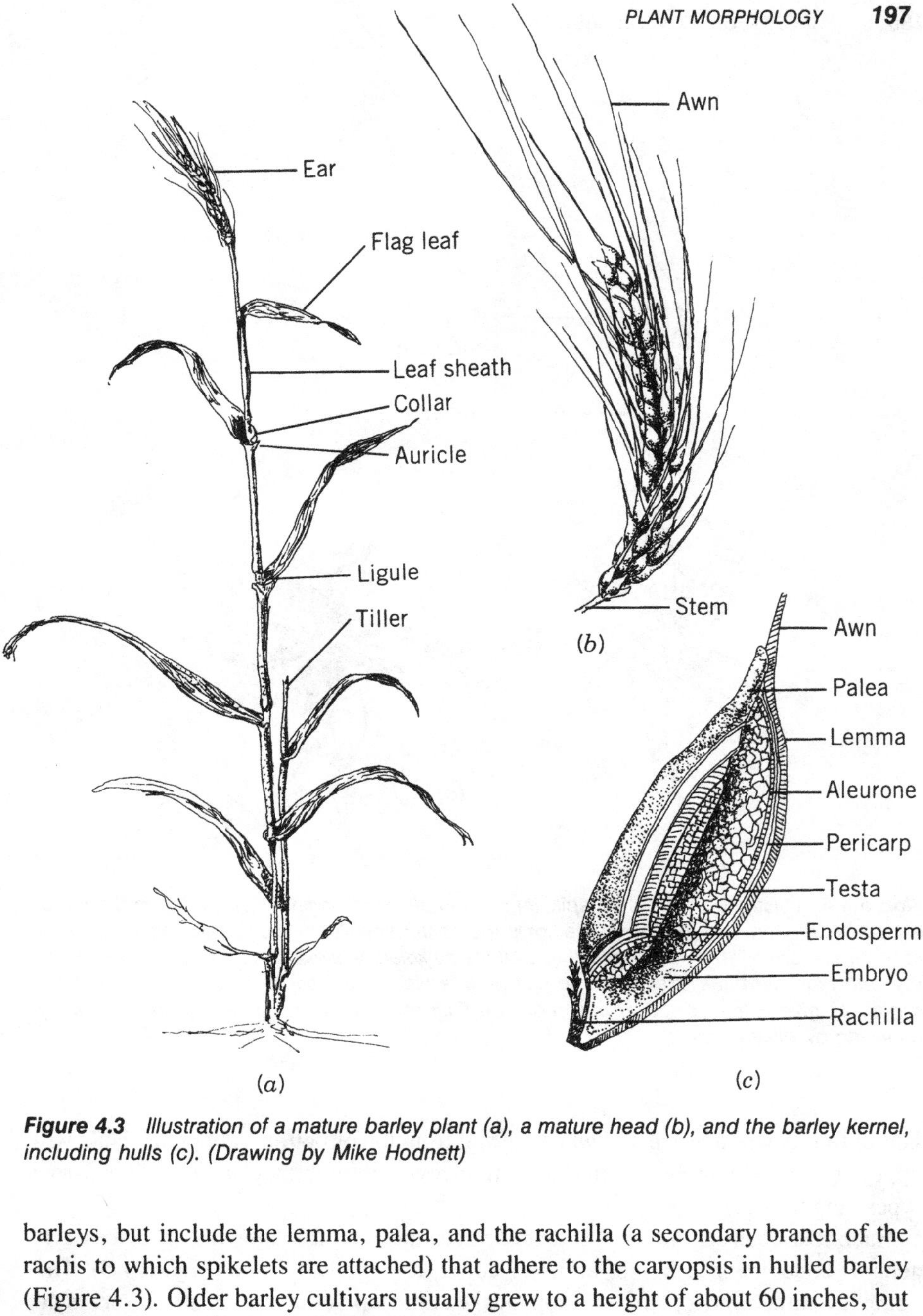

Figure 4.3 Illustration of a mature barley plant (a), a mature head (b), and the barley kernel, including hulls (c). (Drawing by Mike Hodnett)

barleys, but include the lemma, palea, and the rachilla (a secondary branch of the rachis to which spikelets are attached) that adhere to the caryopsis in hulled barley (Figure 4.3). Older barley cultivars usually grew to a height of about 60 inches, but most modern cultivars reach only 36 inches or so.

As noted earlier, all three spikelets at each rachis node may be fertile, resulting in three columns of grain on each side. Barley having only the central spikelet fertile is called two-row, while those with all three spikelets at each node fertile are called six-row barley, so named because when the spike is viewed from the top or bottom, the spikelet arrangement appears to be in two or six columns. Occasionally, the

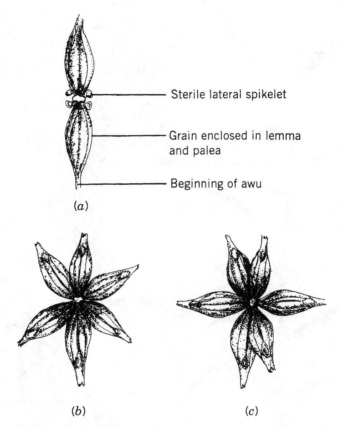

Sterile lateral spikelet

Grain enclosed in lemma and palea

Beginning of awu

(a)

(b)   *(c)*

Figure 4.4 *Illustration of different spikelet configurations in barley as viewed from the top or bottom of the ear. Two-row barley has only the center spikelet of the three spikelets at each rachis node fertile (a); six-row barley has all three spikelets fertile and somewhat evenly spread (b); and four-row barley has all three spikelets fertile but the two outside spikelets overlap slightly to give a four-rowed appearance (c). Four-row barley is technically six-row barley. (Drawing by Mike Hodnett)*

lateral florets will overlap and when viewed from top or bottom, the barley head will appear to be and may be referred to as four-row, although they are actually six-row types (Figure 4.4).

Each leaf consists of a sheath that encloses the stem for some distance above its node of attachment, a ligule, auricles, collar, and the leaf blade (Figure 4.5). The ligule is a thin, almost paper-like appendage at the top of the leaf sheath. It serves no obvious function but is found in most cultivars of barley and wheat, but it is absent in oats. The auricles also occur at the junction of the blade and sheath and consist of two claw-like appendages that appear to clasp the stem. Auricles are very prominent in most barley cultivars, while they are narrow and hairy in wheat and nonexistent in oats.

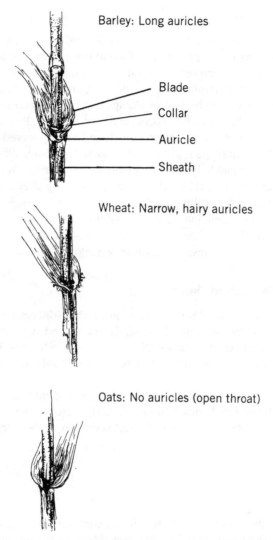

Barley: Long auricles

Blade

Collar

Auricle

Sheath

Wheat: Narrow, hairy auricles

Oats: No auricles (open throat)

Figure 4.5 Comparison of leaves of barley, wheat, and oats. All are composed of blade, collar, sheath, and ligule (not shown). Barley has long auricles, wheat has narrow auricles, and oats have no auricles at the collar. (Drawing by Mike Hodnett)

PRODUCTION PRACTICES

Cultural practices for barley are similar to those for wheat, and therefore will not be repeated here. Producers should be aware that winter types do vary in their winter hardiness and that barley cannot tolerate temperatures as low as can wheat or rye.

Cultivar Choice

Not all ᴜarley cultivars produce grain acceptable for malting. Maltsters prefer grains that are plump, well-matured, and intermediate to low (that is, 10 to 12 percent) in protein content. This lower protein requirement means that producers cannot maximize yields and protein through heavy use of nitrogen fertilizer. Anything such as weathering or damage during harvest or storage that reduces germination below 95 percent will cause the barley to be unacceptable for malting. Producers of malting barley must choose cultivars that have been tested and approved by the malting industry. Of the public cultivars registered in *Crop Science* from 1980 through 1989, only 13 of 44 were acceptable for malting (Table 4.11). Johnson, Klondike, Bowan, Heartland, and Lamont were released as feed barleys only after tests revealed that they were unacceptable for malting, and other releases from major malt-producing states also could have been intended for malting cultivars. Most of the barley cultivars in the United States are feed types, and have six-row spikes with awns, hulls that adhere to the caryopsis, and white or yellow aleurone.

Planting and Harvesting Dates

Spring barley is planted after April 1 in Wisconsin, the Northern Great Plains states of Minnesota, North Dakota, South Dakota, Colorado, and Wyoming (Figure 4.6). Planting spring barley in the Pacific Northwest usually occurs before April 1. Harvesting of spring barley usually begins between mid-July and late August (Figure 4.7).

Winter barley, generally confined to the southern and middle states, is planted from about September 1 in Pennsylvania until early October in South Carolina and parts of the Southwest and California. Harvest occurs from about mid-June through July.

BIOTIC PESTS

Barley is attacked by essentially the same insect complex that attacks wheat, and so the reader is referred to Chapter 2 of this text. Many of the diseases of barley are also common in wheat; however, a number of the more important diseases of barley are listed in Table 4.12.

GRADES AND STANDARDS

Barley is classified as six-row, two-row, or barley by the USDA Federal Grain Inspection Service (USDA-FGIS) (Table 4.13). The six-row and two row classes are further divided into subclasses: six-row malting, six-row blue malting, and six-row barley for the class six-row barley, and two-row malting and two-row barley for the two-row barley class (Tables 4.14–4.16).

TABLE 4.11 Barley Cultivars Registered with Crop Science, 1980–1989.

Name	Year Registered	State Released[3]	Purpose	Spike	Awns[4]	Hulled	Growth Habit	Aleurone
Bonanza	1980	Canada	malt	6-row	smooth	yes	spring	blue
Kimberly	1978	Idaho	malt		rough	yes	spring	white
Milton	1981	N. Carolina	feed	6-row	rough	—	winter	—
Bedford	1981	Canada	feed	6-row	smooth	yes	spring	white
Johnson[1]	1981	Canada	feed	6-row	smooth	yes	spring	white
Klondike[1]	1981	Canada	feed	6-row	smooth	yes	spring	white
Tambar 402	1982	Texas	feed	6-row	rough	yes	winter	blue
Azure	1982	N. Dakota	malt	6-row	smooth	yes	spring	blue
Redhill	1982	S. Carolina	feed	6-row	rough	yes	winter	white
Wintermalt[2]	1982	New York	malt	6-row	rough	—	winter	white
Dawn	1983	Georgia	feed	6-row	rough	—	winter	white
Anson	1983	N. Carolina	feed	6-row	awnlete	—	winter	—
Robust	1983	Minnesota	malt	6-row	smooth	yes	spring	white
Dundy	1983	Nebraska	feed	6-row	rough	yes	winter	white
Abee	1983	Canada	feed	2-row	rough	yes	spring	yellow
Empress	1983	Canada	feed	6-row	rough	yes	spring	yellow
Advance	1983	Washington	malt	6-row	awnless	yes	spring	white
Sussex	1984	Virginia	feed	6-row	awnless	—	winter	white
Hazen	1984	N. Dakota	malt	6-row	smooth	yes	spring	white
Karla	1985	Idaho	malt	6-row	rough	yes	spring	white
Lewis	1985	Montana	malt	2-row	rough	yes	spring	white
Post	1984	Oklahoma	feed	6-row	rough	yes	winter	white
Clark	1985	Montana	malt	2-row	rough	yes	spring	white
Bowan[1]	1985	N. Dakota	feed	2-row	smooth	yes	spring	colorless

(continued)

TABLE 4.11 (Continued)

Name	Year Registered	State Released[3]	Purpose	Spike	Awns[4]	Hulled	Growth Habit	Aleurone
Hitchcock	1985	Nebraska	feed	6-row	rough	yes	winter	blue
Andre	1985	Washington	duel	2-row	rough	yes	spring	white
Heartland[1]	1985	Canada	feed	6-row	smooth	yes	spring	white
Diamond	1985	Canada	feed	6-row	smooth	—	spring	yellow
Kline	1985	Georgia	feed	6-row	rough	—	winter	yellow
Samson	1986	Canada	feed	6-row	rough	—	spring	yellow
Cougbar	1986	Washington	malt	6-row	rough	yes	spring	white
Showin	1986	Washington	feed	6-row	rough	yes	winter	white
Wysor	1987	Virginia	feed	6-row	awnlete	—	winter	white
Gallatin	1987	Montana	feed	2-row	rough	yes	spring	white
Venus	1988	Georgia	feed	6-row	rough	—	winter	yellow
Russell	1988	Idaho	malt	6-row	smooth	—	spring	white
Lamont[1]	1988	Idaho	feed	2-row	rough	—	spring	white
Virden	1988	Canada	feed	6-row	smooth	yes	spring	white
Pennco	1988	Pennsylvania	feed	6-row	awnlete	—	winter	white
Ray	1988	Ohio	feed	6-row	rough	yes	winter	yellow
Preamble	1989	Maryland	feed	6-row	awnlete	—	winter	—
Willis	1989	New York	feed	6-row	rough	—	winter	white
Noble	1989	Canada	feed	6-row	smooth	yes	spring	yellow
Seco	1989	Arizona	conservation	6-row	rough	—	spring	white

1. Released as feed barleys after being unacceptable by malting trade.
2. First winter barley released in U.S. meeting malting specifications.
3. Some releases resulted from joint efforts by two or more states or government agencies. Only the first state listed in registration shown for brevity.
4. Refers to lemma awns and not glumes.

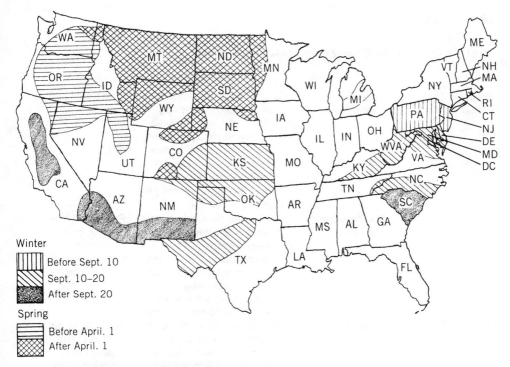

Figure 4.6 *Usual beginning planting dates for winter and spring barley in the continental United States.* Source: *USDA-SRS, 1984. (Redrawn by Mike Hodnett.)*

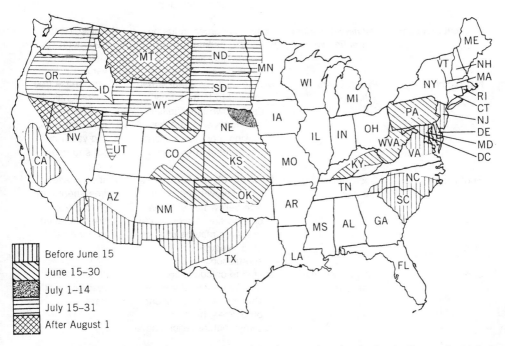

Figure 4.7 *Usual beginning harvest dates for winter and spring barley in the continental United States.* Source: *USDA-SRS, 1984. (Redrawn by Mike Hodnett.)*

TABLE 4.12 Common Diseases and Nematodes of Barley.

Disease	Cause/Source	Symptoms/Injury	Control
Bacterial Leaf Blight	*Pseudomonas syringae*	Usually on uppermost leaves only, especially the flag leaf; small, water-soaked lesions turning gray-green then tan or white; lesions coalesce and entire leaf may become necrotic; may appear on kernel, primarily on lemma.	Field sanitation.
Ergot	*Claviceps purpurea*	Infected florets exude sugary slime that attracts insects; kernel is replaced by purple-black, hornlike sclerotia (ergots) that protrude from glumes; fungi produces alkaloids that are toxic in sufficient amounts.	Plant ergot free seed; crop rotation; deep tillage; clean cultivation.
Scab or Fusarium Head Blight	*Fusarium* spp.	Brownish spots in middle or base of spikelet glume or on rachis; heads may be dwarfed; pink to reddish dust-like mycelial may be seen at the margins or base of glumes; kernels may be shrunken and gray-brown.	Field sanitation; do not plant barley in fields having abundant residue; chemical seed treatment.
Helminthosporium Stripe	*Helminthosporium gramineum*	Long, pale-green strips on leaves, turning brown; leaves may split along stripe.	Fungicidal seed treatment; resistant cultivars.
Net Blotch	*Pyrenophora teres*	Minute spots or streaks that expand to form narrow, dark brown longitudinal and transverse streaks making a net like pattern on leaves and sheaths.	Plant disease free seed; fungicide treated seed; field sanitation; two year rotations to nonhost.
Spot Blotch	*Cochliobolus sativus*	Round to oblong lesions uniformly brown and often with yellow halos; spots enlarge and coalesce; older leaves may be olive colored because of sporulation.	Plant disease free seed; fungicide treated seed; rotation to non-host.
Scald	*Rhynchosporium secalis*	Round to oblong lesions usually on leaf blades and sheaths but may appear on glumes, hulls and awns; mature lesions have	Field sanitation; resistant cultivars.

TABLE 4.12   (*Continued*)

Disease	Cause/Source	Symptoms/Injury	Control
		light gray, tan or white centers with dark brown edges, often with chlorotic halo.	
Septoria Blotch	see wheat		
Powdery mildew	see wheat		
Covered Smut	*Ustilago hordei*	Kernels replaced with black fungal spores enclosed in persistent membrane; smutted heads emerge late or remain enclosed in flag leaf sheath.	Plant disease free seed; fungicidal seed treatment; resistant cultivars.
Nuda or True Loose Smut	*U. nuda*	Diseased heads are blackened with dry, olive-black spore masses that blow away leaving only the rachis.	Systemic fungicidal seed treatments.
Nigra or Semi-loose Smut	*U. nigra*	Same as for Nuda Smut but spores may remain within membrane longer.	Systemic fungicidal seed treatments; resistant cultivars.
Stripe Rust	see wheat		
Stem Rust	see wheat		
Leaf Rust	*Puccinia hordei*	Small, round, light orange-brown uredinia on leaf sheaths and blades; occasionally on heads.	Foliar fungicides; resistant cultivars.
Barley Yellow Dwarf Virus	see wheat		
Barley Stripe Mosaic Virus	Seed transmitted virus; no known vectors.	Yellow to white mottling, spotting and streaking; brown, necrotic stripes, often in a V shape on leaves; plants are stunted; some florets sterile; poorly developed heads or kernels.	Plant virus free seed; resistant cultivars.
Oat Blue Dwarf Virus	Virus vectored by aster leafhopper, *Macrosteles fascifrons*, only.	Plants are stunted with stiffened and shortened leaves; excessive tillering; heads sterile.	None.
Cereal Cyst Nematode	*Heterodera avenae*	Infected roots divide near invasion point resulting in stunted, knotted, much proliferated root system; white cyst found on roots at heading.	Rotation with non-host; resistant cultivars.

TABLE 4.13   Classes, Subclasses, and Grades for Barley.

Class	Subclass	Grades[1]
Six-row	Six-row Malting	U.S. No. 1–U.S. No. 3
	Six-row Blue Malting	U.S. No. 1–U.S. No. 3
	Six-row	U.S. No. 1–U.S. No. 5
		U.S. Sample Grade
Two-row	Two-row Malting	U.S. No. 1 choice, and
		U.S. No. 1–U.S. No. 3
	Two-row	U.S. No. 1–U.S. No. 5
		U.S. Sample Grade
Barley	——	U.S. No. 1–U.S. No. 5
		U.S. Sample Grade

Source: USDA-FGIS, 1988.

1. Bleached, blighted, bright, ergoty, garlicky, smutty, stained, tough and/or weevily are special grades that may be assigned across classes and subclasses.

Terms Used in Grading by the USDA-FGIS

Definition of Barley Grain that, before the removal of dockage, consists of 50 percent or more of whole kernels of cultivated barley (*Hordeum vulgare* L.) and not more than 25 percent of other grains for which standards have been established under the United States Grain Standards Act. The term "barley" as used in these standards does not include hull-less barley or black barley.

Definitions of Other Terms

(a) **Black barley.** Barley with black hulls.

(b) **Broken kernels.** Barley with more than one-fourth of the kernel removed.

(c) **Classes.** There are three classes for barley: six-row barley, two-row barley, and barley.

    (1) **Six-row Barley.** Barley of the six-row type with white hulls that contains not more than 10 percent of two-row barley. This class is divided into the following three subclasses:

        (i) **Six-row malting barley.** Six-row barley of a suitable malting type that has 90 percent or more of kernels with white aleurone layers; that contains not more than 1.9 percent of injured-by-frost kernels that may include not more than 0.4 percent of frost-damaged kernels; that is not more than 0.2 percent of injured-by-heat kernels that may include not more than 0.1 percent of heat-damaged kernels; that is not blighted, ergoty, garlicky, infested, or smutty; and that otherwise meets the grade requirements of the subclass six-row malting barley; and may contain unlimited amounts of injured-by-mold kernels; however mold-damaged kernels are scored as damaged kernels and against sound barley limits.

(ii) **Six-row blue malting barley.** Six-row barley of a suitable malting type that has 90 percent or more of kernels with blue aleurone layers; that contains not more than 1.9 percent of injured-by-frost kernels that may include not more than 0.4 percent of frost-damaged kernels; that is not more than 0.2 percent of injured-by-heat kernels that may include not more than 0.1 percent of heat-damaged kernels; that is not blighted, ergoty, garlicky, infested, or smutty; and that otherwise meets the grade requirements of the subclass six-row blue malting barley; and may contain unlimited amounts of injured-by-mold kernels; however, mold-damaged kernels are scored as damaged kernels and against sound barley limits.

(iii) **Six-row barley.** Any barley of the class six-row barley that does not meet the requirements of the subclass six-row malting barley or six-row blue malting barley.

(2) **Two-row barley.** Barley of the two-row type with white hulls that contains not more than 10 percent of six-row barley. This class is divided into the following two subclasses:

(i) **Two-row malting barley.** Two-row barley of a suitable malting type that contains not more than 1.9 percent of injured-by-frost kernels that may include not more than 0.4 percent frost-damaged kernels; that is not more than 1.9 percent of injured-by-mold kernels that may include not more than 0.4 percent of mold-damaged kernels; and not more than 0.2 percent of injured-by-heat kernels that may include not more than 0.1 percent of heat-damaged kernels; that is not blighted, ergoty, garlicky, infested, or smutty; and that otherwise meets the grade requirements of the subclass two-row malting barley. Injured-by-frost kernels and injured-by-mold kernels are not scored against sound barley.

(ii) **Two-row barley.** Two-rowed Barley that does not meet the requirements of the subclass Two-row Malting Barley.

(3) **Barley.** Barley that does not meet the requirements for the classes six-row barley or two-row barley.

(d) **Damaged kernels.** Kernels, pieces of barley kernels, other grains, and wild oats that are badly ground-damaged, badly weather-damaged, diseased, frost-damaged, germ-damaged, heat-damaged, injured-by-heat, insect-bored, mold-damaged, sprout-damaged, or otherwise materially damaged.

(e) **Dockage.** All matter other than barley that can be removed from the original sample by use of an approved device according to procedures described in FGIS instructions. Also, underdeveloped, shriveled, and small pieces of barley kernels removed in properly separating the material other than barley and that cannot be recovered by properly rescreening or recleaning.

(f) **Foreign material.** All matter other than barley, other grains, and wild oats that remains in the sample after removal of dockage.

(g) **Frost-damaged kernels.** Kernels, pieces of barley kernels, other grains, and wild oats that are badly shrunken and distinctly discolored black or brown by frost.

(h) **Germ-damaged kernels.** Kernels, pieces of barley kernels, other grains, and wild oats that have dead or discolored germ ends.

(i) **Heat-damaged kernels.** Kernels, pieces of barley kernels, other grains, and wild oats that are materially discolored and damaged by heat.

(j) **Injured-by-frost kernels.** Kernels and pieces of barley kernels that are distinctly indented, immature, or shrunken in appearance or that are light green in color as a result of frost before maturity.

(k) **Injured-by-heat kernels.** Kernels, pieces of barley kernels, other grains, and wild oats that are slightly discolored as a result of heat.

(l) **Injured-by-mold kernels.** Kernels and pieces of barley kernels containing slight evidence of mold.

(m) **Mold-damaged kernels.** Kernels, pieces of barley kernels, other grains, and wild oats that are weathered and contain considerable evidence of mold.

(n) **Other grains.** Black barley, corn, cultivated buckwheat, einkorn, emmer, flaxseed, guar, hull-less barley, nongrain sorghum, oats, Polish wheat, popcorn, poulard wheat, rice, rye, safflower, sorghum, soybeans, spelt, sunflower seed, sweet corn, triticale, and wheat.

(o) **Plump barley.** Barley that remains on top of a $6/64 \times 3/4$ slotted-hole sieve after sieving according to procedures prescribed in FGIS instructions.

(p) **Sieves.**

(1) $5/64 \times 3/4$ **slotted-hole sieve.** A metal sieve 0.032 inch thick with slotted perforations 0.0781 ($5/64$) inch by 0.750 ($3/4$) inch.

(2) $5.5/64 \times 3/4$ **slotted-hole sieve.** A metal sieve 0.032 inch thick with slotted perforations 0.0895 ($5.5/64$) inch by 0.750 ($3/4$) inch.

(3) $6/64 \times 3/4$ **slotted-hole sieve.** A metal sieve 0.032 inch thick with slotted perforations 0.0937 ($6/64$) inch by 0.750 ($3/4$) inch.

(q) **Skinned and broken kernels.** Barley kernels that have one-third or more of the hull removed, or with the hull loose or missing over the germ, or broken kernels, or whole kernels that have a part or all of the germ missing.

(r) **Sound barley.** Kernels and pieces of barley kernels that are not damaged, as defined under (d) of this section.

(s) **Stained barley.** Barley that is badly stained or materially weathered.

(t) **Suitable malting type.** Varieties of malting barley that are recommended by the American Malting Barley Association as being suitable for malting purposes. The recommended varieties are listed in FGIS instructions.

(u) **Thin barley.** Six-row barley which passes through a $5/64 \times 3/4$ slotted-hole sieve and two-row barley which passes through a $5.5/64 \times 3/4$ slotted-hole sieve after sieving according to procedures prescribed in FGIS instructions.

(v) **Wild oats.** Seeds of *Avena fatua* L. and *A. sterilis* L.

All Other Determinations Each determination of heat-damaged kernels, injured-by-heat kernels, and white or blue aleurone layers in six-row barley is made on pearled, dockage-free barley. Other determinations not specifically provided for under the general provisions are made on the basis of the grain when free from dockage, except the determination of odor is made on either the basis of the grain as a whole or the grain when free from dockage.

Special Grades and Special Grade Requirements

(a) **Blighted.** Barley that contains more than 4 percent of fungus-damaged and/or mold-damaged kernels.

(b) **Ergoty.** Barley that contains more than 0.10 percent ergot.

(c) **Garlicky.** Barley that contains three or more green garlic bulblets, or an equivalent quantity of dry or partly dry bulblets in 500 grams of barley.

(d) **Smutty.** Barley that has kernels covered with smut spores to give a smutty appearance in mass, or which contains more than 0.20 percent smut balls.

TABLE 4.14 Grades and Grade Requirements for the Subclasses Six-Row Malting Barley and Six-Row Blue Malting Barley.

| | Minimum Limits of— | | | Maximum Limits of— | | | | |
Grade	Test Weight per Bushel	Suitable Malting Type	Sound Barley[1]	Damaged Kernels	Foreign Material	Other Grains	Skinned and Broken Kernels	Thin Barley
	(lbs.)	(%)	(%)	(%)	(%)	(%)	(%)	(%)
U.S. No. 1	47.0	95.0	97.0	2.0	1.0	2.0	4.0	7.0
U.S. No. 2	45.0	95.0	94.0	3.0	2.0	3.0	6.0	10.0
U.S. No. 3	43.0	95.0	90.0	4.0	3.0	5.0	8.0	15.0

Source: USDA-FGIS, 1988.

1. Injured-by-frost kernels and injured-by-mold kernels are not considered damaged kernels or scored against sound barley.

Note: Six-row barley that meets the requirements of U.S. No. 1 to U.S. No. 3, inclusive, for the subclasses six-row malting barley and six-row blue malting barley is classified and graded according to those requirements. Otherwise, it will be graded according to the requirements in Table 4.16.

TABLE 4.15 Grades and Grade Requirements for the Subclass Two-Row Malting Barley.

Grade	Minimum Limits of—				Maximum Limits of—		
	Test Weight per Bushel	Suitable Malting Types	Sound Barley[1]	Wild Oats	Foreign Material	Skinned and Broken Kernels	Thin Barley
	(lbs.)	(%)	(%)	(%)	(%)	(%)	(%)
U.S. No. 1 Choice	50.0	97.0	98.0	1.0	0.5	5.0	5.0
U.S. No. 1	48.0	97.0	98.0	1.0	0.5	7.0	7.0
U.S. No. 2	48.0	95.0	96.0	2.0	1.0	10.0	10.0
U.S. No. 3	48.0	95.0	93.0	3.0	2.0	10.0	10.0

Source: USDA-FGIS, 1988.

1. Injured-by-frost kernels and injured-by-mold kernels are not considered damaged kernels or scored against sound barley.

Note: Two-row barley that meets the requirements of U.S. No. 1 Choice to U.S. No. 3, inclusive, for the subclass two-row malting barley is classified and graded according to those requirements. Otherwise, it will be graded according to the requirements in Table 4.16.

TABLE 4.16 Grades and Grade Requirements for the Subclasses Six-Row Barley, Two-Row Barley and the Class Barley.

Grade	Minimum Limits of—		Maximum Limits of—				
	Test Weight per Bushel	Sound Barley[1]	Damaged Kernels	Heat-Damaged Kernels	Foreign Material	Broken Kernels	Thin Barley
	(lbs.)	(%)	(%)	(%)	(%)	(%)	(%)
U.S. No. 1	47.0	97.0	2.0	0.2	1.0	4.0	10.0
U.S. No. 2	45.0	94.0	4.0	0.3	2.0	8.0	15.0
U.S. No. 3	43.0	90.0	6.0	0.5	3.0	12.0	25.0
U.S. No. 4[2]	40.0	85.0	8.0	1.0	4.0	18.0	35.0
U.S. No. 5	36.0	75.0	10.0	3.0	5.0	28.0	75.0

U.S. Sample grade shall be barley that

(a) Does not meet the requirements for the grades U.S. Nos. 1, 2, 3, 4, or 5; or

(b) Contains 8 or more stones or any number of stones which have an aggregate weight in excess of 0.2 percent of the sample weight, 2 or more pieces of glass, 3 or more crotalaria seeds (*Crotalaria* spp.), 2 or more castor beans (*Ricinus communis* L.), 4 or more particles of an unknown foreign substance(s) or a commonly recognized harmful or toxic substance(s), 8 or more cocklebur (*Xanthium* spp.) or similar seeds singly or in combination, 10 or more rodent pellets, bird droppings, or equivalent quantity of other animal filth per 1 1/8 to 1 1/4 quarts of barley; or

(c) Has a musty, sour, or commercially objectionable foreign odor (except smut or garlic odor); or

(d) Is heating or otherwise of distinctly low quality.

Source: USDA-FGIS, 1988.

1. Includes heat damaged kernels; injured-by-frost kernels and injured-by-mold kernels are not considered damaged kernels or scored against sound barley.

2. Barley that is badly stained or materially weathered shall be graded not higher than U.S. No. 4.

GLOSSARY

Acrospire: The shoot that is formed in the germination process and grows to about the length of the kernel. The part that extends from under the hull is broken off and becomes a feed by-product along with the rootlet.

Aleurone: The barley kernel is composed of the lemma and palea, a short portion of the rachilla and the caryopsis itself. Just inside the seed coat of the caryopsis are a few layers of very important nitrogen-rich cells comprising the aleurone layer. These cells, when properly activated, secrete hydrolytic enzymes into the endosperm that disintegrate the structure of this starch-containing storage material and accomplish what is called "modification."

Country elevator: These facilities are located in production areas and serve as the primary outlet for off-farm sales. Country elevators make unofficial determinations of grain grades and weights. They generally take title to the grain they handle, but in some cases may provide the source of handling grain previously contracted to another buyer.

Dense: Refers to barley with short rachis internodes giving the inflorescence a dense or compact appearance.

Enzyme potential: Barley, when malted, produces two known starch-splitting enzymes, alpha amylase and beta amylase. The combination of these two enzymes results in more rapid and complete hydrolysis of starch to dextrin and fermentable sugars. Only barley, wheat, and rye have both alpha and beta enzymes. The potential for transforming starch to fermentable sugars is not the same for all varieties; thus, the careful selection of malt varieties.

Facultative winter type: Winter barley requiring very little cold temperature for vernalization, which therefore may reproduce when spring-planted; as opposed to an obligate winter type that requires longer periods of cold temperatures to initiate flowering.

Highgrade: Highgrading is a process whereby low- and high-quality products are separated. In the barley industry, poorer quality kernels not suited for malting are separated out and sold to feed manufacturers, resulting in a higher-quality malting product.

Hooded barley: The inflorescence has a modified lemma awn that resembles a hood over the floret. Hoods may have both male and female organs and in rare instances produce viable seed.

Hull-less barley: Barley having the hull thrash free of the caryopsis during harvest. Most U.S. barley cultivars are not hull-less.

Inland terminal elevator: A facility located at a point of accumulation and distribution in the movement of grain. An inland terminal elevator procures a large share of its grain from other elevators rather than directly from farmers. They have facilities for establishing official grades and weights, and may store grain for others.

Kilning: Drying malted barley in a kiln, or oven, with hot, dry air.

Lax: Refers to barley with long rachis internodes giving the inflorescence an open, lax appearance.

Malt sprouts: Primarily rootlets separated from malt before kilning. Malt sprouts average about 27 percent protein, 2.5 percent fat, and 12 percent fiber, and are commonly used in dairy feeds.

Malt: Final product after kilning that is cleaned to remove the dried rootlets and given a degree of polishing (may be called final malt). Casual observation indicates that the final product varies little in appearance from the original barley, but on close inspection it is obvious that the kernels are somewhat larger in size, and the dried acrospires noticeably budge under the husks, and husks do not adhere as tightly to the main body of the kernel.

Midwestern six-row Manchurian: Relatively small-kerneled barleys that are medium-high in protein, vigorous in germination, and produce high enzymatic activities during malting; used for the production of brewer's and distiller's malts.

Modification: The limited breakdown of the starch endosperm of barley by enzymes during the germination phase of malting. Modified starch is readily converted to dextrin and fermentable sugars when ground and mixed with water to 60 to 70°F.

Naked barley: Refer to hull-less barley.

Pearled barley: A barley product remaining after the hull, the kernel coating, practically all of the embryo, and part of the outer layers of the starchy endosperm have been removed by a grinding process. One hundred pounds of barley yields approximately 35 pounds of pearled barley.

Pipeline stocks: The grain or grain products that are not in storage awaiting a buyer or in inventory as stored grain. These stocks may be in transit or may be held in working space.

Port terminal elevator: An elevator located along waterways and designed to load out vessels with grain and other products. A port terminal elevator receives most of its grain from subterminal elevators or inland terminals. Port terminals have facilities for establishing official grades and weights.

Rachilla: An inconspicuous pedicel or branch attached to the rachis and to which the florets are attached.

Six-row barley: The axis of the barley head has nodes throughout its length with spikelets at these nodes alternating from side to side. For six-row barley, three kernels develop at each node, a central kernel and two lateral kernels.

Spent grain: The hulls and other solids remaining in the brewer's mash tub. Spent grains are dried and sold as a feed by-product.

Terminal market: A large concentration of wholesale grain handlers, commission merchants, and grain brokers that may be complemented by a Grain Exchange or Board of Trade, which in turn houses an association organized for the purpose of providing a place where buyers and sellers may conduct trading in both the cash and futures market.

True winter type: Barley requiring a prolonged period of cold for vernalization that, therefore, usually will not flower when spring-planted.

Two-row barley: The axis of the barley head has nodes throughout its length with spikelets at these nodes alternating from side to side. For two-row barley, only the central kernel develops, both laterals being sterile. (Compare with six-row barley.)

Vernalization: The stimulation of reproductive growth of a plant by its passing through a time period of temperatures below 50°F. A true winter barley, for example, will not mature if planted in the spring and not exposed to temperatures below 50°F. Length of vernalization period is cultivar- and temperature-dependent.

Western six-row: Brewing barleys grown primarily in California. Large, hulled kernels medium protein content, rather slow physical and chemical modification potential, and low enzymatic activities after malting are the characteristics of this type of barley. These are used for brewing in the West Coast and Rocky Mountain areas, or for blending with midwestern type malts for brewing.

Western two-row: Grown primarily in the Northwest and intermountain areas of the United States. They have medium sized, uniform, plump kernels with a thin hull. They are generally low in protein and high in starch with vigorous germination and intermediate enzymatic activity during malting. They are used by the brewing industry both alone and blended with midwestern six-row barley.

Primary source: Heid and Leath, 1978.

BIBLIOGRAPHY

Atkins, I. M. 1980. *A History of Small Grain Crops in Texas: Wheat, Oats, Barley, Rye. 1582–1976.* Texas A&M Univ. Agri. Exp. Sta. B-1301.

Atkins, I. M., J. H. Gardenhire, M. E. McDaniel, and K. B. Porter. 1969. *Barley production in Texas.* Texas A&M Univ. Agri. Exp. Sta. B-1087.

Bar-Yosef, O., and M. Kislev. 1989. "Early Farming Communities in the Jordan Valley." pp. 632–642. *In* D. R. Harris and G. C. Hillman (eds.) *Foraging and Farming: The Evolution of Plant Exploitation.* Unwin Hyman Press, London, England.

Bender, B. 1975. *Farming in Prehistory.* St. Martin's Press, New York, NY.

von Bothmer, R., and N. Jacobsen. 1985. "Origin, Taxonomy, and Related Species." pp. 19–56. *In* D. C. Rasmusson (ed.) *Barley.* Am. Soc. Agro., Madison, WI.

Briggs, D. E., J. S. Hough, R. Stevens, and T. W. Young. 1981. *Malting and Brewing Science.* Chapman and Hall, London, England.

Briggs, D. E. 1978. *Barley.* Chapman and Hall, London, England.

Brown, A. R., J. W. Johnson, C. S. Rothrock, and P. L. Bruckner. 1988. "Registration of Venus barley." *Crop Sci.* 28:718–719.

Brown, A. R., D. D. Morey, J. W. Johnson, and B. M. Cunfer. 1985. "Registration of Kline barley." *Crop Sci.* 25:706–707.

Brown, A. R., D. D. Morey, J. W. Johnson, and B. M. Cunfer. 1983. "Registration of Dawn barley." *Crop Sci.* 23:597.

Church, D. C. 1986. *Livestock Feeds and Feeding.* Prentice-Hall, Englewood Cliffs, NJ.

Croissant, R. L., and J. W. Echols. 1990. *Barley Production in Colorado.* Colorado St. Univ. Coop. Ext. Ser. Service in Action No. 120.

Dickson, A. D. 1979. "Barley for Malting and Food." pp. 112–120. *In Barley: Origin, Botany, Culture, Winter Hardiness, Genetics, Utilization, Pests.* USDA-SEA Agri. Handbook No. 338.

Edney, M. J., R. Tkachuk, and A. W. MacGregor. 1992. "Nutrient composition of the hullless barley cultivar, Condor." *J. Sci. Food and Agri.* 60:451–456.

Edwards, L. H., E. L. Smith, H. Pass, and G. H. Morgan. 1985. "Registration of Post barley." *Crop Sci.* 25:363.

Fidanza, F. 1979. "Diets and dietary recommendations in ancient Greece and Rome and the School of Salerno." *Prog. Food and Nut. Sci.* 3:79–99.

Florell, V. H. 1927. "A comparison of selections of Coast barley." *Agron. J.* 19:660–674.

Foster, A. E., J. D. Franckowiak, V. D. Pederson, and R. E. Pyler. 1984. "Registration of Hazen barley." *Crop Sci.* 24:1210.

Foster, A. E., J. D. Franckowiak, V. D. Pederson, and R. E. Pyler. 1982. "Registration of Azure barley." *Crop Sci.* 22:1083.

Franckowiak, J. D., A. E. Foster, V. D. Pederson, and R. E. Pyler. 1985. "Registration of Bowman barley." *Crop Sci.* 25:883.

Gardenhire, J. H., M E. McDaniel, and N. A. Tuleen. 1982. "Registration of Tambar 402 barley." *Crop Sci.* 21:1259.

Graham, W. D., Jr., B. C. Morton Jr., and G. C. Kingsland. 1982. "Registration of Redhill barley." *Crop Sci.* 22:1083.

Hancock, J. F. 1992. *Plant Evolution and the Origin of Crop Species.* Prentice Hall Press, Englewood Cliffs, NJ.

Harlan, J. R. 1979. "On the Origin of Barley." pp. 10–36. *In Barley: Origin, Botany, Culture, Winter Hardiness, Genetics, Utilization, Pests.* USDA-SEA Agri. Handbook No. 338.

Harlan, J. R. 1976. "Barley." pp. 93–98. *In* N. W. Simmonds (ed.) *Evolution of Crop Plants.* Longman Press, London, England.

Harlan, J. R., J. M. J. de Wet, and E. G. Price. 1973. "Comparative evolution of cereals." *Evol.* 27:311–325.

Harlan, J. R., and D. Zohary. 1966. "Distribution of wild wheats and barley." *Sci.* 153:1074–1080.

Harlan, H. V., and M. L. Martini. 1936. "Problems and Results in Barley Breeding." pp. 303–346. *USDA Yearbook of Agriculture.* Government Printing Office, Washington, D.C.

Harlan, H. V. 1925. *Tests of Barley Varieties in America.* USDA Dep. Bul. 1334.

Hartmann, H. T., W. J. Flocker, and A. M. Kofranek. 1981. *Plant Science: Growth, Development and Utilization of Cultivated Plants.* Prentice-Hall, Englewood Cliffs, NJ.

Heid, W. G., Jr., and M. N. Leath. 1978. *U.S. Barley Industry.* USDA, Ec., Stat., and Coop. Ser. Agri. Ec. Rep. No. 395.

Helm, J. H., D. F. Salmon, D. H. Dyson, and W. M. Stewart. 1989. "Registration of Noble barley." *Crop Sci.* 29:235.

Helm, J. H., D. H. Dyson, and W. M. Stewart. 1986. "Registration of Samson barley." *Crop Sci.* 26:384.

Helm, J. H., D. F. Salmon, D. H. Dyson, W. M. Stewart, J. D. M. Skramstad, J. W. Mitchell, T. R. Duggan, and S. M. Hand. 1983. "Registration of Abee barley." *Crop Sci.* 32:1217.

Hendry, G. W. 1931. "The adobe brick as a historical source." *Agri. Hist.* 5:110–127.

Hensleigh, P. F., T. K. Blake, and L. E. Welty. 1992. "Natural selection in winter barley composite cross XXVI affects winter survival and associated traits." *Crop Sci.* 32:57–62.

Hillman, D., and Z. Helsel. 1978. *Cereal Grain Forages for Dairy Cattle.* Michigan St. Univ. Coop. Ext. Ser. Bul. E-1263.

Hockett, E. A., K. M. Gilbertson, G. D. Kushnak, G. R. Carlson, V. R. Stewart, and G. F. Stallknecht. 1987. "Registration of Gallatin barley." *Crop Sci.* 27:815.

Hockett, E. A., K. M. Gilbertson, H. F. Bowman, C. F. McGuire, D. E. Baldridge, and A. L. Dubbs. 1985. "Registration of Clark barley." *Crop Sci.* 25:197.

Hockett, E. A., K. M. Gilbertson, C. F. McGuire, J. W. Bergman, L. E. Weisner, and G. S. Robbins. 1985. "Registration of Lewis barley." *Crop Sci.* 25:571.

Hoffman, L., M. Ash, W. Lin, and S. Mercier. 1990. *U.S. Feed Grains: Background for 1990 Farm Legislation.* USDA-ERS Agri. Inf. Bul. 604.

Jensen, N. F., L. H. Edwards, E. L. Smith, and M. E. Sorrells. 1982. "Registration of Wintermalt barley." *Crop Sci.* 22:157.

Kaufmann, M. L., and S. Kibite. 1985. "Registration of Diamond barley." *Crop Sci.* 25:706.

Lafever, H. N. 1988. "Registration of Ray barley." *Crop Sci.* 28:187.

Mathre, D. E. 1982. *Compendium of Barley Diseases.* Am. Path. Soc., St. Paul, MN.

Matthews, R. H., and J. S. Douglass. 1978. "Nutrient content of barley, oats, and rye." *Cer. Foods World* 23:606–610.

Matz, S. A. 1991. *The Chemistry and Technology of Cereals as Food and Feed.* Van Nostrand Reinhold/AVI Pub. Co., New York, NY.

Moseman, J. G. 1991. "Use of Introduced Germplasm in the USDA-ARS National Barley Collection in Barley Cultivars." pp. 49–68. *In* H. L. Shands and L. E. Wiesner (eds.) *Use of Plant Introductions in Cultivar Development Part I.* Crop Sci. Soc. Am., Madison, WI.

Murphy, C. F. 1983. "Registration of Anson barley." *Crop Sci.* 23:181.

Murphy, C. F. 1981. "Registration of Milton barley." *Crop Sci.* 21:474.

Newman, R. K., and C. W. Newman. 1991. "Barley as a Food Grain." *Cer. Foods World* 36:800–805.

Nilan, R. A., A. J. Lejeune, C. E. Muir, and S. E. Ullrich. 1983. "Registration of Advance barley." *Crop Sci.* 23:1218–1219.

Poehlman, J. M. 1985. "Adaptation and Distribution." pp. 1–18. *In* D. C. Rasmusson (ed.) *Barley.* Am. Soc. Agron., Madison, WI.

Ramage, R. T. 1989. "Registration of Seco barley." *Crop Sci.* 29:487.

Ranaweera, N. 1989. "Feed Value of Barley." pp. 113–116. *In* G. M. Wright and R. B. Wynn (eds.) *Barley: Production and Marketing.* Agron. Soc. New Zealand, Palmerston North, New Zealand.

Rasmusson, D. C., and R. W. Wilcoxson. 1983. "Registration of Robust barley." *Crop Sci.* 23:1216.

Reid, D. A. 1985. "Morphology and Anatomy of the Barley Plant." pp. 73–101. *In* D. C. Rasmusson (ed.) *Barley.* Am. Soc. Agron., Madison, WI.

Reid, D. A., and G. A. Wiebe. 1979. *Taxonomy, Botany, Classification, and World Collection.* pp. 78–104. *In Barley: Origin, Botany, Culture, Winter Hardiness, Genetics, Utilization, Pests.* USDA-SEA Agri. Handbook No. 338.

Renfrew, J. M. 1973. *Palaeoethnobotany: The Prehistoric Food Plants of the Near East and Europe.* Columbia Univ. Press, New York, NY.

Risius, M. L., H. G. Marshall, and J. H. Frank. 1988. "Registration of Pennco barley." *Crop Sci.* 28:186–187.

Sammons, D. J., and R. J. Kratochvil. 1989. "Registration of Preamble winter barley." *Crop Sci.* 29:1568.

Sauer, J. D. 1993. *Historical Geography of Crop Plants: A Selected Roster.* CRC Press, Boca Raton, FL.

Schmidt, J. W., A. F. Dreier, and S. M. Dofing. 1985. "Registration of Hitchcock barley." *Crop Sci.* 25:1123.

Schmidt, J. W., A. F. Dreier, and S. M. Dofing. 1983. "Registration of Dundy barley." *Crop Sci.* 23:1217.

Shellenberger, J. A. 1980. "Milling Technology." pp. 227–270. *In* Y. Pomeranz (ed.) *Advances in Cereal Science and Technology.* Vol. 3. Am. Assoc. of Cer. Chem., Inc., St. Paul, MN.

Shroyer, J. P., R. E. Lamond, E. B. Nelson, W. B. Willis, H. L. Brooks, M. E. Mikesell, and J. K. Brotemarkle. 1986. *Winter Barley in Kansas.* Kansas St. Univ. Coop. Ext. Ser. C-677.

Smart, J. G. 1983. "Malting Quality-What is it?" pp. 109–112. *In* G. M. Wright and R. B. Wynn (eds.) *Barley: Production and Marketing.* Agron. Sco. New Zealand, Palmerston North, New Zealand.

Sorrells, M. E., and N. F. Jensen. 1989. "Registration of Willis winter barley." *Crop Sci.* 29:1086.

Starling, T. M., C. W. Roane, and H. M. Camper Jr. 1987. "Registration of Wysor barley." *Crop Sci.* 27:1306–1307.

Starling, T. M., C. W. Roane, and H. M. Camper Jr. 1984. "Registration of Sussex barley." *Crop Sci.* 24:617.

Therrien, M. C., R. B. Irvine, K. W. Campbell, and R. I. Wolfe. 1988. "Registration of Virden barley." *Crop Sci.* 28:374.

Therrien, M. C., R. B. Irvine, K. W. Campbell, and R. I. Wolfe. 1985. "Registration of Heartland barley." *Crop Sci.* 25:1124.

Ullrich, S. E., C. E. Muir, and R. A. Nilan. 1986. "Registration of Cougbar barley." *Crop Sci.* 26:1079.

Ullrich, S. E., C. E. Muir, and R. A. Nilan. 1986. "Registration of Showin barley." *Crop Sci.* 26:1079–1080.

Ullrich, S. E., R. A. Nilan, A. J. LeJeune, and C. E. Muir. 1985. "Registration of Andre barley." *Crop Sci.* 25:1123–1124.

United States Department of Agriculture-Federal Grain Inspection Service. 1988. *U.S. Standards for Grains.* U.S. Government Printing Office, Washington, D.C.

United States Department of Agriculture-Statistical Reporting Service. 1984. *Usual Planting and Harvest Dates for U.S. Field Crops.* USDA Agri. Handbook No. 628.

United States Department of Agriculture. 1936, 1942, 1952, 1961, 1972, 1982 and 1992. *Agricultural Statistics.* U.S. Government Print Office, Washington, D.C.

Weaver, J. C. 1950. *American Barley Production.* Burgess Pub. Co., Minneapolis, MN.

Weaver, J. C. 1943. "Barley in the United States: A historical sketch." *Geograph. Rev.* 33:56–73.

Wendorf, F., R. Schild, N. El Hadidi, A. E. Close, M. Kobusiewicz, H. Wieckowska, B. Issawi, and H. Haas. 1979. "Use of barley in the Egyptian Late Paleolithic." *Sci.* 205:1341–1347.

Wesenberg, D. M., and G. S. Robbins. 1988. "Registration of Lamont barley." *Crop Sci.* 28:373.

Wesenberg, D. M., J. C. Whitmore, G. S. Robbins, and B. L. Jones. 1988. "Registration of Russell barley." *Crop Sci.* 28:574.

Wesenberg, D. M., G. S. Robbins, and W. C. Burger. 1985. "Registration of Karla barley." *Crop Sci.* 25:570.

Wesenberg, D. M., R. M. Hayes, N. N. Standridge, W. C. Burger, and E. D. Goplin. 1980. "Registration of Kimberly barley." *Crop Sci.* 20:413.

Wiebe, G. A. 1979. "Introduction of Barley into the New World." pp. 2–9. *In Barley: Origin, Botany, Culture, Winter Hardiness, Genetics, Utilization, Pests.* USDA-SEA Agri. Handbook No. 338.

Wolfe, R. I. 1981. "Registration of Bedford barley." *Crop Sci.* 21:143.

Wolfe, R. I. 1981. "Registration of Johnston barley." *Crop Sci.* 21:143.

Wolfe, R. I. 1981. "Registration of Klondike barley." *Crop Sci.* 21:144.

Wolfe, R. I., K. W. Campbell, and W. H. Johnston. 1980. "Registration of Bonanza barley." *Crop Sci.* 20:822.

5

Rice
(Oryza sativa L.)

INTRODUCTION

Rice, also called paddy rice, common rice, lowland or upland rice, not including Americana wild rice, *Zizania palustris* L., is the major calorie source for a large proportion of the world's population, especially in parts of southern and east Asia and Indonesia. As an average of 1989 to 1991, rice comprised 27 percent of all cereal grains produced worldwide, second only to wheat at 30 percent and slightly more than corn, which comprised 25 percent of total cereal production. More rice may be consumed without processing, other than pearling, than any other cereal grain, and may exceed wheat in the amount of production consumed by humans in any form or product, as very little rice is used for products other than food.

ORIGIN

Rice is believed to have evolved within a 2,000-mile-long belt that stretches underneath the Himalaya Mountains, along the Ganges River plains of India, across Bangladesh and Bhutan, through northern Bruma, Thailand, Laos, and Vietnam, and into southern China (Figure 5.1). Because of the wide distribution of related species, cultivars, and intermediates throughout this belt, it is impossible to identify any specific areas where the evolution and domestication of *O. sativa* took place. The hot and humid environment and paucity of archaeological and paleo-ethnobotanical searches, compared with other parts of the world, also have hindered more precise identification of the sites of origin and domestication. Indeed, domestication probably took place at many sites and many times within this area of

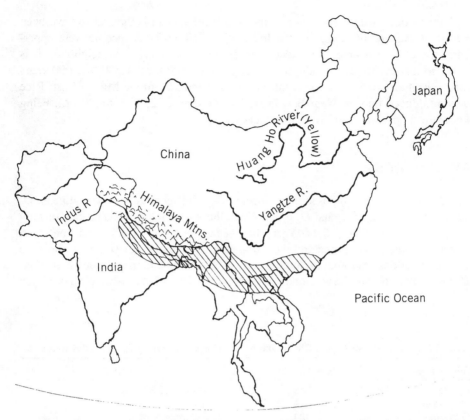

Figure 5.1 Oryza sativa *evolved from an annual ancestor within a broad belt south of the* Himalaya Mountains (hatched area) before 6000 B.C. *Rice culture spread to the outlet of the* Yangtze River by 5000 B.C.; *Indus River Valley of Pakistan by 2300* B.C.; *and Japan by 300* B.C.

Southeast Asia. One theory has been proposed that Japonica rice, a type of rice that will be discussed subsequently, evolved on the northern slopes of the Himalayas in present-day China while Indica evolved south of the Himalayas. It doesn't appear that this theory has unified support.

MOVEMENT FROM ORIGIN

Archaeological discoveries of rice remains, fossil imprints of rice glumes, or pottery imprints at sites within rice's evolutionary belt date as early as 4000 B.C., while remains from the Hemudu excavation in eastern China date to about 5000 B.C. (Figure 5.1). After being cultivated for perhaps millennia, rice culture was introduced into Japan apparently from about 1000 to 300 B.C. The culture of rice spread westward across India by 2000 B.C. and southward into Malaysia, reaching the Philippines by 1400 B.C.

Introduction of domesticated rice into modern Pakistan is thought to have been by the Harappan civilization of the Indus River Valley. These people were apparently a seafaring people that traded with both African and Asian neighbors. It is believed that the Harappan introduced rice around 2300 B.C., 1,500 to 2,000 years after wheat and barley had been introduced into India from the Middle East. Rice was introduced into the Near East in Hellenistic times, 323 B.C. to 23 B.C., being known to both Greek and Roman writers.

SYSTEMATICS

The genus *Oryza* is composed of 22 species, 20 wild species and two cultigens, *O. sativa*, the Asiatic race, and *O. glaberrima*, the African domesticated rice of only local importance (Table 5.1). There has been much controversy and confusion concerning the native home and most immediate progenitors of *O. sativa*, because feral populations are found in South America, Africa, Southern and Southeast Asia, China, Australia, and New Guinea. This New and Old World distribution of *Oryza*

TABLE 5.1   Species of *Oryza*, Chromosome Numbers, and Geographical Distributions.

Species	x = 12 / 2n =	Distribution
O. alta	48	Central and South America
O. australiensis	24	Australia
O. barthii	24	West tropical Africa
O. brachyantha	24	West tropical and central Africa
O. eichingeri	24, 28	East and central Africa
O. glaberrima	24	West tropical Africa
O. grandiglumis	48	South America
O. granulata	24	South and Southeast Asia
O. glumaepatula	24	South America and West Indies
O. latifolia	48	Central and South America, West Indies
O. longiglumis	48	New Guinea
O. longistaminata	24	Africa
O. meridionalis	24	Australia
O. meyeriana	24	Southeast Asia, southern China
O. minuta	48	Southeast Asia, New Guinea
O. nivara	24	South and Southeast Asia, southern China, Australia
O. officinalis	24	South and Southeast Asia, southern China, New Guinea
O. punctata	48, 24	Africa
O. ridleyi	48	Southeast Asia, New Guinea
O. rufipogon	24	South and Southeast Asia, southern China, New Guinea
O. sativa	24	Asia
O. schlechteri	——	New Guinea

Source: Adapted from Chang, 1976.

has presented problems and opportunities for taxonomists, botanists, and others interested in crop evolution.

Te-Tzu Chang of the International Rice Research Institute (IRRI) in the Philippines proposed that rice, both *O. sativa* and *O. glaberrima,* evolved from a common ancestor during the early Cretaceous Period, perhaps 130 million years ago. About 200 million years ago, the super land mass called Pangaea broke apart, forming Laurasia and Gondwanaland. Laurasia split into North America and Eurasia while Gondwanaland broke into the present continents of Africa, Antarctica, Australia, and South America, and the Indian subcontinent, and various islands. Over the course of, say, 20 to 65 million years ago, the Indian subcontinent collided with the mainland Asian plate that had broken away from Laurasia, forming the Himalayan Mountain range. These events, if accurate, would explain the occurrence of rice below the Himalayas and its wide distribution across the tropics and subtropics.

Chang further supports this theory by noting that the "distribution of *Oryza* in the southern hemisphere is nearly identical to the northern boundary" of fossilized remains of seed ferns of the genus *Glossopteris* that is believed to be the progenitor of angiosperms. He postulates that Japonica rice was selected, cultivated, and domesticated in China, because Gondwanic *Glossopteris* flora have been found on the northern slopes of the Himalayas in present-day China.

One cannot argue with this explanation as the first flowering plants (i.e., angiosperms) are believed to have evolved during the Cretaceous Period. However, the same scenario could be developed for all crop plants and does little to identify early or more immediate progenitors of cultivated and domesticated rice.

There are several criteria used by taxonomists to determine botanical and taxonomic relationships among species within a genus. These criteria may include (1) similarities among morphological and physiological traits—e.g., organ size or secondary allelochemical production; (2) similarities in gene frequencies between populations of the postulated species; (3) similarities in chromosome structure such as location of centromeres, arm lengths, and existence or number of knobs; (4) hybridization and post-hybridization barriers such as F_1 sterility, F_1 inviability, or amount of chromosome pairing failure during meiosis in F_1's or later generations; (5) geographical distribution and ecological requirements; and (6) frequency of natural hybridizations and establishment of intermediate forms.

Given these considerations, it appears that the most immediate progenitor of *O. sativa,* or Asiatic rice, is *O. rufipogon,* a wild perennial. *Oryza rufipogon* is synonymous with *O. perennis, O. fatua,* and *O. perennis* subsp. *balunga.* The most immediate progenitor of *O. glaberrima,* African rice, is *O. breviligulata,* or *O. barthii* in Table 5.1.

EVOLUTION OF TYPES

The genus *Oryza* contains eight genomic groups, both diploid and tetraploid species, and both annual and perennial forms. African rice was selected and domesticated in the swampy areas of the upper Niger River, about 1500 B.C., and perhaps again near the coast of Guinea to the west at about 1000 B.C., approximately 4,000 years after the adoption of grain sorghum in that area. This species has been and continues to be limited to those areas, and therefore very little variation in biotype exists for *O. glaberrima*.

This is not the case for Asian rice, *O. sativa*. As one would expect from the vast and diverse area within which it evolved, and the diverse climates into which it apparently was taken by migratory man, *O. sativa* is composed of an array of biotypes defined by geographic isolation, adaptability to submerged or dryland growth, adaptability to water depth, and resulting from cultural traditions.

Ecotypes

Scientists reported in the late 1920s that hybridization of rice cultivars from China, Japan, Korea, India, Sri Lanka, and Indonesia resulted in morphological differences and sterility in F_1 progeny. This and subsequent studies led to the classification of cultivars of tropical and subtropical origin in China as Indica types (apparently referring to India as its origin), and those of Chinese, Japanese, and Korean origin were classified as Japonica types (apparently referring to Japan). These distinctive types have been proposed as subspecies but have never been given that status by most taxonomists.

It turns out that the Chinese used the terms "hsien" (non-sticky) to refer to Indica types and "keng" (sticky) to refer to Japonica types as early as the third century A.D. This brings us back to the proposal of Chang, that Japonica, also erroneously referred to as temperate (climate) rice, evolved from a common ancestor with Indica, 20 to 65 million years ago, and was first gathered, cultivated, and domesticated along the northern slopes of the Himalayas, while Indica, also occasionally and erroneously referred to as tropical rice, evolved, and was gathered, cultivated, and domesticated below the Himalayas. Although this is an intriguing line of thought, one must remember that the same end could have been achieved by Indica rice being carried by migratory people from southeast Asia into China where it would have been exposed to different evolutionary forces that could have brought about natural and human-assisted selection of the Japonica biotype. It is also known today that Japonica types are distributed into parts of the tropics. A third possible hypothesis is that rice evolved somewhere in the Himalayan Mountains and was then distributed by humans both north into China and south into India and southeast Asia.

A third ecotype, Javanica, was suggested in 1952 to describe tropical cultivars having large grains similar to the Japonica ecotype. Further studies have failed to garner unified support of Javanica as an ecotype although some data clearly establish differences. The most convincing data supporting Javanica as a distinct ecotype come from studies showing that crossing Javanica with some Indica and Japonica genotypes results in partially sterile hybrids. Other authorities suggest that Javanica-type rice should be considered a subgroup of Japonica.

The most accepted or common characteristics used to classify rice as being Indica or Japonica are grain length and lodging potential. Indica rice generally has

TABLE 5.2 Characteristics of Indica, Japonica, and the Proposed Javanica Ecotypes.

Character	India	Javanica (Bula)	Japonica
Grain shape	Long	Large	Short
Lemma and palea pubescence	Sparse, short	Long	Dense, long
Foliage	Light green	Light green	Dark green
Tillering propensity	Profuse	Low	Moderate
Plant height	Tall	Tall	Short
Low temperature	Sensitive	Tolerant	Tolerant
Drought	Tolerant	Variable	Sensitive
Grain shattering	Easy	Low	Low
Grain Amylose	23–31%	20–25%	10–24%

Sources: Adapted from Takahashi, 1984; Holder and Grant, 1979; Chang et al., 1990.

long grains and tall, weak stems that will tend to lodge, especially under heavy nitrogen fertilization. Japonica, on the other hand, has short grains and short, stiff stalks that resist lodging. These criteria are not absolute as there is overlap in the ranges of both grain length and plant height. Other criteria used to differentiate Indica and Japonica (and Javanica) are listed in Table 5.2.

Subecotypes

Some authorities recognize several biotypes within the Indica and Japonica ecotypes, which this author prefers to call subecotypes for continuity. Other authors have referred to these biotypes as ecotypes within subspecies, or classifications based on "hydro-edaphic-cultural-seasonal regimes." Indica has the most subecotypes, while Japonica has only four, including two that are sometimes placed in the proposed Javanica ecotype.

Not only is there controversy as to the geographical origin of rice, differences arise as to whether rice was first cultivated/domesticated as a crop requiring flooded conditions or as a dryland crop. Both subecotypes occur in both Indica and Japonica. Some authorities suggest that rice evolved as a dryland plant on the slopes of the Himalayas. However, most authorities opt for an Indian or tropical origin, suggesting that rice began as a plant adapted to low-lying shallow water habitats. As this swamp plant was transported by migratory people and planted in different environments, biotypes adapted to non-flooded (i.e., dryland) culture and deep-water culture were selected. Subecotypes are dryland or dry paddy, lowland or wet paddy, and deep-water or floating.

True dryland rice, also called upland, although this term is confused with elevation such that rice grown "up land" from "low land" may require flooded conditions, is direct-seeded into a dry, prepared seedbed and grows to maturity without being subjected to natural or artificial, prolonged, planned flooded conditions. Dryland rice is grown on unleveed, flat or sloping fields and is totally dependent on rainfall for moisture, just as one would grow corn or wheat. This production method is generally restricted to tropical areas receiving at least seven inches of rain per month, especially during early and mid-season. Although it is estimated that about one-sixth of the world's rice acreage is devoted to dryland rice production, its culture and the development of improved cultivars for dryland production have received less attention than flooded rice. However, the IRRI at Los Baños, Philippines, has had an ongoing effort in dryland rice research since 1962, and has developed and released several improved cultivars. Dryland rice culture also is referred to as shifting paddy, indicating a slash/burn/farm/abandon system, or as dryland paddy.

Lowland or wet paddy rice is rice as we know it in the United States. It is grown in near-level fields divided by levees into paddies, often referred to as cuts or sections in the United States rather than paddies. In the United States, lowland rice is grown under artificial (i.e., irrigated) flood that may be continuous from seeding until drained for maturation and harvest. However, in most of the tropics, lowland or wet paddy production is entirely dependant on seasonal rainfall. About 80 percent of the world's wet paddy rice acreage is grown without irrigation. Most of Southeast Asia receives over 75 inches of rain per year, but variability in amount and distribution renders rainfed, wet paddy rice production a somewhat tenuous operation. In the great delta and basin areas of Southeast Asia that are associated with large river systems such as the Mekong, Irrawaddy, Chao Phraya, Brahmaputra, Ganges, and Indus, drainage becomes more of a problem than flood water. Desirable water depth for wet paddy rice production in Southeast Asia varies from about two inches to four or more, with about four inches being more common.

The deep-water or floating subecotype, found only in the Indica ecotype, is a most fascinating botanical and agronomic plant character. Although not grown in the United States, this subecotype provides food for subsistence farmers in parts of Bangladesh, India, Thailand, Burma, Vietnam, and Indonesia who live near and farm in areas subject to expected, annual flooding. Deep-water rice cultivars have strong seedling vigor, late maturity, produce grain with a 20-to-30-day dormancy, and are strongly photoperiodic. Deep-water rice plants have a singularly unique trait of rapid internode elongation as flood water deepens. Some cultivars of deep-water rice are capable of reaching 16 feet. For purposes of this text, only a few salient points need to be elaborated:

1. Culms have buoyant pith and air spaces that cause plants to tend to be erect even in moving water;
2. Deep-water cultivars are strongly vertical (i.e., even in flowing water the top three leaves tend to be above water), referred to as "kneeing";
3. Basal roots are shallow, such that plants are uprooted in turbulent water, avoiding stem breakage;
4. Uprooted plants grow normally under flooded conditions, absorbing nutrients from flood water through nodal roots, as do rooted plants;
5. The main culm may have as many as 25 nodes when subjected to deepening water;

6. As the flood subsides, plants collapse, nodal roots quickly establish in the soft mud, and nodal tillers grow that produce panicles and subsequently grain;

7. Some deep-water cultivars are relatively early, and grain is produced before flood waters subside and is harvested from boats; and

8. The dense mat of biomass remaining after final harvest is usually burned during the dry season.

Some authorities suggest other subecotypes, within the Indica ecotype, based on morphological similarities, growing season, and/or degree of fertility of their F_1 hybrids. These are aus, aman, and boro, seasonal cultivars grown in summer, autumn, and winter, respectively. Bulu, a large-grained subecotype of Japonica is found in Indonesia and may be classified as the Javanica ecotype.

Preference

The culinary preferences of a people dictate the variants of a food crop grown in any given area when variations exist within that crop. For example, certain regions of Southwest Asia and the Middle East prefer durum wheat flour for local bread as opposed to most people of the United States who prefer bread made from common or bread wheat. Within rice, the primary culinary variants are stickiness and aroma. Japonica rice softens and becomes sticky upon cooking, will become mushy if overcooked, and tends to remain soft if held for the next meal. Indica rice is more difficult to overcook, and properly cooked grains are separated easily and are not sticky. This is the preferred rice in the United States.

One of the many interesting variants found in *O. sativa* is that of aroma. All rice has some aroma, although it is hardly noticeable unless the uncooked grains are ground. However, the Basmati cultivars produced in the Punjab area along the Indus River in Pakistan and India are highly prized for their aroma. These cultivars produce grain with a distinct "mousy" odor when boiled, which is prized by some but generally considered offensive to western palates. Thailand also produces some aromatic rice.

Aromatic rice cultivars usually are low yielding but command higher prices. The Louisiana Agriculture Experiment Station (AES) released a long-grain, aromatic cultivar, Della, about 1970. The source of Della's aromatic genes was Delitus, a selection out of a 1904 introduction, Bertone, from France. Della is marketed also as Texmati, Pecan Rice, and Della Aromatic. Yields are low compared with non-aromatic cultivars, but market price has been about double that received for conventional rice. Because the U.S. consumer equates aroma in rice with contamination, and as the aroma would linger in machinery, larger U.S. mills will not accept Della or other aromatic cultivars. Small, consumer-sized packages were sold to foreign markets. Della's aroma is described as being similar to "roasted popcorn or nuts."

Glutinous rice has a different starch composition than normal rice and cooks to a gelatinous mass with indistinguishable grains. Glutinous rice, also called waxy or sweet rice, is normally used for desserts but is considered a staple in parts of Thailand, Laos, and Kampuchea. Consumers in these areas claim that "it tastes better and sticks to the stomach," a claim supported by research. Glutinous rice absorbs only half as much water as regular rice when cooked, thereby providing greater bulk in the digestive tract; the starch of glutinous rice also is more difficult to digest. In areas of low income and deficit calorie intake, glutinous rice would produce a "full" feeling longer and with less volume consumed than normal rice. A small quantity of sweet rice is produced in the United States.

RICE AS A SUBSISTENCE CROP

It is obvious that any plant species that was cultivated and domesticated by early humans had to be plentiful, provide more calories than it required to gather or produce, and provide adequate nutrition for human growth and development, albeit only as a part of the diet. When humans began to farm, productivity of the land became an important factor. With non-legume crops, humans tilled the earth until it became unproductive and then moved on. Later, they learned to fallow, to rotate with legumes, and to add animal manure. Monocropped rice was an exception. Nutrients were supplied, albeit in small amounts, by flood water, river flooding, or rainfall. Rice is different also in that atmospheric nitrogen is fixed by blue-green algae and anaerobic bacteria associated with flooded conditions. These factors made long-term monocropped subsistence production possible in the tropical areas of India, Southeast Asia, and Indonesia, providing gatherer-hunters, then gatherer-farmers, and then farmers with a nutritious and tasty grain.

BOTANICAL DESCRIPTION

Rice belongs to the *Gramineae* family, subfamily *Oryzoideae,* tribe *Oryzeae,* genus *Oryza.* As noted earlier, *Oryza* contains 22 species, both annual and perennial. Cultivated rice, both Asiatic and African, is an annual grass with culms terminated by a loose panicle inflorescence having single, perfect-flowered spikelets. Plants may reach 16 feet (deep-water types in Southeast Asia) but most modern U.S. cultivars reach only 3 to 4 feet. Culms are round and hollow except for nodes, and leaves are narrow, joined to sheaths with collars. Leaf auricles are well defined, sickle shaped, and hairy in cultivated rice. Ligules may be acute to acuminate or may be biclefted (Figure 5.2).

MOVEMENT TO THE UNITED STATES AND CULTIVAR DEVELOPMENT

Colonists in Virginia planted rice, perhaps dryland, on an experimental basis as early as 1609, but commercialization of the crop apparently began in South Carolina in the 1680s, North Carolina about 1730, and Georgia around 1760. Rice culture was introduced into Alabama, Mississippi, and Florida after 1760, but the exact dates are not available in commonly circulated documents. Rice culture was abandoned in Virginia, tobacco being more profitable and easier and less laborious to produce. Settlers in Pennsylvania attempted to produce rice but found it unproductive and unprofitable. Rice was produced in relatively small amounts in Kentucky, Tennessee, Illinois, Missouri, and Virginia before 1840.

The colony of Carolina, originally called the Province of Carolina (Latin for Charles, therefore meaning land of Charles), was granted to Sir Robert Heath by England's King Charles I in 1629. After Heath had made no attempts to settle the territory, the king's son, Charles II, awarded the colony to eight of his court favorites, called lord proprietors, in 1663. These eight men planned to produce food stuffs in the Carolina colony to supply the sugar plantations on the isle of Barbados. These lord proprietors also hoped to produce other items for England, such as wine, silk, and olive oil. The first English settlement in present-day South Carolina was at Albemarle Point in 1670. In 1680, the settlement relocated to Oyster Point and was renamed Charles Town, later to be Charleston.

South Carolina was bought by England from the lord proprietors in 1719 and was made a royal colony of Great Britain, partly as a southern defense against Spain and France. England bought North Carolina in 1729 and established the southern part of South Carolina as the colony of Georgia in 1732.

These southern territories or colonies were well suited to agriculture with relatively hot summers, mild winters, and reasonable rainfall and distribution. Early commercial agriculture was established along the coast by necessity, because transportation depended on river and sea traffic. Charleston would become the wealthiest city in the South during the early colonial period because of its harbor and its location between the West Indies and the New England colonies.

But settlement was slow. Those who settled away from the coast were pioneers, independent and subsistence farmers. Those who would settle near the coast had a number of different problems, not the least of which was malaria because of the expanse of low, swampy, mosquito-infested wetlands. The mosquito became the vector for malaria, a disease introduced into the colonies by European settlers that would devastate the Amerindian population and keep most Anglo settlers on higher ground, away from the mosquito and the disease. By 1708, 38 years after the first English settlement in South Carolina, the Carolina territory boasted only 3,500 free white settlers, but just over 4,000 black slaves, the Negro race being more resistant to malaria.

A number of crops were grown on an experimental basis in the Carolinas, but rice soon emerged as the most suited to production along the settled, however sparsely, coastal areas. Rice was grown dryland until about 1720 when a "reserve system" of wetland production became popular. This was essentially a reservoir system where rainwater was trapped and released into fields by gravity flow and then drained by gravity flow at the lower-elevation side of the field. Where possible, water from streams or swamps was used. Apparently very few acres were suitable to this system of flooding. In some instances, water was manually added to paddies, one bucket at a time, obviously requiring a lot of manual labor.

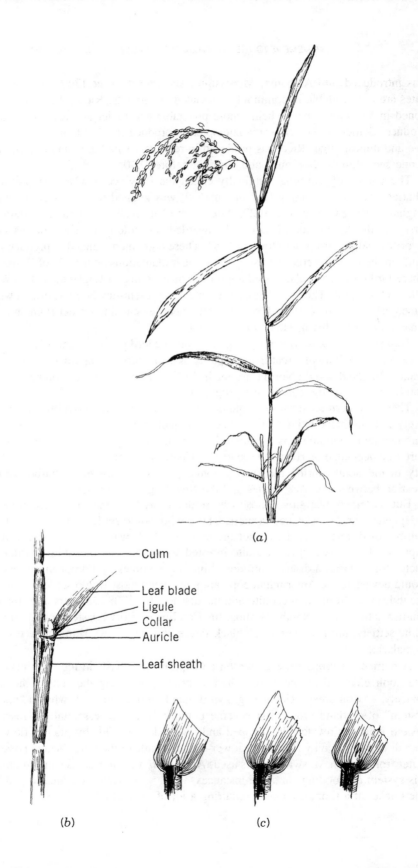

(a)

Culm

Leaf blade
Ligule
Collar
Auricle

Leaf sheath

(b)

(c)

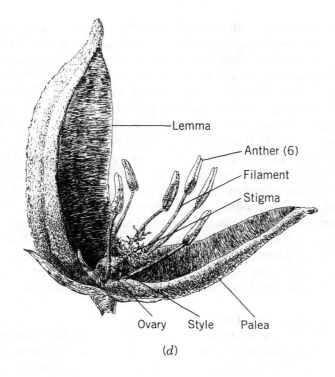

Lemma

Anther (6)

Filament

Stigma

Ovary   Style   Palea

(d)

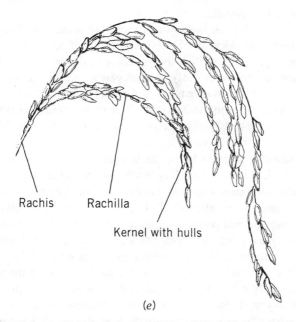

Rachis   Rachilla

Kernel with hulls

(e)

Figure 5.2 *Distinctive morphological characteristics of rice: (a) mature rice plant; (b) rice culm showing leaf blade, leaf sheath, collar, auricle, and ligule; (c) ligules may be acute, acuminate or bicleft; (d) the rice flower; and (e) mature rice panicle. (Drawings by Mike Hodnett)*

In 1750, a planter named McKewn Johnstone began experimenting with a system of flood irrigation that became known as the "tidal-flow" system by the time of its perfection in 1758. This system would become the basis of the expansion and success of the colonial and antebellum rice industry in the United States.

Although rice was "the" agricultural commodity of South Carolina and Georgia during this time period, it was a minor crop in the United States. So minor, in fact that not a single new cultivar was introduced from about 1690 until 1889. Lyman Carrier suggested that the first rice culture attempted in South Carolina was by Sir Nathaniel Johnson, the Provincial Governor in 1688. Others cast doubt on the accuracy of the year, suggesting that rice was well established in Carolina before 1690. Local farmers petitioned the lord proprietors that very year to allow them to pay their land rent in commodities, one of which was rice. This petition was granted in 1695. Several authorities agree that a significant event occurred in either 1685 or 1694, that being the introduction of a rice cultivar from which Carolina White and Carolina Gold cultivars were selected. These were the only rice cultivars produced in any significant quantity for the ensuing 200 years. All seem to agree that this introduction was the significant event, but they disagree on the particulars of the event.

One account has a John Thurber, captain of an English brigantine, docking at Charleston Harbor in or before the year 1686. During this fortuitous occurrence, there being no mention of why the ship was docked at Charleston, Dr. Henry Woodward obtained "about a peck" of rice seed that had been placed on board the ship in Madagascar. Dr. Woodward planted this seed, found it to be productive, and distributed seed to his friends.

The second account has an English or Dutch ship homeward bound from Madagascar being forced to dock at Charleston because of rough seas in 1694. The captain took the opportunity to visit the governor of the province, Thomas Smith, whom the captain had met previously in Madagascar. Smith learned that the ship's stores contained rice from Madagascar, and since he had a "low, moist patch of ground" similar to the sites in Madagascar where he had observed rice growing, he petitioned and the captain obliged by giving Smith a "small bag" of this Madagascar rice.

The first account appears to be the most accurate, but regardless of how this Madagascar, identified later as Carolina White, rice found its way to South Carolina, it set that state on a course toward a stable and booming economy. Charleston, because of rice, and indigo after 1741, became the wealthiest city in the South during much of the colonial period in America. As a footnote of history, the South Carolina Assembly awarded John Thurber a memorial of 100 English pounds in 1715 "for bringing the first Madagascar rice to this province."

Carolina White was well adapted to the coastal lowlands of the South Atlantic states and was an early-maturing, long-grain rice with good table quality, apparently of the Indica ecotype. From Carolina White, Carolina Gold was later selected, although it may have been a separate introduction shortly after the introduction of Carolina White. Carolina Gold was also a long-grain rice with good table quality that matured a few days later than Carolina White (Table 5.3). This possibly acci-

TABLE 5.3   Several U.S. Cultivars of Rice from Direct Introductions or Selections within Introductions, Their Origin, and Characteristics, 1609–1932.

Cultivar	Origin	Date Introduced	Characteristics	Probable Ecotype
Carolina White	Madagascar	1685	Early maturing; long grain, stiff straw; good table quality.	Indica
Carolina Gold	Madagascar	1685	Later maturing than Carolina White, long grain; stiff straw; good table quality	Indica
Honduras	Honduras	1890	Early maturing; long grain; stiff straw.	Indica
Kiushu	Japan	1899	Short grain; good field and mill yield.	Japonica
Shinriki	Japan	1902	Late maturing; short grain, stiff straw; good quality.	Japonica
Wataribune	Japan	1908	Late maturing; short grain.	Japonica
Omachi	Japan	1910	Similar to Wataribune	Japonica
Early Wataribune	Japan	1913	Similar to Wataribune but matures 10 days earlier.	Japonica
Onsen	Japan	1918	Early maturing; short grain.	Japonica
Blue Rose	Japan (?)	1907	Late maturing; medium grain; poor table quality.	Japonica
Early Prolific	unknown	1915	Early maturing; medium grain; poor table quality.	?
Caloro	Japan	1921	Mid-season maturity; short grain.	Japonica
Colusa	Italy	1917	Early maturing; short grain.	Japonica
Lady Wright	?	1920	Early maturing; long grain; stiff straw; poor table quality.	Indica
Rexoro	Philippines	1926	Late maturing; long grain; good table quality.	Indica
Fortuna	Taiwan	1918	Late maturing; long grain; shatters slightly.	Indica
Edith	?	1916	Early maturing; long grain; good table quality.	Indica
Nira	Philippines	1932	Late maturing; long grain; good yield.	Indica
Shoemed	Philippines	1932	Early maturing; medium grain; shatters somewhat.	Indica

dental occurrence provided perhaps the longest-lived cultivars of any row crop in colonial and U.S. history. Carolina White and Carolina Gold were the vastly predominate cultivars of rice grown in the United States from the 1690s until about 1890, an amazing 200 years.

About 1890, the cultivar Honduras was introduced by a commercial seedsman whose name is lost in history. This cultivar was well adapted to the deltas of Louisiana, Mississippi, and Arkansas. Honduras was an early-maturing, long-grain cultivar, suggesting that it too was of Indica origin. The United States Department of Agriculture (USDA), through the efforts of S. A. Knapp, began to introduce foreign accessions with the introduction of Kiushu in 1899 and Shinriki in 1902. Both are short-grain rice, probably of the Japonica ecotype. Four other short-grain cultivars of Japanese origin were introduced from 1908 until 1918 by U.S. farmers: (1) Wataribune by S. Sabaira of Webster, Texas; (2) Omachi by an unknown rice producer from Crowley, Louisiana; (3) Early Wataribune in 1913 by W. K. Brown of Butte City, California; and (4) Onsen by an unknown farmer near Briggs, California. These four Japanese introductions probably were of the Japonica ecotype.

The first commercial rice seedsman in the United States was S. L. Wright of Crowley, Louisiana, who selected and marketed several cultivars from 1912 until his death about 1930. Wright developed a cultivar called Blue Rose that was a selection from an unknown Japanese cultivar growing in 1907 near Jennings, Louisiana. Blue Rose became his most productive release, and by 1934, was planted to 50 percent of U.S. rice acreage. The first cultivar as the result of planned hybridization, Calady, was produced by J. W. Jones and L. L. Davis of the USDA at Briggs, California. Calady was the first medium-length-grain cultivar in U.S. history.

Because of the reliance on foreign germplasm, the lack of widespread knowledge of ecotype distinctions until well into the 20th century, and a strong commitment to scientific principles of plant breeding (i.e., hybridization and selection), the ecotype distinction has little meaning in the United States today. As will be noted subsequently, U.S. rice is classed by length of grain and not by Old World ecotype. In fact, medium-grain rice as we know it is a U.S.-developed class. However, even though our marketing system doesn't take note of it, California rice was distinctly short-grain Japonica ecotype as late as the 1980s, Caloro and Colusa (Table 5.3) being the then-leading cultivars until 1982 when producers began to produce Tebonnet, L20, and L202, all long-grain cultivars. This was the case because the Japonica ecotype has strong, hardy seedlings that can emerge through several inches of cold water, which is typical of California's spring irrigation water. Japonica plants also can tolerate the extremes in temperature prevailing in the Sacramento Valley where most California rice is grown.

However, California lost export markets for its short- and medium-grain rice in 1981, causing a buildup of stocks and a concomitant lowering of market prices. The rising value of long-grain rice relative to medium-grain rice and the development and release of new, adapted, long-grain cultivars further encouraged California to switch from their short- and medium-grain production to long-grain production.

Breeding objectives in the United States are (1) higher yield potential; (2) early maturity; (3) quality; (4) semi-dwarfism; and (5) biotic and abiotic stress resistance. Higher yield potential has been possible over the past 20 years because of the introduction of germplasm carrying the semi-dwarfing gene, sd_1. Cultivars carrying sd_1 will respond with higher grain yield, as much as 25 percent more than tall cultivars, to added inputs of nitrogen without lodging. The sd_1 gene was introduced into most recent U.S. cultivars from three Indica ecotype cultivars, IR8 from the Philippines, and Taichung Native 1 (TN1), and Deo-Geo-Woo-Gen (DGWG) from Taiwan. The worldwide distribution and use of the sd_1 gene from these three cultivars gave impetus to the so called "green revolution" in rice. Additional yield potential may be possible through the use of a semi-dwarf cultivar from China designated Gui-Chou and a Taiwanese cultivar, Tainan-iku 487.

Early maturity is a common goal of all rice development programs in the United States. Benefits of early maturity are obvious and include such things as reduced water requirement, shorter time frame for protection from predation, double-cropping, ratooning, a wider planting date range that allows producers to spread their harvest, thereby making better economic use of machinery and other facilities.

Quality is an ambiguous term when applied to rice, or almost any agricultural commodity, meaning different things to people in different marketing avenues of that commodity. Quality in rice likewise depends on the ultimate consumer, but usually means cooking or table quality to the final consumer of whole-grain rice. Even this varies by consumer, some preferring rice that will cook moist and sticky while others prefer a rice that will cook dry and fluffy. Most U.S. consumers prefer the latter.

U.S. rice has long had the reputation of being a high-quality product, a fact that keeps U.S. rice competitive in world markets. We produce three primary types of rice based on grain length: long, short, and medium. Typical milled, long-grain rice in the United States is 0.26 to 0.28 inches (6.7 to 7.0 mm) in length while medium- and short-grain, milled kernels are 0.22 to 0.23 inches (5.5 to 5.8 mm) and 0.20 to 0.21 inches (5.2 to 5.4 mm), respectively (Table 5.4). The United States also produces very small amounts of specialty rices such as waxy and aromatic.

The consistency of cooked rice depends on several factors, some not well understood, but paramount among these factors in amylose and amylopectin content. Amylose, the straight-chained starch fraction, and amylopectin, the branched-chain fraction, make up the vast majority of starch molecules in rice endosperm. Based on amylose content, worldwide rice cultivars are grouped into four categories: (1) glutinous, or waxy, 1 to 2 percent; (2) low, 8 to 20 percent; (3) intermediate, 21 to 25 percent, and (4) high, 25+ percent. Long-grain rice produced in the United States will have 23 to 27 percent amylose, while medium- and short-grain types will have 15 to 20 percent and 18 to 20 percent, respectively. Water absorption and grain expansion are positively correlated with amylose content. Amylopectin, on the other hand, is positively related to tenderness and stickiness, and, of course, nega-

TABLE 5.4 Typical Rice Kernel Lengths by Class as Recognized by the USDA and the United Nations, Food and Agriculture Organization.

	Typical U.S. Rice			FAO Classification
Grain	Rough	Brown	Milled	Milled
	-- (mm) --			
Extra long-grain	–	–	–	>7.0
Long-grain	8.9–9.6	7.0–7.5	6.7–7.0	6.0–7.0
Medium-grain	7.9–8.2	5.9–6.1	5.5–5.8	5.0–5.9
Short-grain	7.4–7.5	5.4–5.5	5.2–5.4	<5.0

Source: Adapted from Adair, 1955 and Matz, 1991.

Note: One inch = 25.4 mm.

tively correlated with amylose level. So the endosperm starch of milled, U.S. long-grain rice will be about 23 to 26 percent amylose, 74 to 77 percent amylopectin, therefore absorbing a relatively higher amount of water during cooking. The cooked product will have dry, flaky grains that will tend to remain separate and will tend to become hard upon cooling. Medium- and short-grain rice will tend to be moist and sticky when cooked, will remain soft upon cooling, and may split and disintegrate if overcooked. There are some rice cultivars that are regarded as intermediate in amylose content that have been developed for tropical production that will have the dry fluffiness of high-amylose types when cooked, but will remain soft when cooled.

It has been estimated that at least 100,000 cultivars of *Oryza* had existed world-wide before a concentrated effort began by IRRI in 1962 to collect and maintain current and obsolete cultivars, and breeding lines or strains. By 1983, IRRI had acquired 68,000 *O. sativa* and 2,600 *O. glaberrima* cultivars and breeding lines. A duplicate set of these accessions is held at Fort Collins, Colorado, by the USDA. To illustrate the paucity of research and to underscore the relative position of rice as a U.S. crop commodity, all public cultivars grown in the United States in 1990 could be traced to only 23 plant introductions.

Rice is attacked by a number of fungi, bacterial leaf blight, and several insects. Resistance to many of these pests has been identified in several plant introductions by U.S. scientists and/or reported by scientists in other countries. Breeders continually try to identify useful genes for host plant resistance and incorporate them into commercially acceptable genotypes.

Tolerance to early-season cool temperatures and cool irrigation water has been an objective of scientists working to develop cultivars for California. Useful germplasm has been identified in accessions adapted to high altitudes in Europe and Korea. In the southern-rice producing states, tolerance to cool temperatures during maturation has been desirable. Early maturity provides an escape from such stress by maturing before the onset of cool temperatures.

HISTORICAL EVENTS

Husbandry Advances in the Orient

In most other crops discussed in this text, some of the historical events that directly affected the production or marketing of that crop in the United States are noted. Many of these are monumental events, such as wars and economic depressions. With rice, however, there is the opportunity to document some of the more minor events or inventions in the evolution of rice culture, since the Chinese kept written records for centuries before the birth of Christ. A few of these inventions or events as cited by Chang (1985) are as follows: (1) domestication of the water buffalo dates to about 1500 B.C.; (2) the hoe has been used since 1122 B.C.; (3) flood control of wetland rice was begun between 600 and 500 B.C.; (4) by 400 B.C., the iron plow, spade, and scythe were used; (5) deep plowing and mid-season cultivation was practiced by the time of Christ; (6) transplanting appeared about 155 A.D.; (7) foot-pedalled pumps to move water were used by 900 A.D.; and (8) spike-tooth harrows for cultivation and rollers to compact and firm seedbeds have been used since about 1000 A.D.

Before 155 A.D., farmers began to germinate seeds in nursery beds and transplant seedlings to their paddies. This establishment technique led to enhanced productivity by allowing for better weed control, more efficient use of limited water, and more intense land management for rice and other crops. Chang contends that the development of the spike-tooth harrow greatly facilitated transplanting because of its impact in tilling puddled soils.

Wetland Production 1720 to 1750

Rice production through 1720, as briefly noted earlier, was dryland, called "providential production" by some, and depended on natural rainfall. From 1720 through the early 1750s, wetland production was practiced by those fortunate enough to have a plot of land suitable for gravity-flow irrigation. Surely this could not have involved many producers or many acres of rice. Although yield-per-acre data are not available, rice exports from the United States rose from about 9,000,000 lbs. in 1721 to 39,000,000 lbs. in 1751, apparently reflecting increased acreage of rice, although wetland production and improved cultural techniques could have played a role. Vastly expanded acreage is indicated by the number of imported African slaves during this time period. In 1710, about 130 slaves a year were being brought into the Carolinas. The yearly average increased to 600 by 1720 and had reached 1,000 a year by 1725.

Johnstone Tidal-Flow

McKewn Johnstone began experimenting with irrigating rice growing on land adjacent to rivers and swamps whose water level was affected by the rising and falling tides of the Atlantic Ocean. By devising a system of levees and gates, he was able to flood his rice fields with fresh water at high tide and drain them at low tide. The essential elements of this system were as follows:

1. A high embankment was constructed along the river or stream to prohibit water from entering the field at high tide.
2. The field to be flooded was then encircled by smaller embankments of soil, called levees.
3. The leveed fields had to be nearly level; in some cases the fields could be subdivided into a series of subfields that were near level or high spots that could be leveed and excluded from planting.
4. Flood gates were installed then in the outer embankment, consisting of two doors constructed such that when the outer door was raised, the water pressure from the rising tide pushed water through the inner door. As the tide

receded, the inner gate would close under outward water flow pressure such that no water would be lost from the field.

5. To drain the field, the inner gate needed only to be held open during low tide.

But tidewater land was scarce, limited to a strip of land extending only 10 to 20 miles inland from the Atlantic Coast and along rivers and swamps. South Carolina and Georgia had the most land suitable for lowland rice, and by 1850 accounted for 90 percent of the rice produced in the United States (Table 5.5). There was another, more insidious problem with rice production. Just as with cotton and sugarcane, those who would be entrepreneurial and aggressive in attaining wealth in southern agriculture did so at the expense of slavery. This unbelievable and dispiteous way of life and fortune allowed men to own and operate large plantations in the tidewater country. The average size of rice plantations in Georgia in 1860 was 885 acres, while the average farm size for the whole state was about half that, 430 acres. In 1850, there were 551 rice farms in Georgia, each averaging 226 slaves, while the state boasted 74,031 cotton plantations, averaging 24 slaves per farm. Sugar plantations "required" only 55 slaves per farm. While cotton would become king for the masses in the Southeast, for those with substantial acres suitable to tidewater rice production, cotton was no equal. Tremendous power, wealth, and cultural affluence were to be found in the rice growing areas of South Carolina and Georgia.

Rice culture in the Carolinas and Georgia was a labor-intensive operation, even before the perfection of the tidewater system of flooding. Everything was accomplished by hand labor, from planting to cultivation to milling. Workers frequently worked in standing water, surrounded by mosquitoes, snakes, and possibly alligators. Work was compartmentalized and accomplished by a "gang" whose number depended on the size of the field or task. For example, fields were subdivided into sections that a gang could hoe or plant in one week. There was always something to do on rice plantations in the tidewater country. Fields to be planted were hoed, later plowed and harrowed, and seeded by hand. Seeding would consist of some slaves creating a trench, or drill, others placing seed in the drill, and still others coming behind covering the seed. Obviously, there was a critical number of people required for an efficient operation, as well as a critical number of suitable acres to be profitable. Sometimes the seed would be soaked, or even embedded in a mud ball, called "muddied," before planting to keep the seed from floating out when the field was flooded.

After planting, fields, if possible, were flooded for two to four days, drained and hoed or hand weeded. The weeding operation was repeated in two weeks, followed by flooding for three to four weeks before being drained for a final hoeing. Fields were then flooded until such time that they were drained in preparation for harvest.

Rice plants were hand-harvested with sickles, bundled, allowed to dry, flailed (see Chapter 2, Wheat), winnowed, and grain held for milling. Even when the rice was under flood, there was plenty of work. Embankments and flood gates had to be continually maintained. Corrosion, alligators, crayfish, muskrats, and snakes were capable of destroying gates and cutting levees, to say nothing about the endangerment to workers. Crews tending levees and gates were called "trunkminders," while

TABLE 5.5 Rough Rice Production by State in the United States, 1850–1990.

State	1850	1860	1870	1880	1890	1900	1910	1920	1930	1940	1950	1960	1970	1980	1990
									—(×1,000 cwt)—						
SC	1,599.3	1,191.0	323.1	520.8	303.4	473.6	160.6	78.8	—	—	—	—	—	—	—
GA	389.5	525.1	222.8	253.7	145.6	111.8	39.6	46.8	—	—	—	—	—	—	—
NC	54.7	75.9	20.6	56.1	58.5	78.9	12.2	—	—	—	—	—	—	—	—
LA	44.3	63.3	158.5	231.9	756.5	1,727.3	5,746.0	11,340.0	8,617.0	8,442.0	10,882.0	13,053.0	20,397.0	20,768.0	26,469.0
MS	27.2	8.1	3.8	17.2	6.8	7.4	37.8	41.8	—	—	189.0	1,350.0	2,244.0	9,226.0	14,250.0
FL	10.8	2.2	4.0	13.0	10.1	22.5	8.6	32.4	—	—	—	—	—	—	—
AL	23.1	4.9	2.2	8.1	4.0	9.3	11.2	14.0	—	—	—	—	—	—	—
AR	0.6	0.2	0.7	—	0.1	0.1	1,080.0	3,858.8	3,676.5	4,314.6	7,780.0	12,639.0	21,024.0	52,615.0	60,000.0
TX	0.9	0.3	0.6	0.6	1.1	71.9	3,932.1	4,299.3	4,396.0	7,490.2	11,568.0	12,927.0	21,015.0	24,814.0	21,180.0
CA	—	—	—	—	—	—	1.4	3,717.9	3,272.0	4,248.0	8,270.0	13,248.0	18,867.0	36,386.0	30,429.0
MO	—	—	—	—	—	—	—	—	—	—	—	139.0	207.0	2,341.0	3,760.0
US[1]	2,150.4	1,871.0	736.3	1,101.4	1,286.1	2,502.8	11,029.5	23,429.8	19,961.5	24,494.8	38,689.0	53,356.0	83,754.0	146,150.0	156,088.0

Source: Adapted from Adair et al., 1962; USDA Agricultural Statistics, 1942 and following years.

1. Totals do not include minor production in other states.

other crews called "birdminders" were responsible for keeping blackbirds, ducks, and such out of the fields.

Slaves

I have alluded already to the cruel and barbaric institution of slavery in colonial and early U.S. history. Dethloff makes a case that African slaves actually contributed in ways other than as forced labor to the American rice industry. Of Africans enslaved and imported to the United States during the 1700s, 43 percent were from regions where rice was an important crop. These slaves brought with them experience and know-how in producing rice. An interesting footnote is that in some African cultures it was the men who grew rice, while in others women did. Regardless of who taught whom, production of rice in the United States became uniquely American, with borrowed technology and business organization from Asia, Africa, and Great Britain.

One should note here that there likely were rice farms in the Southeast that were not attended to or dependent on slave labor. However, these would be in most cases very small operations because of labor intensity of production, and they probably contributed only a fraction of total production.

Pirates

Although a thorough discussion of this subject is beyond the intent of this text, pirates are an interesting footnote of agriculture history during the early shipping trade with England, the New England Colonies, and the West Indies. Privateers were legal pirates, at least to the commissioning country, and used as such by England, France, and even the United States, among other nations. In fact, George Washington was part owner of at least one privateer vessel. Privateers were used to plunder and/or sink merchant ships belonging to the enemy, and as such operated along the Atlantic coast during the on-again, off-again colonial wars among England, France, and Spain from 1640 through 1783. A number of vessels carrying Carolina rice were plundered between 1717 and 1721, prompting the colonists to appeal to England for relief. England sent a small fleet of warships commanded by a Captain Rogers, who quickly trapped the pirates on Providence Island. Conditional amnesty was offered and accepted by most, probably because many lived respectable lives in Charleston and other cities.

But the few who refused amnesty continued to control the seas just outside Charleston harbor. Governor Johnson, obviously fed up and displeased at the unwillingness of England to bring the matter to a close, outfitted a ship for battle and authorized the pursuit of the pirates. On one such outing, Johnson was apparently commanding the vessel when it fought with and captured Richard Worley, a pirate of some local fame. Unimpressed with Worley's fame, Johnson ordered a quick return to Charleston with the wounded Worley and one other surviving pirate, where they were quickly tried, convicted, and hanged before they could die of their battle wounds. One other such excursion was recounted by Dethloff, that of the capture of one Steed Bonnett and 40 of his men. They were all tried, convicted, and hanged. One would suspect that the plundering of merchant ships carrying Carolina rice from Charleston harbor dropped precipitously after these events.

Lucas's Rice Mill

As noted earlier, rice in colonial and early U.S. times was hand-harvested by sickle, bundled, and transported to a central location where it was stored until the harvest was completed and/or the grain was dry. At such time the bundles were thrashed by flailing. The resulting mass of stems, chaff, and rough rice typically was tossed into the air to allow the wind to blow away the chaff while the heavier grains fell back to earth, a process called winnowing. This operation completed, the producer was left with rough rice (i.e., rice kernels with the lemma and palea, sometimes called flower or inner glumes) still tightly adhering to the grain. Rough rice was then milled, a process of removing these flower glumes. This was accomplished in all cultures across the world by use of a mortar and pestle. In colonial America, until the mid 1700s, producers typically used a hollowed log as the mortar and a shaped wooden pestle. Milling was done during winter months when each male slave was required to pound out or mill three pecks of rough rice per day, and each female slave was assigned two pecks per day. A grinding system was occasionally employed utilizing two large blocks of wood, measuring 20 inches in diameter and 6 inches thick.

By the 1780s, two types of mills were used that utilized animal power, those being the pecker mill and the cog mill. Both mills were based on the mortar-and-pestle principle of milling rice and could turn out no more than six barrels, 3,600 pounds, per day, certainly an improvement over the "one-man-three-pecks" system. Then in 1787, Jonathan Lucas built a water-powered mill for a Mr. Brown on Peach Island Plantation, and by early 1793 he had constructed four more mills. Lucas's mill combined both grinding and mortar-pestle techniques into one mill that could be operated by only a few people.

Lucas's early models required water holding tanks to supply the kinetic energy upon its release and movement across the wheel activating the milling process. In 1791 or 1792, he had devised a means to take advantage of the tide-induced movement of water in streams and rivers in the tidewater region of South Carolina, just as McKewn Johnstone had harnessed for irrigation about 40 years before. In fact, Lucas used the same one-way gate system to allow rising water upstream, but which would not allow the water back through as the tide receded, thus storing water, which he released at will to turn his water wheel. When the water level was rising, the water was allowed to flow through the one-way gate in the coffer or dam constructed across the river or stream. This movement also activated the water wheel and therefore the mill. As the water was allowed to flow out during low tide, the water wheel was turned in the opposite direction, again activating the mill. These mills were capable of producing 100 barrels, 60,000 pounds, of cleaned rice per day and required only three people for operation.

Lucas's mill was a marvel of engineering for its day. The mill included rolling screens, mortars and pestles, and even elevators for moving the rice through the milling process. By 1820, almost all cog and pecker mills had been replaced with Lucas mills. By the 1840s, steam powered many mills, but those located on streams with good tidal flow continued to operate Lucas mills into the 1850s. There is no doubt that Jonathan Lucas was to rice what Eli Whitney was to cotton.

Other inventions would complement the Lucas water mill, inventions typical of the industrial revolution that swept the nation and Europe during the early 1800s. Improvements in the Lucas mill, animal- and then steam-powered harvesting and thrashing, and other such mechanizations of agriculture led to expanded production that led to expanded markets and improvements in the quality of farming life.

Civil War and Wheat Farmers

The outcome of the great war between the states rightly destroyed the slave system in the United States, but created a shortage of capital, facilities, and labor in the Atlantic tidewater region. The carefully planned and built systems of embankments, flood gates, and levees were most likely impaired and nonfunctional or completely destroyed through neglect during the war years. Improved rice land that was valued at up to $300/acre in 1860 could be bought for $30/acre or less in 1866. Production in 1870, five years after the cessation of hostilities, was about 74,000,000 lbs., only 34 percent of production in 1850. Nevertheless, South Carolina and Georgia continued to be the leading producers of rice, producing 70 percent of U.S. production in 1870. But production already had begun to move westward. By 1890, Louisiana was the leading producer state and, by 1930, rice was no longer produced along the Atlantic coast (Table 5.5).

Although the fortunes of war hastened the decline of rice in the tidewater region, it encouraged its production in Louisiana. While the eastern tidewater rice farmers were confronted with rice production problems because of the war, eastern Louisiana sugar cane growers were faced with similar problems. These sugar cane plantations had easily available and ample surface water for irrigation and vast expanses of nearly flat land, so producers turned to rice, a crop that had been grown previously in eastern Louisiana for local consumption but never on a large, commercial scale. Producers generally rented land once devoted to sugar cane, produced rice until annual grasses made the land no longer productive, then moved on to other land. In 1860, Louisiana produced 6,000,000 lbs. of rough rice; 16,000,000 by 1870; 23,000,000 in 1880; and 76,000,000 lbs. in 1890, over twice as much as South Carolina. Most of the Louisiana rice was produced similarly to South Carolina rice until the 1880s, when some Midwest wheat farmers settled in southwest Louisiana and brought modern production technology to rice.

In 1884 and 1885, these wheat farmers settled in the Great Southern Prairie that stretches along the Gulf of Mexico for about 140 miles east of the Louisiana and Texas border and for about 100 miles west into Texas. These innovative producers realized rather quickly that rice was well suited to the flat, open prairie with an abundance of surface water for irrigation. They set about adapting the equipment that they used to produce wheat in the Northwest Prairie to rice production. The gang plow, disc harrow, drill, broadcast seeder, and twine binder had been modified sufficiently by 1886.

These entrepreneurs bought with them the "big country" outlook. They would quickly throw up a levee around any size field that was near level and that had a dissecting stream. Fields of 40 to 80 acres were common. Size and mechanization made their cost-per-unit-of-production much lower than in the still labor-intensive tidewater area. Rice production had caught up with the 20th century. From here, rice production moved to east Texas, the Arkansas Grand Prairie and California.

Canals and Wells

After successive drought years, 1897 and 1898, in Louisiana in which many smaller streams dried up, producers living near major rivers turned to building pumping plants to provide a dependable water supply. Gradually, private enterprise installed larger pumps and dug larger canals that could service from 1,000 to 30,000 acres at a cost to the producer in 1899 of 324 pounds of rough rice per watered acre.

The success of the commercial canal system had scarcely been demonstrated before an abundant supply of underground water was discovered. Producers quickly moved from flooding their rice with surface water to "putting down" wells on their own property, making them dependent on no one for this necessary input.

U.S. PROCESSING INDUSTRY

To appreciate the processing of rice, one must have a working knowledge of the rice caryopsis or fruit (Figure 5.3). Rough rice grain, also called paddy or paddy rice, as it comes from the combine will be composed of the caryopsis, the edible portion, a short section of the rachilla (or secondary panicle branch), perhaps segments of the sterile lemma, and the overlapping and interlocking palea and lemma. The caryopsis is composed of the ovary wall, or pericarp, seed coat, nucellus tissue, aleurone layer, embryo, and endosperm. During the milling process, some of the outer layer of the endosperm will be abraded, producing rice flour. The protein content of rice flour may be as high as 20 percent, while the protein content of the entire milled kernel ranges from 6 to 10 percent. One possible explanation for this phenomenon is that the rice flour from the outer endosperm will be contaminated with aleurone cells that are high in protein. The aleurone layer in rice is from one to seven cells thick and varies in thickness by location on the kernel and by grain size, with thicker grains having more layers of cells than thin grains.

As the vast majority of rice is consumed as whole grain, albeit with the outer portion removed, the rice-processing industry is rather straightforward. The goal of the industry is to remove the germ and the outer layer of the rough rice kernel, those being the lemma, palea, pericarp, seedcoat and aleurone layer. Because the American consumer prefers unbroken kernels, great care is taken by millers to produce a high percentage of unbroken kernels. Whole, milled kernels are called "head rice" in milling parlance.

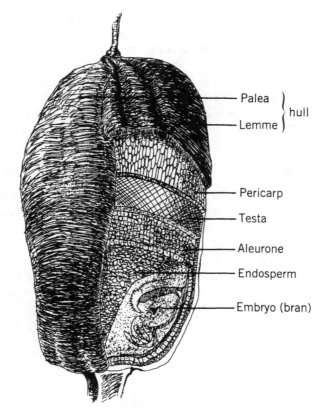

Palea ⎫
⎬ hull
Lemme ⎭

Pericarp

Testa

Aleurone

Endosperm

Embryo (bran)

Figure 5.3 *Illustration of the rough rice kernel. (Drawing by Mike Hodnett)*

Rice milling is composed of several distinct processes. Succinctly stated, these are as follows:

1. Removal of foreign matter: Rough rice is conveyed through four cleaning machines, plus a magnet to remove ferrous contaminants. The cleaning operation removes pieces of straw, soil particles, sticks, metal, and seeds of other plants. Removing seeds of other plants is probably the most critical part of the cleaning operation, because the USDA allows only one non-rice seed/18 ounces of milled rice for their highest grade, U.S. No. 1, regardless of other quality criteria. It is obvious that, since rice is consumed as whole grains, the occurrence of seeds of other plant species would detract rapidly from product acceptance.

2. Removal of hulls: Most cleaned rough rice is hulled by use of a rubber-roll sheller or huller. This machine works by passing the rough rice kernels between two rubber-covered rollers turning in opposite directions, thus having the same directional movement at the point of closest setting. One roller turns faster than the other, creating a shearing action on the rough rice kernels

as they pass between the rollers, thus breaking the lemma and palea away from the resulting brown rice. Other types of hullers have been used that employ emery-faced wheels and are based on the grindstone principle. In these types of hullers, rough rice enters through the center of a top, rotating, emery-faced surface. Centrifugal force causes the grains to be discharged to the outside of this apparatus.

The rubber roll is considered the more desirable of the two machines because it causes less kernel breakage than the abrasive disc-type huller. The desired minimum efficiency is 92 percent whole kernels of brown rice after hulling.

3. Hulls are removed by aspiration; rough rice is returned to the huller; broken, hulled grains are separated. The resulting product is clean, brown rice. Some brown rice is consumed in the United States.

4. Milling: Dry calcium carbonate is added, about ⅓ pound /100 pounds brown rice, that acts as an abrasive during removal of the bran. In some mills a small amount of water may be added. Brown rice plus calcium carbonate is conveyed into a pearling chamber where a mixing roller causes the grains to rub against each other, as well as against the chamber walls, thus abrading off the bran.

   The product of the pearling chamber is whole and broken white rice, and bran that includes embryonic tissue. The miller takes great care in pearling because most breakage occurs here. It has been estimated that a 1 percent increase in breakage during pearling could cause a $100,000 difference annually in profit margin for an average-size mill. Whole, milled kernels are referred to as "head" or "first-head" rice. Broken, milled rice is classed as "second-head," "screenings," or "brewers," depending on the amount of broken kernels and their size.

5. Some mills also polish head rice by conveying it through a rotating cylinder having strips of leather attached. Although additional bran is removed, some mills have discontinued the use of polishers because of the increase in broken grains.

After polishing, head rice is separated from broken kernels by use of screens or aspiration. Typical products from a rice mill are 70 percent head and broken kernels, 20 percent hulls, 8 percent bran, and 2 percent polish. The compositions of these constituents are shown in Table 5.6.

Rice hulls are more of a nuisance than a usable by-product of milling. The most common use is to mix them with bran to produce rice millfeed, with a typical composition of 61 percent hull, 35 percent bran, and 4 percent polish. Hulls are used also as bedding, litter, mulch, or in some cases, burned as industrial fuel. It is estimated that only two-thirds of the hulls produced are utilized, and that the disposal of the other one-third poses a continuing problem for rice millers.

Rice bran, polish, and germ are used in a number of products, including food, pharmaceuticals, soaps, fuel, and feed. However, these by-products are not used to the extent one might expect.

TABLE 5.6   Typical Composition[1] of Rice Milling Products.

Constituent	Brown Rice	Milled Rice	Polish	Bran	Hulls
			(%)		
Carbohydrates	87.2	86.0	66.8	46.6	34.0[2]
Protein	9.2	7.2	13.3	13.3	3.0
Ash	1.5	0.5	10.1	10.3	20.0
Fat	2.5	0.5	12.9	13.2	2.0
Crude fiber[2]	1.3	0.4	4.8	12.4	41.0

Source: Adapted from Luh, 1980; Matz, 1991; Hoseney, 1986.

1. Values may total more than 100 because each value is an average of more than one source.
2. Mostly undigestible.

Rice oil can be extracted from brown rice, whole or broken, and from rice bran, which includes the embryo or germ. After refining, bleaching, and deodorizing, the oil is used for cooking and frying, salad oils, and mayonnaise. Less-refined oil (i.e., without removal of saturated glycerides that cause the oil to become solid at low temperatures) is used in specialty margarine and other emulsified products. Crude rice oil may be used in glycerine, soap, antirust or anticorrosive products, and for lubricants to be used in textile or leather finishing. Rice oil contains a natural wax that is relatively easy to separate from the oil by cooling in settling tanks followed by filtration or centrifugation. This wax is used in cosmetics and has other, although limited, industrial and food uses.

Rice bran and polish are excellent sources of B vitamins and protein, but they are largely ignored by the food industry. The major problems are (1) a high fat content; (2) high level of lipase that, if not inactivated, will cause the fat to hydrolyze and oxidize, therefore becoming rancid and unpalatable; and (3) if damp they become an excellent growth media for bacteria and fungi that could produce mycotoxins, which could cause the product to be unsafe as a food or feed. Another major problem is that rice bran and polish are highly variable products, especially if purchased from different mills.

RICE PRODUCTS

As noted already, long-grain rice produced in the United States will cook dry and fluffy, with grains being easily separated. This rice is preferred for quick-cooking rice, canned rice, convenience foods containing rice, and soups. The medium- and short-grain rices that cook moist and sticky are typically used for breakfast foods, baby foods, and brewing. The vast majority of rice is consumed as whole kernels.

Rice kernels will undergo some poorly understood biochemical changes, called after-ripening, during storage as dry, sound kernels. The primary effect or expression of these changes is that the kernels will have a greater tendency to be flaky and less sticky when cooked. These changes are desirable in the United States but are not considered so in the Far East and other cultures desiring sticky rice.

During the cooking process, water must penetrate the entire kernel with sufficient heat to gelatinize the endosperm starch. This process requires about 30 minutes because undamaged rice kernels have tightly packed cells with no air spaces or other avenues for water penetration. Quick-cooking rice has been treated in such a manner as to create channels for water penetration.

Quick-Cooking Rice

Rice may be precooked to about 60 percent moisture, followed by drying to 8 percent moisture. Properly precooked rice will have a porus structure that will allow for quick penetration of water and heat. Precooked rice also may be handled in such a manner as to flatten the grain, effectively decreasing the distance water and heat must travel to complete the gelatinization process. Freezing, thawing, and then drying or puffing by rapid pressure change in precooked rice, or rapid heating of dry rice are other ways that producers of quick-cook rice create fissures for water penetration.

Parboiled Rice

Although little parboiled rice is consumed in the United States, an estimated 25 percent of world production is so treated. This treatment is ancient and probably started as a process to aid in removing the hull without breaking the kernel. Parboiling also improves the nutritional aspects of hulled rice by moving minerals and vitamins concentrated in the outer layers of the endosperm and/or in the aleurone layer into the endosperm where they are trapped during gelatinization. Paddy or rough rice is steeped in water at about 140°F, then drained and heated with steam to sterilize the rough rice and gelatinize the endosperm starch. The rice is then dried rapidly to about 20 percent moisture and then dried more slowly to a safe storage moisture in order to prevent cracking or fissuring. Parboiled rice is not quick-cooking, actually requiring a slightly longer cooking time. Parboiling improves the nutritional value of milled rice, especially vitamin B_1, makes the kernel more resistant to breakage and insects, and makes the kernels less sticky when cooked. Parboiling does, however, impart a slightly darker color to the milled kernel, changes the flavor slightly, and increases the probability of rancidness during storage.

RICE ENRICHMENT

A nutrient deficiency disease, beriberi, was once common in countries such as China and the Philippines where many people's diet consisted primarily of milled rice that contained very low levels of vitamin B_1, or thiamine. Beriberi is characterized by stiffness of the legs, paralysis, pain, and eventual breakdown of muscle tissue. Dietary changes, higher standards of living, and the availability of synthetic thiamine have reduced the occurrence of beriberi in these cultures.

Although consumers in the United States have a much more varied diet and beriberi is not likely to be encountered, milled rice is enriched with thiamine, niacin, also a B-complex vitamin, and iron. This is accomplished by coating the rice in a rinse-resistant mix of the desired additives.

U.S. AND WORLD PRODUCTION

Although rice is the second leading grain produced worldwide, it ranks fifth among grain crops produced in the United States (see Chapter 3, Grain Sorghum, Tables 3.1 and 3.2). Average annual U.S. production for 1989 to 1991 was 15,496,324,000 pounds, while corn production averaged 428,082,614,200 pounds; wheat 135,082,455,800 pounds; grain sorghum 33,005,266,600 pounds; and average barley production was 20,652,692,800 pounds. Total production doubled from 1940 to 1960, and 1990 production was almost three times that of 1960, partly resulting from the abolition of acreage control (Table 5.7). Acres planted to rice increased from 1,815,000 in 1970 to 3,312,000 in 1980 and then dropped slightly to about 2,823,000 in 1990.

TABLE 5.7 Production Statistics for Rough Rice, *Oryza sativa*, for the United States, 1900–1990.

Year	Harvested Acres	Average Yield	Total Production	Avg. Price Received	Exports[1]
	(×1,000)	(cwt./ac.)	(×1,000 cwt.)	(dollars/cwt.)	(%)
1900	361	12.20	4,407	——	9
1905	457	15.80	7,217	2.10	9
1910	666	16.70	11,129	1.51	4
1915	740	15.89	11,748	2.01	17
1920	1,299	17.91	23,242	2.62	31
1925	853	17.42	14,866	3.30	5
1930	966	20.93	20,218	1.74	23
1932	874	21.43	18,729	0.93	15
1935	817	21.73	17,753	1.60	8
1940	1,069	22.91	24,495	1.80	26
1945	1,499	20.46	30,668	3.98	37
1950	1,637	23.71	38,820	5.09	36
1955	1,826	30.61	55,902	4.81	43
1960	1,595	34.23	54,591	4.55	56
1965	1,793	42.55	76,281	4.93	57
1970	1,815	46.18	83,805	5.17	55
1973	2,170	42.74	92,765	13.80	54
1975	2,818	45.58	128,437	8.35	44
1980	3,312	44.13	146,150	12.80	63
1985	2,492	54.14	134,913	6.53	44
1990	2,823	55.29	156,088	6.70	45

Source: Adapted from USDA Agricultural Statistics, 1936 and following years.

1. Exports are percent of annual domestic production of rough rice equivalent for 1900–1935 and 1965–1990 and percent of domestic production of milled rice for 1940–1960.

Rice yields and amounts are referenced usually in the U.S. in 100 pound units, or hundred weights (cwt). Average yield/acre in the United States was below 20 hundred weights before 1930 and remained at about that level until 1955 when it reached 30.61 hundred weights/acre (Table 5.7). Yield since has increased rather dramatically, reaching 55.29 hundred weights/acre in 1990. The inclusion of the sd_1 semi-dwarf gene in more recent cultivars has allowed producers to increase nitrogen input with reduced risk of lodging and thereby increasing yields. Better management techniques, pest and stress resistance, power machinery, etc., also have contributed to higher yields.

Prices received by producers have fluctuated over the years just as with other crop commodities (Table 5.7). During this century, the lowest price received was just less than one cent/pound during the height of the Great Depression in 1932, while the record high was $13.80/hundred weights during the inflationary years of the 1970s.

Most rice is consumed in the country of production with the exception of the United States and Thailand, which were the only countries in 1990 exporting more than 2,000,000 pounds of milled rice. The United States has exported from 36 to 63 percent of its production since the 1950s (Table 5.7).

Fifty-nine percent of U.S. production in 1980 was long-grain, with 5 percent being the short-grain type (Table 5.8). By 1990, 69 percent was long-grain and less than 1 percent was short-grain. The major reasons for this shift were the loss of export markets in the early 1980s for California short- and medium-grain rice, and the introduction of improved long-grain cultivars that could tolerate the low irrigation water temperatures in California. Newer, long-grain cultivars also have been developed that outcompete medium-grain cultivars in Arkansas, Mississippi, and Missouri. Of course, prices received for any agricultural commodity are the major driving force behind any change in production practices.

Arkansas was the number-one rice-producing state in the United States in 1990, producing 38 percent of all U.S. rice (Table 5.8). California ranked second, followed by Louisiana and Texas. Arkansas produced more long-grain rice than any other state in 1990, followed by Texas, Louisiana, and Mississippi. California

TABLE 5.8   Production of Rough Rice, *Oryza sativa*, by Grain Length and by State for 1980 and 1990.

State	1980			1990		
	Long	Medium	Short	Long	Medium	Short
	---(×1000 cwt)---			---(×1000 cwt)---		
Arkansas	42,480	9,073	1,062	53,034	6,912	54
Louisiana	8,875	11,893	——	14,805	11,664	——
Mississippi	9,086	140	——	14,250	——	——
Missouri	2,100	191	50	3,713	47	——
Texas	24,310	504	——	20,690	490	——
California	——	29,606	6,780	1,314	28,215	900
U.S. Total	86,851	51,407	7,892	107,806	47,328	954

Source: Adapted from USDA Agricultural Statistics, 1982 and 1992.

**TABLE 5.9 Production of Rice, Milled Basis, in Specified Countries
Producing at Least 5,000,000 cwt., 1990–91.**

Country/Continent	Harvested Acres (×1,000)	Yield (cwt./ac.)	Total Production (×1,000 cwt.)
Cuba	331	19.55	6,459
Dominican Republic	242	25.26	6,107
United States	2,821	39.81	112,391
North and Central America	4,276	32.94	140,808
Argentina	272	24.28	5,049
Brazil	11,239	12.67	142,417
Colombia	1,074	23.21	24,912
Ecuador	373	21.24	7,937
Peru	457	28.83	13,184
Uruguay	272	29.36	7,981
Venezuela	299	19.19	5,732
South America	14,600	14.91	218,079
Italy	531	40.34	21,429
Spain	225	39.63	8,818
Europe	1,099	26.87	37,985
Soviet Union (former)	1,514	20.44	30,997
Egypt	1,074	43.56	46,782
Guinea	1,210	6.16	7,452
Ivory Coast	1,556	6.60	10,229
Madagascar	3,063	11.07	33,951
Nigeria	1,606	7.41	11,905
Tanzania	852	11.87	10,141
Africa	13,076	11.34	148,568
Afghanistan	519	13.21	6,834
Bangladesh	25,774	15.26	393,565
Burma	11,849	15.26	181,152
Cambodia	4,298	6.78	29,167
People's Republic of China	81,668	35.79	2,921,800
India	105,212	15.62	1,644,389
Indonesia	25,940	24.99	647,403
Iran	1,494	22.15	33,069
Japan	5,123	41.15	210,627
North Korea	2,174	39.18	85,186
South Korea	3,073	40.25	123,590
Laos	2,124	9.37	19,841
Malaysia	1,625	16.16	26,235
Nepal	3,335	15.44	51,367
Pakistan	5,222	13.75	71,980
Philippine	8,480	16.69	141,646
Sri Lanka	2,045	18.57	38,051
Taiwan	1,121	32.67	36,640
Thailand	21,741	11.51	250,156
Vietnam	15,141	17.23	260,451
Asia	328,359	21.87	7,180,426
Australia	217	49.27	10,714
World Total	363,142	21.42	7,767,577

Source: Adapted from USDA Agricultural Statistics, 1992.

produced the most medium-grain rice, producing just over 28,000,000 hundred weights, with Louisiana the next largest producer at about 11,664,000 hundred weights.

Just over 363,000,000 acres were devoted to rice production worldwide during 1990 and 1991 (Table 5.9). Average yield/acre of milled rice ranged from a low of only 6.16 hundred weights/acre in Guinea to 49.27 hundred weights/acre in Australia. World production was 7,767,577,000 hundred weights, or 776,757,700,000 pounds, of milled rice. Other than the Asian countries, only Brazil produces more rice than the United States.

STAGES OF GROWTH

Emergence

The radicle or primary root and coleoptile, the protective cover for the first leaf, both emerge from the imbibed seed in five to seven days after planting, assuming appropriate conditions. A second negative geotropic structure called the mesocotyl may be found at the base of the coleoptile that has elongated to push the coleoptile out of the soil. The coleoptile stops elongating when it encounters light and emergence has occurred.

Primary Leaf

The coleoptile splits at its tip, allowing the exertion of the primary leaf of the rice plant. This leaf does not have a typical leaf blade and grows to only about one inch in length. This primary leaf serves as a protective covering for the first complete leaf.

First Leaf

The first leaf emerges through the primary leaf and is composed of the three necessary structures of complete leaves, a sheath, a collar, and blade. The blade is the first part of the leaf to emerge from within the coleoptile and the primary leaf, followed by the rigid collar that is the juncture of the blade and sheath. Only the blade and collar are completely visible until internode elongation (Figure 5.4).

Second Leaf

The tip of the second leaf will be visible at first leaf stage, as growth and development are continuous from the meristematic crown located at the base of the coleoptile. The second leaf stage is established when the collar of the second leaf is visible above the collar of the first leaf.

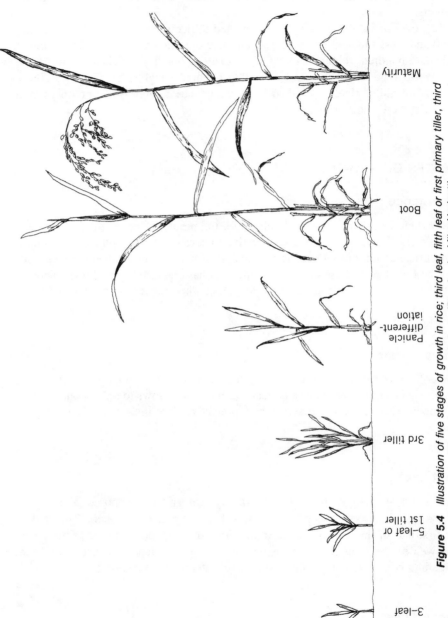

Figure 5.4 *Illustration of five stages of growth in rice; third leaf, fifth leaf or first primary tiller, third primary tiller, panicle differentiation, and early boot. (Drawing by Mike Hodnett)*

Third Leaf and Continuing

Additional leaf stages are identified in like manner as the second leaf. Leaves appear opposite and alternately along the rice culm, including tiller culms, with each subsequent leaf being wider and longer than the preceding leaf until about the ninth leaf when each decreases in size relative to the preceding leaf. Leaf stages overlap with tiller stages.

First Primary Tiller

Tillers originate from buds found inside leaf sheaths and located just above the point of attachment of the sheath to the main culm. Since a leaf blade tip is the first visible sign of a tiller, it first appears as an emerging complete leaf with two blades. The first primary tiller stage is established when the first complete leaf of the tiller is visible—i.e. sheath, collar, and blade.

Second Primary Tiller and Continuing

Subsequent primary (i.e., the main culm) tillers are identified as indicated for the first primary tiller. These will emerge, therefore, from the main culm of the plant in an opposite and alternate pattern.

Secondary Tillers

When plant density is less than about 10 plants/foot^2, secondary tillers (i.e., originating from primary tiller culms) may occur. These tillers develop and grow in the same manner as primary tillers and main stems.

Development and maturity of both primary and secondary tillers will lag behind the main stem. Later-emerging tillers may be slower-maturing if adverse growing conditions such as late-season cool temperatures are encountered after the main stem approaches maturity. More tillers are usually initiated than will mature and produce grain, with later tillers dying before maturity.

There will normally not be more than five or six complete, mature green leaves on the main culm of a rice plant at any point during its life. And, because each emerging leaf is younger than the preceding leaf, it should always be those leaves below the top five or six leaves that die back, beginning at the leaf tip and progressing toward its base.

To understand rice morphology and development, the student of agriculture must understand the structure referred to as the crown. This structure occurs at the juncture of the shoot and secondary root system that develops at the base of the coleoptile. The crown may be almost at the seed if seed placement was shallow and no mesocotyl elongation occurred, or it will be located where the mesocotyl joins the coleoptile. Inside the crown, nodes form simultaneously with each early leaf, with successively younger nodes forming above older nodes. All shoot growth (i.e., primary leaf, complete leaves, and primary tillers) originates from these nodes within the crown until internode formation begins. Older nodes and internodes within the crown will coalesce and will cease being individually discernable until internode elongation begins. Once internodes begin to lengthen, the node from which a leaf originates is easily discernable. This process of internode elongation is sometimes referred to as "jointing." Internode number, albeit a small range of differences, and internode length are cultivar-dependent. Although it is possible for roots to be produced at stem nodes above the crown, most will originate at crown nodes. Primary tillers also originate from crown nodes, although they can emerge from main stem nodes above the crown.

In most crop plants, vegetative progression is measured by the number of main stem nodes. However, in rice, it is internodes, although each must be associated with a node, that are used to identify vegetative progression. Stem formation, as it is commonly thought of, begins with internode elongation.

First Internode or First Green Ring

The first stem node and the upper most crown node become clearly separated by an internode. Because chlorophyll accumulates in tissue immediately below the first stem node, a green ring is discernable in properly dissected stems. This green ring may be observed in early stages of each additional internode. Rice plant stages are assessed using main stem development only. Internode length varies by cultivar and management, from one to ten inches for the base internode to the uppermost internode, respectively, in semi-dwarf cultivars and from two to fifteen inches for tall cultivars.

Plant age at the appearance of the first internode depends on cultivar, early-season heat unit accumulation, and cultural practices. However, very-early-maturing cultivars reach first internode about 40 days after emergence, and early-season cultivars require approximately 50 days, while 55 days may be required for mid-season cultivars in Louisiana.

Second Internode and Continuing

A second node and its supporting internode will be visible above the first stem node in properly dissected stems before the first internode has completely elongated. This process will continue and stages will be **third internode, fourth internode,** etc. Each additional node and internode will be apical to the preceding one, and internodes will become progressively longer and thinner. As noted, there will be four to six internodes on the main culm of a mature rice plant with the last stem node supporting the panicle. The leaf originating from the uppermost node is the flag leaf, recognizable by its more horizontal blade architecture relative to the vertical stem axis.

Panicle Initiation

The panicle, or inflorescence, can be detected microscopically 35 to 40 days before it is visible above the flag leaf sheath. Panicle initiation, interestingly, is associated with other growth parameters and varies with cultivar maturity differences. In very-early-season cultivars, panicle initiation occurs about the same time as first internode. In early-season cultivars, initiation occurs at second internode, while it is most closely associated with third internode in mid-season cultivars.

Panicle Differentiation

The panicle reaches one-eighth of an inch long and appears as a small tuft of hairs inside the stem, just above the crown. At panicle differentiation the rachis, rachilla, and florets are microscopically visible.

Panicle Stages

Panicle growth is identified in one-inch increments until the flag leaf is visually identifiable as described above for leaf stages. The panicle is about four inches long and still inside the stem at this time, and this is therefore referred to as the **four-inch panicle** stage.

Many stages of the rice plant development overlap in time. For example, panicle stages will overlap with internode elongation stages because the panicle progresses up the culm as each subsequent internode elongates. However, all other designated stages are completed by the boot stage in the maturing plant.

Early Boot

The developing panicle continues to be completely enclosed in the flag leaf sheath. This stage begins when the flag leaf sheath first appears above the collar of the penultimate leaf (i.e., the immediately preceding leaf) and ends when it extends two inches above.

Middle Boot

The flag leaf sheath is from two to five inches above the collar of the penultimate leaf.

Late Boot

The flag leaf sheath extends more than five inches above the collar of the penultimate leaf. The entire panicle remains fully enclosed by the flag leaf sheath.

Panicle development is completed during the boot stages of growth. Since floret development proceeded from top to bottom within the panicle, pollen development also progressed top to bottom, with production of mature pollen being maximum during early boot. As the panicle develops, its size becomes sufficient to cause a bulge in the upper stem while it is still enclosed within the flag leaf sheath and the flag leaf's penultimate leaf sheath. By the end of the boot stages, leaf development is complete and, except for the top two internodes, all internodes are fully elongated.

First Head

Rice is at first head when the tip of the panicle is visible above the flag leaf sheath on the main culm. Pollination may commence about this time.

Full Head

The entire panicle has been pushed through the flag leaf sheath by the full elongation of the topmost internode. Pollination in some cultivars and under some climatic conditions may not occur until after full head. Pollination is obvious in rice because the white or yellow anthers, or pollen sacs, are extruded when the flowers open. Although the rice flower opens and pollen is observable outside of the flowers, rice is a self-pollinating plant species with pollination occurring usually about the time of flower opening.

Grain Fill

Once ovule fertilization occurs, the process of grain filling begins. Most of the photosynthate that is stored in rice grains is produced by the top three leaves with the majority produced by the flag leaf. This assimilate is stored in rice grains, as well as in all monocot seeds, as starch in the endosperm. When the starch or endosperm within kernels is in a highly liquified state, grains are considered to be in the **milk stage** of development. As the kernels reach maximum grain fill and begin to dry, the consistency becomes similar to that of bread dough, and is referred to as the **dough stage.** Some authorities divide the dough stage into **soft dough** and **hard dough.** These terms are self-descriptive.

Maturity

The focus for identification now turns from individual grains to the average of the panicle or of the field. At **physiological maturity,** grain fill is complete and the average moisture content is in the range of 30 percent. At **harvest maturity,** average moisture of main stem grain should be between 15 and 18 percent, with the entire crop averaging about 18 to 21 percent moisture.

PRODUCTION PRACTICES

Cultivar Choice

Rice cultivars in the United States are developed to fit into one of three market classifications—i.e., long-grain, medium-grain, or short-grain. Cultivars developed for the southern states are predominately long-grain types with a few medium-grain cultivars. There is essentially no active development of short-grain cultivars for the southern states. California, on the other hand, has historically grown short- and medium-grain cultivars because much of their irrigation water comes from the Sacramento and Feather Rivers that are fed by mountain snow melt. Water temperature in early spring is rarely over 65°F and stand establishment and early seedling development are slower at these temperatures. The Japonica ecotype is much more tolerant of cool temperatures than Indica and therefore California cultivars have been predominately of the Japonica ecotype that has short- to medium-grain lengths. New, long-grain cultivars have been developed for California but they accounted for only 4 percent of California's production in 1990. Within each grain length classification, cultivars are categorized as having very-early, early, or mid-season maturity in the southern states, with California cultivars divided into very-early, early, inter-mediate, or late.

Producers should consider a number of characteristics when choosing a cultivar. These include yield potential, mill acceptance (which will be based on grain length and the turnout of whole grains after milling), seedling vigor when planting early in the spring or in California, days to maturity, and lodging potential. Cultivars should be selected and planted appropriately to ensure harvest at the most opportune time—that is, to avoid extremely high summer temperatures if possible during grain maturation and to harvest before the onset of inclement fall weather patterns.

The expected time of harvest can be spread over several days or weeks by planting cultivars from more than one maturity group. A good rule of thumb is to never plant more of a particular maturity group in a week than you can harvest in one week. Spreading planting dates also will spread harvest dates but producers should consider the disadvantages of planting too early or too late. Planting too early may result in poor emergence and poor seedling performance if low temperatures are encountered. A three-week spread in planting dates under adverse spring temperatures may result in only a few days' difference in maturity dates. Planting the same maturity group cultivars later than optimum to spread harvest can result in poor performance because of low maturation temperatures or undesirable seasonal rainfall during harvest.

Producers should look to their state's agricultural experiment station and cooperation extensive service for information on rice cultivars. Typical data are shown from Arkansas in Table 5.10.

TABLE 5.10   Cultivar Performance Data from the Arkansas Agricultural Experiment Station, 1990.

Cultivar	Grain Type	Yield Index[1]	Days to Heading	Days to Maturity	Days Flooded	Stem Strength[2]	Percent Mill Yield Head	Percent Mill Yield Total
		(%)						
Newbonnet	Long	110	95	130	85	2	59	72
Gulfmont	Long	108	89	124	79	1	52	72
Rexmont	Long	97	86	121	76	1	55	70
Millie	Long	102	82	117	72	2	60	72
Alan	Long	105	76	111	66	2	55	71
Mars	Med.	115	90	135	90	4	59	71

Source: Adapted from Univ. of Arkansas, 1990.

1. Percent of experimental standard cultivar.
2. Stem strength: 0 = very strong; 9 = very weak.

TABLE 5.11   Usual Dates of Planting and Harvesting Rice in the United States.

State	Planting	Harvest Dates Begin	Harvest Dates Most Active	Harvest Dates End
Arkansas	10 Apr.–5 June	1 Sept	15 Sept.–15 Oct.	25 Oct.
Louisiana	15 Mar.–25 May	20 July	1 Aug.–15 Sept.	5 Oct.
Mississippi	10 Apr.–31 May	25 Aug.	20 Sept.–20 Oct.	10 Nov.
Missouri	1 May–20 May	1 Oct.	5 Oct.–25 Oct.	1 Nov.
Texas	20 Mar.–15 May	20 July	10 Aug.–10 Sept.	15 Nov.
California	15 Apr.–1 June	1 Sept.	1 Oct.–5 Nov.	30 Nov.

Source: Adapted from USDA Agri. Handbook 628.

Date of Planting

With all other crops discussed in this text, the soil temperature at seeding depth is the essential criteria determining planting date. However, as rice is often flooded immediately after planting, or water seeded in California, southwest Louisiana, or southeast Texas, producers of rice are concerned with air temperature because the water can be warmed sufficiently to ensure a soil temperature more conducive to germination and seedling development. Louisiana recommends an air temperature of at least 65°F; Arkansas and California recommend that producers should wait until daily air temperatures, or water temperatures in California, reach a high of 70°F. Obviously, long-term expected weather patterns figure into the planting date equation. Usual dates of planting are shown in Table 5.11.

Seeding and Seedbed Preparation

Rice may be drilled directly into a dry seedbed, broadcast dry seeded, or broadcast into standing water. As with any crop, obtaining an optimum stand of healthy, fast growing plants is the first step toward optimizing yield. The optimum plant density for rice is 15 to 20 plants/foot^2, with about 8 being the minimum and about 30 being the maximum. This wide acceptable range is due to the tillering capacity of rice. The producer should be aware that exceptional thick stands can lead to more severe disease pressure and thin culms that may lodge easily. Excessively thin stands may lead to uneven maturity because of excessive tillering.

Producers usually plant on a pound basis, 150 pounds/acre being recommended in California where water seeding is normally practiced, 120 to 150 pounds/acre recommended for water seeding or broadcast dry seeding in Louisiana and Texas, and 90 to 110 pounds/acre when direct drilling seed in Louisiana, Arkansas, and Texas. Again, as for other row crops, producers should be cognizant that cultivars vary in the number of seeds per pound and this could influence plant density (Table 5.12). Producers in all rice-growing areas should strive to plant 40 seeds/foot^2 under the best of conditions, because as many as half of these are normally lost to seedling diseases, insects, birds, and other environmental stresses. When planting early,

TABLE 5.12   Number of Seeds per Pound and Number of Seeds per Square Foot at Different Seeding Rates for Several Cultivars.

Cultivar	No. Seeds/lb.	No. Seeds/ft² when Seeding Rate (lbs./ac.) is:								
		80	90	100	110	120	130	140	150	160
L202	17,379	32	36	40	44	48	52	56	60	64
Mars	17,719	33	37	41	45	49	53	57	61	65
Tebonnet	18,900	35	39	43	48	52	56	61	65	69
Rexmont	19,895	37	41	46	50	55	59	64	69	73
Katy	21,808	40	45	50	55	60	65	70	75	80

Source: Adapted from Univ. of Arkansas, 1990.

planting into a poorly prepared seedbed, or planting associated with some other environmental stress, seeding rates should be increased from 10 to 50 percent, depending on the severity of the existing or expected stress.

An appropriately prepared seedbed is essential not only for stand establishment, but also to provide for uniform water depth, a topic that will be discussed subsequently. When direct seeding, the seedbed should be smooth, free of vegetation, firm, and level within tolerances. The seedbeds should be disturbed only deep enough to accomplish these objectives. Seedbeds are often "rolled" with a roller device attached behind a field cultivar as the last seedbed operation before drilling. Rolling firms the soil and decreases the degree of depressions caused by tillage equipment, thus encouraging more uniform water depth across the field. Rice is then drilled within the levees, the levees having been broadcast seeded just ahead of the last pass with levee-building equipment. Producers should be aware that semi-dwarf cultivars should not be planted more than one inch deep and the old, taller cultivars not more than two inches deep.

With broadcast dry seeding, seedbed preparation may be the same as with direct drill. Seeds are broadcast by either ground equipment or airplane and then covered with a harrow or similar tillage tool. Increased seeding rates are required because some seed will be covered too deeply to emerge and some will not be covered at all, or covered very shallowly, and will desiccate.

Seedbed preparation for broadcast seeding into water is somewhat different. Here, the seedbed should be left rough, with clods or with groves. Preparation of a seedbed with clods may be obtainable only with certain soil types and in certain years. Otherwise, the field is prepared as for direct drilling, but the final operation will be to put groves in the seedbed. These groves, say two inches deep on seven-inch centers, help minimize seed drift in the water and facilitate seedling root development, or pegging, and therefore provide for rapid emergence and uniform stands. Seeding into water should be completed as soon as possible after flooding to allow for maximum oxygen concentration at the water-soil interface and to avoid slick soil surfaces that contribute to seed drift. Seeds are usually soaked, or pregerminated, before aerial seeding in California, Louisiana, and Texas. Research has shown that there is not an advantage of one seeding method (i.e., dry drilled, dry broadcast, or water broadcast) over another if adequate stands are obtained.

TABLE 5.13 Time Required to Pump One Acre-Inch of Water onto Fields of Varying Size with Varying Pump Capacity.

Pump Discharge	Field Size (ac.)				
	20	40	80	160	240
(gpm)	------------------------------ (hr.) ------------------------------				
200	45	91	(182)[2]	(364)	(540)
400	23	45	91	(182)	(276)
600	15	30	60	120	(180)
800	11	23	45	91	(132)
1000	——[1]	18	36	72	108
2000	——	——	18	36	54
3000	——	——	12	24	36
4000	——	——	——	18	27
5000	——	——	——	——	22

Source: Adapted from Univ. of Arkansas, 1990.

1. Less than 12 hours.
2. Parenthesis indicates an impractical time requirement—i.e., requires more than five days to add one acre-inch.

Water Management

An adequate water supply is a necessity in rice production. A producer must be able to (1) flush a field in four days—that is, put water on a field to add soil moisture but not to create flooded conditions; (2) flood a field to at least two inches deep in five days; (3) drain, dry, and flood in 17 days; or (4) maintain flooded conditions all season long. Producers should know the amount of water that their pumps will deliver, whether well or surface water is used. Not only is pumping capacity important, but soil type also will dictate the rice acreage that can be reasonably watered. A pump that will deliver 1,000 gallons/minute will provide enough water for 67 acres of rice on a silt loam soil, but only 40 acres on a sandy loam soil. Table 5.13 indicates the time required for each acre-inch of water supplied to varying field sizes across several pump capacities. Another rule of thumb is that watering capacity should be about 30 gallons/minute/acre.

Producers should strive to have their rice under flooded conditions as much as possible. Fields, once flooded, should only be drained to apply pesticides to control insects, diseases, or weeds, and to help control straighthead. Straighthead is a physiological disorder that is most often associated with sandy soils, or fields where a relatively large amount of crop residue was incorporated into the soil before flooding, thereby having anaerobic decomposition occurring. Arsenic-containing residue also is linked to the occurrence of straighthead.

When rice is planted into a well-prepared seedbed by the use of a grain drill, producers often flush the field to ensure uniform emergence, taking care that surface water is drained completely within three days. Herbicides and part of the seasonal nitrogen requirement, as well as other nutrients if not applied preplant, are applied before the rice plants reach six inches, usually three to four weeks after planting. Water is then added and a continuous flood is desirable until the field is ready to drain in preparation for harvest. A flood of two to four inches is desirable unless weeds become a problem. If rice plants are tall enough, the water level can be raised slowly, about one inch per day, to help suppress weed growth. This is the most common method of rice culture in Arkansas, Mississippi, and North Louisiana. This system works best when red rice is not a severe problem.

When rice is planted as dry seed (i.e., not pregerminated) into standing water, the field is drained after 24 hours and is not flooded for about four days to allow enough oxygen for the seed to germinate and "peg down." This system is referred to as "pinpoint flood." The flood is then maintained until harvest unless straighthead becomes a problem.

"Continuous flood" is where flooded conditions are maintained throughout the season. With this system, pregerminated seed are broadcast into standing water. This is the predominate system for California and is used to a much lesser degree in Louisiana and Texas. Seeds are soaked for 18 to 24 hours and allowed to drain for 24 to 48 hours before seeding. If seeding is delayed, seeds should be aerated with cool air or flushed with cold water to prevent damage from the heat buildup during the germination process. Continuous flood is the most common method of rice production in California, while pinpoint flood is preferred in most rice-producing areas of Louisiana and Texas.

The advantages of producing rice under continuous flood are (1) less management required; (2) less water use; (3) reduced seedling stress from cool air temperatures; (4) elimination of early-season bird problems; (5) reduced salt death in seedlings; (6) maximum suppression of red rice and other weedy grasses; and (7) increased nitrogen efficiency when nitrogen is applied to dry soil before flooding. With either continuous or pinpoint flood, a portion of the rice plant must be above the water by the time plants reach the four-leaf stage.

Although water can be used to suppress weeds and grasses, certain herbicides can be applied by air to flooded rice. In some instances, water depth is reduced to expose weeds and grasses. Weed control is important in all cropping endeavors, but more difficult in some than others. Johnsongrass in grain sorghum for example is a major problem because they are both in the same plant genus. The same occurs in rice. Red rice, *O. sativa,* occurs in many rice growing areas and is a major problem. Red rice can be controlled if there is sufficient height differential such that it can be submerged without submerging the rice crop. Otherwise, herbicides must be relied on. Potential effects of several weeds are shown in Table 5.14.

TABLE 5.14 Potential Yield Losses to Weeds Commonly Found in Rice Fields in the Midsouth.

Weed Species	Density (Plants/ft²)	Potential Yield Loss (%)
Barnyardgrass	5	50
Signalgrass	15	40
Red rice	2	50
Ducksalad	100	30
Hemp sesbania	1	50

Source: Adapted from Univ. of Arkansas, 1990.

Tillage

Producers have three major concerns relative to tillage in rice production: (1) destruction of previous crop residue, (2) field levelness, and (3) levee construction.

Producers should begin to prepare for the next rice crop as soon as harvest of the current crop is complete. Fields should be disked or plowed, and sometimes rolled, to incorporate residue into the soil or make sure that all residue is in contact with the soil to encourage decomposition. In some operations, residue is shredded and may be removed before tillage commences. Burning has been practiced for years as a means of controlling rice residue but this practice is dangerous and wasteful of organic matter that should be returned to the soil to maintain good tilth and structure. Early tillage operations, whether in fall or spring, are designed first to manage residue and second to encourage weed seed germination and subsequent destruction. Later tillage is to prepare the proper seedbed discussed previously.

Once residue management is complete and before final seedbed preparation, rice fields are often "floated" with a land plane to remove pot holes that have developed as a result of tillage, equipment, or soil shifts under flood. Today, laser guided planes ensure much more precision in the leveling operation. Rice fields should have no high or low areas that would result in water more shallow or deeper than desired, a condition that could result in reduced yield and complicated weed control.

After leveling, levees are constructed such that the change in elevation between levees is not more than 0.2 feet per 100 horizontal feet. California recommends a slope not more than 0.05 feet per 100 feet if the field will be monocropped with rice. Laser guidance for identifying levee position has greatly simplified levee construction. Water is added to the rice field at the highest elevation. Gates are placed in each levee so that the paddies at lower elevations can be watered by gravity flow from upper levees. Gates also serve to drain fields. Levees must be maintained during the growing season with leaks or breaks repaired as soon as possible.

Fertility

Soil tests should be taken every three years on non-problem fields and every year on fields with production problems that may be related to fertility. As with other crops, nitrogen, phosphorus, and potassium are most often limiting. However, rice grown as an aquatic plant—it can be grown without being under a continuous flood but yield is severely curtailed—presents some unusual management opportunities. Without going into all possible scenarios, it seems desirable for the student of agriculture to be aware of the effects of flooded soil conditions on applied and native primary nutrients.

Nitrogen is the most common limiting factor in rice production. Whereas nitrogen is often applied to soils for upland crops as nitrate, only reduced forms such as urea and ammonium forms of nitrogen are stable under anaerobic conditions. Under anaerobic conditions, soil bacteria will use NO_3^- as a source of oxygen for respiration, thus releasing N_2 that volatilizes into the atmosphere. There may be some nitrates in the top half-inch or so of soil, referred to as an oxidized layer, but these are subject to leaching. When ammonium nitrogen is added to flooded soils, the ammonium ion, NH_4^+, is quickly adsorbed onto cation exchange sites until an equilibrium is established between these sites and the soil solution. If nitrate is added to flooded soils, up to 70 percent will be lost to leaching or denitrification to nitrites and then to gaseous nitrogen that is lost to volatilization. Draining fields already fertilized with nitrogen and allowing the soil to dry will create oxidative conditions that will allow the conversion of ammonium to nitrite and nitrate, $NH_4^+ + O_2 \rightarrow NO_2^{-2}$ and NO_3^-, that may be lost to leaching and denitrification to gaseous nitrogen when soils are reflooded. If such draining is necessary, fields should be reflooded as quickly as possible.

In dry seeded production, up to 65 percent of the total nitrogen needs may be applied before permanent flood. Usually, one-half of this 65 percent is incorporated just ahead of planting and the other one-half is applied topically just ahead of flooding. The remainder, essentially one-third of total requirements, is applied to the flood water at mid-season—i.e., from panicle initiation to panicle differentiation. Under the pinpoint or continuous flood production systems, nitrogen may be all preplant incorporated since the permanent flood will maintain anaerobic conditions and prohibit nitrogen losses.

Nitrogen needs may vary from 70 to 200 pounds/acre. Medium-grain cultivars will require the lower amount while the newer, long-grain cultivars carrying the semi-dwarf gene will require the higher amount for maximum yields. Information on appropriate levels to apply can be obtained from state agricultural experiment stations and cooperative extension services.

Soil phosphorus is generally more available under flooded conditions because of the reduction of ferric phosphate to the ferrous state and the hydrolysis of iron and aluminum phosphates. Nevertheless, phosphorus and potassium are usually added as a part of a complete preplant incorporated fertilize program.

Zinc is the most often limiting micronutrient, especially where high pHs are encountered. Soil test recommendations should be followed for all required nutrients. Plant analysis also may prove a valuable guide for correcting in-season deficiencies.

Ratoon Rice

The probable success of a ratooned or second crop of rice continues to improve as newer cultivars require fewer days to maturity than the cultivars that they replace. However, high to moderate fall temperatures that the producer has no control over will continue to dictate success. Low temperatures and early frost can devastate a potentially excellent second crop. Among the items a producer should consider before attempting a second crop are (1) date of harvest of the first crop and therefore the remaining days of expected appropriate weather; (2) a field severely rutted from the first harvest will be low yielding and difficult to harvest; (3) rolling or clipping the remaining stubble may be advisable in some locations as regrowth will be initiated from the base of the plant and clipping therefore will not affect regrowth; (4) extra nitrogen will be required for maximum yields; (5) weeds, especially broadleafs, can be a major problem; and (6) only certain cultivars have good regrowth potential.

BIOTIC PESTS

Not only is rice attacked by the usual plethora of insects and disease pathogens, but California producers must also contend with tadpole shrimp that chew off seedling roots and coleoptiles, and uproot seedling with their constant digging. California and other areas must contend with crayfish that burrow into and weaken levees and irrigation ditches, and muskrats and beavers that dam irrigation ditches and canals. Many areas must also contend with birds, especially blackbirds. These will consume rice seed left on top of the soil and will pull emerging seedlings from the soil. The best control practices are to cover seeds with soil, flush dry-seeded fields, and maintain continuous flood.

Insects and disease pathogens attacking rice are shown in Tables 5.15 and 5.16, respectively.

ABIOTIC DISORDERS

There are a number of disorders of rice and other crop species that are not caused by insects or disease pathogens. The most common causes are herbicide or insecticide toxicities and nutrient deficiencies. Symptoms are usually not uniform across a field except in the case of nitrogen deficiency, which will be uniform or streaked as a result of non-uniform application. Soil variation such as changes in types, elevation, pH, and salinity will affect nutrient availability or toxicity.

Deficiencies of iron, zinc, manganese, boron, and copper will appear usually in the younger or upper leaves with older leaves remaining green because these elements are immobile once assimilated into plant tissue (see Chapter 1, Corn, Table 1.16 for functions of the necessary elements for plant growth). However, upper leaves can become normal if the deficiency is relieved by natural conditions such as roots growing deeper into the soil where the element is more plentiful.

Symptoms of nitrogen, potassium, phosphorus, or sulfur deficiency will appear on older leaves because these elements are more mobile—i.e., tissues and compounds in older leaves will be broken down and the elements recycled into developing plant structures. The deficiency symptoms of the most commonly lacking nutrients are shown in Table 5.17.

Boron and salt toxicity are not uncommon in rice production, especially in areas of poor water quality in terms of sodium. Salt toxicity is manifested usually by the tips of old leaves turning brown and then white, along with stunted growth and reduced tillering. In extreme cases, salt may actually accumulate on stems at the water line and on soil along the water line on levees. Boron toxicity is recognized by necrosis of leaf tips and margins of older leaves. These necrotic areas later have an elliptical appearance.

Straighthead is a physiological disorder of rice that apparently is associated with sandy soils, arsenate-containing residue, or fields with a large amount of old crop residue undergoing anaerobic decomposition. Panicles of affected plants are upright at maturity because the kernels do not fill, and panicles may not emerge from the boot. Lemmas and paleas may be reduced in size, missing, and/or distorted and discolored. Florets that are distorted have a characteristic "hook" on the distal end, a condition called "parrot beak." Control of straighthead is with resistant cultivars or draining fields at first internode and allowing the soil to dry to surface dryness before reflooding.

HARVESTING

The rice kernel is more difficult to thrash free from its panicle than are other grain crops considered in this text. Therefore, the rice combine is fitted with a spike-tooth cylinder that is more aggressive in its thrashing action than in combines used exclusively for such grains as wheat or sorghum. The rice combine also may differ from other grain combines in having wide tires for flotation to prevent rutting in moist soil and with deep lugs for traction. Such tires are usually referred to as "rice-cane specials." California producers may use combines fitted with tracks rather than rubber tires.

Timing of harvest may be more critical in rice than in many other grains. If rice remains in the field too long and dries to less than 18 percent moisture, the cyclic nature of moisture content caused by day-night relative humidity differences in the South will cause stress cracks, sometimes called checks, that will reduce head rice yields and increase the yield of broken kernels at milling. The aggressive thrashing action of the spike-tooth cylinder also cracks more grain at moisture levels below 18 percent. Rice should be harvested at 18 to 22 percent moisture or when grains on the lower part of the panicle are in the hard dough stage in southern states and the soft dough stage in California. Grain harvested at higher moisture levels may have an excess of immature, chalky kernels.

Producers must time field drainage to coincide with appropriate grain moisture for harvest without sacrificing grain fill of kernels on the lower portion of the panicles. Irrigation is discontinued several days ahead of the anticipated harvest date, but not early enough to allow the soil to dry before lower panicle maturity. Rice produced on fine sandy loams in Texas is drained ten days ahead of the targeted harvest date, while clay and clay loam soils are drained 15 days ahead of harvest. These drainage dates are approximately 25 and 20 days, respectively, after heading.

Once harvested at 18 to 22 percent moisture, rice should be dried to 12 to 13 percent moisture for safe storage and for maximum head rice yields. Harvested rice should be dried to 16 percent or lower moisture in seven days or less to prevent mold and fungi growth. Milled grain discolored by mold or fungus organisms is called "heat damaged" or "stackburn" grain and is, obviously, low quality. Rice dried too quickly will fissure, and thus be more likely to break during milling, hence reducing head rice yield and profit for both farmer and mill.

TABLE 5.15 Insect Pests of Rice.

Insect	Description	Symptoms/Injury	Control
Rice Water Weevil *Lissorhoptrus oryzophilus* (Kuschel)	Adults brown, 1/8-inch long; white larvae without legs.	Slit-like scars on leaves from adult feeding; larvae feed on and prune roots.	Chemical; cultural.
Chinch Bug *Blissus leucopterus* (Say)	Adults black to 1/6-inch long; white forewings with black, triangular spot; nymphs are red with yellow band on abdomen.	Damage rice seedling by sucking plant sap.	Flooding; chemical.
Rice Stink Bug *Oebalus pugnax* (Fabricius)	Tan-colored adults to 1/2-inch long; nymphs are first red and black, later tan with red and black pattern.	Feeding on florets and developing kernels may result in discolored, "pecky," grains.	Chemical.
Fall Armyworm *Spodoptera frugiperda* (Smith)	Moth dark gray to brown; males with white bar near wing tip; larvae light green to black with white strips along body.	Larvae feed on plants, chewing out large sections of leaf; may prune seedlings to ground.	Flooding; chemical.
Rice Leaf Miner *Hydrellia scapularis* (Loew)	Adults dark-bodied, clear wings, and metallic blue-green to gray thorax; larvae becoming yellow to light green to 1/4-inch long.	Larvae tunnel through leaves, killing leaves closest to water first.	Chemical; water depth.

Pest	Description	Damage	Control
Rice Stalk Borer *Chilo plejadellus* (Zincken)	Moths to 1 inch, pale white wings with row of gold scales with black dots; larvae brown with 1 dark and 1 light brown stripes.	Larvae tunnel inside stem and may kill developing leaves-called deadheart; may result in lodging; panicles may be black.	Cultural; chemical.
Sugarcane Borer *Diatraea saccharalis* (Fabricius)	Moths are straw-colored, forewings with black dots in V-shape; larvae to 1 inch, yellowish white with brown spots.	Same as Rice Stalk Borer.	Cultural; chemical.
Grasshoppers *Melanoplus* spp. other spp.	Obvious.	May attack blooming rice, blank heads may result.	Chemical.
Grape Colaspis *Colaspis brunnea* (Fabricius)	Larvae small, white and legless.	Larvae eat into stem just above seed.	Rotation.
Western Yellowstriped Armyworm *Spodoptera praefica* (Grote)	Larvae vary in color with many distinct, very thin yellow stripes on sides and an intense black spot on side of its first legless segment.	Larvae may defoliate plant; leaves chewed in angular pieces; feeds day and night.	Chemical.
Armyworm *Pseudaletia unipuncta* (Haworth)	Larvae vary in color with distinct yellow stripes but no intense black spot.	Same as Western Yellowstriped Armyworm; feeds at night.	Chemical.

TABLE 5.16 Common Diseases of Rice.

Disease	Cause/Source	Symptoms/Injury	Control
Watermold	*Achlya* spp. *Pythium* spp. *Fusarium* spp.	Seeds rotted, copper or greenish-brown spots on soil or gelatinous mass coating seed after water drainage.	Plant good quality seed; water management.
Seedling Disease	*Bipolaris oryzae*, *Fusarium* spp, *Rhizoctonia solani*, *Sclerotium rolfsii*, others	Seedlings 1–4 inches; may have brown spots on coleoptile; may turn white and have fluffy, white mycelium; may have small, round, tan sclerotia at plant base; seedling death.	Plant good quality seed; water management.
Stem Rot	*Magaporthe salvinis*	Crown area decayed with black or dark brown streaks extending to lower stem internodes, and/or black angular lesions on sheaths at water level; tillers die singularly; roots die and turn black; adventitious roots growing from lower stem node(s); outer sheaths die; culms discolored with raised areas of dark fungus mycelium.	Resistant cultivars; water management.
Brown Leaf Spot	*Cochiobolus miyabeamus*	Round, dark brown lesions with yellowish margins; centers become necrotic with brown margins.	Cultural; seed fungicide.
Narrow Brown Leaf Spot	*Cercospora oryzae*	Reddish brown, elongated lesions, 1/5th to 1-inch in length, on leaf blades, usually parallel with veins; may occur on sheaths.	Resistant cultivars; foliar fungicides.
Sheath Blight	*Thanatephorus cucumeris*	Alternating wide bands of white, greenish gray or tan with narrow bands of reddish brown to brown on leaf blades; water-soaked, gray-green lesions on sheaths at water line, becoming oval, white to straw colored with reddish brown margins 1/2 to 1 inch wide and 1 to 1 1/2 inches long; lesions may form a continuous band.	Resistant cultivars; cultural chemical.
Leaf Scald	*Gerlachia oryzae*	Lesions of wide bands of gray, dying tissue alternating with narrow, reddish brown bands forming chevrons from leaf tip or leaf margins.	Control not recommended; cultivars vary in degree of susceptibility.
Leaf Smut	*Entyloma oryzae*	Small, black, linear lesions that may have dark gold or light brown halos.	Cultivars vary in degree of susceptibility.
Alternaria Leaf Spot	*Alternaria padwickii*	Round to oval lesions, 1/5th to 1/2 inch, with white or pale tan centers and reddish brown margins; adjacent lesions may coalesce to form an oval double spot.	Cultivars vary in degree of susceptibility.

Disease	Causal Agent	Symptoms	Control
White Tip	*Aphelenchoides besseyi*	Leaf tips white with yellow band between healthy green areas; flag leaf may be twisted with poor panicle emergence; kernels may be aborted, poorly filled and/or discolored.	Cultivars vary in degree of susceptibility.
Sheath Rot	*Sarocladium oryzae*	Flag leaf sheath uniformly reddish or reddish brown with yellow-tan lesions; dark, irregular ring patterns within lesions; panicles twisted and poorly emerged; florets may be reddish to dark brown; kernels poorly or not filled.	Cultivars differ in degrees of susceptibility.
Blast: Leaf; Flag Leaf; Nodal; Panicle	*Pyricularia oryzae*	Leaf: small, dark brown, round to oval lesions or with gray or white center and dark or reddish margin; lesions may elongate and become diamond shaped. Flag Leaf: collar discolors brown to chocolate brown, blade may become detached. Nodal: stem nodes turn black, shrivel and turn gray toward maturity, or may appear blue-gray with conidia. Panicle: branches and florets may have gray brown lesions or florets light brown, straw colored, or gray; grain development stops.	Leaf: flood if field drained. Leaf and others: water management and resistant cultivars.
Neck Blight	*Bipolaris oryzae* or *Cercospora oryzae*	Stem above and below node immediately below the panicle becomes light brown or tan-brown, dies and shrivels; kernels in lower panicle do not fill.	Resistant cultivars; early-maturing cultivars.
Pecky Rice	*Bipolaris oryzae, Curvularia* spp., other fungi; feeding by rice stink bug	Florets, kernel or hulls of kernels with brown, reddish brown, purple, or white surrounded by purple-brown spots.	Chemical control of stink bug; foliar fungicide; field sanitation.
Panicle Blight	unknown	Single to several florets on a panicle branch light brown to straw-colored to gray; grain development stops.	None.
Brown Spot	*Cochiobolus miyabeanus*	Grains partially filled, shriveled, chalky with black mass on grain or at junction of lemma and palea that will not easily rub off on finger tips.	Plant good quality seed; field sanitation.
Kernel Smut	*Neovossia barclayana*	Grains partially filled; black mass on grain or at juncture of lemma and palea that rubs off easily onto fingers.	Control not recommended; cultivars vary in degree of susceptibility.
False Smut	*Ustilaginoidea virens*	Orange fruiting structures on maturing grains that rupture to expose a mass of green-black spores; grain replaced by sclerotia.	Chemical.

TABLE 5.17   Typical Symptoms Associated with the Most Commonly Deficient Plant Nutrients.

Element	Mobility	Symptoms
Iron	Immobile	First appears as interveinal chlorosis on young leaves; entire leaf becoming pale yellow to whitish; older leaves remain greener.
Zinc	Immobile	Midribs of younger leaves turning yellow and drooping, i.e. not upright but may float in water; stunted growth and delayed maturity.
Nitrogen	Mobile	Older leaves becoming pale green to yellow; younger leaves greener but may be narrow, short and erect; plants are stunted with few tillers.
Phosphorus	Mobile	Leaves becoming dark green to purple, narrow, short and erect; later becoming necrotic; plants are stunted with few tillers.
Potassium	Mobile	Leaves becoming dark green followed by interveinal chlorosis and small necrotic spots coalescing from tip; leaves will be shortened and only slightly reduced.
Sulfur	Mobile	Similar to nitrogen.

GRADES AND STANDARDS

Definitions

Rough rice: Rice (*Oryza sativa* L.) that consists of 50 percent or more of paddy kernels of rice.

Brown rice for processing: Rice (*Oryza sativa* L.) that consists of more than 50 percent of kernels of brown rice, and which is intended for processing to milled rice.

Milled rice: Whole or broken kernels of rice (*Oryza sativa* L.) from which the hulls and at least the outer bran layers have been removed and that contain not more than 10 percent of seeds, paddy kernels, or foreign material, either singly or combined.

Classes:

1. Long-grain rice
2. Medium-grain rice
3. Short-grain rice
4. Mixed rice

The following additional three classes apply to milled rice and shall be based on the percentage of whole kernels and of broken kernels of different size:

1. Second-head milled rice

2. Screenings milled rice

3. Brewers milled rice

Classes shall be based on the percentage of whole kernels, large broken kernels, and types of rice.

Additional Definitions

(1) **Long grain rough rice** shall consist of rough rice that contains more than 25 percent of whole kernels and which, after milling to a well-milled degree, contains not more than 10 percent of whole or large broken kernels of medium- or short-grain rice.

(2) **Medium grain rough rice** shall consist of rough rice that contains more than 25 percent of whole kernels and which, after milling to a well-milled degree, contains not more than 10 percent of whole or large broken kernels of long-grain rice or whole kernels of short-grain rice.

(3) **Short grain rough rice** shall consist of rough rice that contains more than 25 percent of whole kernels and which, after milling to a well-milled degree, contains not more than 10 percent of whole or large broken kernels of long-grain rice or whole kernels of medium-grain rice.

(4) **Mixed rough rice** shall consist of rough rice that contains more than 25 percent of whole kernels and which, after milling to a well-milled degree, contains more than 10 percent of "other types" as defined in paragraph 20.

(5) **Long-grain brown rice for processing** shall consist of brown rice for processing that contains more than 25 percent of whole kernels of brown rice and not more than 10 percent of whole or broken kernels of medium or short-grain rice.

(6) **Medium-grain brown rice for processing** shall consist of brown rice for processing that contains more than 25 percent of whole kernels of brown rice and not more than 10 percent of whole or broken kernels of long-grain rice or whole kernels of short-grain rice.

(7) **Short-grain brown rice for processing** shall consist of brown rice for processing that contains more than 25 percent of whole kernels of brown rice and not more than 10 percent of whole or broken kernels of long-grain rice or whole kernels of medium-grain rice.

(8) **Mixed brown rice for processing** shall be brown rice for processing that contains more than 25 percent of whole kernels of brown rice and more than 10 percent of "other types."

(9) **Long-grain milled rice** shall consist of milled rice that contains more than 25 percent of whole kernels of milled rice and in U.S. Nos. 1 through 4 not more than 10 percent of whole or broken kernels of medium or short-grain rice. U.S. Nos. 5 and 6 long-grain milled rice shall contain not more than 10 percent of whole kernels of medium or short-grain milled rice (*broken kernels do not apply*).

(10) **Medium-grain milled rice** shall consist of milled rice that contains more than 25 percent of whole kernels of milled rice and in U.S. Nos. 1 through 4 not more than 10 percent of whole or broken kernels of long-grain rice or whole kernels of short-grain rice. U.S. Nos. 5 and 6 medium-grain milled rice shall contain not more than 10 percent of whole kernels of long- or short-grain milled rice (*broken kernels do not apply*).

(11) **Short-grain milled rice** shall consist of milled rice that contains more than 25 percent of whole kernels of milled rice and in U.S. Nos. 1 through 4 not more than 10 percent of whole or broken kernels of long-grain rice or whole kernels of medium-grain rice. U.S. Nos. 5 and 6 short-grain milled rice shall contain not more than 10 percent of whole kernels of long or medium-grain milled rice (*broken kernels do not apply*).

(12) **Mixed milled rice** shall consist of milled rice that contains more than 25 percent of whole kernels of milled rice and more than 10 percent of "other types." U.S. Nos. 5 and 6 mixed milled rice shall contain more than 10 percent of whole kernels of "other types" (*broken kernels do not apply*).

(13) **Second-head milled rice** shall consist of milled rice that contains
   (i) Not more than (a) 25 percent of whole kernels, (b) 7 percent of broken kernels removed by a 6 plate, (c) 0.4 percent of broken kernels removed by a 5 plate, and (d) 0.05 percent of broken kernels passing through a 4 sieve (*southern production*); or
   (ii) Not more than (a) 25 percent of whole kernels, (b) 50 percent of broken kernels passing through $6^1/_2$ sieve, and (c) 10 percent of broken kernels passing through a 6 sieve (*western production*).

(14) **Screenings milled rice** shall consist of milled rice that contains
   (i) Not more than (a) 25 percent of whole kernels, (b) 10 percent of broken kernels removed by a 5 plate, and (c) 0.2 percent of broken kernels passing through a 4 sieve (*southern production*); or
   (ii) Not more than (a) 25 percent of whole kernels, (b) 15 percent of broken kernels passing through a $5^1/_2$ sieve, (c) more than 50 percent of broken kernels passing through a $6^1/_2$ sieve, and (d) 10 percent of broken kernels passing through a 6 sieve (*western production*).

(15) **Brewers milled rice** shall consist of milled rice that contains not more than 25 percent of whole kernels and which does not meet the kernel-size requirements of the class second head milled rice or screenings milled rice.

(16) **Damaged kernels.** Whole or large broken kernels of rice that are distinctly discolored or damaged by water, insects, heat, or any other means, and whole or large broken kernels of parboiled rice in nonparboiled rice. "Heat-damaged kernels" (see paragraph 17) shall not function as damaged kernels.

(17) **Heat-damaged kernels.** Whole or large broken kernels of rice that are materially discolored and damaged as a result of heating, and whole or large broken kernels of parboiled rice in nonparboiled rice that are as

dark as, or darker in color than, the interpretive line for heat-damaged kernels.

(18) **Milling yield.** An estimate of the quantity of whole kernels and total milled rice (whole and broken kernels combined) that are produced in the milling of rough rice to a well-milled degree.

(19) **Objectionable seeds.** Seeds other than rice, except seeds of *Echinochloa crusgalli* (commonly known as barnyard grass, water grass, and Japanese millet).

(20) **Other types.** (1) Whole kernels of (i) long-grain rice in medium- or short-grain rice, (ii) medium-grain rice in long- or short-grain rice, (iii) short-grain rice in long- or medium-grain rice, and (2) large broken kernels of long-grain rice in medium- or short-grain rice and large broken kernels of medium- or short-grain rice in long-grain rice.

NOTE: Large broken kernels of medium-grain rice in short-grain rice and large broken kernels of short-grain rice in medium-grain rice shall not be considered other types.

(21) **Paddy kernels.** Whole or broken unhulled kernels of rice.

(22) **Red rice.** Whole or large broken kernels of rice in which there is an appreciable amount of red bran.

(23) **Seeds.** Whole or broken seeds of any plant other than rice.

(24) **Smutty kernels.** Whole or broken kernels of rice that are distinctly infected by smut.

(25) **Types of rice.** The following three types:

1. Long grain
2. Medium grain
3. Short grain

Types shall be based on the length/width ratio of kernels of rice that are unbroken and the width, thickness, and shape of kernels of rice that are broken as set forth in the Rice Inspection Handbook.

(26) **Ungelatinized kernels.** Whole or large broken kernels of parboiled rice with distinct white or chalky areas that are due to incomplete gelatinization of the starch.

(27) **Whole and large broken kernels.** Rice (including seeds) that (1) passes over a 6 plate (*for southern production*), or (2) remains on top of a 6 sieve (*for western production*).

(28) **Whole kernels.** Unbroken kernels of rice and broken kernels of rice that are at least three-fourths of an unbroken kernel.

Grades and grading requirements as established by the USDA for rough rice, brown rice, milled rice, second head milled rice, screenings, and brewers rice are shown in Tables 5.18 through 5.23.

TABLE 5.18 Grades and Grade Requirements for the Classes of Rough Rice.

| | Maximum Limits of— | | | | | | |
| | Seeds and Heat Damaged Kernels | | | Chalky Kernels | | | |
Grade	Total (Singly or Combined) Number in 500 g.	Heat Damaged Kernels and Objectionable Seeds (Singly or Combined) Number in 500 g.	Red Rice and Damaged Kernels (Singly or Combined) (%)	In Long Grain Rice (%)	In Medium or Short Grain Rice (%)	Other Types[1] (%)	Color Requirements
U.S. No. 1	4	3	0.5	1.0	2.0	1.0	Shall be white or creamy.
U.S. No. 2	7	5	1.5	2.0	4.0	2.0	May be slightly gray.
U.S. No. 3	10	8	2.5	4.0	6.0	3.0	May be light gray.
U.S. No. 4	27	22	4.0	6.0	8.0	5.0	May be gray or slightly rosy.
U.S. No. 5	37	32	6.0	10.0	10.0	10.0	May be dark gray or rosy.
U.S. No. 6	75	75	15.0[2]	15.0	15.0	10.0	May be dark gray or rosy.
U.S. Sample grade	U.S. Sample grade shall be rough rice which (a) does not meet the requirements for any of the grades for U.S. No. 1 to U.S. No. 6, inclusive; (b) contains more than 14.0 percent of moisture; (c) is musty, or sour, or heating; (d) has any commercially objectionable foreign odor; or (e) is otherwise of distinctly low quality.						

Source: USDA-FGIS, 1983.

1. These limits do not apply to the class mixed rough rice.
2. Rice grade U.S. No. 6 shall contain not more than 6.0 percent of damaged kernels.

TABLE 5.19 Grade and Grade Requirements for the Classes of Brown Rice for Processing.

			Maximum Limits of—							
			Seeds and Heat-Damaged Kernels					Broken Kernels Removed by a 6 Plate or a 6½ Sieve[1]		
Grade	Paddy Kernels		Total (Singly or Combined)	Heat-Damaged Kernels	Objectionable Seeds	Red Rice and Damaged Kernels	Chalky Kernels		Other Types[2]	Well-Milled Kernels
	(%)	No. in 500 g.	No. in 500 g.	No. in 500 g.	No. in 500 g.	(%)	(%)	(%)	(%)	(%)
U.S. No. 1	—	20	10	1	2	1.0	2.0	1.0	1.0	1.0
U.S. No. 2	2.0	—	40	2	10	2.0	4.0	2.0	2.0	3.0
U.S. No. 3	2.0	—	70	4	20	4.0	6.0	3.0	5.0	10.0
U.S. No. 4	2.0	—	100	8	35	8.0	8.0	4.0	10.0	10.0
U.S. No. 5	2.0	—	150	15	50	15.0	15.0	6.0	10.0	10.0

U.S. Sample Grade U.S. Sample grade shall be brown rice for processing which (a) does not meet the requirements for any of the grades from U.S. No. 1 to U.S. No. 5, inclusive; (b) contains more than 14.5 percent of moisture; (c) is musty, or sour, or heating; (d) has any commercially objectionable foreign odor; (e) contains more than 0.2 percent of related material or more than 0.1 percent of unrelated material; (f) contains live weevils or other live insects; or (g) is otherwise of distinctly low quality.

Source: USDA-FGIS, 1983.

1. Plates should be used for southern production rice and sieves should be used for western production rice, but any device or method which gives equivalent results may be used.

2. These limits do not apply to the class mixed brown rice for processing.

TABLE 5.20 Grades and Grade Requirements for the Classes Long-Grain Milled Rice, Medium-Grain Milled Rice, Short-Grain Milled Rice, and Mixed Milled Rice.

Grade	Seeds, Heat-Damaged and Paddy Kernels (Singly or Combined) Total No. in 500 g.	Heat Damage Kernels and Objectionable Seeds No. in 500 g.	Red Rice and Damaged Kernels (Singly or Combined) (%)	Chalky Kernels[1] In Long Grain (%)	Chalky Kernels[1] In Medium or Short Grain (%)	Broken Kernels Total (%)	Broken Kernels Removed by a 5 Plate[2] (%)	Broken Kernels Removed by a 6 Plate[2] (%)	Broken Kernels Through a 6 Sieve[2] (%)	Other Types[3] Whole Kernels (%)	Other Types[3] Whole and Broken Kernels (%)	Color Requirements[1]	Milling Requirements-Minimum[4]
U.S. No. 1	2	1	0.5	1.0	2.0	4.0	0.04	0.1	0.1	—	1.0	Shall be white or creamy	Well milled
U.S. No. 2	4	2	1.5	2.0	4.0	7.0	0.06	0.2	0.2	—	2.0	May be slightly gray	Well milled
U.S. No. 3	7	5	2.5	4.0	6.0	15.0	0.1	0.8	0.5	—	3.0	May be light gray	Reasonably well milled
U.S. No. 4	20	15	4.0	6.0	8.0	25.0	0.4	2.0	0.7	—	5.0	May be gray or slightly rosy	Reasonably well milled
U.S. No. 5	30	25	6.0[4]	10.0	10.0	35.0	0.7	3.0	1.0	10.0	—	May be dark gray or rosy	Lightly milled
U.S. No. 6	75	75	15.0[5]	15.0	15.0	50.0	1.0	4.0	2.0	10.0	—	May be dark gray or rosy	Lightly milled

U.S. Sample grade U.S. Sample grade shall be milled rice of any of these classes which (a) does not meet the requirements for any of the grades from U.S. No. 1 to U.S. No. 6., inclusive; (b) contains more than 15.0 percent of moisture; (c) is musty, sour, or heating; (d) has any commercially objectionable foreign odor; (e) contains more than 0.1 percent of foreign material; (f) contains live or dead weevils or other insects, insect webbing, or insect refuse; or (g) is otherwise of distinctly low quality.

Source: USDA-FGIS, 1983.

1. For the special grade parboiled milled rice, see below.
2. Plates should be used for southern production rice, and sieves should be used for western production rice; but any device or method which gives equivalent results may be used.
3. These limits do not apply to the class mixed milled rice.
4. For the special undermilled milled rice, see below.
5. Grade U.S. No. 6 shall contain not more than 6 percent of damaged kernels.

TABLE 5.21 Grades and Grade Requirements for the Class Second Head Milled Rice.

Grade	Maximum Limits of—				Minimum Milling Requirements[2]
	Seeds, Heat-Damaged, and Paddy Kernels (Singly or Combined)		Red Rice and Damaged Kernels (Singly or Combined)	Chalky Kernels[1]	Color Requirements[1]
	Total	Heat-Damaged Kernels and Objectionable Seeds			
	Number in 500 grams	Number in 500 grams	(%)	(%)	
U.S. No. 1	15	5	1.0	4.0	Shall be white or creamy.	Well milled.
U.S. No. 2	20	10	2.0	6.0	May be slightly gray.	Well milled.
U.S. No. 3	35	15	3.0	10.0	May be light gray.	Reasonably well milled.
U.S. No. 4	50	25	5.0	15.0	May be gray or slightly rosy.	Reasonably well milled.
U.S. No. 5	75	40	10.0	20.0	May be dark gray or rosy.	Lightly milled.
U.S. Sample Grade	U.S. Sample Grade shall be milled rice of this class which (a) does not meet the requirements for any of the grades from U.S. No. 1 to U.S. No. 5, inclusive; (b) contains more than 15 percent of moisture; (c) is musty, or sour, or heating; (d) has any commercially objectionable foreign odor; (e) contains more than 0.1 percent of foreign material; (f) contains live or dead weevils or other insects, insect webbing, or insect refuse; or (g) is otherwise of distinctly low quality.				

Source: USDA-FGIS, 1983.

1. For the special grade parboiled milled rice (see below).
2. For the special grade undermilled milled rice (see below).

TABLE 5.22 Grades and Grade Requirements for the Class Screenings Milled Rice.

| Grade | Maximum Limits of— Paddy Kernels and Seeds | | Chalky Kernels[1] | Color Requirements[1] | Minimum Milling Requirements[2] |
	Total (Singly or Combined) Number in 500 g.	Objectionable Seeds Number in 500 g.			
			(%)		
U.S. No. 1[3,4]	30	20	5.0	Shall be white or creamy	Well milled
U.S. No. 2[3,4]	75	50	8.0	May be slightly gray	Well milled
U.S. No. 3[3,4]	125	90	12.0	May be light gray or slightly rosy	Reasonably well milled
U.S. No. 4[3,4]	175	140	20.0	May be gray or rosy	Reasonably well milled
U.S. No. 5	250	200	30.0	May be dark gray or very rosy	Lightly milled
U.S. Sample Grade	U.S. Sample Grade shall be milled rice of this class which (a) does not meet the requirements for any of the grades from U.S. No. 1 to U.S. No. 5, inclusive; (b) contains more than 15 percent of moisture; (c) is musty, or sour, or heating; (d) has any commercially objectionable foreign odor; (e) has a badly damaged or extremely red appearance; (f) contains more than 0.1 percent of foreign material; (g) contains live or dead weevils or other insects, insect webbing, or insect refuse; or (h) is otherwise of distinctly low quality.				

Source: USDA-FGIS, 1983.

1. For the special grade parboiled milled rice (see below).
2. For the special grade undermilled milled rice (see below).
3. Grades U.S. No. 1 to U.S. No. 4, inclusive, shall contain not more than 3 percent of heat-damaged kernels, kernels damaged by heat and/or parboiled kernels in nonparboiled rice.
4. Grades U.S. No. 1 to U.S. No. 4, inclusive, shall contain not more than 1 percent of material passing through a 30 sieve.

TABLE 5.23   Grades and Grade Requirements for the Class Brewers Milled Rice.

| | Maximum Limits of―― | | | |
| | Paddy Kernels and Seeds | | | |
Grade	Total (Singly or Combined)	Objectionable Seeds	Color Requirements[1]	Minimum milling Requirements[2]
	(%)	(%)		
U.S. No. 1[3,4]	0.5	0.05	Shall be white or creamy	Well milled
U.S. No. 2[3,4]	1.0	0.1	May be slightly gray	Well milled
U.S. No. 3[3,4]	1.5	0.2	May be light gray or slightly rosy	Reasonably well milled
U.S. No. 4[3,4]	3.0	0.4	May be gray or rosy	Reasonably well milled
U.S. No. 5	5.0	1.5	May be dark gray or very rosy	Lightly milled
U.S. Sample Grade	U.S. Sample Grade shall be milled rice of this class which: (a) does not meet the requirements for any of the grades from U.S. No. 1 to U.S. No. 5, inclusive; (b) contains more than 15.0 percent of moisture; (c) is musty, or sour, or heating; (d) has any commercially objectionable foreign odor; (e) has a badly damaged or extremely red appearance; (f) contains more than 0.1 percent of foreign material; (g) contains more than 15.0 percent of broken kernels that will pass through a 2½ sieve; (h) contains live or dead weevils or other insects, insect webbing, or insect refuse; or (i) is otherwise of distinctly low quality.			

Source: USDA-FGIS, 1983.

1. For the special grade parboiled milled rice (see below).
2. For the special grade undermilled milled rice (see below).
3. Grades U.S. No. 1 to U.S. No. 4 inclusive shall contain not more than 3 percent of heat-damaged kernels, kernels damaged by heat and/or parboiled kernels in nonparboiled rice.
4. Grades U.S. No. 1 to U.S. No. 4 inclusive, shall contain not more than 1 percent of material passing through a 30 seive. This limit does not apply to the special grade granulated brewers milled rice.

Special Grades, Requirements, and Designations

A special grade, when applicable, is supplemental to the grade assigned. Such special grades for milled rice are established and determined as follows:

Coated milled rice. Coated milled rice shall be rice that is coated, in whole or in part, with substances that are safe and suitable according to commercially accepted practice.

Granulated brewers milled rice. Granulated brewers milled rice shall be milled rice that has been crushed or granulated so that 95 percent or more will pass through a 5 sieve, 70 percent or more will pass through a 4 sieve, and not more than 15 percent will pass through a 2½ sieve.

Parboiled milled rice. Parboiled milled rice shall be milled rice in which the starch has been gelatinized by soaking, steaming, and drying. Grades U.S. No. 1 to

U.S. No. 6, inclusive, shall contain not more than 10 percent of ungelatinized kernels. Grades U.S. Nos. 1 and 2 shall contain not more than 0.1 percent, grades U.S. Nos. 3 and 4 not more than 0.2 percent, and grades U.S. Nos. 5 and 6 not more than 0.5 percent of nonparboiled rice. If the rice is (1) not distinctly colored by the parboiling process, it shall be considered "parboiled light"; (2) distinctly but not materially colored by the parboiled process, it shall be considered "parboiled"; (3) materially colored by the parboiling process, it shall be considered "parboiled dark." The color levels for parboiled light, parboiled, and parboiled dark shall be in accordance with the interpretive line samples for parboiled rice.

Note: The maximum limits for "chalky kernels," "heat-damaged kernels," "kernels damaged by heat," and the "color requirements" are not applicable to the special grade "parboiled milled rice."

Undermilled milled rice. Undermilled milled rice shall be milled rice that is not equal to the milling requirements for "well-milled," "reasonably well-milled," and "lightly milled" rice. Grades U.S. Nos. 1 and 2 shall contain not more than 2 percent, grades U.S. Nos. 3 and 4 not more than 5 percent, grade U.S. No. 5 not more than 10 percent, and grade U.S. No. 6 not more than 15 percent of well-milled kernels. Grade U.S. No. 5 shall contain not more than 10 percent of red rice and damaged kernels (singly or combined) and in no case more than 6 percent of damaged kernels.

Note: The "color and milling requirements" are not applicable to the special grade "undermilled milled rice."

The grade designation for coated, granulated brewers, parboiled, or undermilled milled rice shall include, following the class, the word(s) "coated," "granulated," "parboiled light," "parboiled," "parboiled dark," or "undermilled," as warranted.

GLOSSARY

Adventitious: plant structures arising from an unusual place.

Anaerobic: without oxygen.

Auricles: narrow extensions of the collar in grasses; may completely clasp the culm; may be absent.

Blank: describes individual florets of rice producing no grain.

Boot: describes growth stage identified by the bulge within the flag leaf sheath caused by the enlarging panicle.

Borrow pits: depressions on each side of a rice levee created when soil is embanked by a levee disk.

Brewers: the smallest size of broken milled rice, less than one-fourth of a whole kernel.

Broken yield: broken grain resulting from milling 100 lbs. of rough rice and equals total mill yield minus head yield.

Brokens: milled rice kernels that are less than three-fourths of whole kernels. Includes second heads, screenings, and brewers.

Brown rice: rice with the hull removed and bran and embryo remaining.

Carbohydrate: organic chemicals composed of C, H, and O—e.g., photosynthetically produced sugar and starch in plants.

Chlorosis: yellowing of normally green tissue resulting from chlorophyll destruction or failure to develop.

Coleoptile: a sheath-like structure enclosing the shoot of a grass seedling for a short time after germination.

Collar: juncture of leaf blade and sheath; may be used to identify grass species.

Commingled rice: blend of rice of similar grain type, quality and grade.

Culm: stem of a grass plant; in rice there are main culms and tiller culms.

Dough stage: stage of grain maturity when the grain turns from a thick liquid to a soft dough consistency.

Embryo: a microscopically small plant attached to the endosperm of a seed; often called the germ.

Endosperm: the stored food, primarily carbohydrate in the form of starch, comprising most of a monocot seed, such as rice, that serves as food for the embryo and developing seedling during germination and early growth.

Flag leaf: the top-most leaf of the rice plant immediately below the panicle.

Floret: a grass flower, including the lemma, palea, and sex organs.

Green rice: high moisture (18.5 to 22.5 percent) rough rice.

Head milling yield: pounds of head rice milled from 100 pounds of rough rice.

Head rice: milled rice kernel that is more than three-fourths of the whole kernel.

Hulling: the process of removing the lemma and palea (husks or hulls) from rough rice.

Hulls: lemmas and paleas of rough rice that are usually a waste product but may be used in rice millfeed and as filler in feed products.

Inflorescence: the flowering structure of a plant.

Instant rice: milled rice that is cooked, cooled, and dried under controlled conditions and packaged in a dehydrated form and requiring very little preparation time before eating; enriched with thiamine, riboflavin, niacin, and iron.

Internode: the area of a stem between two nodes.

Jointing: the process of internode elongation in grasses.

Lemma: the larger of two structures enclosing the rice seed that, together with the palea, form the hard, outer covering of rough rice.

Ligule: a short membranous projection on the inner side of a leaf blade at the junction of blade and sheath; present in many grasses.

Long-grain rice: rice that is long and slender, and when milled measures typically from 0.26 to 0.28 inches (6.7 to 7.0 mm) or longer and with a length:width ratio from 3.27:1 to 3.41:1.

Main culm or shoot: the stem or aboveground portion, respectively, of a plant originating directly from the seed.

Medium-grain rice: rice that when milled measures typically from 0.22 to 0.23 inches (5.5 to 5.8 mm) with a length:width ratio of 2.09:1 to 2.49:1.

Milk stage: early stage of grain development when milky liquid fills the grain.

Milled rice: grain with husks, bran, and germ removed.

Milling: process of converting rough rice into milled or brown rice.

Necrotic: dead.

Nematode: unsegmented, round, threadlike worms, generally microscopic that may be free-living or parasitic of plants and animals. Soil-inhibiting nematodes may be parasitic on rice.

Node: the solid portion of the rice stem from which leaves arise.

Paddy: a term used to describe a subdivision, bounded by levees, of a rice field.

Panicle: a many-branched inflorescence composed of few to many spikelets that are composed of one to several florets.

Parboiled rice: rough rice steeped in warm water under pressure, steamed, dried, and milled; improves head milling yield.

Palea: the smaller of two enclosing structures forming a hard outer covering of unmilled rice.

Pathogen: a living organism that causes an infectious disease.

Precooked rice: milled rice processed to make it quick-cooking.

Processed rice: rice used in breakfast cereals, soups, baby foods, packaged mixes, etc.

Ratoon crop (second crop): regrowth from the rice stubble of the first crop that grew from seed.

Residue management: the management of rice straw and stubble after harvest, or residue of previous crop before planting.

Rice bran: the outer layer of the rice caryopsis just beneath the hull containing the outer bran layer and parts of the germ; rich in protein and vitamin B; used as livestock feed and in vitamin concentrates.

Rice polish: a layer sometimes removed as the final stage of milling and composed of the inner white bran and perhaps aleurone cells that are high in protein and fat; used in livestock feed and baby food.

Rough or paddy rice: rice grains with hulls but without any stalk parts; consists of 50 percent or more whole or broken unhulled kernels.

Screenings: broken milled rice kernels less than one-half but more than one-quarter of the whole kernel.

Second heads: broken milled rice that is more than one-half of the whole kernel but less than three-fourths.

Senescence: plant or plant part stage of growth from maturity to death.

Sheath: the portion of a complete grass leaf below the collar that may enclose the stem.

Shoot: the aboveground portion of a plant; in rice the portion above the crown.

Short-grain rice: rice that is almost round and when milled measures typically from 0.20 to 0.22 inches (5.2 to 5.4 mm) with a length:width ratio of 1.66:1 to 1.77:1.

Spikelet: the subdivision of a spike consisting of one to several flowers.

Tiller: a vegetative shoot in rice, not the main culm.

Total milling yield: pounds of head, second heads, brewers, and screenings milled from 100 pounds of rough rice.

White rice: kernel remaining after hulls, bran layer, and germ are removed; includes head rice and brokens.

Y-leaf: the most recently matured leaf.

BIBLIOGRAPHY

Adair, C. R., M. D. Miller, and H. M. Beachell. 1962. "Rice Improvement and Culture in the United States." pp. 61–108. *In* A. G. Norman (ed.). *Advances in Agronomy.* Acad. Press, New York, NY.

Adair, C. R. 1955. "The Irrigation and Culture of Rice." *USDA Yearbook of Agriculture.* U.S. Government Printing Office, Washington, D.C.

Bailey, T. A., and D. M. Kennedy. (not dated). *The American Pageant.* (no pub. listed)

Bender, B. 1975. *Farming in Prehistory: From Hunter-Gatherer to Food-Producer.* St. Martin's Press, New York, NY.

Carrier, L. 1923. *The Beginnings of Agriculture in America.* McGraw-Hill Book Co., New York.

Chang, T. T. 1989. "Domestication and Spread of the Cultivated Rices." pp. 408–417. *In* D. R. Harris and G. C. Hillman (eds.) *Foraging and Farming: The Evolution of Plant Exploitation.* Unwin Hyman Pub., London, England.

Chang, T. T. 1976. "The origin, evolution, cultivation, dissemination, and diversification of Asian and African rices." *Euphytica* 25:425–441.

Chang, T. T. 1976. "Rice." pp. 98–104. *In* N. W. Simmonds (ed.) *Evolution of Crop Plants.* Longman Press, London, England.

Chang, T. T., Y. Pan, Q. Chu, R. Peiris, and G. C. Loresto. 1990. "Cytogenetic, electrophoretic, and root studies of javanica rices." *Rice Genetics II.* Proc. Sec. Intern. Rice Gen. Sym. 1990:21–32.

Chang, T. T. 1976. "The rice cultures." *Phil. Trans. R. Soc. Lond.* B. 275:143–157.

Chang, T.T. 1985. "Crop history and genetic conservation: Rice-a case study." *Iowa State J. Res.* 59(4):425—495.

Coale, F. J., and D. B. Jones. 1988. "Biology and Control of Rice Stink Bug." *Univ. of Florida Coop. Ext. Ser. Everglades Rice Newsletter* Vol. 1 No. 2.

Coale, F. J., and D. B. Jones. 1988. "Weed Control in Rice." *Univ. of Florida Coop. Ext. Ser. Everglades Rice Newsletter* Vol. 1 No. 3.

Coale, F. J, and D. B. Jones. 1989. "Ratoon Crop Management." *Univ. of Florida Coop. Ext. Ser. Everglades Rice Newsletter* Vol. 2 No. 2.

Coale, F. J., and D. B. Jones. 1990. "Rice Variety Performance Trials." *Univ. of Florida Coop. Ext. Ser. Everglades Rice Newsletter* Vol. 3 No. 3.

Coale, F. J., and D. B. Jones. 1990. "Water Seeding of Rice." *Univ. of Florida Coop. Ext. Ser. Everglades Rice Newsletter* Vol. 3, No. 4.

Cobley, L. S., and W. M. Steele. 1976. *An Introduction to the Botany of Tropical Crops.* Longman Press, London, England.

Cowan, C. W., and P. J. Watson. 1972. *The Origins of Agriculture: An International Perspective.* Smithsonian Instit. Press, Washington, D.C.

Crawford, G. W. 1992. "The Transitions to Agriculture in Japan." pp. 117–129. *In* A. B. Gebauer and T. D. Price (eds.) *Transitions to Agriculture in Prehistory.* Prehistory Press, Madison, WI.

Dethloff, H. C. 1988. *A History of the American Rice Industry, 1685–1985.* Texas A&M Press, College Station, TX.

Dilday, R. H. 1990. "Contribution of ancestral lines in the development of new cultivars of rice." *Crop Sci.* 30:905–911.

Drees, B. M., A. R. Gerlow, G. Fipps, A. D. Klosterboer, J. D. Krausz, R. Smith and T. D. Valco. 1992. *Rice Production Guidelines.* Texas A&M Univ. Coop. Ext. Ser. D-1253.

Efferson, J. N. 1985. "Rice Quality in World Markets." pp. 1–14. *In Rice Grain Quality and Marketing.* Intern. Rice Res. Inst., Manila, Philippines.

Food and Agriculture Organization of the United Nations (FAO). 1992. *Production Yearbook.* Vol. 45. FAO, Rome, Italy.

Grant, W. R., and S. H. Holder Jr. 1975. *Recent Changes and the Potential for U.S. Rice Acreage.* USDA-ERS RS-26.

Grant, W. R., E. M. Rister, and B. W. Brorsen. 1983. *The Value of Rice Quality.* USDA-ERS RS-42.

Grist, D. H. 1986. *Rice.* Longman Inc., New York.

Groth, D. E., M. C. Rush, and C. A. Hollier. 1991. *Rice Diseases and Disorders in Louisiana.* Louisiana St. Univ. Expt. Sta. Bul. 828.

Harlan, J. R., J. M. J. de Wet, and E. G. Price. 1973. "Comparative evolution of cereals." *Evol.* 27:311–325.

Hill, L. D. 1990. *Grain Grades and Standards.* Univ. of Ill. Press, Urbana and Chicago, IL.

Holder, S. H., Jr., and W. R. Grant. 1979. *U.S. Rice Industry.* USDA Ag. Ec. Rep. No. 433.

Holmes, G. K. 1912. *Rice Crop of the United States, 1712–1911.* USDA Bur. Stat. C. 34.

Hoseney, R. C. 1986. *Principles of Cereal Science and Technology.* Am. Assoc. Cer. Chem., St. Paul, MN.

Houston, D. F. 1972. *Rice Chemistry and Technology.* Am. Assoc. Cer. Chem., St. Paul, MN.

Hutchinson, J. 1976. "India: Local and introduced crops." *Phil. Trans. R. Soc. London B.* 275:129–141.

Ikehashi, H., H. Araki, and S. Yanagihara. 1990. "Screening and analysis of wide compatibility loci in wide crosses of rice." *Rice Genetics II.* Proc. Sec. Intern. Rice Gen. Sym. 1990:33–43.

International Rice Research Institute. 1975. *Major Research in Upland Rice.* Los Banos, Philippines.

Jennings, P. R., W. R. Coffman, and H. E. Kauffman. 1979. *Rice Improvement.* Int. Rice Res. Inst. Los Banos, Philippines.

Jones, J. W. 1936. "Improvement in Rice." pp. 415–450. *USDA Yearbook of Agriculture, 1936.* U.S. Government Print Office, Washington, D.C.

Jones, J. W. 1952. *Rice Production in the Southern United States.* USDA Farmers' Bul. No. 2043. Washington, D.C.

Juliano, B. O. 1980. "Rice: Recent Progress in Chemistry and Nutrition." pp. 409–428. *In* G. E. Inglett and L. Munck (eds.) *Cereals for Food and Beverages: Recent Progress in Cereal Chemistry.* Acad. Press, New York, NY.

Keith, T. L., G. Rutz, and F. Laws. 1994. *Delta Agricultural Digest—1994.* Farm Press Pub., Clarksdale, MS.

Knapp, S. A. 1899. *The Present Status of Rice Culture in the United States.* USDA Div. Bot. Bul. 22.

Louisiana State University. 1987. *Rice Production Handbook.* Louisiana St. Univ. Coop. Ext. Ser. Pub. 2321.

Luh, B. S. 1980. *Rice: Production and Utilization.* AVI Pub., Westport, CT.

MacNeish, R. S. 1992. *The Origins of Agriculture and Settled Life.* Univ. of Oklahoma Press, Norman, OK.

Martin, J. H., W. H. Leonard, and D. L. Stamp. 1976. *Principles of Field Crop Production.* MacMillan Pub. Co., New York.

Matz, S. A. 1991. *The Chemistry and Technology of Cereals as Food and Feed.* Van Nostrand Reinhold/AVI Pub., New York.

Metcalf, R. L., and R. A. Metcalf. 1993. *Destructive and Useful Insects: Their Habits and Control.* McGraw-Hill, New York.

Melville, R. 1966. "Continental drift, mesozoic continents and the migrations of the angiosperms." *Nature* 211:116–120.

Morishima, H., K. Hinata, and H. I. Oka. 1963. "Comparison of modes of evolution of cultivated forms from two wild rice species, *Oryza Breviligulata* and *O. perennis.*" *Evol.* 17:170–181.

Oka, H. I. 1988. *Origin of Cultivated Rice.* Japan Sci. Soc. Press and Elsevier Sci. Pub., Tokyo, Japan.

Peake, H. 1979. *The Origins of Agriculture.* Ernest Benn Ltd., London, England.

Rutger, J. N., and C. N. Bollich. 1991. "Use of Introduced Germplasm in U.S. Rice Improvement." pp. 1–14. *In* H. L. Shands and L. E. Wiesner (eds.) *Use of Plant Introductions in Cultivar Development Part I.* Crop Sci. Soc. Am., Madison, WI.

Rutger, J. N., and D. M. Brandon. 1980. "California rice culture." *Sci. Am.* 244(2):42–51.

Sauer, J. D. 1993. *Historical Geography of Crop Plants: A Selected Roster.* CRC Press, Boca Raton, FL.

Smith, J. F. 1985. *Slavery and Rice Culture in Low Country Georgia, 1750–1860.* Univ. of Tennessee Press, Knoxville, TN.

Takahashi, N. 1984. "Differentiation of Ecotypes in *Oryza sativa* L." pp. 31–37, *In* S. Tsunoda and N. Takahashi (eds.) *Biology of Rice.* Japan Sci. Soc. Press and Elsevier Sci. Pub., Tokyo, Japan.

United States Department of Agriculture. 1984. *Usual Planting and Harvesting Dates for U.S. Field Crops.* USDA Agri. Handbook 628.

University of Arkansas. 1990. *Rice Production Handbook.* Univ. of Arkansas Coop. Ext. Ser. MP 192.

University of California. 1983. *Integrated Pest Management for Rice.* Univ. California Div. Agri. Pub. No. 3280.

United States Department of Agriculture-Federal Grain Inspection Service. 1983. *United States Standards for Rice.* U.S. Government Printing Office, Washington, D.C.

Vernon, A. W. 1993. *African Americans at Mars Bluff, South Carolina.* Louisiana St. Univ. Press, Baton Rouge, LA.

Wailes, E. J., S. H. Holder, and J. Luebkemann. 1985. "Potential impact of California long grain rice production in U.S. milled rice flow patterns." pp. 12–17. *In* USDA-ERS. *Rice Outlook and Situation Report.* Washington D.C.

Zohary, D., and M. Hopf. 1993. *Domestication of Plants in the Old World.* Clarendon Press, Oxford, England.

6

COTTON
(Gossypium Hirsutum L.)

INTRODUCTION

The origin, evolution, and domestication of cotton remains somewhat of a mystery even to this day. There are 43 species, both diploid (2n = 2x = 26 chromosomes) and tetraploid (2n = 4x = 52) types, classified in the genes *Gossypium,* to which cotton belongs. Of these, 37 are diploid and distributed predominately across Africa, Asia Minor, Mexico, and Australia. Two diploid species, *G. arboreum* and *G. herbaceum,* were domesticated by humans; *G. arboreum* is still grown to some extent in India where the environment is too hostile for modern tetraploid cultivars. The remaining six species of cotton are tetraploid and their origins are believed to be in the New World, specific countries being Mexico, Peru, Brazil, Hawaii, and the Galapagos Islands. Two of these, *G. hirsutum* and *G. barbadense,* are cultivated types (Table 6.1). Only species of *Gossypium* producing seed hairs can accurately be called cotton.

The oldest known word for cotton is "karpasa-i," a term from the Sanskrit language of the ancient inhabiters of India. "Karapas," an apparent derivation, either before or after karpasa-i, is used in early manuscripts of the Bible (Ester 1:6). The English word "cotton" comes from the Arabic "al-or el-kutum," "gutum," or "kutum."

Cotton is unique among cultivated crops in that four distinct species were domesticated by humans and in that two each evolved in the Old World and New World. Also setting cotton apart is the fact that the two Old World cultigens have 26 chromosomes while the two New World cultivated species have 52 chromosomes. This being the case, we must consider cotton's origins, movement from those origins, and the domestication of Old World and New World species separately.

TABLE 6.1   Diploid and Tetraploid Species of the Genus *Gossypium*.

Species	Genomic Group	Distribution
DIPLOIDS (2n = 2x = 26)		
G. herbaceum L.	A_1	Old World cultigen
G. arboreum L.	A_2	Old World cultigen
G. anomalum Wawr. & Peyr.	B_1	Africa
G. triphyllum (Harv. & Sand.) Hochr.	B_2	Africa
G. capitis-viridis Mauer	B_3	Cape Verde Islands
G. trifurcatum Vollesen	——[1]	Africa
G. bricchettii (Ulbri.) Vollesen	——	Africa
G. benadirense Mattei	——	Africa
G. sturtianum J.H. Willis	C_1	Australia
G. nandewarense (Derera) Fryx.	C_{1-n}	Australia
G. robinsonii F. Muell.	C_2	Australia
G. australe F. Muell.	——	Australia
G. costulatum Tod.	——	Australia
G. cunninghamii Tod.	——	Australia
G. nelsonii Fryx.	——	Australia
G. pilosum Fryx.	——	Australia
G. populifolium (Benth.) Tod.	——	Australia
G. pulchellum (C.A. Gardn.) Fryx.	——	Australia
G. thurberi Tod.	D_1	Mexico, United States (Arizona)
G. armourianum Kearn.	D_{2-1}	Mexico
G. harknessii Brandg.	D_{2-2}	Mexico
G. davidsonii Kell.	D_{3-d}	Mexico
G. klotzschianum Anderss.	D_{3-k}	Galapagos Islands
G. aridum (Rose & Standl.) Skov.	D_4	Mexico
G. raimondii Ulbr.	D_5	Mexico
G. gossypioides (Ulbr.) Standl.	D_6	Mexico
G. lobatum Gentry	D_7	Mexico
G. trilobum (Moc. & Sess. ex DC.) Skov. emend. Kearn.	D_8	Mexico
G. laxum Phillips	D_9	Mexico
G. turneri Fryx.	——	Mexico
G. schwendimanii Fryx.	——	Mexico
G. stocksii Mast. ex Hook.	E_1	Arabia
G. somalense (Gurke) Hutch.	E_2	Arabia
G. areysianum (Defl.) Hutch.	E_3	Arabia
G. incanum (Schwartz) Hillc.	E_4	Arabia
G. longicalyx Hutch. & Lee	F_1	Africa
G. bickii Prokh.	G_1	Australia
Allotetraploids (2n = 4x = 52)		
G. hirsutum L.	$(AD)_1$	New World cultigen
G. barbadense L.	$(AD)_2$	New World cultigen
G. tomentosum Nutt. ex Seem.	$(AD)_3$	Hawaii
G. mustelinum Miers ex Watt	$(AD)_4$	Brazil
G. darwinii Watt	$(AD)_5$	Galapagos Islands
G. lanceolatum Tod.	$(AD)_7$	Mexico

Source: Adapted from Endrizzi et al., 1984.

1. Dash indicates that genome designation has not been determined. Reproduced from *Cotton*, R. J. Kohel and C. F. Lewis (eds.), p. 63, 1984, American Society of Agronomy.

Figure 6.1 *Origin and early movement of diploid cotton. Ancient gold traders in Zimbabwe and Mozambique (a) transported* Gossypium herbaceum *race* africanum *from its origin in South Africa (b) into Ethiopia, Arabia, and Iran (c) where G. herbaceum race acerifolium apparently evolved without assistance from mankind. The oldest evidence of cotton being used by humans was found at Mohenjo-Daro (d) in Pakistan. These remains were estimated to date from about 2300 B.C. (Drawing by Mike Hodnett)*

MOVEMENT FROM ORIGIN—DIPLOIDS

To understand the movement and domestication of Old World cotton, we must consider the movements of primitive peoples on two continents and the prevailing climates in northern Africa and Asia Minor (Figure 6.1). Ancient trade routes apparently existed along the eastern coast of Africa, across southern Arabia and Iran, and into Pakistan and down the western coast of India. There appears to be evidence that gold was mined in present-day Zimbabwe, carried eastward into Mozambique, and transported overland up the East African Coast, or perhaps even transported by "dhow" seamen that sailed with the reliable and alternating winds of the northeast and southwest monsoon seasons. Such ocean trade routes predate the time of Christ.

Perhaps the wild South African cotton, *G. herbaceum* var. *herbaceum* race *africanum*, was taken northward by people as packing or wadding to protect breakables such as pottery or for dressing wounds. This species/race was moved northward along the east coast of Africa to southern Arabia, probably during a time when the entire area that today encompasses the Sahara desert received considerable rain and therefore was much more conducive to plant life than we find it today. The movement to and establishment of this feral cotton in southern Arabia could have occurred during the pluvial periods of 10000 and 6000 B.C. and/or between 5000 and 2500 B.C. Archaeologists believe that agriculture—that is, the planting and harvest of crops and herding of animals by people living sedentary life styles—was practiced at least 7000 years B.C. in the Iraqi-Kurdistan area and that at least a "decrue" system of agriculture was practiced in northern Africa by 5000 B.C. Given these circumstances, *G. herbaceum* var. *herbaceum* race *africanum* could have become established as a perennial plant in northern Africa and Arabia during these times since the winter season is not sufficiently cold to kill the plants. This human-assisted movement from about 25° South latitude to about 20° North latitude led to the establishment of a new race of *G. herbaceum*, that being *acerifolium*, the race considered to be the most primitive cultivated form of diploid cotton.

Although these events could explain the presence of cotton in northern Africa and Arabia, there is no evidence that ancient people of this region ever used the fiber for spinning and weaving. The last piece of the puzzle relative to the domestication of cotton as a textile may lie with the Indus Civilization of 2500 B.C. to 1500 B.C., perhaps the first of the world's four great early civilizations (the other three occurring along the Nile River Valley in Egypt, between the Tigris and Euphrates rivers of Mesopotamia, and in the Huang Ho Valley of China). Archaeological excavations have revealed an advanced people in the Indus Valley, a civilization with brick construction, elaborate drainage systems, grain warehouses and even a system of weights and measures. Mesopotamian artifacts found at dig sites indicate that the Indus people traded with the people of northern Africa and the Mediterranean area. Evidence of a textile industry was found at the Mohenjo-Daro excavation site in the Indus River Valley, and estimated to be from about 2300 B.C. Archaeologists found fabric and string made from coarse cotton fibers, indicating the existence during that time period of a well-established textile craft, if not an industry.

From *G. herbaceum* race *acerifolium* evolved *G. arboreum*, probably a mutation selected by humans. This species is found today only under production, or as escapes from production. There is no evidence that *G. arboreum* exist in the wild today except as escapes from cultivation. From this beginning, more annualized (actually phenotypes that will produce a crop in less than a year—all *Gossypium* species are perennial) biotypes of *G. herbaceum* and *G. arboreum* evolved that spread cotton culture west and north of southern Arabia, and east of the Indus River Valley into China and Burma.

ORIGIN OF NEW WORLD TETRAPLOIDS

The origin of the two cultivated New World species of *Gossypium* is less understood than that of the Old World diploids. Cytological and molecular evidence indicates that the New World tetraploid species are allo-tetraploids, which means that their chromosomal makeup combines the chromosomes (i.e., genomes) of two distinct diploid species. The two genomes were sufficiently different that their chromosomes would not synapsis, or pair, during meiosis, meaning that the initial hybrid was sterile. Therefore, natural doubling of each set of chromosomes had to occur for the natural hybrid to be fertile and produce offspring, an extremely rare event in nature. But even more mysterious is the fact that one of the tetraploid genomes (designated the A genome) is from *G. herbaceum,* an Old World species, and the other (designated the D genome), is from a New World species, most likely *G. raimondii.* This mystery leaves many intriguing questions: How and when did a chance hybridization occur between two species separated in distribution by vast oceans, and how many chance hybridizations had to occur before a chance doubling of the two genomes occurred that would ensure offspring? Percival and Kohel (1988) recount three possibilities.

1. Ancient Origin Theories:
   a. Since New World tetraploids are found in the feral state in the Micronesian and Polynesian islands, it has been suggested that the A and D genomes were brought together briefly (by geologic time scale) in a trans-Pacific land bridge about 60,000,000 years ago—i.e. in the late Cretaceous and Tertiary time periods.
   b. The A and D genome diploid species were native in what is today eastern South America before the land mass separated and continental drift resulted in the formation of the African and South American continents.

2. Recent Origin Theories: Several people have proposed a recent origin for the development of the American tetraploid species. The earliest such proposal was in 1935 and the latest was in 1975. The commonality in all recent origin theories is that *G. herbaceum* was transported to the Americas by people, or resulted from human activity, via the Pacific Ocean. It seems, to this author, that this system would mandate feral *G. herbaceum* plants in the Americas today, yet they are not known.

3. Unknown Origin: No one has developed a theory or presented acceptable data on how, when, or where the natural hybridization of *G. herbaceum* and *G. raimondii,* and of course the subsequent doubling of the hybrid's chromosome complement, occurred. Recent cytological and molecular DNA studies do support that tetraploid cotton is relatively recent in origin, being placed at about 1 million years ago.

An interesting tangent to the study of the origin of tetraploid cotton is that they seem to be native to littoral habitats—i.e., found near seashores. Three characteristics of feral tetraploids, those being an impermeable seed coat, a relatively long-lived embryo, and highly developed seed hairs or fibers (effective flotation device) condition their seed for dispersal by sea. Geographical distribution of feral tetraploid cotton, especially *G. hirsutum,* suggests that transport by sea was a means of dispersal before movement by humans.

MOVEMENT FROM ORIGIN—TETRAPLOIDS

The apparent distribution of the tetraploid cotton species in the New World at the time of its discovery by Columbus in 1492 is shown in Figure 6.2. The center of diversity for *G. barbadense* appears to be northwestern South America. Its first use was as cordage for fishing nets by a settled people in what is present-day Peru. Archaeological excursions at Huaca Prieta, northern coast of Peru, and La Galgada, in north sierra Peru, have yielded cotton seed, fibers, boll parts, yarn, cordage, fishing nets and/or fabric dating to 2400 to 1500 B.C. Similar artifacts recovered from the Ancon-Chillon coastal site in central Peru could be as old as 2500 B.C. From this center, *G. barbadense* dispersed by whatever means to most of South America, the West Indies, and the Galapagos Islands. Movement overland could have and may have occurred without aid by humankind. Movement from the South American continent to the West Indies and the Galapagos Islands could have been human-assisted or accomplished simply by ocean currents. Cottonseed has been shown to survive in salt water for as long as two months, sufficient time to reach these islands.

G. hirsutum apparently evolved in southern Mexico and was dispersed throughout Mexico, Central America, the Caribbean Islands, the southern tip of Florida, the Florida Keys, and southern New Mexico. The oldest archaeological specimens of this species were found in Tehuacan, Oaxaca, Mexico and are tentatively dated at 3400 to 2300 B.C. While nonhuman-assisted dispersal could account for some of this distribution, as noted above for *G. barbadense,* we do know that the Maya people (250 to 900 A.D.) were seafarers and could have been responsible for the distribution of *G. hirsutum* to the West Indies. The Maya were using two kinds of cotton at the time of the Spanish conquest, one to weave clothing called manta, a cloak or wrap, and ceiba (piim), which was used for pillows. The ceiba fibers were reportedly a better quality fiber and came from a sacred tree that the Maya believed held up the heavens. This better quality fiber could have been *G. barbadense* obtained through trade with Amerindians in South America.

Because there are no written records and little documentation of the movements of early people in the Western Hemisphere, one can only speculate as to the association of humans in the movement of *G. barbadense* and *G. hirsutum.* We do know, however, that the Inca civilization that dominated the west coast of South America at the time of Spanish conquest in 1532 used cotton that was derived by trade or barter from people of the upper Amazon region. This cotton was famous for its long, naturally brown-colored fibers. We know also that tribes of Amerindians traded with other tribes. We do not know how extensive this trade was before the

G. hirsutum

G. barbadense

Figure 6.2 *Approximate distribution of* G. barbadense *(South American and West Indies) and* G. hirsutum *(Central America, Southwest U.S., and West Indies) during pre-Colombian times. (Drawing by Mike Hodnett)*

discovery of the New World by the Europeans. But given the knowledge that corn, or maize, had been dispersed and had become a standard of Amerindian diets on two continents by the time Columbus discovered the New World, it seems reasonable to assume that trade occurred in very ancient times, at least over short, overlapping distances.

In both the New and Old Worlds, cotton was probably used by humankind for reasons other than spinning into yarn and then weaving into fabric. As noted earlier, cotton could have been used as wadding, packing, or for dressing wounds. The use of cotton as a spinnable fiber probably occurred in societies that already spun flax or animal hair. Ancient selection of better fiber types in both the Old and New Worlds seems obvious as there appears no existing adaptational advantage for abundant seed hairs unless they evolved naturally as a flotation device for New World species or as a device to prevent inhibition of water. The latter would have some natural "fitness" in that it would ensure that germination would occur only after the rainy season had begun in the tropics.

Humans probably drove the selection process for day neutral and annualized forms that were necessary to transform cotton from an occasional home-based enterprise to a commercial enterprise. All commercial cotton species are perennial when grown under conditions suitable for continued growth, those conditions being moisture, adequate nutrients, and sufficiently warm temperatures. By annualized forms, we mean biotypes that will produce an economic product, in this case fibers, during one season, be that the warm, summer season in temperate climates or a wet-dry season in tropical climates.

BOTANICAL CLASSIFICATION

The genus *Gossypium* was named by Linnaeus in the middle of the 18th century. The genus has been classified in both the *Malvaceae* and the *Bombacaceae* families and in both the *Hibisceae* and *Gossypieae* tribes. Today, the genus seems firmly placed in the *Malvaceae* family, tribe *Gossypieae*, because of the uniqueness of the lysigenous glands found throughout species within the genus. These glands contain a number of sesquiterpenes, collectively called gossypol.

MOVEMENT TO THE UNITED STATES AND CULTIVAR DEVELOPMENT

The first cotton planted in the United States, other than that planted by the Amerindians, was planted by the Spanish in Florida around 1556. English colonists first planted cotton near Jamestown in 1607. Cotton has been grown annually as a commercial crop in the United States since 1621. Colonists living in major seaports had the opportunity to buy articles of clothing imported from England, but those living in the interior or colonists of lesser economic means spun and wove wool and/or utilized leather for clothing. Early production by the colonists therefore was not as a production venture but simply as a means to provide a small amount of cotton fiber to supplement wool and leather. As late as 1774 there was such a small amount of cotton produced in the colonies that when eight sacks of lint, approximately 400 pounds, were shipped from Norfolk, Virginia, to Liverpool, England, the clerk refused to validate the receipt, thereby identifying it as contraband, because he did not believe that there was that much cotton in all of America.

Although England protected its wool industry and its spinning and weaving monopolies by passing laws in 1700 and 1721 that forbade the use of and wearing of cotton, the colonists were not officially discouraged from producing cotton. Farms as far north as New Jersey and Pennsylvania had small fields of cotton. England, however, was more interested in other American products such as tobacco, corn, lumber, and even silk. In the Carolinas, for example, there was a fine of £10 for any colonist owning at least ten acres of land who did not plant at least ten mulberry trees, the trees being used as a source of food for silkworms. No one, in England or the American colonies, foresaw the vast quantities of cotton that would one day be produced by America, and purchased by England.

Nonetheless, cotton culture was a home craft industry in the early days of the American colonies. Having a few garments made of cotton was a convenience enjoyed in the middle and New England colonies, the weather there being mild summers and cold winters. But the southern colonies were a different story. Imagine leaving a country that had summer high temperatures in the 60 to 75°F range and moving to a southern outpost where summer temperatures might reach 100°F on any given day from June through September. Cooler clothing was not a luxury but, rather, a near necessity. The small garden plot of cotton soon was not enough to supply an expanding population, and raw cotton began to be imported from the West Indies during early colonial times.

The early introductions of cotton into the colonies were probably from the Levant, i.e. the lands along the eastern end of the Mediterranean Sea. Most were likely reintroductions of G. hirsutum, the New World species having spread to many Old World countries during the century following Columbus's discovery of the Americas. Some of these early introductions surely were G. arboreum or G. herbaceum, but those were unproductive and were soon abandoned. White Siam (probable form Thailand), and Chinese Nankeen cotton cultivars were brought into the Louisiana Territory during the middle 1700's. Other introductions were New World tetraploids from the West Indies, Mexico, Central and South America, or reintroductions into the New World of tetraploid cotton taken to Europe and Asia by earlier explorers.

The Old World species probably were annualized biotypes but likely did not produce well half a world away from their area of adaptation. Too, they possessed inferior fiber compared with G. hirsutum and G. barbadense. The New World introductions were probably photoperiodic or were perennial types that did not "set" fruit that would mature before frost. As noted earlier, feral or domestic cotton growing in the tropics behaved as perennials and were harvested whenever bolls matured, probably in the early weeks of the "dry" season. These biotypes were not generally adapted to annual production nor could the plants, as a general rule, tolerate frost. But the immediate isolation when grown in the Colonies, and severe selection pressure for the day neutral and annualized growth habits greatly sped the selection of biotypes that became the first true cottons of the U.S. The progenitors of these biotypes were doubtless G. hirsutum from eastern Mexico and Central America.

There were two types of G. hirsutum cotton that became established in the colonies and territories, black or naked seed types (introduced by the French about 1730 and known also as Creole Black Seed), and fuzzy green seed (later known as Georgia Green Seed), the latter introduced from the West Indies by botanist Phillip Miller. As will be discussed later, the fibers of cotton are extensions of seed epidermal cells, and many of these cells produce fibers long enough to be spun into yarn. Other cells produce fibers that are very short, and these, along with the stubs of the longer fibers that have been broken in the ginning process, are called fuzz or linters. Black-seed or naked-seeded types produce fiber cells that become detached from the seed coat and may produce varying amounts of linters, from none to sparse to tufts of linters at the chalazal cap. Fibers of the black-seed types can be easily removed from the seed by passing the fibers between two rollers, effectively squeezing the seeds from the fibers. Fibers on the green-seeded biotype, however, were very laborious and time-consuming to remove since they adhered tightly to the seed. Nonetheless, it was green or fuzzy-seeded type that predominated, and still does. It was simply a matter of economics; the green-seed cottons produced more lint.

The other New World species, *G. barbadense,* was introduced into the colonies in 1786 when Frank Levett, spelled Leavet by Watt (1907), and family moved to Sapelo, an island off the coast of Georgia. That year, Levett received three sacks of Pernambuco cotton seed from Patrick Walsh of Kingston, Jamaica. This cotton probably originated in Brazil, as Pernambuco is a seaport in that country. It is unclear whether Levett selected plants of Pernambuco or if the Pernambuco plants were the only plants to become established. At any rate, by 1789 this biotype of *G. barbadense* had become accepted because of the superior quality of its fibers. The selected Pernambuco cotton had longer, stronger, and thinner fibers than other *G. barbadense* types being grown in the colonies and certainly better quality than *G. hirsutum* or either of the Old World species. (*G. barbadense* is characterized by short, coarse fibers in and near its center of origin in South America.) This high-quality cotton was most suited to the climate and soils of the coastal regions of South Carolina, Georgia and Florida, and the islands off their coasts, and in time became known as sea island cotton. *G. hirsutum,* being better adapted inland or upland from the coast, became known as upland cotton. Those names became fixed in the American lexicon and remain so even today.

Few records exist relative to the introduction of cotton into the United States. One of the early anecdotal accounts of a very significant importation was first reported in writing in 1854. The importation of an upland cotton biotype from Mexico in 1806 is arguably the most important factor in cultivar development in the United States. This cotton became known as Mexican Highland Stock, and the reported story of its importation is a fascinating footnote of history. A cotton producer named Walter Burling from present-day Mississippi traveled to Mexico City under the official guise of attempting to settle a boundary dispute between the Spanish territory of Mexico and the Louisiana Territory, now U.S. territory, having been purchased from France in 1803. To this end, Burling sought and was granted an audience with the Viceroy of Mexico, Jose De Iturrigaray. After addressing the boundary issue, Burling requested seed of a cotton reportedly grown by the Mogui Amerindians of Mexico. This cotton was reported to be very productive and of excellent quality. Perhaps because the Viceroy was under orders not to allow the exportation of Mexican products, and certainly not to a territory recently a part of

Spain's long-term enemy, France, Burling's request was officially denied. However, apparently with the blessings of the Viceroy, Burling smuggled this cottonseed out of the country in a number of Mexican dolls.

The following year, Burling gave the seed to William Dunbar, who apparently received favorable reports on the fiber from textile experts in England. Between 1807 and 1810, Dunbar increased the contraband seeds to over 3,000 pounds, and by 1820, this Mexican introduction had out-crossed with the Georgia green seed and Creole black, creating a plethora of new genotypes of cotton. History suggests that other introductions of the Mexican cotton were made in the early 1800s. However, definite proof is lacking, because the appearance of the Mexican phenotype in Georgia and the Carolinas about 1825 could have originated from seed of the 1806 introduction.

Dr. Rush Nutt of Rodney, Mississippi, and Henry W. Vick, son of the founder of Vicksburg, Mississippi, developed cultivars called Petite Gulf and One Hundred Seed, respectively, apparently as selections from the Mexican Highland Stock. These cultivars spread across southern Mississippi and beyond, out-crossing with other commercial cottons and being reselected by farmers in neighboring communities. By 1880, there were 58 cultivars of cotton grown in the U.S. Cotton Belt, which stretched from the Atlantic to Texas; 118 cultivars by 1895; and over 600 by 1907.

Cotton breeding and cultivar development was practiced in the United States through selection for seed and plant characteristics from 1807 through the remainder of that century. Some hybridizations probably were made along the way, but scientific cotton breeding began with the appointment of Dr. H. G. Webber in 1897 by the USDA to oversee cotton cultivar development activities of that agency. Although some cotton breeding had already taken place at several state experiment stations, including Alabama, Georgia, South Carolina, and Tennessee, the programs were of little permanent value from a cultivar development viewpoint. The USDA stepped up its role in cotton cultivar development in 1911 by hiring H. B. Brown and E. C. Ewing to conduct cotton breeding activities at Stoneville, Mississippi.

Commercial cotton breeding, utilizing the rediscovered Mendelian laws of genetics, began in 1914 with D. R. Coker at Hartsville, South Carolina. In 1915, E. C. Ewing resigned from the USDA and began a commercial program with the Delta and Pine Land Company of Scott, Mississippi, and H. B. Brown left the USDA in 1922 to initiate a cotton breeding program with the Stoneville Pedigreed Seed Company at Stoneville, Mississippi. These three companies dominated the old south cotton cultivar market until the 1980s when the Coker Pedigreed Seed Company was purchased by the parent company owning the Stoneville Pedigreed Seed Company. Cultivars of the Stoneville and Deltapine Companies continue to dominate acreage from Texas to the East Coast.

In order to understand the remaining details of cultivar development in the United States we must consider two aspects of U.S. cotton culture that shaped that development. First, the Mexican boll weevil, *Anthonomus grandis* Boheman, reached the southern tip of Texas in 1892 and spread across all of the eastern U.S. Cotton Belt by 1933. Second, cotton is grown in the United States in a wide range of environments, from subtropical to temperate to desert.

The boll weevil arguably has caused more dollar value losses and caused changes in more production practices than any other insect in the history of agriculture. Reports of 90 percent losses from infestations of the boll weevil were heard from South Texas in the 1890s. Producers across the Cotton Belt hoped to no avail that the weevil could not overwinter in the Mid-South and Southeastern states. In 1892 there were essentially no insecticides, so producers tried all sorts of control measures: picking and burning infested squares (i.e., flower buds) from plants and the ground; planting rows in an east-west direction to increase soil temperatures that would kill developing larvae in cotton squares that had fallen to the ground; planting wide rows with a deep furrow between rows so that fallen squares would slide/roll to the furrow where they would be exposed to direct sun. At best, all of these were marginally successful. Destruction of stalks very soon after harvest, a cultural recommendation around the turn of the century, would become a cornerstone of weevil management in later years, but in 1900 it was impractical without power machinery.

However, producers had one viable option with which to combat the weevil—earlier maturity. The natural intermating of different Mexican Highland Stocks with each other and probably with the green and black seeded types, followed by farmer-seedsman selections, had resulted in over 100 cultivars by 1895. Some of these cultivars were grown in the northern reaches of the Cotton Belt where the growing season was much shorter than in South Texas. By the turn of the 20th century, cotton was grown commercially as far north as Missouri, Tennessee, Virginia, and Ohio. Cultivars from these areas would produce a crop of cotton in fewer days than cultivars grown at the time in South Texas. Producers in the Southern United States quickly imported and began growing these earlier-maturing types because they realized that earlier-maturing fields produced more cotton than later-maturing fields when the boll weevil was present. The great diversity in cultivars grown in the United States allowed cotton producers to remain in cotton production through the early years of the boll weevil era. Too, the weevil further spurred the development of new cultivars, reaching 600+ by 1907.

The boll weevil also caused farmers to move north and west in efforts to escape the destruction of their only cash crop by this persistent and devastating insect. Some producers moved to the Rolling and High Plains of Texas and Oklahoma, while others moved west with the opening up of the southwestern states. Winters are too cold most years for weevil migration to the High Plains, and in far West Texas, New Mexico, Arizona, and California, producers were protected somewhat by the expanse of desert to their east. Producers and seedsmen took seeds of cultivars that they had been growing, imported others, and selected still others that fit their new environments; very early types that tolerated the cool nights of the High Plains were developed, as well as types that produced under the desert conditions of the Southwest United States.

While producers in Texas and east of Texas have produced short staple (i.e.,

TABLE 6.2   Comparison of Types of Cotton Produced in the United States.

Biotype	Range in Fiber Properties			Production Area
	Micronaire	Length	Strength	
	(units)	(in.)	(g./tex.)	
Upland short staple[1]	2.5–4.5	<1.00	23.0–30.0	Northern Cotton Belt, High Plains
Upland medium staple	3.2–5.0	1.00–1.10	25.0–30.0	Southeast, Mid-south, parts of Southwest, High Plains
Upland long staple (Acala)	3.5–4.9	1.11–1.28	28.0–37.0	New Mexico, Arizona and California
Pima (Extra Long Staple or ELS)	3.5–4.9	1.29–1.45	40.0–51.0	Arizona and California

1. Modern plant breeding efforts have decreased the amount of short staple cotton produced in the United States in recent years.

length of fibers) and medium staple upland cotton, New Mexico and California also produced a long staple upland cotton referred to as Acala (Table 6.2). This type was first collected in 1906 by USDA scientists looking for boll weevil-resistant cottons near Ocosingo, Chipas, Mexico. Those seed were inviable, and the area was scouted again the next year. Plants resembling the phenotypes of a year earlier were found near the town of Acala, Chipas, Mexico. Those seeds eventually gave rise to the Acala cultivars of today. There are two types of Acalas, New Mexico Acalas and California Acalas.

G. barbadense, also called sea island, American-Egyptian, Egyptian, Pima, or extra long staple (ELS), has been grown on limited acres in the United States since colonial times. The first *G. barbadense* grown in the United States was on the coastal islands of South Carolina and Georgia, and later, on limited acres interior from the coast in south Georgia and north Florida, and occasionally along coastal areas of Texas. The lint from this cotton is exceptionally long, strong, and very fine relative to *G. hirsutum,* or upland cotton. Sea island fibers went to spinners of fine yarns for sewing thread, laces, and other delicate products, as well as to products requiring structural strength such as sail cloths and, at one time, for cords in rubber pneumatic tires.

Sea island cotton was introduced into the United States from Jamaica, and the West Indies just after the Revolutionary War. Its exact origin is unknown, although there is the suggestion that the cotton originated in Brazil as noted earlier. Another account has it originating in Persia, or present-day Iran. If that is the case, sea island probably resulted from the intermating of *G. hirsutum* and *G. barbadense* that had been carried to the Levant from the New World between, say, 1500 and 1750.

The U.S. crop of sea island reached 50,000 bales (probably less than 480 pound bales) by 1890 and over 100,000 bales by 1897. By the early 1920s essentially all production had ceased; the soft carpel wall, or boll wall, and a much slower growth habit than upland made this type of cotton extremely vulnerable to the boll weevil. Since practically all sea island was grown along the east coast of the United States, the impact of the weevil was not felt until the crops of 1918 and 1919. In 1917, the United States produced 92,619 bales of sea island; 52,208 in 1918; 6916 in 1919; and 1868 in 1920. Sea island cotton ceased to be of economic importance in the United States.

An effort was made around 1936 to revive the sea island industry, but it declined to the point of nonexistence by 1943. Another attempt was made by the USDA and the Georgia Experiment Station to revive the industry, and scientists developed and released a cultivar of sea island called Coastland in 1953. This attempt failed and the program was discontinued in 1962. Today, the Carolinas and Georgia are free of the boll weevil because of the boll weevil eradication program that has as its goal the elimination of the boll weevil from the continental United States, perhaps opening the way for another attempt to resurrect sea island cotton.

Other *G. barbadense* having lower fiber quality than sea island, but still exceptional when compared with upland cotton, was introduced into the southwest United States about 1900. This cotton became known as American-Egyptian or Egyptian. It is believed that the parental stock that gave rise to this cotton was from the hybridization of a sea island and a *G. barbadense* tree cotton (suggesting only that it was allowed to grow for several years) called Jumel's tree. Jumel, a French engineer, observed this tree growing in a garden near Cairo. An Egyptian cultivar called Mitafifi was developed from such hybridizations in 1887 and was introduced into the United States in 1900. The USDA released the cultivar Yuma in 1908, probably a direct reselection from Mitafifi. This was the first modern *G. barbadense* or extra long staple grown in the United States. This type of cotton soon became known as Pima, named for the county having that name in Arizona, which was doubtless named for the Pima Amerindians that inhabited the southern part of the state. Five cultivars were released by 1949 from this Egyptian germplasm base. Pima S-1, released in 1951, was developed from a series of crosses involving sea island, Pima, and Tanguis, all *G. babadense* types, and an upland or *G. hirsutum* cultivar developed by the Stoneville Pedigree Seed Company in Stoneville, Mississippi. Today that program is run by an association of Pima cotton growers, called Supima, with limited help from the USDA. Pima production makes up approximately 2 percent of U.S. production.

SPINNING AND WEAVING

The beginnings of spinning and weaving of cotton are lost in antiquity. It is generally believed that the earliest efforts at spinning and weaving cotton were by people that already spun and weaved some product such as flax. Fiber flax, as opposed to seed flax for the production of linseed oil, is believed to have been produced in Syria and Turkey as early as 7000 B.C. Fibers from the flax plant were probably one of earliest weaving fibers in the Old World. Of course, many ancient cultures in both the Old and New Worlds spun and weaved animal hairs—e.g., sheep's wool in the Old World and llama hair in the New World.

Spinning

Spinning is a process whereby twisting is applied to a supply of overlapping fibers to produce a yarn. The ancient form of spinning employs a spinning stick, also called a spindle, that is rotated by one hand to take up the yarn that is produced as fibers are twisted between the thumb and forefinger of the other hand. The origin of this technology is unknown, but it is common to all ancient cultures in both hemispheres. The process is used to this day in some cultures—e.g., the Bedouin in the Middle East and by some cultures in Central America.

The spinning wheel was the next step toward developing the modern methods of mass-producing yarns. It is believed to have been invented in India between 500 and 1000 A.D. The earliest versions had a wheel, of sorts, turned by hand that rotated a band or small belt running through a groove in the spindle, thereby rotating the spindle taking up the yarn. The yarn still was made by twisting fibers together with the other hand. Foot pedals for turning the wheel were later added as was a distaff to hold the unwoven fibers. These improvements freed both hands of the spinner for twisting the fibers together.

Cleaning batches of cotton fibers, or wool for that matter, of unwanted particles of plant matter would have been of obvious importance to even the earliest spinners. By 1750, the process took on the name carding, because the home spinner used two pieces of wood, each called a card because of its size, about 6 × 8 inches, having inset rows of teeth or bristles. The cotton fibers were gently pulled through one card by the forward action of the other card. This had three advantageous outcomes: (1) the cotton fibers were untangled, allowing pieces of trash to fall out; (2) short fibers also tended to fall out of the resulting airy mass of fibers; and (3) the remaining longer fibers were somewhat parallel. This treatment resulted in a thin sheet of cotton fibers to which a slight twist was applied to hold the individual fibers together. This slightly twisted rope of cotton fibers is called a sliver, from the old English word "slifan" meaning to split or cleave. When more twist is added to the sliver, it is called roving. Additional twisting and stretching of the roving results in the finished product, yarn.

By the middle 1700s, the expanding commerce of the British Empire had created a demand for cotton yarn that could not be met by use of single-yarn spinning wheels (weaving was consuming yarn faster than it was being produced). James Hargraves in 1767 was the first to invent an improvement in the single-yarn spinning wheel by adding multiple spindles turned by a single wheel. This machine was called the Spinning Jenny, supposedly because spinning was traditionally a female task. The Spinning Jenny produced yarn by the additional pulling and twisting of roving supplied independently to each spindle. Hargraves's invention was not looked on favorably by English spinners, as it replaced the work of eleven people by only one person. On one occasion, disgruntled spinners broke into Hargraves's home and destroyed his machine, while on another occasion a mob scoured the countryside looking into homes and businesses and destroying every Spinning Jenny they could find. Hargraves relocated to another part of the country, assisted in the building of other Spinning Jennies, but died without receiving just profits from his invention. Yarn produced on the Spinning Jenny, also called the Common Jenny, was not suitable for warp (i.e., the lengthwise yarns) being too weak and lacking in uniformity. This problem was alleviated by Sir Richard Arkwright in 1769. Arkwright improved the Spinning Jenny by adding a set of rollers turning at different speeds to draw out and therefore straighten cotton fibers in the roving prior to the final twist and winding onto spindles. Hargraves's Spinning Jenny stretched the cotton fibers at the same time as twisting was added but did not sufficiently draw the fibers (i.e., did not make the roving, or rope, of fibers longer and finer) before the final twisting into yarn. This improved Jenny is often referred to as the Water Spinning Frame as it was the first spinning machinery to be powered by water. Yarn produced on these frames was well suited for warp, often called Water Twist yarn, and thus good quality, 100 percent cotton fabric became possible.

Samuel Crompton in 1779 combined the stretching system developed by Hargraves and the drawing rollers of Arkwright into one machine, dubbed the Mule Jenny and later called the muslin wheel as its yarn was widely used to produce this type of material. The original Mule Jenny had 48 spindles and produced a yarn much finer and more uniform throughout its length. By 1938, the mule spinning frame had been enlarged to carry as many as 1,300 spindles.

In 1828, James Thorpe, an American, invented the Ring Spindle, or ring spinning frame. By the early 1900s, the ring frame was used almost exclusively by U.S. industry, although the mule frame continued to be the spinning frame of choice in England. There are three systems of spinning cotton in the United States today: ring, rotor, and air jet (Figure 6.3).

In ring spinning, roving is drawn or stretched through a series of rollers, each set to turn faster than the previous set. This effectively reduces the size of the roving and stretches the roving such that the fibers become more parallel. The reduced roving is then passed through a traveler that is attached to a ring and onto a spool that is attached to a spindle. The spool turns at several thousand revolutions per minute (rpm), much faster than the traveler. This difference in speed puts drag on the roving and gives it a final twist to produce yarn. The frame that houses the ring moves up and down the spool, evenly distributing the yearn over the length of the spool (Figure 6.3a).

Air jet spinning may replace ring spinning because it too is a roller drafted system that producers yarn much faster than ring spinning frames. Sliver is drafted through a series of rollers, after which the fibers on the outer edge of the drafted sliver are wrapped around a central core of sliver fibers by the action of air forced through a series of jets (Figure 6.3b).

The rotor spinning process is very different from ring spinning in that it is not a roller drafted system. The sliver is fed directly into a rotor that is revolving at several thousand rpm. The sliver is cleaned of trash particles a final time by a combing roller because minute trash particles can cause significant damage to the collecting grove of the rotor. The centrifugal action of the rotor wraps fibers around a central core of fibers as they are "pulled" out of the rotor (Figure 6.3c).

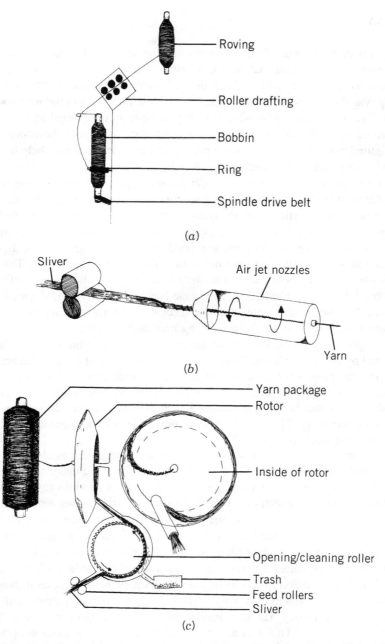

Roving

Roller drafting

Bobbin

Ring

Spindle drive belt

(a)

Sliver

Air jet nozzles

Yarn

(b)

Yarn package
Rotor

Inside of rotor

Opening/cleaning roller

Trash
Feed rollers
Sliver

(c)

Figure 6.3 *Illustrations of the three types of spinning machinery used in the United States. The ring spinning frame (a) and the air jet system (b) are roller drafting systems whereby the cotton fibers are stretched and twisted. The rotor system (c) is not a roller drafting system but utilizes centrifugal forces to wrap fibers into a yarn. (Drawing by Mike Hodnett)*

Weaving

The process of producing fabric by interlacing one set of yarn with another set is termed weaving. The yarns that run the length of the fabric are called warp, while the crosswise yarns are called filling or weft yarn. It is believed that the first weaving was the intertwining of relatively stiff plant material to produce mats and baskets. The lengthwise strips were probably laid out on the ground and the cross strips woven through without the use of any mechanical assistance. Someone eventually figured out that weights or some other stabilizing device would help hold the lengthwise strips parallel and stationary, thereby making weaving easier. Thus, the ground loom was invented. The earliest depiction of a ground loom is on the side of a bowl found at Badari, Egypt, and archaeologically dated to about 4000 B.C. The simple ground loom is still in use today by the Bedouin people of the Middle East.

Warp-weighted looms, where weights are fixed to the ends of warp yarns and the yarns passed over a bar such that the weights do not touch the ground, are depicted on Greek vases estimated to be from the sixth to the fourth centuries B.C. This type of loom was used in parts of Europe into the 20th century A.D. Illustrations of looms from about 1900 B.C. in Egypt depict a second bar used as the weight device for the warp. The Navaho Amerindians still use vertical looms by erecting two vertical posts with horizontal bars for hand weaving rugs and blankets.

The simplest loom is the backstrap loom, where the tension on the warp yarns is determined by the weaver's body position. Warp yarns are laid out parallel across a bar, then brought together (probably with notches or some other device to keep them from getting crossed) and tied to a tree or post. The other end of the warp yarn is attached to a bar that has a strap attached such that the weaver can put the strap around his or her body. The weaver then has only to lean back to put tension in the warp, making it ready for the weft.

Putting the weft through the warp in an over and under pattern obviously was accomplished originally by hand (i.e., over one warp yarn and under the next); the next row of weft went over and under in the reverse of the previous row. Unnamed innovators realized that attaching the weft to a stick, or using a long needle with the weft threaded through, would make the process much easier. Thus was devised the shuttle. The next anonymous innovator deduced that a flat stick inserted through the warp yarns in an over-under pattern would, when turned up on its side, create an opening, later to be called the shed, through which the shuttle could be passed unimpeded. The evolution of the loom probably stopped here for several thousand years. Indeed, the backstrap loom with a crude shuttle and one-way shed is still used in parts of the world today. The next improvement came with the use of a heddle bar to create counter-shed. This was accomplished by attaching alternate warp to the heddle bar that could be lifted to raise odd-numbered warp and the shed stick could still be used to raise and lower the even-numbered warp yarns. Thus, the alternating use of the huddle bar and shed stick allowed the weaver to pass the shuttle through in both directions without interference from the warp yarns.

In 1788, John Kay advanced the cause of mass production of woven fabric by developing and patenting the fly shuttle, also called the flying shuttle. For untold centuries, weavers had laboriously weaved and later thrown the shuttle across or through the warp. The fly shuttle operated on a series of springs and a lathe that allowed the weaver to send the shuttle "flying" through the warp with a simple pull of the lathe. Increased productivity resulted.

The size of fabric woven was limited by the size of the loom, the weaver being able to reach only so far. Excessively long ground looms or excessively tall vertical looms were obviously impractical. This limitation was removed with the use of a large spool, called a warp beam, holding parallel wrapped warp yarns to supply warp on a continuous basis, with woven fabric collected on another spool called the cloth beam.

The steps in the modern processing of cotton fibers into yarn and yarn into fabric are:

1. *Blending:* Several bales of cotton are opened in the opening room of modern mills and the fibers thoroughly blended to provide a large lot of a more uniform mixture of the desired quality. This ensures a uniform product.

2. *Cleaning:* Cotton lint is cleaned of trash such as bits of leaves, stems, seeds, soil, etc., by passing through a series of small-tooth saws that pick the fibers apart so that trash can fall out of the lint.

3. *Picking:* A continuation of the cleaning process. Here the cotton fibers are picked or fluffed to the point that they barely hold together in a continuous sheet called the picker lap.

4. *Carding:* The card machine converts the picker lap into a thin, mist-like sheet that is then twisted slightly to produce a loose rope of fibers called a sliver.

5. *Drawing:* Several slivers are brought together and drawn out or stretched to form a single sliver with fibers more parallel with each other. The drawn-out sliver is also more uniform.

6. *Roving:* The sliver is further drafted and twisted to produce roving. The roving is wound on to a bobbin to await spinning.

7. *Spinning:* The roving is further drafted into small strands of fibers that are then twisted into yarn.

8. *Warping:* Thousands of yarns are wound parallel on a giant spool called a warper beam. Warp yarns run the length of the woven fabric.

9. *Slashing:* Warp yarns are coated with starch to give temporary added strength and resistance to abrasion that is needed for the weaving process.

10. *Weaving:* The process of forming a fabric on a loom that interlaces the warp yarns with filling or cross yarns called weft.

HISTORICAL EVENTS

The contribution of agricultural scientists, farmers, seedsman, and especially Walter Burling in collecting, evaluating, relocating, and breeding outstanding germplasm/cultivars of upland and Pima cotton has been noted. Naming all of the individuals who made great contributions to any crop species is impossible, but doubly so for cotton as it is the only crop species that involves four botanical species of different ploidy levels in both hemispheres. However, there are certain events in history that are of such magnitude that students of agriculture must be cognizant of them. For cotton, these events, in addition to crop/cultivar evolution/development, are (1) the development of a market economy in England and Europe; (2) the flying shuttle developed by John Kay in 1738; (3) Hargraves's invention of the Spinning Jenny in 1764; (4) the improvements in the Spinning Jenny by Arkwright and Crompton between 1765 and 1775; (5) the cotton gin, patented by Eli Whitney in 1794; (6) the boll weevil; and (7) mechanical harvesting.

The Middle Ages (i.e., the years between the end of the Roman Empire and about 1500 A.D.) saw the development and decline of the feudal system in Europe and England. A single family, or individual, owned and controlled large tracts of land called manors. Each manor had a landowner, or lord, and a number of tenant managers responsible for the production of subdivisions of each manor. Most of the people, however, were peasants that actually farmed the land. The tenant managers and ultimately the landlord had complete control over the land; they received what it produced, collected taxes, held court, etc. In return, the landlord maintained an army and police of sorts for the protection of the peasants. The peasants were a part of the manor and were not much different from slaves, except that they could not be bought or sold apart from the land. However, if land holdings did change hands, the peasants were part of the transaction.

The number of social events that drove the decline of the feudal system are beyond the agricultural scope of this text. However, an economic revival in Europe around 1000 A.D. began to shape a new order, an order where money was accepted for goods and services. Peasants were allowed to sell wool and other production from their small allotments of land; goods were traded over ever-englarging distances; and trade between countries evolved. The ideas of democracy, and individual rights and responsibilities of government began to take hold in England as evidenced by the Magna Carta of 1215.

By 1200, England's wool trade had become quite sophisticated. The most visible part of this trade involved the export of the very best quality wool to the expert weavers of Flanders, present-day Belgium, and to France. Some of this wool no doubt came back into England in the form of cloth or finished goods.

But there was another, quieter and less visible wool trade that over the course of, say, 500 years, established England as the world's leading textile producer. With growing freedoms for the peasants, who soon became renter-farmers, and an increasing middle class, small-scale trading among the populace of villages began to increase. Home-based spinning and weaving became a source of income for small sheep-rearing farmers with wives, and perhaps daughters, who could spin the raw fleece into yarns and then weave those yarns into cloth. Even in large villages or towns, spinning wool into yarn became the most characteristic livelihood for a woman living alone, hence the term "spinster" that was used later to describe the unmarried, older woman in the United States.

The contributions of Key, Hargraves, Arkwright, and Crompton have been noted already. Add to these inventors and entrepreneurs a growing demand for cotton clothing by the populace on two continents and the stage was set for one Eli Whitney, an unemployed school teacher from Massachusetts.

Cotton Gin

In the new country called the United States of America, there was a relatively small labor force, there being only 4,600,000 inhabitants by the late 1700s. There were not enough craftsman, apprentices, and smiths to supply manufactured goods in the new country as there were in England and Europe where skilled laborers abounded. Forward-thinking Americans had to rely on new and novel methods of production. Eli Whitney was such a person, remembered as the inventor of the cotton gin, but also often referred to as the "Father of Interchangeable Manufactured Parts," a mainstay of manufacturing today.

Whitney, after graduating from Yale, accepted a position as tutor for the children of one Major Dupont of South Carolina. Whitney had been hired by Phineas Miller, also a Yale graduate, who had served as tutor for the children of General Nathanael Greene and had continued on at Mulberry Grove to manage the estate for Catherine Greene after the death of the General in 1786. As for hiring Whitney, Miller had been acting as emissary for Major Dupont.

Whitney traveled south with Miller, Catherine Greene, and her children in 1792. He never did tutor the Dupont children, but he remained at Mulberry Grove where he was encouraged to put his mechanical ability to work resolving the problem of separating cotton lint from the seed of the green-seeded type of cotton, later to be called upland cotton. Whitney, in writing to his father, outlined the problem and indicated that he built the first model in "about 10 days . . . for which I was offered, if I gave up all right and title to it, a hundred guineas . . . turned my attention to perfecting the machine. I made one before I came away which required of one man to turn it and with which one man will clean ten times as much cotton as he can in any other way . . . may be turned by water or horse . . . it makes the labor 50 times less . . . ."

Whitney's method for removing the lint from the seeds was, in retrospect, a simple idea. Teeth or hooks to catch the lint were embedded into a wooden cylinder because, as Whitley would state ten years later, he had no iron plates from which to fashion circular saws with teeth (Whitney successfully defended his claim as the sole inventor of the cotton gin, including the use of saws, in federal court in 1806). The hooks, when pulling the lint from the seeds, were passed through slits in an iron plate wide enough to allow the hooks with lint to pass but which were too narrow to allow the seed to pass through, thus pulling the lint from the seeds which were caught in a box setting below the roller. To keep the cotton lint from remaining on the hooks or teeth, a rear cylinder with brushes attached rotated faster and in the opposite direction such that it swept the lint from the teeth.

Eli Whitney, in his own words to his father, expected to "make a fortune by it." That never happened. The cotton gin, so long hoped for and sought, was so simple in design and workings that any blacksmith, upon seeing the machine work, could copy it. Although Whitney received the patent on March 14, 1794, he never received his just rewards. An accounting might show that he made a few thousand dollars between 1793 and 1807 when the patent expired, but at the cost of extended and recurring illness and perhaps even an early death as a result of borderline solvency for his company and some 60 or more court battles over infringements. As Whitney would declare, "An invention can be so valuable as to be worthless to the inventor."

Whitney was right. In 1790, three years before Whitney's cotton gin, the United States produced only 3,159 bales (one bale = 480 pounds of lint, either upland or sea island). The sea island could be ginned on a roller gin because the lint grows free of the seed, but the upland had to be ginned by pulling the seed by hand from the mass of lint, a task that would yield about one pound of lint per day per person. In seven years after Whitney's machine, production of cotton reached 73,145 bales, a 23-fold increase. By the time of Whitney's death in 1825, the United States produced 168 times more cotton than it did in 1793. By 1890, the U.S. produced 8.6 million bales, reached a high of about 17 million bales in the 1930s, and today produces 13 to 17 million bales annually, depending on weather, biotic pests, and federal government subsidy programs. At the time of the Civil War, cotton accounted for over 50 percent of U.S. export dollars.

An interesting footnote of history is the use of the term "gin." A popular theory is that Whitney patented a cotton "engine" and that Southerners coined the word gin from engine. This assumption is incorrect; Whitney's patent is for a "cotton cleaning machine." The root word from which gin was derived is from the Latin word *ingenium,* meaning inborn talent or skill. From *ingenium,* the Old French language derived the word *engin,* from which the Middle English took the word *engine* that is defined as we think of it today, a machine that converts energy to mechanical motion. However, Middle English also derived the word *gin* (also spelled *gyn, gynne,* and such), meaning a snare or trap, from the Old French *engin.* It was this Middle English word that was used to mean the trapping of lint from the green-seeded cotton of Whitney's day. The term appears to have been in common usage during the late 1700s. Whitney filed his patent application with the Secretary of State, Thomas Jefferson, on June 20, 1793. Jefferson responded to Whitney in a letter dated November 16, 1793, inquiring " . . . Has the machine been thoroughly tried in the ginning of cotton?" Whitney, as a precaution against others patenting his machine before he had done so, appeared before Elizur Goodrich, an alderman and notary public in New Haven, Connecticut, on October 28, 1793. Goodrich certified a deposition by Whitney that he, Whitney, was ". . . the true inventor and discoverer of the machine for ginning cotton. . . ." In fact, the first public notice of Whitney's cotton machine appeared in an advertisement in the *Gazette of the State of Georgia* newspaper, placed there by Phineas Miller, who had become Whitney's business partner. The ad, which read in part, "the subscriber will engage a gin," appeared on March 6, 1794, eight days before Whitney was granted his patent. It appears that everyone except Whitney used the term "gin" as a verb meaning to separate seed and lint and as a noun to identify any machinery to accomplish the task. (The roller gin has been used since antiquity in India and had been imported into the colonies as early as 1734. Improvements in the basic roller gin reportedly were made by M. Dubreuil about 1750.)

Boll Weevil

Cotton production was almost unaffected by biotic pests from 1621 until 1892, except for occasional predation by bollworms and leaf-feeding caterpillars, when the Mexican Boll Weevil, *Anthonomus grandis* Boh., migrated into Texas. Early reports, probably exaggerated, from South Texas were of 90 percent crop loss. (Later, more temperate analysis of the weevil's affect in Texas showed losses to be more on the order of 40 percent.) Producers north of the Rio Grande Valley hoped that the insect could not withstand the winters further north. Their hopes were unrealized and by 1921 the boll weevil could be found in most cotton-growing regions of the southern United States.

Producers tried any and all suggested remedies to combat the weevil: planting patterns, row direction, row widths, even physically removing and burning squares that showed weevil egg deposition. One town in Georgia even went so far as to declare that all businesses and schools would close one day a week to encourage everyone to go to the cotton fields to remove weevil-infested squares, demonstrating the importance of cotton to local economies. However, early on, agricultural scientists such as Mally, Townsend, Howard, and Bennett realized that two production practices would decrease losses to the boll weevil: prompt stalk destruction and early-maturing cultivars.

Adequate stalk destruction would await power machinery, but, as already noted, early-maturing cultivars were championed as a means to survive the boll weevil. The genetic variability that allowed the expansion of cotton cultivars from 2 to 118 between 1806 and 1895 can probably be traced to two events: Walter Burling's smuggled Mexican Highland Stock and Whitney's cotton gin.

One other change occurred that perhaps was caused by the boll weevil, and that was the movement of cotton producers from more southern growing regions of the Cotton Belt to the Rolling Plains and High Plains of Texas and Oklahoma. Because of cultural practices and cold winters, and in some areas the lack of overwintering habitat, producers in these areas today have only sporadic problems with the boll weevil. The weevil did not become established in the Rolling Plains of Texas until the 1960s.

Cotton Harvesters

With the advances in spinning, weaving, and ginning, harvesting became the limiting factor in the supply of cotton worldwide. Some people were adept at handpicking cotton, harvesting upwards of 500 pounds per long day, while others could pick only 100 pounds or so per day. Just as men dreamed about and finally realized a machine to separate lint and seed, they also sought ways to speed up cotton harvest and reduce dependency on hand labor.

The first patents for a mechanical picker were issued in 1850 to two Memphis, Tennessee, businessmen, Samuel Rembert and Jedediah Prescott. Between 1850 and 1935, an average of almost ten patents per year were issued for mechanical cotton harvesters. The first significant amount of cotton mechanically harvested was with a sled stripper harvester near Lubbock, Texas, in 1926. A simple device, the sled was equipped with a box with a V-shaped front that was simply pulled along the row, stripping off bolls, both mature and immature. This idea is still utilized today with "finger strippers" and to a lesser extent with more modern brush strippers that utilize brushes and "batter bars" to remove both mature and immature bolls.

The stripper technique works well in Oklahoma and parts of Texas, but it was not suitable for taller-growing cotton grown under irrigation or in the U.S. Rain Belt. John Rust, an itinerant mechanic, developed the first successful cotton picker in 1931. Rust's machine used barbed spindles that would catch the fluffed seed cotton as they rotated, pulling the seed cotton from the plant as the bar holding the spindles revolved away from the plant as the machine moved forward. The spindle was slightly wetted before it traveled through the cotton plants. The water acts as a lubricant that allows the seed cotton to slide off of the spindle when the spindle travels very close to a rubber pad, called a doffer. The seed cotton was then moved pneumatically into a container called a basket. This basic design is still used today. In 1931, Rust's machine could harvest one bale per day; five bales per day by 1934; and thirteen bales by 1937.

DEVELOPMENT OF U.S. INDUSTRY

The colonial farmers along the Atlantic coast had little need for growing cotton because they could buy textiles imported from the mother country. Consequently, they preferred to produce a number of food crops for personal and local consumption, and to produce rice, tobacco, and indigo as cash crops to be exported to England. Further inland, settlers produced only enough cotton for home use, the difficulty of hand ginning, spinning, and weaving being noted earlier. Nevertheless, cotton was being produced in Virginia in 1621, in South Carolina by 1664, and in Georgia in 1735. During the Revolutionary War, textiles and raw cotton supplies were cut off from the coastal inhabitants, creating a shortage of clothing; a shortage exacerbated by the need to clothe an army. Cotton culture agreed with the climate in the southern colonies and spread also to coastal areas of Maryland, Delaware, and New Jersey, and even into eastern Pennsylvania. Roller gins (suggesting that *G. barbadense* or Old World types were being grown), were set up in Philadelphia during the Revolutionary War and lint distributed to those who could spin and weave. Philadelphia remained a center of seedcotton trade until the invention of the cotton gin established the interior of the southeast United States as the center of production of mass quantities of lint cotton.

The improvements in textile machinery that occurred in England 1738 and 1775 made England the undisputed leader in quantity of woven textiles produced. The designs of these improvements were closely guarded and textile artisans were prohibited by law from immigrating to other countries after they had worked with the modern machinery. However, in 1770, plans for building a Hargraves's Spinning Jenny were smuggled into the colonies, and commercial production of woven fabric by the United Company of Philadelphia began in 1788, operating 26 looms by the close of the war.

The first mill devoted to cotton was established in Beverly, Massachusetts, in 1789 by John and Andrew Cabot, Moses Brown, and others. The Beverly Manufacturing Company could produce 10,000 yards of cotton cloth a year, most, if not all, sold in Beverly. George Washington visited the factory in 1789 and noted in his diary, "In short, the whole seemed perfect, and the cotton stuffs which they turn out excellent of their kind; warp and filling [weft] both row of cotton." But it was not "perfect," at least not as good as that produced in England where manufacturers enjoyed the improvements of Arkwright. That coveted quality of cotton fabric awaited the arrival of one Samuel Slater from England.

Samuel Slater, at age 14, became an indentured servant to Jedediah Strutt, a partner of Richard Arkwright. After his seven-year indentureship, he remained at Strutt's mill as an overseer and eventually supervised the construction of a new spinning mill using Arkwright's advancements. Having heard and read of the fortune awaiting anyone who could reproduce Arkwright's machinery in America, Slater disguised himself as a farmer to elude English authorities, boarded a ship, and sailed for America, arriving in New York City in 1789. By December of 1790, Slater and business partner Moses Brown opened shop in Providence with a 72-spindle Arkwright spinning frame. High-quality cotton yarn production had reached America, and England's much guarded secrets were out. By 1792, the cost of cotton cloth had dropped from 50 cents per yard when produced by a simple spinning wheel and loom to only 10 cents when made by the newer technology.

Within three years of Slater's first mill, there were ten mills operating in Rhode Island and one in Connecticut. The census of 1810, 18 years after the first Slater mill and 17 years after Whitney's cotton gin, identified 226 cotton mills scattered from Tennessee to Massachusetts (Table 6.3).

F. W. Dawson, a southern newspaper editor, wrote in 1880: "The point in which we lay the most stress is that to the extent in which cotton . . . produced in South Carolina is manufactured in the state, the whole of the profit upon the state, from the first stage to the last, remains in some form within the state for the benefit of its people. Where the cotton is produced here and manufactured elsewhere, South Carolina is in the position of furnishing the elements which make other communities rich . . . we know that the wealth of New England is due to the profit made upon the manufacture of the raw material which the South supplies, and which the South . . . buys back from New England at a high price in its manufactured state." This realization and its impact on rebuilding the economies of the southern states following the Civil War resulted in an International Cotton Exposition held in Atlanta, Georgia, in 1881 that is credited with opening the eyes of northern industrialists to the field of investment opportunities in the South. Reduced shipping costs meant reduced expenditures for raw cotton, and cheap and abundant labor fueled the expansion of mill activity in the South after 1880.

TABLE 6.3   Number of Cotton Mills Operating in the United States in 1810, 1880, 1925, and 1990.

State	Mills (Spinning and/or Weaving)			
	1810	1880	1930	1990[1,2]
Pennsylvania	64	59	4	46
Massachusetts	54	175	83	19
Rhode Island	28	115	34	8
New York	26	36	9	26
Kentucky	15	3	5	1
Connecticut	14	82	22	3
Maryland	11	19	8	2
New Jersey	4	17	2	28
Tennessee	4	16	0	15
Delaware	3	8	0	4
Ohio	2	4	0	4
Vermont	1	7	0	0
Alabama	0	16	86	48
Arkansas	0	2	5	1
California	0	0	4	16
Florida	0	1	0	5
Georgia	0	40	131	73
Indiana	0	4	3	1
Kansas	0	0	0	1
Illinois	0	2	2	3
Louisiana	0	2	2	1
Maine	0	24	6	2
Michigan	0	1	1	0
Mississippi	0	8	13	3
Missouri	0	3	0	1
New Hampshire	0	36	8	3
Nebraska	0	0	0	1
North Carolina	0	49	334	238
Oklahoma	0	0	2	0
Oregon	0	0	0	1
South Carolina	0	14	174	137
Texas	0	2	0	10
Utah	0	1	0	0
Virginia	0	8	0	15
Washington	0	0	0	1
Wisconsin	0	1	0	1
Totals	226	755	938	718

Source: Adapted from American Cotton Handbook, 1949; Cotton Trade Journal, 1931; U.S. Dept. Int., 1983; Davison's Textile Blue Book, 1992.

1. Not official census data and therefore subject to reporting discrepancies.
2. Includes knitting mills.

TABLE 6.4   A Few of the Many Uses of Cotton Fibers and Competing Fibers in the United States in 1990.

Items	Total Pounds		% Cotton*
	Cotton	Total	
Trousers, slacks, jeans	540,000	716,320	76
Shirts, blouses	175,200	283,200	62
Underwear, diapers	279,360	352,320	79
Dresses, skirts	158,400	380,160	42
Gloves, mittens	51,360	62,880	82
Coats, jackets	45,600	175,200	26
Towels, wash cloths	556,320	573,600	97
Sheets, pillowcases	249,600	420,000	59
Drapery, upholstery	286,560	778,080	37
Bedspreads	49,920	87,840	57
Medical supplies	69,600	154,560	45
Tarpaulins, tents, awnings	44,640	125,280	36
Thread	39,840	200,160	20
Rope, cordage	19,680	217,920	9
Automotive	11,040	971,040	1
Luggage, handbags	4,320	32,160	13

Source: Adapted from James Howell and Melanie Gordon, *Cotton Counts Its Customers* (Memphis, TN: The National Cotton Council, 1991): reprinted with permission, registration #TX 3260229.

* Percent of the total pounds of all fibers used that were cotton.

USES OF COTTON

The fiber, excluding linters, is the primary product of the cotton plant, although the seed (meal, oil, and seed coats) are used in a number of products and/or as animal feed. Fibers are used mostly to produce yarns for weaving a number of products, from baby diapers to fashionable suits and dresses to NASA space suits (Table 6.4). Man-made fibers gained wide acceptance in the 1960s in the United States because of their easy-care, wrinkle-free appeal and by 1973 cotton's share of the apparel and home fabrics market dropped to only 18 percent. Since then, stronger consumer demand for apparel made from absorbent, soft cotton, along with advances in easy-care treatment of cotton fabrics and in the blending of cotton with man-made fibers to produce easy care plus comfort, have resulted in cotton regaining a 54 percent share of that market. Cotton fibers made up only 1 percent of all fibers used by the automotive industry in 1990 but near 100 percent of the towel and wash cloth textiles. From one bale of U.S. cotton, 480 pounds, industry produces 200 full-size flat sheets or 22,000 ladies handkerchiefs (Table 6.5).

Cotton seeds, usually thought of as a by-product of cotton lint production, are used in a plethora of consumer products, including margarine, mayonnaise, salad and cooking oils, salad dressing, and shortening. In all of these products, the gossypol, a biochemical constituent of cotton foliage and seed leaves, or cotyledons, must be chemically removed, as it is toxic to nonruminant animals. Gossypol, composed of a number of polyphenolic compounds, is compartmentalized by the

TABLE 6.5   Approximate Number of Some Common Items That Can Be Made from One Bale (480 lbs.) of Cotton.

Item	Number Per Bale
Ladies' handkerchiefs	22,000
Mens' handkerchiefs	8,000
Mens' knee socks	3,400
Diapers	3,000
Ladies' blouses/shirts	850
Mens' dress/business shirts	800
Mens'/boys' dress/sport trousers	450
Ladies' knit/work dress	350
Mens'/boys' jeans	325
Sheets, flat, full size	200

Source: Adapted from James Howell and Melanie Gordon, *Cotton Counts Its Customers* (Memphis, TN: The National Cotton Council, 1991): reprinted with permission, registration #TX 3260229.

plant in small vacuoles or glands. These glands are black in color and are referred to as gossypol glands. The gossypol is removed by solvent extraction from the oil following oil extraction from whole cotton seeds by mechanical and/or solvent extraction.

The remaining cottonseed meal is used in feed rations, and seed hulls are often used as feed roughage. Whole cottonseeds are often fed free-choice to cattle, but caution must be exercised because the rumen is not fully developed in calves weighing less than about 400 pounds, and therefore young calves are susceptible to gossypol toxicity. Gossypol toxicity has been reported in mature dairy cattle that are fed cottonseed free-choice plus rations using cottonseed meal as a protein supplement, but problems in mature cattle are rare.

Upland cotton seeds are covered with short fuzz fibers and with the stubble of longer fibers that are broken off during the ginning process. Portions of these very short fibers are removed by additional saw ginning at the oil mill before seeds are crushed. These short fibers are used in the manufacture of at least 30 products in the United States, including rayon, plastics, paper, twine, automobile upholstery, and photographic film.

S. C. McMichaels, in 1953, found a cotton plant that did not contain gossypol in either foliage or seed, in an old variety-type grown by the Hopi Amerindians of Arizona. This discovery has led to the development of a number of genetically improved cultivars and strains having lint quality and yield potential similar to their glanded counterparts. A few of these cultivars have been grown commercially on the Texas High Plains, but their general use has not occurred to date. The glandless trait, obviously, renders these seeds edible by humans, and in a world where many people do not have enough calories to consume, let alone protein, it is difficult to understand this lack of interest. The first biotypes developed in the 1960s and 1970s were highly susceptible to the usual array of insect pests of cotton and also were consumed by rabbits, field mice, crickets, and deer. Although later types continue to be consumed by the latter list, they are less susceptible to the former group than

TABLE 6.6 Comparison of Cottonseed Flour with Three Common Sources of Protein for Human Consumption, per 100 Grams.

Nutrient	Cottonseed Flour	Defatted Soy Flour	Hamburger	Nonfat Dry Milk Solids
Calories	356	356	268	363
Protein (%)	48.1	43.4	17.9	35.9
Carbohydrate (%)	33.0	36.6	0	52.3
Fat (%)	6.6	6.7	21.2	0.8

Source: Adapted from Watt and Merrill, 1975. With permission from Watt and Merrill, Handbook of the Nutritional Contents of Food, 1975, Dover Publishing.

earlier glandless types. Continued development of glandless cotton may one day result in cottonseed and/or its constituents comprising a significant part of human diets around the world, especially in developing nations where land that must be devoted to food crops also could produce cotton fiber for clothing and other products.

There are approximately 30,000,000 tons of cotton seeds produced worldwide each year from about 80,000,000 bales of cotton, with about 6,000,000 tons produced in the United States. Cottonseeds account for approximately 18 percent of the world's vegetable oil production, making cotton one of the five major oilseed crops for the production of oil for human consumption. The cottonseed is 17 percent crude oil, 45 percent meal, 10 percent linters and 28 percent hulls.

Although oil accounts for about 50 percent of the total value of cottonseed in the United States, it is the protein content that could become vital to an expanding world population. Cottonseed flour contains the same level of calories as defatted soybean flour and nonfat dry milk solids, yet contains 1.4 times as much protein as defatted soybean flour and 1.8 times as much as nonfat dry milk solids (Table 6.6). One hundred grams of cottonseed flour have 2.5 times more protein than hamburger, but more than 14 times less fat. After the oil is removed through crushing, cottonseed meal is 41 percent protein, whereas peanut, soybean, and linseed meal are 50, 44, and 34 percent, respectively. Seed of grains such as corn and sorghum generally have 8 to 12 percent protein.

PLANT GROWTH AND MORPHOLOGY

The seed of the cotton plant is typical of dicotyledonous plants. Placed in the proper environment, the cottonseed imbibes moisture predominately through the chalazal cap, and the germination process is set into motion. Water imbibition is rapid during the first 12 hours indicating initial wetting, thereafter becoming more steady state. Germination is optimum at about 85°F and the process slows and stops for most cultivars around 58°F, although seeds of some cultivars will germinate at temperatures as low as 53 to 55°F. Producers are encouraged not to plant until the soil temperature at planting depth is at least 65°F for three consecutive days, and the five-day weather forecast is for sunshine and high temperatures of 70°F or higher.

TABLE 6.7   Effects of Prolonged below Normal Temperatures after Planting Cotton.

Days of Chill	Days Delayed Flowering	Fiber Maturity	Percent 1st Harvest	Final Plant Height
0	0	3.9	60	65
2	3	3.8	59	61
4	6	3.6	54	59
6	10	3.4	46	54

Source: Adapted from Christiansen and Thomas, 1969.

Emergence should occur in five to seven days under reasonable soil and air temperatures. Delayed emergence because of low temperature stress at and following planting can cause decreased yield, shorter plants, and delayed maturity (Table 6.7). Temperature is the most critical factor in getting an appropriate stand for commercial production, as one would expect when a perennial species, requiring a long growing season, is produced as an annual in the temperate zone.

Once the plant has emerged, it may remain in the cotyledonary stage of development (i.e., having only the two seed leaves) for seven to nine days, again depending usually on temperature, before the unfolding of the first true leaf. Once this first true leaf has developed, a new main stem leaf will develop every 2.5 to 3 days. Main stem leaves will appear at node positions in a $^3/_8$'s phyllotaxy until growth is slowed and finally stopped by fruit load, temperature, soil moisture, photoperiod, or a combination of these factors. The mature cotton plant usually has 18 to 24 main stem nodes.

Staging a cotton plant or crop is a simple matter of counting main stem nodes. A node is considered established when its attending leaf has unrolled sufficiently that the edges are not touching. Stages are divided into vegetative and reproductive, and these do not overlap in time as stages do in some other crops dealt with in this text. A plant is at stage V-C from emergence until the first true leaf has unrolled, at which time the plant will be at stage V-1 (Figure 6.4). Additional vegetative stages occur until the first reproductive or fruiting limb appears on the plant. Once fruiting begins, branches at main stem nodes above the first fruiting limb will normally be reproductive limbs also. Stages are then referred to as R-1 through R-i, where i is the number of fruiting branches arising from the main stem of the plant. For example, a plant with four true leaves unrolled on the main stem will be at stage V-4; a plant that has three fruiting limbs arising from the main stem will be at stage R-3.

The first fruiting limb, or sympodium, having flower buds called squares, will occur at the fifth through the ninth main stem node. This will occur about 30 days after planting in most fields. The cotton plant fruits vertically via the main stem and horizontally via the extension of fruiting limbs. A new flower bud is visible to the unaided eye at 2.5- to 3-day intervals at equivalent fruiting limb nodal positions as the plant develops vertically, and at 4.5- to 6-day intervals along individual fruiting limbs (Figure 6.5).

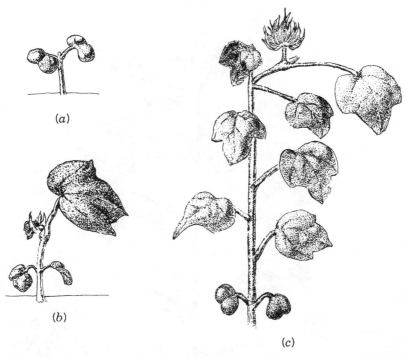

Figure 6.4 *Illustrations of three stages of growth of cotton. Plant (a) is at stage V-C having no true leaves above the cotyledons. Plant (b) is at stage V-2 because the second main stem leaf has unrolled. Plant (c) is at stage R-3 because two main stems have leaves unrolled sufficiently that the edges are not touching above the node with the first fruiting limb so they too should have fruiting limbs at those main stem nodes (although on back side and hidden from view). (Drawing by Mike Hodnett)*

Flower buds mature to an open flower in 20 to 25 days after they can be observed by the unaided eye, depending on temperature and other growth factors (Figure 6.6). The petals of upland cotton flowers are white, while petals of Pima cotton are cream to yellow with a deep red "blood" spot at the inside base of each of the six petals. Buds open predominately during the morning hours, fade to pink by late afternoon, and are red by the next day. Late on the second day, blooms begin to wither and are dead by the third day. Peak bloom will occur about three weeks after first bloom and blooming may continue for six weeks or so. The ovary of the cotton flower is termed a boll on the second day of bloom, anthesis having taken place the previous day. The boll will enlarge for about 20 days and will open in about 50 days after anthesis, exposing the seed cotton for harvest. Cotton being an indeterminate plant species, the crop will be 90 percent open in 150 to 195 days in the rain belt and irrigated west, respectively, but may reach 90 percent opened bolls in 130 days under drought conditions in parts of Texas and Oklahoma (Table 6.8).

All plant metabolic functions are temperature-dependent. Hence, the concept of

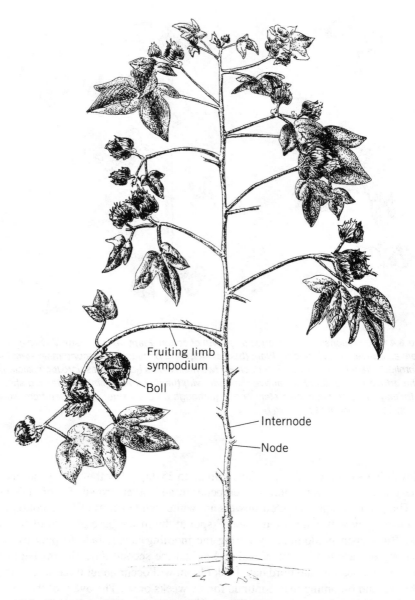

Figure 6.5 *Illustration of the growing cotton plant. The plant grows vegetatively and reproductively at the same time. New main stem nodes are established every 2.5 to 3 days during
the active growing phase. Fruit forms are discernible at equivalent fruiting limb nodes every 2.5
to 3 days as the plant grows vertically. Fruit forms are discernible at sequential fruiting limb
nodes every 4.5 to 6 days. This plant is in growth stage R-10. (Drawing by Mike Hodnett)*

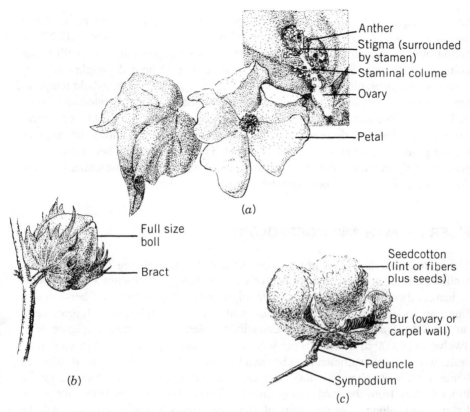

Figure 6.6 *Illustration of cotton plant morphology: (a) the cotton flower and subtending leaf; (b) a full size boll before maturity; and (c) an open boll of cotton ready for harvest. (Drawing by Mike Hodnett)*

TABLE 6.8 Range in Number of Days Required for Upland Cotton Plant Development in Different Cotton Growing Regions of the United States and the Range in GDD$_{60}$ Units Required in All Regions.

Stage[1]	Rainbelt[2]	Plains[3]	Irrigated[4] West	GDD$_{60}$[5]
Planting to emergence	5–20	5–20	5–20	45–130
Emergence to first square	27–38	33–38	40–60	350–450
Square to bloom	20–27	20–27	20–27	250–500
First bloom to peak bloom	26–45	26–45	26–45	——
Bloom to open boll				910–950
Early–mid season	45–55	45–55	45–65	
Late season	55–75	55–75	65–85	
Season	130–170	120–150	180–210	2550–4600

1. Temperature and cultivar dependent.
2. Southeast and Mid-South states.
3. Texas High Plains, Texas Rolling Plains, Texas Blacklands, Texas Coastal Bend, and Oklahoma.
4. Arizona, California, New Mexico and Far West Texas.
5. Growing Degree Days with 60°F base temperature adapted from *Cotton Physiology*, 1986.

growing degree days (GDD) has been applied to cotton's development, as it has with most crops. A base temperature of 60°F is usually used in calculating GDDs in cotton, with no maximum temperature limit. Growing degree days, also called heat units, are calculated as {[(minimum temperature in a 24-hour day-night cycle) + (maximum temperature in the same 24 hours)] ÷ 2} − 60 (the threshold temperature). These are calculated for each day after planting. Values for cotton are shown in Table 6.8. The use of GDD has not gained wide acceptance in the United States, since its use does not decrease the degree of variability in estimating the range of time required for a given morphological event. This is presumably because cotton is grown mostly in areas where the vast majority of morphological benchmarks occur during summer months when daily temperatures vary little.

FIBER GROWTH AND MORPHOLOGY

The cotton fiber is unique in development and structure among vegetable fibers. Unlike other plant fibers that are part of the morphological structure of plant stems or leaves, thereby being composed of multiple cells (often called bast fibers), cotton fibers are unicellular extensions of epidermal seed cells that begin to elongate early on the day of anthesis, a phenomenon independent of fertilization (Figure 6.7). Twelve to 20,000 of these cells per seed, approximately 20 percent of its epidermal cells, will elongate sufficiently to be spun into yarn. Other epidermal seed cells will begin to elongate about six days after anthesis and will not elongate enough to be broken away from the seed during ginning. These short, thicker fibers are called linters, and, along with the stubs of spinnable fibers left after ginning, will be removed at the oil mill and used in a number of products such as rayon, paper, and automobile upholstery. The cotton fiber is approximately 94 percent cellulose; about 1 percent protein; 1 percent pectic substances; and less than 1 percent wax or sugars.

The elongating seed epidermal cell, now called a fiber, grows rapidly in length for 16 to 19 days after anthesis. At this point in development, the fiber consists of a primary wall, a small secondary wall, a layer of protoplasm, and a central vacuole called the lumen (Figure 6.7). Rate of secondary wall development increases at about 19 days after anthesis, and development is rapid until about 40 days after anthesis, assuming suitable temperature, radiant energy, nutrients, and moisture. The secondary wall grows by daily deposits of cellulose microfibrils on the inside of the cotton fiber, daily growth rings being observable under proper magnification. At maturity, the fiber dries and the lumen collapses, giving the cotton fiber its characteristic twisted, ribbon-like appearance.

FIBER QUALITY

Quality is often an elusive character in agricultural commodities because it depends on the final consumer of the product. The same is true in cotton; fibers that would be

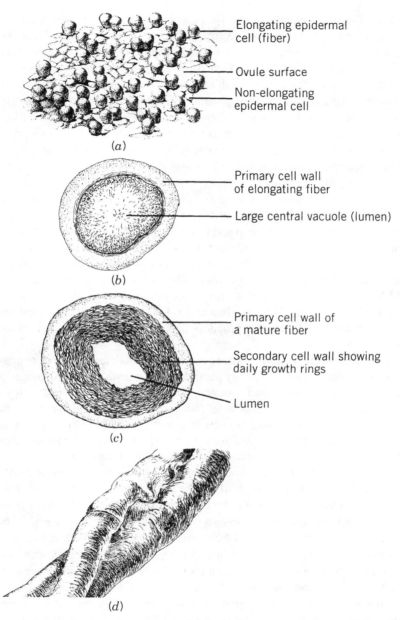

Figure 6.7 *Illustrations of the cotton fiber: (a) seed epidermal cells beginning to elongate about the time of anthesis; (b) cross-section of newly elongating fiber showing a large lumen and thin primary cell wall; (c) cross-section of mature fiber showing a small lumen and a well developed secondary wall interior to the primary wall; (d) collapsed, mature cotton fiber. (Drawing by Mike Hodnett)*

considered low- or medium-quality for production of delicate fabrics may be of perfect quality for producing larger yarns that are woven into denim. Fibers appropriate for nonwoven fabric could be of such quality as to be unsuitable even for denim. Yarn quality also may be affected by the type of spinning machinery employed as well as by the quality of the lint used on that machinery. For example, fiber length is the most critical fiber parameter for spinning certain yarns on the ring spinning frame, while fiber bundle strength is considered more important in rotor spinning machines. Be that as it may, cotton fiber quality is defined in terms of its length, its strength, and its diameter/maturity. Whiteness and trash content also are important identifiers of quality to a mill. These enter into grade considerations that will be considered later in this chapter.

In the drawing and twisting process of producing yarns by hand, or with machinery other than rotor technologies, fiber length plays a pivotal role. Longer fibers are easier to twist because they overlap and fewer long fibers will twist together to form a yarn than will be required of shorter fibers. Yarn strength also is enhanced by longer fibers in twisted yarns because of fiber-to-fiber friction properties. The average length of the longest 50 percent of fibers of upland cotton produced in the United States ranges from about 1 inch for cotton produced under drought conditions to 1.12 inches in the U.S. rain belt and as long as 1.20 inches for Acala cotton in the irrigated west. Pima cotton produces fibers averaging as high as 1.35 to 1.40 inches.

Stronger fibers tend to produce stronger yarns, but fiber length and fiber diameter interact significantly with this quality parameter. Ring-spun yarn using longer fibers of a given strength will produce a stronger yarn than would be produced with shorter fibers because of the friction characteristic. This interaction, however, is not as dramatic with rotor-spun yarn, because the twisting of individual fibers is not accomplished under drawing conditions; therefore, a better correlation exists between fiber strength and yarn strength when fibers are rotor-spun.

The diameter of fibers determines the number of fibers that can be twisted or wrapped to form a yarn of a given size. As a generalization, a yarn produced with more fibers in its cross-section will be stronger than that same size yarn produced with larger but fewer fibers per cross-section. At the present time, the U.S. cotton classing offices of the Agricultural Marketing Service, along with spinning mills, use a value called micronaire that reflects the linear density of fibers, expressed as ug/25.4 mm length, or in g/tex, where a tex is a bundle of fibers equivalent to 1,000 m of fiber. Mills further use micronaire as an indicator of fiber maturity. A mature fiber is one that has a well-developed secondary wall and a lumen of sufficient size that will result in the characteristic twist when collapsed. Mature fibers are necessary to ensure uniform dyeing, which in turn is necessary for end-use product acceptance.

Other determinants of fiber quality include the degree of whiteness, friction properties, resilience, and elongation before break. Whiteness is a determinant of grade that will be discussed later, but the other three are not used to determine the quality of cotton at the producer level.

U.S. PRODUCTION

Cotton has been produced in the present-day continental United States each year since 1621. Because it was a very labor-intensive crop to harvest and prepare for spinning before Whitney's cotton gin, the United States produced only 3,000 bales in 1790 (Table 6.9). Production increase 24-fold within seven years of the invention

TABLE 6.9 Production Statistics for All Cotton Produced in the United States, 1790–1990.

Year	Harvested Acres	Average Yield	Total Production	Average Price Received	Exports[7]
	(×1000)	(lbs./ac.)	(×1000 bales)	(cents/lb.)	(%)
1790	——	——	3	25.0[4]	——
1800	——	——	73	—	——
1810	——	——	178	15.0	70
1820	——	——	334	16.6	75
1830	——	——	731	9.8	76
1840	——	——	1,346	10.2	79
1850	——	——	2,134	12.1	87
1860	——	——	3,837	11.1	16
1863	——	——	449	52.8	5
1865	——	——	2,091	30.8	62
1866	7,666	121	2,097	32.2	63
1870	9,238	208	4,352	17.0	66
1880	15,921	191	6,606	9.8	67
1890	20,937	196	8,653	8.6	68
1900	24,886	195	10,124	9.2	67
1910	31,508	176	11,609	14.0	69
1920	34,408	187	13,429	15.9[5]	44
1926	44,616[1]	193	17,978	12.5	63
1930	42,454	157	13,932	9.5	51
1931	38,705	212	17,096	5.7	54
1937	33,623	270	18,946[3]	8.4	32
1940	23,861	253	12,566	9.9	9
1950	17,843	269	10,012	40.1[6]	43
1960	15,309	446	14,272	30.2	48
1963	14,212	517[2]	15,334	32.2	38
1970	11,155	438	10,192	22.0	38
1976	10,914	465	10,581	64.1	43
1980	13,215	404	11,122	74.7	51
1990	11,732	634	15,505	62.2	48

Source: Adapted from Ware, 1951; USDA Agricultural Statistics, 1936 and following years.

1. Highest acreage in U.S. history.
2. First year United States averaged at least one bale (480 lbs.) per acre.
3. Highest one-year production.
4. Price received is average export price for 1790—1865, season average price on New York or New Orleans Cotton Exchange for 1866–1870, and actual average price received by producers for 1880–1900.
5. Average price received for 1917–1919 was 30.5 cents/lb.
6. Average price received for 1942–1945 was 20.0 cents/lb.
7. Current year exports as a percent of current year production; minus imports for 1976–1990.

of the cotton gin and had increased 59-fold by 1810. Production rose steadily until the outbreak of the war between the states in 1860, and then reached a low of only 449,000 bales in 1863. Average price received jumped from 11.1 cents/pound in 1860 to a record 52.8 cents in 1863, a price not matched until 1976.

Production continued to increase until the Great Depression of the 1930s. Record production of over 17,000,000 bales along with the general economic decline associated with the Great Depression pushed the averaged price received to a record low of 5.7 cents/pound in 1931, the lowest average price recorded by the USDA. Record production, again in 1937, depressed prices below 10 cents/pound, but the demand during and after World War II pushed prices to 40.1 cents/pound by 1950. Competition from man-made fibers in the 1960s and early 1970s depressed production and prices received, but strong consumer demand since has resulted in record prices received by historical standards.

Cotton is produced in 17 states across the southern United States with over a million bales produced in each of five states (Table 6.10). Texas is the leading producer of upland cotton, producing about 5,000,000 bales in 1990, while Kansas produced the fewest bales at less than 1,000. The United States produced about 15,000,000 bales of upland cotton in 1990. Of these 15,000,000 bales, 7,488,000 were exported and over 8,000,000 bales were used domestically, the difference in total used plus exported and production being carryover stocks from previous years. The United States exported about 46 percent of upland cotton produced from 1980

TABLE 6.10 Harvested Acreage, Average Yield per Acre and Total Production of Upland Cotton Lint by State in 1990.

State	Harvest Acres	Average Yield	Total Production
	(×1,000)	(lbs./ac.)	(×1,000 bales)
Alabama	378	476	375
Arizona	348	1,119	811
Arkansas	750	692	1,081
California	1,090	1,204	2,734
Florida	36	640	48
Georgia	350	555	405
Kansas	1	280	<1
Louisiana	790	715	1,177
Mississippi	1,220	728	1,850
Missouri	235	641	314
New Mexico	62	735	95
North Carolina	200	631	263
Oklahoma	370	496	382
South Carolina	154	452	145
Tennessee	515	461	495
Texas	5,000	477	4,965
Virginia	5	562	6
TOTAL	11,505	632	15,147

Source: Adapted from USDA Agricultural Statistics, 1992.

TABLE 6.11   Harvested Acreage, Average Yield per Acre and Total Production of Pima Cotton Lint by State in 1990.

State	Harvested Acres	Acreage Yield	Total Production
	(×1,000)	(lbs./ac.)	(×1,000)
Arizona	124	751	194
California	26	1,180	57
Mississippi	1	591	2
New Mexico	19	609	25
Texas	57	682	81
TOTAL	227	758	358

Source: Adapted from USDA Agricultural Statistics, 1992.

through 1991, with annual exports ranging from 14 to 68 percent of production. Japan and the republic of Korea each bought over a million bales of U.S. cotton in 1989, 1990, and 1991. Other countries buying at least 200,000 bales in any one year during that time period were Germany, Italy, Egypt, Hong Kong, Indonesia, Taiwan, and Thailand. Annual domestic use averaged about 5,000,000 bales from 1980 through 1984 and around 7,000,000 bales from 1985 through 1989.

Pima cotton was produced in five states in 1990 with total production reaching 350,000 bales (Table 6.11). The vast majority of Pima has been produced historically in Arizona, but favorable price structure and more adapted cultivars encouraged producers in California, Mississippi, and Texas to experiment with Pima production during the late 1980s. The long growing season requirement for Pima, along with its sensitivity to the boll weevil, again forced producers in Mississippi and Texas to abandon production of this species of cotton. Very little Pima was produced in states east of New Mexico by 1993. However, California appears entrenched as the leading producer of Pima cotton, producing 293,000 bales in 1992 while Arizona produced only 137,900 bales. The majority of Pima produced in the United States is exported, with annual domestic consumption averaging only 57,000 bales during the 1980s.

The U.S. Upland Cotton Belt can be divided into seven production regions (Table 6.12). Acreage in the southeastern states declined from about 1950 through the late 1980s because of (1) the expense of and difficulty in controlling insects, especially the boll weevil, (2) the increasing cost of and decreasing availability of hand labor, and (3) the increase in profitability of alternate crops, primarily peanuts, and later soybeans.

The USDA-ARS, cooperatively with State Agriculture Experiment Stations, began a program in 1987 designed to eradicate the boll weevil from the United States. The basis of this program is that the boll weevil is a near obligate feeder of and reproducer on cotton. The program started in the northeastern corner of the Cotton Belt, and by 1993 had eradicated the weevil from Virginia, North Carolina, South Carolina, Georgia, and parts of Alabama. This program and favorable governmental farm policies have resulted in an increase in cotton acreage and production in these

TABLE 6.12   U.S. Upland Cotton Production by Regions, 1990.

Region	States	Production[4]	Rainfed	Irrigated
			(%)	
Southeast	Virginia	0.04	90	10
	North Carolina	1.74	90	10
	South Carolina	0.96	90	10
	Georgia	2.67	60	40
	Florida	0.22	70	30
	Alabama	2.48	90	10
Subtotal		8.11	—	—
Deltas[1]	Tennessee	3.27	95	5
	Missouri	2.07	50	50
	Arkansas	7.14	55	45
	Mississippi	12.21	60	40
	Louisiana	7.77	60	40
	Texas	3.19	60	40
Subtotal		35.65	—	—
Lower Rio Grande	Texas	2.19	50	50
Blacklands[2]	Texas	2.22	95	5
Plains[3]	Texas	24.96	60	40
	Oklahoma	2.52	75	25
Subtotal		31.89	—	—
Southwest	Texas	0.32	0	100
	New Mexico	0.63	0	100
	Arizona	5.35	0	100
Subtotal		6.30		
West	California	18.05	0	100

Source: Adapted from USDA Agricultural Statistics, 1992; Texas Agricultural Statistics, 1991; personal communication.

1. Delta areas of the Mississippi, Arkansas, Red, Brazos, Colorado, Trinity and other rivers plus other acreage with similar production practices and adapted cultivars.
2. Blacklands and Coastal Bend regions.
3. Includes the Edwards Plateau, Cross Timbers, Northern and Southern Low Plains, and Northern and Southern High Plains.
4. Percent of total U.S. production for 1990.

states. The removal of this pest from these states has reduced dramatically the expense of insect control.

WORLD PRODUCTION

China was the leading producer of cotton in 1990, producing over 20,000,000 bales, followed by the United States, producing over 15,000,000 bales, and India, Pakistan, and Brazil at 9,084,000, 7,492,000 and 3,200,000 bales, respectively (Table 6.13). Total world production of all cotton types (i.e., upland, Pima, and *G. arboreum*) was 86,573,000 bales in 1990.

TABLE 6.13   Area Planted, Average Yield and Total Production of All Cotton in Countries Producing at Least 3,000,000 Bales and Continental and Worldwide Statistics, 1990.

Continent/Country	Number Acres	Average Yield	Total Production[2]
	(×1,000)	(lbs./ac.)	(×1,000 bales)
N. America			
United States	11,728	633	15,466
Mexico	459	813	777
Others (5+)[1]	272	——	403
Total	12,459	640	16,646
S. America			
Argentina	1,556	417	1,352
Brazil	4,907	313	3,200
Paraguay	1,358	429	1,214
Others (5)	1,017	——	1,055
Total	8,838	369	6,821
Europe			
Greece	697	664	964
Spain	207	856	369
Other (4+)	77	——	101
Total	981	700	1,434
Former USSR	7,832	728	11,878
Africa			
Zimbabwe	674	239	336
Sudan	484	376	379
Egypt	1,030	640	1,373
Other (13+)	6,198	——	3,575
Total	8,386	325	5,663
Asia			
China	13,802	718	20,645
India	18,167	240	9,084
Pakistan	6,575	547	7,492
Turkey	1,583	909	2,997
Other (9+)	2,192	——	2,189
Total	42,319		42,407
Australia	667	1,428	1,984
World Total	81,480	510	86,573

Source: Adapted from USDA Agricultural statistics, 1992.

1. Numbers in parenthesis are other countries within a given continent "producing" at least 1,000 bales; + indicates islands or countries with less than 1,000-bale production included in continental area and total production.
2. Numbers adjusted for rounding errors.

PRODUCTION PRACTICES

Cultivar Choice

The U.S. cotton producer chooses the cultivar(s) that he or she will plant based on information obtained from advertisements from seed companies, cultivar performance trials conducted by state agricultural experiment stations, and advice from fellow producers. Although there are several distinct types of cotton cultivars developed for specific production areas, producers in all regions demand cultivars with (1) high yield potential, (2) good quality fiber, and (3) early maturity.

Producers of upland cotton choose cultivars that have demonstrated high yield potential for their particular locale and have at least the minimum level of fiber quality demanded by the industry today. Data reported from Mississippi demonstrated that lint yield potential of cultivars grown in the Mississippi River Delta increased from, say, 180 pounds lint/acre in 1920 to 763 pounds lint/acre by the 1960s, an increase in yield potential of 14 pounds lint/acre/year (Figure 6.8). Concurrent with selecting for increased yield potential, breeders have succeeded in reducing the number of days required to produce a crop of cotton by 30 days between 1966 and 1986 among cultivars evaluated at the Delta States Experiment Station at Stoneville, Mississippi. If one considers the U.S. Cotton Belt in general terms, cultivars required probably 175 to 200+ days for production around the turn of this century, while modern cultivars require from 120 to 175 days. Early maturity gives producers more flexibility in planting date and allows producers in certain production areas to harvest before the onset of inclement weather.

Date of Planting

Cotton is planted from late February in the Lower Rio Grande Valley of Texas to early June on the High Plains of Texas and Oklahoma (Figure 6.9). Planting can extend into early July in some years in the more northern limits of production because of inclement weather or replant situations following hail. Since cotton is a botanically indeterminate crop species, requiring a relatively long growing season, and because producers want to take advantage of spring moisture in much of the U.S. Cotton Belt, planting is generally recommended when the soil temperature at seeding depth is at least 65°F for three consecutive days and there is a favorable five-day weather forecast.

Early planting, in addition to the positive attributes just noted, also allows for maximum plant growth and development before certain insect pest populations, most notably the boll weevil and boll worm, build to summertime highs. In some sections of the Cotton Belt, such as the Rolling Plains of Texas, producers practice uniform delayed planting in an effort to deprive the boll weevils that are emerging from winter diapause in the early spring of their only source of food—i.e., cotton flower buds. Other sections of the belt may delay planting until soils have warmed to above 65°F to insure rapid seedling emergence and early seedling growth, or to time harvest to the most favorable weather period.

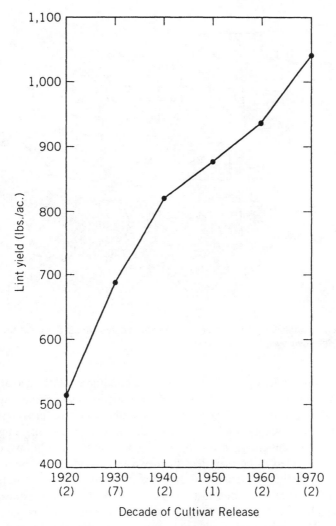

Figure 6.8 *Increase in genetic yield potential of cotton cultivars developed for the Midsouth from 1920 until the 1970s. Adapted from Bridge and Meredith, 1983.*

Seeding Rates

The optimum number of plants per linear foot of row varies surprisingly little across the U.S. Cotton Belt. Under rain belt and irrigated conditions, producers should aim for a final stand of three to four plants/row-foot. Lint yields of Pima are optimized at two to three plants/row-foot.

Producers normally overplant by at least one seed/row-foot (i.e., planting four to five seeds/row-foot when the desired final stand is three to four plants/row-foot), under normal conditions. Planting quality seed, defined as seed having at least 80

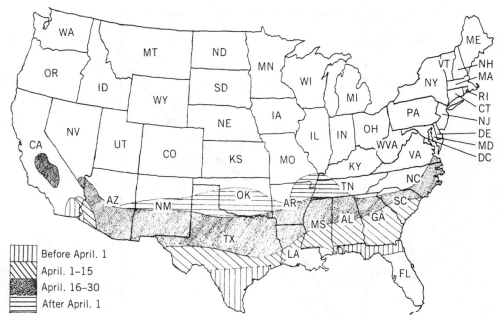

Figure 6.9   *Normal dates for beginning of cotton planting across the U.S. Cotton Belt.* Source: *USDA, 1984. (Redrawn by Mike Hodnett)*

percent germination under optimum or normal test conditions (86°F) and at least 50 percent germination in the cool (65 to 68°F) germination test, on an appropriate planting date should result in 75 percent of the seeds producing productive seedlings. Planting lower-quality seed or planting during cooler soil/air temperatures will decrease the percent emergence. Under marginal planting conditions, or where plant height is expected to be, say, less than two feet, producers should increase seeding rates, since slightly higher-than-optimum plant densities result in less yield loss than slightly lower (i.e., one to two plants/row-foot) plant densities. Where plants are normally two feet tall or less, optimum light interception requires higher plant densities. However, a uniform stand of two plants/row-foot should, in most cases, be kept and the field not replanted. Seeding rates should be decreased if planting later in the spring when soil temperatures at seed depth exceed, say, 75°F and moisture conditions are not a factor. As a general rate, an equivalent number of seeds/acre should be planted regardless of row width such that the number of seeds/row-foot decreases as row width decreases. Cotton producers often gauge seeding rates in terms of pounds of seed planted per acre; however, since seed size varies by cultivar and environmental conditions under which the seeds were produced, producers should calibrate each planter unit to drop the desired number of seed/row-foot.

Tillage

As in most other crops, tillage in cotton production accomplishes the following: (1) shatters hard pans to allow penetration of moisture and cotton roots; (2) incorporation of fertilizer; (3) provides furrows for surface drainage; (4) destruction of weeds and other plant residue; (5) ensures good soil-seed contact for proper germination; (6) incorporates herbicides; and (7) controls in-season weeds and grasses.

In many areas of the Cotton Belt, preparation for next year's cotton crop begins almost immediately after harvest. Stalks are cut out to (1) remove a source of food for the remaining boll weevils, and (2) begin the destruction of crop residue that could harbor disease pathogens as well as make planting next season's crop more difficult. Destruction of stalks by shredding or disking as soon after harvest as possible will decrease the number of boll weevils going into diapause, thus decreasing the number of weevils emerging the following spring.

Following stalk destruction, producers may subsoil or deep plow fields where a traffic or natural hard pan exists in order to shatter such hard pans and allow for percolation of winter rain or melting snow to recharge the subsoil moisture level. In areas of limited rainfall and limited or no irrigation, it is important to have a full soil profile of moisture at the beginning of the season if maximum yields are to be obtained. Cotton, a tap-rooted species, can extract water from five to six feet deep. Except for incorporation of dry fertilizer by disking, fields may be left in this condition until just before planting when disking will be necessary to incorporate herbicides, destroy winter weeds, and prepare a proper seedbed.

Fields to be planted to cotton are often "bedded" in the fall or winter. The bedding equipment consist of large disks, two per row, that produce a raised area of soil that will serve as the drill, or row, for next season's planting pattern. Beds may be as much as ten inches high when first formed and settle to, say, eight inches high from middle furrow to bed apex. Bedding should be completed well ahead of the planting season to allow the soil to settle to ensure proper seed-soil interface at planting. Beds may be reshaped just prior to planting by the same bedding implement to destroy weeds or to restore the old beds that have eroded during the winter.

At planting, these beds are "knocked off" by use of an implement designed to pull down the beds and loosen the top two to four inches of soil remaining. Beds should be about four inches high at this point to maintain furrows for furrow irrigation and surface drainage. Preplant herbicides can be incorporated after beds are knocked off by use of a rotary-hoe, rotary field cultivator, or other implement that will not incorporate the herbicide deeper than two inches. Since the top four to six inches of the beds are knocked off before planting, they provide one other advantage, namely that seeds usually can be placed into moist soil even during dry planting seasons if the beds were put up early enough to have received winter and/or early spring rains.

In-season cultivation is usually necessary to control weeds, ensure rain or irrigation water percolation, and to maintain soil aeration. Plant species vary in the amount of soil oxygen necessary for normal root growth and function, but most crops require 10 percent or more. Not only is oxygen necessary for root growth, but the uptake of nutrients can be severely curtailed at O_2 levels below 10 percent. Potassium uptake appears to be the most sensitive to root zone O_2, followed by nitrogen > phosphorus > magnesium > calcium. If the rhizosphere oxygen is replaced by nitrogen for as little as three minutes, root elongation ceases but will return to normal state with no permanent damage when the oxygen level returns to

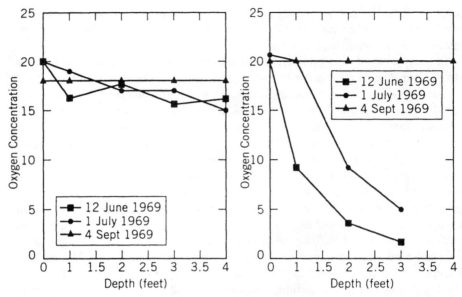

Figure 6.10 *Oxygen concentration in a soil with good internal drainage (left) and a soil with poor internal drainage (right) in the Mississippi River Valley.* Source: *Adapted from Patrick et al., 1973.*

normal. If the oxygen in the rhizosphere is replaced with nitrogen for 30 minutes, some of the root tissue will die but the remaining root tissue will return to normal when the atmosphere returns to 10 percent O_2. However, if the cotton plant root system is deprived of oxygen for three hours, complete root death will occur. Replacing the rhizosphere oxygen with nitrogen is an extreme example, since even under complete water-logged conditions some oxygen is available to plant roots. However, it does serve to point out that some cotton root function will be lost under conditions of depressed rhizosphere oxygen levels.

Although cultivation will increase the rate at which oxygen diffuses into soils, soil type is a major determinate of oxygen distribution at various depths throughout the growing season. Comparison of O_2 concentration in a soil with good internal drainage (Dundee) and a soil with poor internal drainage (Tunica), both found in the Mississippi River Valley of Louisiana, is shown in Figure 6.10. Distribution of and amount of cotton roots at lower soil depths is directly related to soil oxygen concentration throughout the growing season.

Cotton is predominantly grown under full tillage, but economics, concern for the environment, and U.S. government policy dictate reducing tillage in all crops where possible. Limited research and experience with cotton grown without tillage suggest that no-till will not fit every farm situation. The most successful no-till production has occurred when planting into old cotton crop stubble or into small grain stubble. If the trend toward reducing soil erosion by maintaining old crop residue as a mulch continues, then changes in cotton production such as planting, rotation, weed control, fertilizer application, and others are likely to occur in the near future. (See Chapter 7. Soybean for additional discussion of no-till and reduced-till systems.)

Fertility

Since only seed cotton is removed from land dedicated to its production and plant residue is returned to the soil, cotton requires less fertilizer nutrients than many other crops. Each bale (480 pounds of lint) of cotton produced removes 40 pounds of nitrogen (N), 15 pounds potassium (K) as K_2O, 20 pounds phosphorus (P) as P_2O_5, 1 pound of calcium (Ca), 5 pounds of magnesium (Mg), and 2 pounds of sulphur (S). Micronutrients required in trace amounts are boron (B), manganese (Mn), zinc (Zn), chlorine (Cl), copper (Cu), iron (Fe), and molybdenum (Mo). Most soils in the United States where cotton is grown have sufficient amounts of these micronutrients. Producers should determine the fertilizer requirements of individual fields to be planted to cotton by (1) a soil test, (2) estimating nutrient removal by previous crops, (3) past experience, and (4) in-season plant analyses should problems be indicated or suspected.

Cotton, as other nonlegume crops, responds to nitrogen deficiency and excess more rapidly than to other nutrients. Excess nitrogen will lead to excess vegetative growth that will decrease yield and increase the difficulty of insect control. Applying nitrogen at rates required for maximum production for several years while removing less than maximum production will result in excess soil nitrogen. If this situation occurs, or if cotton is being planted after corn or grain sorghum that were heavily fertilized with nitrogen, then nitrogen rates should be reduced to prevent excessive vegetative growth. Additional nitrogen can be added to the soil by side-dress or applied directly to the plants by foliar application during the growing season, should conditions warrant.

Cotton tolerates a wide range of soil pH, 4.5 to 8.5, but production is optimized at 5.7 to 7. Nitrogen, K, P, Ca, Mg, and S become less available as soil pH drops below 6, whereas B, Mn, Zn, Fe, and Cu solubilize and become more available to cotton plants. At low pH, Mn and aluminum may solubilize to the point of concentrations sufficient to be toxic.

Harvest

Cotton harvest will begin in August in south Texas and north Florida with start-up date progressively later in more northerly production areas (Figure 6.11). In most years, harvest will begin in early September for the central areas of the Cotton Belt and after the first of October in the north Mississippi River Valley, the Plains of Texas and Oklahoma, and in western states. In years of late planting or inclement autumn weather, harvesting may not begin until November or December on the High Plains. When autumn rains delay harvest in some northern areas, producers may have to wait for freezing temperatures so that the ground will freeze sufficiently to keep their machines from becoming stuck in mud. Producers in these areas plant cultivars that have bolls that do not fluff enough to be harvested with a spindle machine. This character also allows the mature cotton crop to withstand inclement weather and prolonged field weathering without seedcotton shattering from the plant. The types of machines used to harvest cotton in the United States were noted earlier.

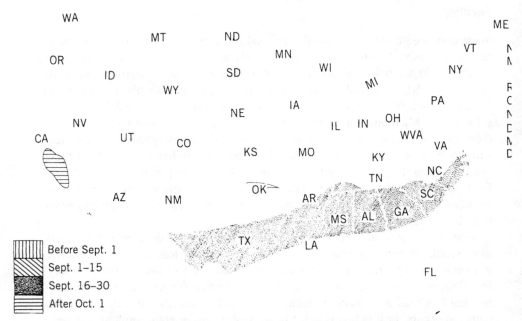

Figure 6.11 *Normal beginning of harvest across the U.S. Cotton Belt.* Source: *USDA, 1984.* *(Redrawn by Mike Hodnett)*

Plants are conditioned for spindle harvest by applying a chemical called a defoliant that induces the development of the abscision zone at the proximal end of the leaf petiole, resulting in leaf drop. Defoliants serve two purposes: (1) removal of leaves to expose the open boll for harvest, and (2) reduction of leaf trash in harvested seedcotton that would reduce grade and prices received.

Plants that are to be harvested by stripper harvesters may be conditioned by applying a defoliant, followed by application of a desiccant to quickly dry the plant. Plants thus conditioned will result in the least amount of leaf trash and stem bark contaminates in the harvested seedcotton. Stem bark is one of the most difficult types of plant material to remove from lint at the gin and spinning mill.

Producers in many cases will apply a desiccant without first defoliating plants. In this situation, no abscision zone develops and killed leaves remain attached to the plants, a condition often referred to as "sticking" leaves. Stuck leaves will eventually be broken off the plant by high winds in many areas of Texas and Oklahoma where stripper harvesting is practiced. The other option for conditioning plants for stripper harvest is to wait for the onset of freezing temperatures to kill the plants. This is a viable option for the High Plains of Texas and Oklahoma, and the northern fringes of the Texas Blacklands and Rolling Plains.

BIOTIC PESTS

Cotton is attacked by a plethora of insects and disease organisms. In addition to the boll weevil that has been noted earlier, cotton can host over 20 insects plus spider mites. Seedling diseases that cause damping off of young seedlings are probably the most widespread disease problem across the Cotton Belt, but other diseases such as bacterial blight, cotton rust, cotton rot root, and verticillium and fusarium wilts are problems somewhere in the United States each year. Major insect and disease pathogens are noted below, along with notations on appearance of damaging stage, symptoms or injury, and general control measures (Tables 6.14 and 6.15).

TABLE 6.14   Common Insect Pests of Cotton.

Insect	Description[1]	Symptom/ Injury	Control
Boll Weevil *Anthonomus grandis* (Boheman)	Hard shelled weevil; 1/4-inch long; grayish to brown with prominent snout.	Feeds on and oviposits eggs in squares and young bolls; fruit with developing larvae fall from plant.	Chemical; some cultural control measures.
Bollworm *Helicoverpa zea* (Boddie)	Up to 1 1/2-inch larvae; green to pink to brown with light and dark bands along back and sides.	Bores into and consumes inside of squares and bolls.	Chemical; some cultural measures.
Tobacco budworm *Heliothis virescens* (Fabricius)	Same as bollworm to the untrained eye.	Same as bollworm.	Same as bollworm.
Pink bollworm *Pectinophora gossypiella* (Saunders)	New larvae are white with brown heads but developing pink back and sides.	Larvae consume developing seeds inside young bolls, obvious entrance and exit holes in bolls.	Chemical; quarantines.
Plant bugs *Lygus Hesperus* (Knight) *L. elisus* (Van Duzec) *Neurocolpus nubilus*	Nymphs are most damaging; greenish; 1/25 to 1/8-inch long; adults light mottled tan to brown; 3/8-inch.	Feed on pinhead squares with piercing mouth parts; squares turn brown and fall from plant.	Chemical.
Fleahopper *Pseudatomoscelis seriatus* (Reuter)	Nymphs and adults light green; adults to 1/8-inch long.	Same as plant bugs.	Chemical.
Thrips *Frankiniella* spp.; *Thrips tabaci* (Linde.); *Caliothrips fasciatus* (Pergande)	Small, elongated, 1/60- to 1/25-inch; yellow to brown to black.	Larvae and adults suck plant sap from young leaves and terminals; characteristic silvery tracks on underside of leaves.	Chemical; maintain good growing conditions.

(continued)

TABLE 6.14   *(Continued)*

Insect	Description[1]	Symptom/ Injury	Control
Spider mites *Tetranychus* spp.	Microscopic size mites generally found on underside of leaves; greenish with reddish hue.	Leaves become reddened with masses of tiny webs on underside.	Maintain beneficial insects; chemical.
Aphids *Aphis gossypii* (Glover)	Soft bodied; bulb shaped; almost clear to green.	Feed on youngest tissue in terminals and young leaves causing curling and crinkling; symptomatic secretion of honeydew.	Maintain beneficial insects; chemical.
Whitefly, *Bemisia tabaci* (Gennadius)	Adults are white, $1/16$-inch long; immature stages resemble small scale insects.	Feed on underside of leaves causing desiccation in sufficient numbers; symptomatic secretion of honeydew.	Field and adjacent field sanitation; control alternate host; chemical.

1. Stage most damaging or most prominent to observer.

COTTON GRADES (CLASSING)

The United States Department of Agriculture, Agricultural Marketing Service (USDA-AMS) is responsible for identifying the quality of every bale of cotton produced and sold in the United States. The first official standards were adopted by Congress in 1909 and consisted of nine grades of upland cotton: middling fair, strict good middling, good middling, strict middling, middling, strict low middling, low middling, strict good ordinary, and good ordinary. (Middling = average, strict = precise or exact, and ordinary = below average.) Congress mandated the use of the cotton standards in 1923 for all cotton sold in interstate or foreign commerce. Also in 1923, American and European cotton trade organizations met and established Universal Standards of Grades for American upland cotton, which were the same as those established by the U.S. Congress. Standards for Pima, American-Egyptian, and sea island cotton were established in 1918.

Although the number of grades and their description have varied over the years, their purpose has not. Standardized grades in any commodity serve to identify and reward the production, harvest, and conditioning of a high-quality product. During colonial and early years of U.S. cotton production, the only standards were those imposed by English mills and the few, but growing, number of U.S. mills. Brokers for these mills often bought cotton from given locales and specified cultivars. As the

TABLE 6.15 Common Diseases and Nematodes of Cotton.

Disease	Cause/Source	Symptom/Injury	Control
Seedling disease complex	*Pythium* spp.; *Fusarium* spp.; *Rhizoctonia solani*; *Thielaviopsis basicola*	Seed rot; pre and post emergence damping off; seedling root rot.	Cultural; chemical.
Bacterial Blight	*Xanthomonas campestris* pv. *malvacearum*	Water-soaked spots on cotyledons, leaves, stems, and/or fruit becoming tan to brown with reddish border.	Acid delinted seeds; resistant cultivars; chemical.
Alternaria Leaf Spot	*Alternaria macrospora*	Small, dull brown circular spots enlarging to near 1/2-inch diameter; dry, dead or grey centers that may fall out.	Resistant cultivars.
Cercospora Leaf Spot	*Cercospora gossypiia*	Small, reddish dots that enlarge to 3/4-inch diameter with narrow red margin and whitish or necrotic center.	Resistant cultivars.
Wet-Weather Blight or Ashen Spot	*Ascochyta gossypii*	Small pale brown spots, enlarging and coalescing into large, irregular dead areas surrounded by dark brown border.	Resistant cultivars.
Southwestern Cotton Rust	*Puccinia cacabata*	Pale green lesions that develop into bright yellow pustules usually on upper leaf surface.	Resistant cultivars; Chemical.
Boll Rot	*Ascochyta gossypii; Glomerella gossypii; Diplodia gossypina; Fusarium* spp.; *Phomopsis* spp.; *Phytophthora capsici; Rhizoctonia solani, Xanthomonas campestris* pv *malvacearum*; + 160 other microorganisms	Vary from water soaked to dark green to brown to black lesions.	Decrease relative humidity in plant canopy by cultural means if possible.

(continued)

TABLE 6.15   (Continued)

Disease	Cause/Source	Symptom/Injury	Control
Fusarium Wilt[1]	*Fusarium oxysporum* f. sp. *vasinfectum*	Leaves wilt, turning yellow or brown and drop from plant; plant death; characteristic dark coloring of vascular system just below bark and in woody tissue.	Resistant cultivars; crop rotation.
Verticillium Wilt	*Verticillium dahliae*	Leaves becoming mottled, chlorotic and finally necrotic; light brown discoloration of woody tissue near pith.	Resistant cultivars; crop rotation.
Phymatotrichum Root Rot (Texas Root Rot)	*Phymatotrichum omnivorum*	Slight yellowing of young leaves followed by sudden wilting and death of plant.	Crop rotation.
Root Knot Nematode	*Meliodogyne incognita*	Stunting and yellowing of foliage; distinctive spindle-shaped galls or knots on roots; often associated with Fusarium wilt.	Crop rotation; resistant cultivars.
Reniform Nematode	*Rotylenchulus reniformis*	Stunting of seedlings followed by death caused by associated seedling blight; surviving plants are stunted.	Crop rotation; resistant cultivars.

cotton industry flourished, cotton exchanges were established in several locations in the United States; Savannah and Atlanta, Georgia; New Orleans, Louisiana; New York, New York; and Mobile, Alabama. Each exchange established its own standards or grades, resulting in considerable differences of opinion. For example, middling grade cotton at Augusta, Georgia, was of higher quality than middling at Savannah, mainly because cotton sold on the Savannah exchange was surely headed for Liverpool, England, to be used in mills requiring lower quality cotton because of their end products and because of the spinning machinery used to obtain those end products.

After 1923, the USDA identified the quality of each bale of cotton based on samples taken from each side of that bale and forwarded to the nearest USDA Cotton Classing Office. Trained classers subjectively assigned a grade to each bale based on the sample's trash or nonlint content, the whiteness of the sample, and its preparation—i.e., how well it was ginned (cotton ginned too wet or with too much heat is inferior to properly ginned cotton).

In 1918, the length of the longest fibers in the sample was added as an additional determinate of fiber quality, and micronaire determination was added in 1966. Micronaire is an indirect measurement of fiber diameter and therefore fiber maturity. The number of grades have been revised and expanded several times since 1909, and today there are 44 grades of upland cotton (Table 6.16). Pima is graded on a different scale, as its fibers are more of a deep yellow than upland and its leaf content and preparation are much different than upland. Pima is ginned on a roller gin, resulting in baled lint that has a stringy or lumpy preparation. A grade of 1 is the highest grade of Pima and the lowest is 6, with 7 being below grade.

Grading or classing cotton was strictly a subjective art, with the exception of micronaire after 1966, until the 1980s when the USDA-AMS began to gradually replace human judgment with objective measurements by machines. By the 1990s, all cotton classed in the United States was done so by high volume instrument (HVI) measurements, except for grade determination. Machinery capability for accurately identifying amounts or kind of trash or nonlint content has not been perfected. Nevertheless, the quality of a bale of cotton as identified by HVI-assisted classing is currently more accurately defined than at any other time in history.

Grade continues to be subjectively determined by trained personal based on comparisons of bale samples with specially prepared physical and descriptive standards when viewed under standardized lighting. Grade continues to be determined by color, trash, and preparation. Lint color affects the bleaching and dyeing properties of fabric, and variation within the fabric can result in uneven color. Trash

TABLE 6.16 Codes for Upland Cotton Grades within Color Groups.

	Color Group						
Name	White	Light Spotted	Spotted	Tinged	Yellow Stain	Light Gray	Gray
Good Middling	11	12	13			16	17
Strict Middling	21	22	23	24	25	26	27
Middling Plus	30						
Middling	31	32	33	34	35 (85)	36	37
Strict Low Middling Plus	40						
Strict Low Middling	41	42	43	44		46	47 (87)
Low Middling Plus	50						
Low Middling	51	52	53	54 (84)			
Strict Good Ordinary Plus	60						
Strict Good Ordinary	61	62 (82)	63 (83)				
Good Orinary Plus	70						
Good Ordinary	71 (81)[1]						

Source: Adapted from USDA, 1993.

1. Numbers in parentheses are codes to indicate Below Grade bales.

TABLE 6.17 Average Non-Lint Content and Trash Meter Readings of Several Grades of Upland Cotton.

Grade Name	Non-Lint Content	Trash Meter[1]
	(%)	(%)
Strict Middling	1.9	0.1
Middling	2.3	0.2
Strict Low Middling	3.1	0.4
Low Middling	4.4	0.7
Strict Good Ordinary	5.6	1.1
Good Ordinary	7.2	1.5

Source: Adapted from Banks et al., 1992; Perkins et al., 1984.

1. Percent of surface of sample covered by non-lint particles; cannot detect embedded non-lint particles. Reproduced from Cotton, R. S. Kohel and C. F. Lewis (eds.), p. 457, American Society of Agronomy.

content correlates with the amount of waste in processing fibers into yarn (Table 6.17). More processing is necessary to clean dirty cotton, which adds to processing costs and results in more damaged fibers—e.g., increase in the amount of fibers less than $1/2$-inch and more tangled fibers. Preparation is reflected in the "smoothness" or homogeneity of the sample. Mills usually prefer bright, white fibers with no foreign material and with no harvesting or ginning damage.

Table 6.17 indicates continued efforts by the USDA-AMS to make grading totally objective. Trash meters are experimentally used to identify nonlint content and other machinery is used to objectively identify the whiteness of the sample. Color of upland cotton can be defined in terms of degree of grayness and degree of yellowness. Degree of gray is reported as Rd (percent reflectance) values and normally range from about 50 percent to, say, 85 percent, with higher numbers desirable. Yellowness is designated by the "Hunter's +b" value that ranges from 5, least yellow, to 18, most yellow. The relationship of grade, with Rd and +b is shown in Figure 6.12.

Other items that are measured by HVI machinery and indicated on the class card for both buyer and seller are (1) fiber length, (2) uniformity, (3) micronaire, and (4) fiber strength. Fiber length is reported on the class card in two ways. First, the average length of the longest one-half of the fibers sampled (upper half mean or UHM) is reported in hundredths of an inch, and then this value is converted to 32nds of an inch using established conversion criteria. The second value is included because it puts length in terms of "staple length," a term used before HVI classing for the subjective length as reported by a classer. Longer fibers generally produce stronger and finer yarns that can be used to produce higher value products such as shirts and sheets versus ropes, blankets, denim, etc.

Length uniformity is reported as uniformity on the class card and is the ratio of the average length of all the fibers in the sample to the UHM length. Lower uniformity, especially below 80 percent, indicates a relative high content of short fibers that may be lost in the spinning process—that is, waste or processing loss. Yarn produced from bales with low uniformity tends to be variable in thickness and weaker than yarns produced from highly uniform bales.

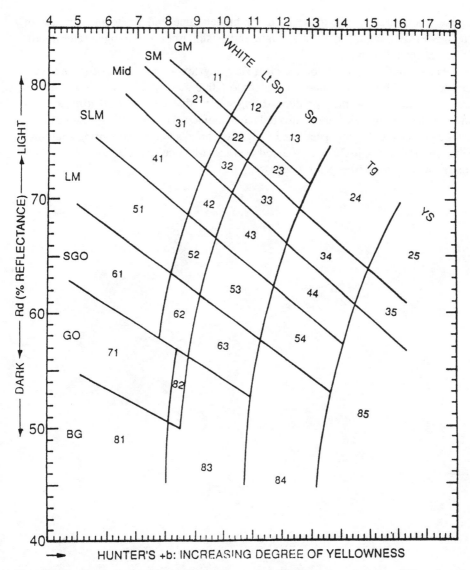

Figure 6.12 *Rd and +b values indicating the grey and yellow color values from HVI testing of cotton quality. Source: USDA, AMS, Cotton Division. See Table 6.16 for code descriptors.*

The micronaire instrument indicates the resistance to air flow of a given weight of fibers that have been compressed to a given volume. Finer fibers will more completely block the air passage when compressed resulting in lower readings, whereas more coarse fibers have the opposite effect. Micronaire values between 3.5 and 4.9 have traditionally indicated mature, well-developed fibers that adequately absorb dye stuffs. Producers receive less for bales having micronaire readings outside of this range, as lower values suggest immature fibers that can result in increased wastage during manufacturing and cause irregular dyeing of fabric, while coarse fibers are too big to be used for higher-value end products. Because of changes in spinning technology, spinning mills using rotor technology prefer micronaire between 3.9 and 4.2.

Fiber strength is reported on the class card in g/tex—i.e., grams of force necessary to break a bundle of fibers equivalent to 1,000 meters of fiber. Stronger fibers are desirable because they produce stronger yarn that can be used to produce a stronger fabric. Stronger fibers also reduce manufacturing costs because machinery can be operated at higher speeds and with fewer breaks in the yarn during spinning. Fiber strength of 24 to 26 is considered average. Most cultivars today will produce fibers having 26 to 30+ g/tex fiber strength.

GLOSSARY

Anther: the upper portion of the stamen that contains pollen grains.

Bast fibers: course fiber made from plant conductive tissue (e.g., phloem fibers).

Boll: the cotton fruit that consists (after fertilization) of the carpel wall, the placental column, seeds, and the cellulose fibers.

Bract: modified leaves, subtending the floral bud and later the boll.

Calyx: the first series of floral organs, immediately external to the flower bud and later the boll, consisting of five sepals.

Carpel: the plant organ bearing ovules along the inside margins; a unit of structure of cotton's compound pistil; may be three to six, usually four, in upland, and three in Pima.

Chalazal cap: the large end of the seed, or ovule, where its parts merge.

Corolla: collectively the petals of the cotton flower; consists of five petals.

Cotyledons: the rudimentary or first leaves found inside the seed; the seed leaves of the embryo.

Filiment: the lower portion of the stamen that supports the anther.

Fruiting branch: branch on which pedicled fruit are borne directly from the branch, terminates in a fruit and subtending leaf; lateral extension is via axillary bud break; also referred to as a sympodium.

Glabrous: having no hairs or pubescence.

Gossypol gland: gland found in all aboveground plant tissue; contains gossypol and related compounds that are toxic to nonruminate animals.

Hirsute: having pubescence or hairs on the stem or leaves.

Hull: the seed coat, removed at milling; the outer covering of a seed.

Lint: cellulose fibers resulting from the elongation of seed epidermal cells.

Linters: fibers too short to be removed during ginning and fiber stumps left on the seed from normal ginning processes. They are removed at oil mills (used in a variety of industrial products and as padding in seats and furniture).

Micropyle: the pointed end of the seed; the opening in the integuments of the ovule.

Nectary: a specialized structure or gland that exudes sugary nectar.

Ovary: the swollen basal portion of the pistil, the part that contains ovules, or seeds.

Ovule: the female egg in plants containing a haploid number of chromosomes.

Peduncle: a stalk supporting the flower, sometimes referred to as a pedicle.

Petiole: a stalk supporting the leaf.

Phyllotaxy: the distribution or arrangement of leaves on a stem, given as a fraction of a circle.

Pistil: female portion of a flower composed of the stigma, style, and ovary.

Raphe: the broad seam (or mid-vein) of the seed.

Sepals: one of the units of the calyx; also a modified leaf.

Stamen: male portion of a flower composed of filaments and anthers.

Stigma: the part of the pistil that receives the pollen grain.

Style: the portion of the pistil between the stigma and the ovary.

Vegetative branch: branch that does not directly bear fruit; reproductive branches that bear fruit directly may originate from vegetative branches.

BIBLIOGRAPHY

Agricultural Research Service. 1978. *Glandless Cotton: Its Significance, Status, and Prospects.* USDA-ARS Conf. Proc. Dec. 13–14, 1977. Dallas, TX.

Aguillard, W., D. J. Boquet, and P. E. Schilling. 1980. *Effects of Planting Dates and Cultivars on Cotton Yield, Lint Percentages, and Fiber Quality.* Louisiana St. Univ. Agri. Expt. Sta. Bul. 727.

The American Cotton Handbook. 1965. D. S. Hamby (ed.) Interscience Pub., John Wiley and Sons, New York, NY.

Banks, J. C., L. M. Verhalen, G. W. Cuperus, and M. A. Karner. 1992. *Cotton Production and Pest Management in Oklahoma.* Oklahoma St. Univ. Coop. Ext. Ser. C. E.-883.

Bennett, R. L. 1904. *Early Cotton.* Texas A&M Univ. Agri. Expt. Sta. Bul. 75.

Blankenship, D. C., and B. B. Alford. 1983. *Cottonseed: The New Staff of Life.* Texas Woman's Univ. Press, Denton, TX.

Blalock, J., and R. Metzer. 1989. *Cotton Production on the Blackland Prairie and Grand Prairie.* Texas A&M Univ. Coop. Ext. Ser. B1628.

Bonner, C. M. 1991. *1991 Cotton Production Recommendations.* Univ. Arkansas Coop. Ext. Ser. Ag-86-3-91.

Bowman, F. H. 1908. *The Structure of the Cotton Fiber.* MacMillan & Co., London, England.

Bridge, R. R., and S. R. Miller. 1989. *The Influence of Plant Populations in Two Cotton Varieties.* Mississippi St. Univ. Agri. and For. Expt. Sta. Res. Rep. 14:7.

Bridge, R. R., and L. D. McDonald, 1987. *Beltwide efforts and trends on development of varieties for short season production systems.* Proc. Beltwide Cotton Prod. Conf. 1987: 81–85.

Bridge, R. R., and W. R. Meredith. 1983. "Comparative performance of obsolete and current cotton cultivars." *Crop Sci.* 23:949–952.

Britton, K. G. 1992. *Bale o'Cotton: The Mechanical Art of Cotton Ginning.* Texas A&M Univ. Press, College Station, TX.

Brown, M. 1988. *Cottonseed Plus Cottonseed Meal Diet May Contain Too Much Gossypol for Cattle.* Texas Agri. Expt. Sta. Res. Rep. 617.

Brown, H. B. 1958. *Cotton.* McGraw-Hill, New York, NY.

Brown, H. B. 1923. *Cotton.* McGraw-Hill Book Co., New York, NY.

Burmester, C. H., and K. Edmisten. 1991. *Cotton Production Guide.* Auburn Univ. Coop. Ext. Ser. CT-91-1.

Christiansen, M. N., and R. O. Thomas. 1969. "Season-long effects of chilling treatments applied to germinating cottonseed." *Crop Sci.* 9:672–673.

Christidis, B. G., and G. J. Harrison. 1955. *Cotton Growing Problems.* McGraw-Hill Book Co., Inc., New York, NY.

Cotton Board. 1992. *Cotton Incorporated Research and Promotion Facts.* Vol. 7, No. 1.

Cotton Disease Council. 1981. *Compendium of Cotton Diseases.* G. M. Watkins (ed.) Am. Phytopath. Soc., St. Paul, MN.

Cotton Trade Journal. 1931. International Trade Edition. "Principal American Cotton Mills and Number of Spindles." Vol. 11, No. 12. New Orleans, LA.

Crawford, J. L., W. R. Lambert, S. M. Brown, D. Shurly, S. C. Hodges, W. M. Powell, and J. F. Hadden. 1991. *1991 Cotton Production Package*. Univ. of Georgia Coop. Ext. Sec. Misc. Pub. 380.

Crawford, M. D. C. 1948. *The Heritage of Cotton*. Fairchild Pub. Co., New York, NY.

Culp, T. W., D. C. Harrell, and J. B. Pitner. 1974. Population Studies with Cotton (*Gossypium hirsutum* L.). South Carolina Agri. Expt. Sta. Bul. 575.

Davison's Textile Blue Book. 1992. Davison Pub. Co., Inc., Ridgewood, NJ.

Dennis, R. E., and R. E. Briggs. (not dated). *Growth and Development of the Cotton Plant on Arizona*. Univ. of Arizona Coop. Ext. Ser. Pub. 8168.

Elsner, J. E., C. Wayne Smith, and D. F. Owen. 1979. "Uniform stage descriptions in upland cotton." *Crop. Sci.* 19:361–363.

Endrizzi, J. E., E. L. Turcotte, and R. J. Kohel. 1984. "Qualitative Genetics, Cytology, and Cytogenetics." pp. 81–129. *In* R. J. Kohel and C. F. Lewis (eds.) *Cotton. Am. Soc. Agron.*, Madison, WI.

Fryxell, P. A. 1984. "Taxonomy and germplasm resources." pp. 27–57. *In* R. J. Kohel and C. F. Lewis (eds.) *Cotton*. Am. Soc. Agron., Madison, WI.

Fryxell, P. A. 1979. *The Natural History of the Cotton Tribe*. Texas A&M Univ. Press, College Station, TX.

Fryxell, P. A. 1965. "Stages in the evolution of *Gossypium* L." *Adv. Front. Plant Sci.* 1:31–56.

Gilroy, C. G. 1845. *The History of Silk, Cotton, Linen, Wool and Other Fibrous Substances*. Harper and Brothers, New York, NY.

Gray, L. C. 1933. *History of Agriculture in the Southern United States to 1860*. Vol. 1. Carnegie Inst., Washington, D.C.

Green, Constance McL. 1956. *Eli Whitney and the Birth of American Technology*. Little, Brown and Co., Boston, MA.

von Hagen, V. W. 1961. *The Ancient Sun Kingdoms of the Americas*. The World Pub. Co., Cleveland, OH.

Hake, K., T. Burch, L. Harvey, T. Kerby, and J. Supak. 1991. "Plant Population." *Physiology Today* Vol. 2 No. 4. Natl. Cotton Coun. Am., Memphis, TN.

Hake, K., K. Cassman, and W. Ebelhar. 1991. "Cotton nutrition—N, P. and K." *Physiology Today* Vol. 2 No. 3. Natl. Cotton Coun. Am., Memphis, TN.

Hake, K., W. McCarty, N. Hopper, and G. Jividen. 1990. "Seed quality and germination." *In Physiology Today Newsletter*. Natl. Cotton Coun. Am., Memphis, TN.

Hake, K., B. Mayfield, H. H. Ramey, and P. Sasser. (not dated). *Producing Quality Cotton*. Cotton Physiology Education Program. Natl. Cotton Coun. Am., Memphis, TN.

Harris, D. G., and C. H. M. van Bavel. 1957. "Nutrient uptake and chemical composition of tobacco plants as affected by composition of the root atmosphere." *Agron. J.* 49:176–181.

Hawkins, B. S., and H. A. Peacock. 1972. *Agronomic and Fiber Characteristics of Upland Cotton (Gossypium hirsutum L.) as Affected by Hill Spacing, Plants per Hill, and Plant Populations*. Univ. of Georgia Agri. Expt. Res. Bul. 101.

Hecht, A. 1989. *The Art of the Loom: Weaving, Spinning and Dyeing Across the World*. Rizzoni Internatl. Pub., Inc., New York, NY.

Helms, Douglas. 1980. "Revision and reversion: Changing cultural control practices for the cotton boll weevil." *Agri. History* 54:108–125.

Hodges, S. C., J. L. Crawford, and C. O. Plank. 1989. *Fertilizing cotton.* Univ. of Georgia Coop. Ext. Ser. Bul. 966.

Holekamp, E. R., E. B. Hudspeth, and L. L. Ray. 1960. *Soil Temperature: A Guide to Timely Cotton Planting.* Texas A&M Univ. Agri. Expt. Sta. Bul. 673.

Howard, L. O. 1896. *Insects Affecting the Cotton Plant.* USDA Off. Expt. Sta. Bul. 33.

Hubhouse, H. 1986. *Seeds of Change: Five Plants that Transformed Mankind.* Harper and Row, New York, NY.

Huck, M. G. 1970. "Variation in taproot elongation rate as influenced by composition of soil air." *Agron. J.* 62:815–818.

Hudspeth, E. B., and A. P. Brashears. 1974. *Dryland Narrow-Row Cotton Production, Texas High Plains.* Texas A&M Univ. Agri. Expt. Sta. Prog. Rep. 3280.

Hutchinson, Sir Joseph. 1962. "The history and relationships of the world's cottons." *Endeavor* 21(81):5–15.

Hutchinson, J. B. 1959. *The Application of Genetics to Cotton Improvement.* University Press, Cambridge, England.

Hutchinson, J. B. 1954. "New evidence on the origin of the Old World cottons." *Heredity* 8:225–241.

Johnson, W. H. 1926. *Cotton and Its Production.* MacMillan and Co., Limited, London, England.

Jones, R. T. 1975. *Cottonseed Protein.* Market Res. Ser. Natl. Cotton Coun., Memphis, TN.

Kennedy, J. 1830. *A Brief Memoir of Samuel Crompton.* Manchester, England.

Kerby, T., and D. Bassett. *1991 San Joaquin Cotton Varieties.* Univ. of California Coop. Ext. Ser. California Cotton Rev. No. 19.

Kerby, T. A., M. Keeley, and S. Johnson. 1987. *Growth and Development of Acala Cotton.* Univ. of California Agri. Expt. Sta. Bul. 1921.

Kirk, I. W., A. D. Brashears, and E. B. Hudspeth. 1969. *Influence of Row Width and Plant Spacing on Cotton Production Characteristics on the High Plains.* Texas A&M Univ. Agri. Expt. Sta. MP-937.

Kittock, D. L., R. A. Selley, C. J. Cain, and B. B. Taylor. 1986. "Plant population and plant height affects on Pima cotton lint yield." *Agron. J.* 78:534–538.

Lee, J. A. 1984. "Cotton as a world crop." pp. 6–26. *In* R. J. Kohel and C. F. Lewis (eds.) *Cotton.* Am. Soc. of Agron., Madison, WI.

Lewton, F. L. 1938. "Historical notes on the cotton gin." *In Annual Report of the Board of Regents of the Smithsonian Institution.* U.S. Government Printing Office No. 3451. Washington, D.C.

Lyman, J. B. 1968. *A Treatise on Cotton Culture.* Orange Judd & Co., New York, NY.

Malm, N. R. 1974. *Heat Units and Upland Cotton Production, New Mexico, 1939–1972.* New Mexico St. Univ. Agri. Expt. Sta. Res. Rep. 275.

Mauney, J. R. 1986. "Vegetative Growth and Development of Fruiting Sites." pp. 11–28. *In* J. R. Mauney and J. McD. Stewart (eds.) *Cotton Physiology.* The Cotton Foundation, Memphis, TN.

Mauney, J. R., and L. L. Phillips. 1963. "Influence of day length and night temperature on *Gossypium.*" *The Bot. Gaz.* 124:278–283.

Meloy, G. S. 1922. *Meade Cotton, an Upland Long-Staple Variety Replacing Sea Island.* USDA Bul. 1030.

Meredith, B., K. Hake, and M. Lange. 1992. "Choosing the Right Cotton Variety." *Physiology Today* Vol. 3 No. 4 Natl. Cotton Coun. Am., Memphis, TN.

Merrill, G. R., A. R. Macormac, and H. R. Mauerserger. 1949. *American Cotton Handbook.* Textile Book Pub., Inc., New York, NY.

Miller, T. S. 1915. *Cotton Trade Guide and Student's Manual.* E. L. Steck Pub., Austin, TX.

Mirsky, J., and A. Neving. 1952. *The World of Eli Whitney.* MacMillan Co., New York, NY.

Mitchell, C., C. Burmester, and K. Edmisten. 1991. *Micronutrients for Cotton in Alabama.* Auburn Univ. Coop. Ext. Ser. Agron. Ser. S-2-91.

Montgomery, J. 1932. *The Carding and Spinning Masters Assistant.* John Niven, Throngate, Whittaker, Treacher, Arnot, and G. Herbert, London, England; Oliver & Boyd, and Stirling & Kennedy, Edinburgh, Scotland.

Moore, J. H. 1956. "Cotton breeding in the old south." *Agri. History* 50:95–101.

Olmsted, D. 1846. *Memoir of Eli Whitney, Esq.* Duorie and Peck, New Haven, CT.

Oosterhuis, D. M. 1990. *Growth and Development of a Cotton Plant. Nitrogen Nutrition in Cotton: Practical Issues.* Am. Soc. Agron., Madison, WI.

Percival, A. E., and R. J. Kohel. 1990. "Distribution, collection, and evaluation of *Gossypium.*" Vol. 44, pp. 225–256. *In* N. C. Brady (ed.) *Advances in Agronomy.* Academic Press, Inc. New York.

Perkins, H. H., D. E. Ethridge, and C. K. Bragg. 1984. "Fiber." pp. 437–509. *In* R. J. Kohel and C. F. Lewis (eds.) *Cotton.* Am. Soc. Agron., Madison, WI.

Phillips, L. L. 1962. "The cytogenetics of *Gossypium* and the origin of New World cotton." *Evol.* 17:460–469.

Quisenberry, J. E., and R. J. Kohel. 1975. "Growth and development of fiber and seed in upland cotton." *Crop Sci.* 15:463–467.

Rasmussen, W. D. 1975. *Agriculture in the United States: A Documentary History.* Random House.

Ray, L. L., E. B. Hudspeth, and E. R. Holekamp. 1959. *Cotton Planting Rate Studies on the High Plains.* Texas A&M Univ. Agri. Expt. Sta. MP-358.

Ryder, M. L. 1968. "The origin of spinning." *Text. Hist.* 1:73–82.

Schubert, A. M., C. R. Benedict, J. D. Berlin, and R. J. Kohel. 1973. "Cotton fiber development-kinetics of cell elongation and secondary wall thickening." *Crop Sci.* 13:704–709.

Shepperson, A. B., and C. W. Shepperson. 1920. *Cotton Facts.* Shepperson Pub. Co., New York.

Silvertooth, J. C. 1991. *Agronomic Guidelines for Pima Cotton Production in Arizona.* Univ. of Arizona Coop Ext. Ser. Pub. 190038.

Smith, C. Wayne. 1992. "History and Status of Host Plant Resistance in Cotton to Insects in the United States." *In* D. L. Sparks (ed.) *Advances in Agronomy* 48:251–296.

Smith, C. Wayne. 1983. *Development characteristics of earliness in upland cotton.* Proc. Beltwide Cotton Prod. Res. Conf. 1983:98.

Smith, C. W., B. A. Waddle, and H. H. Ramey. 1979. "Plant spacings with irrigated cotton." *Agron. J.* 71:858–860.

Stephens, S. G. 1975. "A reexamination of the cotton remains from Huaca Prieta, north coastal Peru." *Am. Antiquity* 40:406–419.

Stephens, S. G., and E. D. Moseley. 1974. "Early domesticated cottons from archaeological sites in Central Coastal Peru." *Am. Antiquity* 39:109–122.

Smith, E. C., and S. G. Stephens. 1971. "Critical identification of Mexican archaeological cotton remains." *Ec. Bot.* 25:160–168.

Stephens, S. G. 1970. "The botanical identification of archaeological cotton." *Am. Antiquity* 35:367–373.

Stephens, S. G. 1966. "The potential for long range oceanic dispersal of cotton seeds." *The Am. Naturalist* 100:199–209.

Texas Agricultural Statistics Service. 1992. *1991 Texas Agriculture Statistics.* Texas Dep. Agri. and USDA Bul. 250.

Turner, J. 1981. *White Gold Comes to California.* Book Pub, Inc., Fresno, CA.

Tyler, F. J. 1910. *Varieties of American Upland Cotton.* USDA Bur. Plt. Ind. Bul. 163.

United States Department of Agriculture. 1993. *The Classification of Cotton.* USDA-ARS Agri. Handbook 566.

United States Department of Agriculture. 1962. *Essential Steps in Cotton Textile Processing.* USDA-ARS, So. Util. Res. and Dev. Div. New Orleans, LA.

United States Department Interior. 1883. *Compendium of the Tenth Census (June 1, 1880).* U.S. Government Printing Off., Washington, D.C.

United States Department of Agriculture. 1936, 1942, 1952, 1961, 1972, 1982 and 1992. *Agricultural Statistics.* U.S. Government Printing Office, Washington, D.C.

Wailes, B. L. C. 1854. "Agriculture." p. 143. *In Report on the Agriculture and Geology of Mississippi.*

Walker, J. K., and C. Wayne Smith. 1995. "Cultural Control." *In* E. King and J. Phillips (eds.) *Cotton Insects.* The Cotton Found. Memphis, TN.

Wanjura, D. F., and D. R. Buxton. 1972. "Water uptake and radicle emergence of cottonseed as affected by soil moisture and temperature." *Agron. J.* 64:427–431.

Wanjure, D. F., E. B. Hudspeth, and I. W. Kirk. 1966. *Effects of Compaction on Cotton Emergence.* Texas A&M Univ. Agri. Expt. Sta. MP-800.

Ware, J. O. 1951. *Origin, Rise, and Development of American Upland Cotton Varieties and their Status at Present.* Univ. of Ark. Mimeograph (unnumbered).

Ware, J. O. 1930. *Cotton Spacing II: Effect of Blooming on Earliness, Fruit Set and Yield.* Ark. Agri. Expt. Sta. Bul. 253.

Watt, B. K., and A. L. Merrill. 1975. *Handbook of the Nutritional Contents of Foods.* Dover Pub., Inc., New York.

Watt, G. 1907. *The Wild and Cultivated Cotton Plants of the World.* Longman, Green, & Co., New York, NY.

Wendel, J. F. 1989. "New World tetraploid cottons contain Old World cytoplasm." *Proc. Natl. Acd. Sci. USA.* 86:4132–4136.

Whitty, E. B., B. J. Brecke, D. H. Teem, and H. A. Peacock. 1981. *Cotton Production in Florida.* Univ. of Florida Coop. Ext. Ser. Agron. Facts No. 111.

7

SOYBEAN
(Glycine max (L.) Merrill)

INTRODUCTION

The soybean is native to eastern Asia, Australia, and several of the Pacific Islands. It has been called soya bean, soja bean, Chinese pea, Manchurian bean, Japan pea, Japan bean, and Japanese fodder plant. According to Morse (1950), soybean culture dates to 2800 B.C. in China and it was probably well established as a food, feed, and medicinal plant by that date. Written recommendations on proper planting rates, dates of planting, variety choice, harvest, storage, and utilization date to 2207 B.C. in China; however, the earliest archaeological evidence of soybean dates its use to only 770 B.C. Other evidence place its domestication during the Shang and Chou Dynasties of 1700 to 700 B.C.

The name soya traces to the Chinese "Chiang-yiu," meaning soy sauce. Soy sauce in Japanese is pronounced "show-yu" and was shortened to "so-ya." The English further contracted the common name of *Glycine max* from soy-a or soya to soy plus bean, or soybean.

The soybean, as with many other plants, was considered to have medicinal value by early producers. Texts written as early as 450 A.D. noted that soybean was a specific remedy for the malfunctioning of the heart, liver, kidneys, stomach, bowels, lungs, and for most any ailment. The yellow and green seeded varieties were regarded as valuable foods that increased lung power and beautified the complexion, while black seeded varieties were used in certain foods and customarily fed to horses to increase their strength and lung power before departing on a long journey.

Like most grains, soybean became a valuable source of human and animal energy because it could be stored to provide food and feed during winter months, and during periods of famine. Apparently, the early Chinese planted five acres of soybean for each member of the family; however, it is unclear as to how rigidly this rule was followed, but one must remember that yields of crops during ancient times were much different than in more recent times in the United States.

SYSTEMATICS

The genus *Glycine* is divided into subgenera *Glycine* and *Soja*. *Glycine Glycine* contains eight species; all eight are indigenous to Australia, with two species also found in Taiwan, south China, and some of the Pacific Islands. *Glycine Soja* contains the cultivated species *Glycine max* (note that subgenera names are not included when identifying a plant by binomial nomenclature) and its wild annual counterpart *G. soja*. *Glycine soja* is distributed throughout parts of China, Taiwan, Japan, Korea, and the former USSR. Current knowledge identifies *G. soja* as the wild ancestor of cultivated soybean *Glycine max*. Some authorities add a weedy form, *G. gracilis*, into the subgenera *Soja* and suggest that it is an intermediate form between the wild ancestor and the domesticated soybean. While some evidence exist that soybean is an allopolyploid with $x = 10$, $n = 20$, and $2n = 4x = 40$, it behaves as a diploid and $2n = 40$ is considered the diploid number of chromosomes.

MOVEMENT FROM ORIGIN

The soybean was domesticated in central China and was exported as a cultigen to Manchuria before recorded history. Early soybean production apparently was for consumption as whole beans or whole bean products only, as normally circulated literature makes no mention of soybean oil. How and when soybean oil became a commodity of interest is not obvious. Soybean remained a local crop in Manchuria until 1895 when Japan began to import soybean oil cake as a fertilizer. Soybean cake became the primary product of the oil mill industry of Manchuria shortly thereafter. Greater attention to this crop was gained during the Russo-Japanese war of 1904–1905, and worldwide attention to the soybean was gained with the introduction and acceptance of soybean products into Europe in 1908. Production in Manchuria expanded to keep pace with demand, and soybean was established as a staple crop in that country.

The soybean had been introduced in Europe as early as 1712 by botanist Englebert Kaempfer. Missionaries in China sent soybean seed back to France in 1740, but efforts to establish soybean as a crop species from 1740 through 1855 failed. The lack of suitable climate in most of Europe and the fact that the imported soy oil and oil meal had to compete with similar products already produced in Europe were the primary reasons for the lack of interest in Europe prior to the 20th century. Deterioration of soy oil and cake during shipment from Manchuria to Europe also added to the early unacceptance of these products.

MOVEMENT TO THE UNITED STATES AND CULTIVAR DEVELOPMENT

Samuel Bowen, a seaman for the East India Company, introduced soybean into North America in 1765 (Hymowitz and Harlan, 1983). Bowen brought soybean from China to his farm near Savannah, Georgia. In 1767, he received patent number 878 for his invention of methods (probably learned in China) to prepare sago powder, vermicelli, and soy sauce from plants grown in North America. The soy sauce was exported to London for consumption.

In 1804, a yankee clipper ship left China with an unspecified amount of soybean seed aboard as inexpensive ballast. These seeds were subsequently dumped in port in the United States to make room for cargo. The fate of these seeds apparently went unrecorded. Although it may be coincidental, soybean was evaluated in Pennsylvania that year and its production recommended for that state's farmers. Soybean was grown in the botanical garden at Cambridge, Massachusetts, in 1829. Nuttall noted that it was nothing more than a luxury crop for the production of soy sauce. Admiral Perry is reported to have brought back two soybean cultivars in 1854, one white seeded and one red seeded, which were given to the U.S. Commissioner of Patents.

The earliest cultivated variety, or cultivar, in the United States was Mammoth Yellow, reportedly collected by missionaries in China in 1873. Mammoth Yellow was cultivated extensively in the southern United States, especially North Carolina, for many years; other soybean cultivars were imported and evaluated by Cook in New Jersey, Georgeson in Kansas, and Brooks in Massachusetts. Probably no more than eight cultivars were grown in the United States, predominately in the southern states, before the United States Department of Agriculture (USDA) began organized introductions of soybean in 1898. By 1907, Ball reported that 65 lots of soybean seeds, representing 20 cultivars had been collected. These introductions and subsequent U.S. breeding efforts have significantly expanded the areas of adaptation and production in this country. By 1937, this program had introduced over 10,000 selections of at least 2,500 distinct types from China, Manchuria, Japan, Korea, Java, Sumatra, and India.

According to Hymowitz and Bernard (1991), three technological advances occurred that catapulted soybean into major crop status in the non-east-Asia world. These advances were:

1. The discovery in Germany that legumes fix atmospheric nitrogen when roots were nodulated by rhizobia, rod shaped bacteria that are widely distributed in soils. The Massachusetts Experiment Station subsequently reported that soybean yields were highest when nodules were most abundant. Commercial inoculant was available in the United States by 1905.
2. The discovery in 1917 that properly heated soybean meal is superior in quality to unheated meal.
3. The understanding, reported in 1920, of the photoperiodic sensitivity of soybean to day length.

Soybean was grown primarily as a forage crop in the United States until World War II. Prior to 1930, acreage of soybean harvested for seed was less than 25 percent of total acres planted. The production of soybean increased in 1934 as a producer response to the severe drought of that year in the upper midwest states. Large acreages of corn, small grains, and hay crops were lost very early on that season, and soybean was planted as an emergency crop. The performance of soybean under drought conditions was better than corn, thereby convincing midwest corn producers to try this relatively new crop. Additionally, in an effort to decrease the supply of corn and thereby increase the price received by producers, the Agricultural Adjustment Administration, USDA, set corn acreage allotments. This restricted the number of acres that producers could plant to corn and increased the amount of land available for production of other, noncontrolled crops. Soybean for seed was one of the nonrestricted alternate crops and acreage harvested for seed increased to 40 percent by 1939. The need in the United States for fats, oils, and oilseeds after war was declared with Japan resulted in 72 percent of the planted acreage in the United States being harvested for seed by 1944.

An enormous number of soybean cultivars exist today. Modern cultivars in much of the world today were developed through the scientific application of genetics. These cultivars are well-defined in morphological and agronomic terms, and many also are well defined biochemically for oil and protein quality and quantity. Prior to modern plant breeding, cultivars—then known as varieties—were much more localized. That is, a cultivar would be selected by farmers or early seedsmen within an ecological community. Separated by some number of miles—a distance that grew ever smaller in industrialized nations as travel became easier and accessible to the masses—another community could have a different but adapted cultivar.

Soybean cultivars in eastern Asia were classified on the basis of grain color, size, and shape, and on time and method of planting. Finding a cultivar that was widely known was virtually impossible. Cultivars in China, Manchuria, and Korea were defined as yellow, green, black, brown, or mixed colored. The yellow seeded beans generally were used as human food, especially soybean curd. The yellow bean also was valued as having a higher oil content. Cultivars having green seed with yellow embryo and cotyledons produce greater quantities of curd, although inferior in quality to that produced from yellow beans. The green bean with green germ was preferred for sprouts. Black beans were used for oil, horse feed, or for salted, fermented beans.

Early cultivars grown in the United States were direct selections from germplasm introduced from China, Japan, and Korea. The first soybean cultivars developed through pure line selection from controlled hybridizations were released for commercial production in 1939–40. Pagoda was developed by R. Kimmock with the Canada Department of Agriculture; Ogden was developed by H. P. Ogden with the Tennessee Agricultural Experiment Station (AES); Chief was released by C. M. Woodworth at the Illinois AES. Between 1939 and 1988, 274 soybean cultivars were released for commercial production by public institutions. All but 15 of these were developed for the seed market.

HISTORICAL EVENTS

There are several notable events in history that led to the development of the soybean industry as we know it today. The six most notable were (1) the development of trade between Manchuria and Japan following the Sino-Japanese war of 1895; (2) the Russo-Japan war of 1904–05; (3) events leading to and during World War II; (4) machine power; (5) the dust bowl years; and (6) Franklin D. Roosevelt's New Deal.

Sino-Japanese War

In 1894, China sent troops to squash a rebellion in Korea (Figure 7.1). Japan, having considerable financial interest in Korea, sent troops to protect those businesses. China crushed the rebellion but Japan refused to withdraw its troops, and fighting broke out between China and Japan. The hostilities ended with independence for Korea and several concessions to Japan. So Japan was introduced to the soybean and began to import soybean meal as a fertilizer, thus expanding its market area.

Russo-Japan War

With major concessions from the Chinese following the 1894–95 hostilities, Japan wanted to expand its empire by controlling Korea and other territory, especially the Liaotung Peninsula of China where two major ports were located (Figure 7.1). The Russians also had been expanding their interest in eastern China, and through treaties and leases had built the Chinese Eastern Railroad across Manchuria, thus gaining partial control of the province. Russia leased the Liaotung Peninsula from China in 1898 and built a naval base at Port Author and a commercial port at Dairen, thereby having considerable influence in Manchuria and Korea by the early 1900s.

Japan, bolstered by its defeat of China in 1895, prepared for war with Russia and attacked Russian ships at Port Author on February 8, 1904. Fighting a short war, compared with most, Russia and Japan signed the treaty of Portsmouth at Portsmouth, New Hampshire, in 1905. The war cost the Russians dearly; they conceded southern Sakhalin Island, Port Author, and Dairen to the Japanese, removed all of their troops from Manchuria, and gave up all interest in Korea.

Food production had to increase in northern China as Japan deployed 200,000 troops there during the short hostilities with Russia. In Manchuria, this meant an increase in soybean production. With the end of the war, troops were withdrawn and surpluses of soybean resulted, and Japan was somewhat forced to initiate efforts to open markets for this excess of Manchurian soybeans. Soybean was introduced into England in 1908 and gained some acceptance as oil for soap manufacture and as meal in some mixed livestock feeds. Germany, Denmark, and the Netherlands imported Manchurian soybean by 1910. This expansion of demand for Manchurian soybean led to Chinese farmers migrating into Manchuria and thereby further exacerbating the overproduction problem. The amount of soybean imported into England and Europe continued to increase into the 1930s, although interrupted briefly

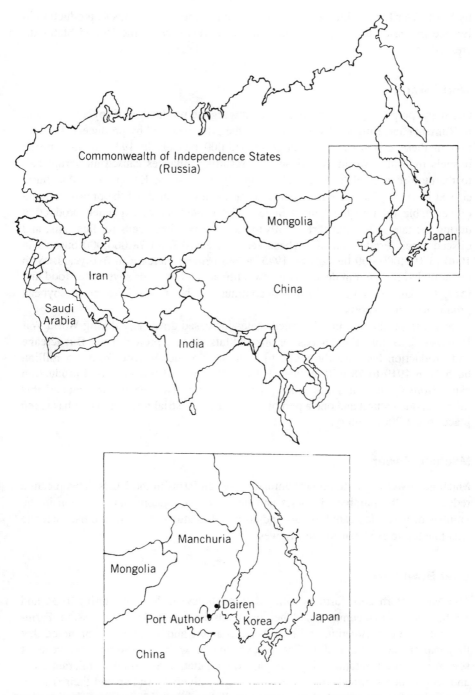

Figure 7.1 *Geography of Central East Asia. Inset shows Japan, China, Manchuria, Korea, and the Liaotung Peninsula including the sea ports of Port Author and Dairen. (Redrawn by Mike Hodnett)*

by World War I. Manchurian exports began to decline in the 1930s as production in Europe increased and imports from other countries such as the United States increased.

World War II

Germany, preparing for war, began encouraging the production of soybean in 1935 in Rumania and Bulgaria by guaranteeing the price received by producers. Production in these countries increased from 375,000 bushels in 1934 to 5.5 million bushels by 1941. World War II and the resulting alliances caused an abrupt halt to theretofore normal trade and shipping. Trade between Europe and Manchuria ceased in 1940. The Far East accounted for about two-thirds of the prewar imports of vegetable oil into the United States. The United States expanded production during the late 1930s and early 1940s to meet the wartime needs for fats, oils, and oilseed meal. Production of soybean seed increased from 78,000,000 bushels in 1940 to 192,000,000 bushels by 1945, a 246 percent increase in five years. With unparalleled research in developing new cultivars and uses for soybean as food and feed, the United States gained and continues to be a world leader in soybean production and exports.

Soybean production in the United States increased drastically during World War II to meet U.S. and allied needs for oils and fats, and for protein meal. But acreage and production had already begun to increase. Production rose from 2.5 million bushels in 1919 to 78 million bushels by 1940. Most of this increased production came from Corn Belt states where corn was king. Two events precipitated this increase, one natural and one a part of the intense industrial revolution that has taken place in the 20th century.

Machine Power

Machine power began to replace animal power on farms in the 1920s. This meant a reduction in the number of horses and mules and a concomitant reduction in the number of acres dedicated to oats and corn as feedstuffs. Most of the oats acreage was planted to soybean in the Midwest.

Dust Bowl Years

The Midwestern corn farmers suffered through severe droughts during 1934 and 1936, droughts that compounded their problems during the Great Depression. Farmers that were experimenting with soybean as a new and alternative crop noted that they experienced less yield decline in soybean than with corn during those years of severe moisture limitation. Soybean gained a reputation as a drought tolerant crop, and so producers turned to soybean as a way to diversify and spread their risks of crop failure from low rainfall and poor rainfall distribution. This diversification also spread their risk of financial ruin resulting from overproduction of only one cash crop.

Franklin D. Roosevelt's New Deal

Acreage control and soil improvement programs were initiated by President Franklin D. Roosevelt in 1933 in an attempt to decrease the oversupply of corn (and other crops) and therefore increase prices received as a way to help the country out of the Great Depression. No acreage controls were placed on soybean, and farmers planted some of their Midwest corn acres to soybean for this reason.

U.S. AND WORLD PRODUCTION

Soybean production in the United States in 1991 was an $11 billion crop, more farm income than any other single crop. Yet soybean was virtually unknown to the American farmer in 1900. As noted earlier, there were only eight cultivars known in 1898 and those were probably forage types. In 1917, the United States planted only 460,000 acres of soybean and less than 37 percent of those acres were harvested for seed. Most of the early acres grown for seed were grown in North Carolina and other Southeastern states, with only a small percentage grown for seed in Corn Belt states. However, between 1917 and 1940, most of the increase in acreage planted to soybean in the Corn Belt states of Illinois, Indiana, Ohio, Iowa, and Missouri was dedicated to seed production, while about half of the soybean acres in other states, particularly the southern states, were grown for hay or green manure (Table 7.1).

U.S. production of soybean was only 4,875,000 bushels (273,000,000 pounds at 56 pounds/bushels) in 1925 (Table 7.2). Production reached 100,000,000 bushels in 1941; 200,000,000 bushels in the late 1940s; the 1 billion bushels mark in 1968; and has been about 2 billion bushels/year since 1979. Production in 1990 was 1,925,947,000 bushels, or 107,853,032,000 pounds.

Average yield was 20 bushels/acres or less through the 1940s but reached the high 20 bushels/acres range by the late 1960s, and today the United States averages about 35 bushels/acres (Table 7.2). Prices received were at an all time low of only 67 cents/bushels (1.2 cents/pound) in 1938 during the Great Depression and reached the $7.00/bushels or higher levels during the inflationary years of the 1970s.

Illinois was the leading producer of soybean in 1990, producing 354,900,000 bushels. The second leading state in production was Iowa with 327,850,000 bushels, followed by Minnesota with 179,400,000 bushels; Indiana with 171,380,000 bushels; Ohio with 135,720,000 bushels; and Missouri with 124,500,000 bushels. These six states produced 67 percent of the soybeans produced in the United States in 1990 (Table 7.3).

The U.S. soybean industry has grown from its meager beginnings in 1765 as a novelty crop for one seaman to the largest producer of soybean in the world today. In 1990/91, the United States produced 1,923,667,000 bushels, 50 percent of the world's supply (Table 7.4). Only three other countries produced more than 5 percent of the world's supply each, those being Brazil, producing 578,025,000 bushels; Argentina, producing 422,050,000 bushels; and the People's Republic of China, which produced 403,700,000 bushels.

TABLE 7.1   Acreage by State Planted to Soybean for 1930 through 1990 in the United States. (1,000 Acres)

State	1930	1940	1952	1960	1970	1980	1990
				($\times$1,000 ac.)			
Alabama			170	155	642	2,200	470
Arkansas	9	63	996	2,448	4,393	4,800	3,400
Delaware			67	194	165	265	200
Florida			14	35	188	475	80
Georgia			128	146	550	2,200	900
Illinois	410	1,995	3,830	5,013	6,848	9,400	9,200
Indiana	151	723	1,782	2,458	3,300	4,400	4,200
Iowa	66	709	1,540	2,615	5,709	8,300	8,000
Kansas	7	26	703	594	1,074	1,550	2,000
Kentucky			227	260	582	1,650	1,250
Louisiana			258	308	1,730	3,450	1,800
Maryland			94	238	216	400	505
Michigan			105	229	539	980	1,150
Minnesota		53	1,197	2,118	3,140	4,800	4,700
Mississippi	28[2]	54[2]	650	996	2,358	4,000	2,050
Missouri	138	109	1,824	2,387	3,546	5,700	4,200
Nebraska			90	165	824	1,830	2,400
New Jersey			36	42	52	200	110
North Carolina	128	190	508	673	981	2,030	1,400
North Dakota			31	182	184	210	500
Ohio	31	561	1,146	1,514	2,434	3,800	3,500
Oklahoma			154	137	200	350	250
Pennsylvania			745	23	30	135	280
South Carolina			185	580	1,550	1,600	800
South Dakota			89	102	253	780	1,950
Tennessee			385	492	1,280	2,450	1,300
Texas	28	23	10	84	170	560	220
Virginia			253	347	372	620	540
Wisconsin			61	102	155	355	440
TOTAL[1]	3,072	10,478	15,048	24,440	43,082	69,930	57,795

Source: Adapted from USDA Agriculture Statistics, 1936 and following years.

1. Acres do not add to total because of minor production in other states.
2. Includes Louisiana.

SOYBEAN PROCESSING INDUSTRY

The first soybean processed in the United States was crushed in a hydraulic press mill in Seattle, Washington, in 1911. The beans were imported from Manchuria. The first domestic soybean processed was in a cotton oil mill in Elizabeth City, North Carolina, in 1915–16, a total of 10,000 bushels.

The soybean processing industry began in 1917 when a plant with both expeller and hydraulic equipment was built in Chicago Heights, Illinois. The plant was

TABLE 7.2   Production Statistics for Soybeans Grown for Seeds in the United States, 1925–1990.

Year	Harvested Acres	Average Yield	Total Production	Average Price Received
	(×1,000)	(bu./ac.)	(×1,000 bu.)	($/bu.)
1925	na	na	4,875	na
1930	na	na	13,929	1.37
1933	1,044	12.9	13,509	0.94
1935	2,915	16.8	48,901	0.73
1938	3,035	20.4	61,906	0.67
1940	4,807	16.2	78,045	0.90
1941	5,889	18.2	107,197	1.55
1942	9,894	19.0	187,524	1.01
1943	10,397	18.3	190,133	1.81
1944	10,245	18.8	192,121	2.05
1945	10,740	18.0	193,167	2.08
1950	13,807	21.7	299,249	2.47
1955	18,620	20.1	373,682	2.22
1960	23,655	23.5	555,085	2.13
1965	34,449	24.5	845,608	2.54
1968	41,104	26.8	1,103,129	2.43
1970	42,056	26.7	1,123,740	2.85
1974	51,341	23.7	1,216,287	6.64
1975	53,617	28.9	1,548,344	4.92
1979	70,566	32.1	2,267,901	6.28
1980	67,856	26.4	1,792,062	7.57
1985	61,599	34.1	2,099,056	5.05
1990	56,512	34.1	1,925,947	5.74

Source: Adapted from USDA Agricultural Statistics, 1936 and following years.

moved to Bloomington, Illinois, in 1924 because not enough beans were produced near Chicago Heights to maintain operations. The first soybean crushing by a major corn processing company was in 1924 in Decatur, Illinois. As other companies opened operations in Decatur, it soon became the center of soybean processing in the United States during the 1920s and 1930s.

Problems that had to be addressed by the soybean industry in the United States were (1) improving processing methods to improve oil yield and therefore reduce cost; (2) convincing farmers to produce enough beans for economical processing; and (3) creating markets for soybean oil and meal. Oil yield was improved by the development of solvent extraction, with the first solvent operation established in 1934. Solvent extraction involves cleaning, cracking, and dehulling the soybean seed, followed by rolling the seed meats into flakes. A solvent, usually hexane, is added to the flakes that dissolves the oil and carries it out of the soybean flakes to be collected and desolventized, a process of heating the mixture sufficiently to vaporize the solvent. Water is then added to aid in the separation of oil and lecithin, which are further refined into a number of products. The flakes are cleaned of solvent by forcing steam through them, which heats and vaporizes the solvent. Flakes are then

TABLE 7.3 Soybean Production in United States by State and Years.

State	1930	1940	1952	1960	1970	1980	1990
				(×1,000 bu.)			
Alabama			1,748	3,192	13,800	31,500	7,480
Arkansas	68	756	13,856	50,589	99,000	65,250	90,450
Delaware			986	4,536	3,276	5,200	6,766
Florida			240	780	5,152	10,120	1,425
Georgia			336	1,275	10,925	23,400	9,800
Illinois	6,970	34,912	85,128	129,298	210,800	313,225	354,900
Indiana	2,114	9,399	38,493	65,205	101,618	157,680	171,380
Iowa	1,023	14,180	37,587	66,274	184,600	318,395	327,850
Kansas	52	312	7,360	12,599	13,950	23,925	46,800
Kentucky			1,767	4,378	14,310	36,000	39,040
Louisiana			594	5,184	40,512	67,000	42,000
Maryland			1,350	5,850	5,112	9,360	17,820
Michigan			1,748	4,420	13,250	32,010	43,320
Minnesota		795	21,945	40,755	78,780	149,940	179,400
Mississippi	246[2]	592[2]	6,142	20,152	58,050	61,600	39,900

State							
Missouri	966	1,417	32,756	50,396	88,358	135,485	124,500
Nebraska			2,288	4,592	17,864	53,100	81,420
New Jersey			410	808	1,350	3,492	3,996
North Carolina	1,344	2,280	4,785	12,262	20,808	34,740	32,400
North Dakota			362	2,288	2,715	3,500	12,870
Ohio	434	8,976	20,680	36,726	72,675	135,360	135,720
Oklahoma			861	2,480	3,330	3,000	4,620
Pennsylvania			361	161	896	3,185	11,275
South Carolina			1,127	9,730	20,200	20,800	13,875
South Dakota			1,275	1,700	4,323	20,020	53,760
Tennessee			3,620	8,865	26,450	45,900	33,750
Texas			450	2,025	4,424	13,860	5,000
Virginia			2,958	7,200	6,780	9,150	16,800
Wisconsin			816	1,536	3,672	10,890	17,630
Total[1]	13,929	78,045	291,682	555,307	1,127,100	1,797,543	1,925,947

Source: USDA Agricultural Statistics, 1936 and following years.

1. Yields do not add to total because of minor production in other states.

2. Includes Louisiana.

TABLE 7.4   Production Statistics for Soybeans for Countries Producing at Least a Million Bushels in 1990/91.

Country/Continent	Harvest Acres	Average Yield	Total Production
	(×1,000)[1]	(bu./ac.)	(×1,000 bu.)[2]
Canada	1,213	39.1	47,416
Guatemala	37	37.8	1,358
Mexico	682	30.5	20,809
United States	56,489	34.1	1,923,667
N. and C. America	58,423	34.2	1,993,324
Argentina	11,856	35.7	422,050
Bolivia	459	28.1	12,918
Brazil	23,836	24.3	578,025
Colombia	252	27.1	6,826
Ecuador	141	26.0	3,670
Paraguay	2,198	21.7	47,710
Uruguay	84	17.6	1,468
South America	34,841	27.7	1,073,071
France	287	31.4	8,992
Italy	1,351	49.6	66,794
Spain	42	36.8	1,541
Hungary	89	19.8	1,615
Romania	469	11.0	5,175
Yugoslavia	225	24.8	5,578
Europe	2,561	30.5	92,264
Former USSR	2,043	15.8	32,296
Egypt	101	38.5	3,890
Nigeria	185	12.9	2,386
South Africa	215	21.6	4,624
Zambia	96	17.1	1,652
Zimbabwe	143	24.8	3,560
Africa	753	21.9	16,405
People's Republic China	18,673	21.7	403,700
India	5,842	15.2	88,777
Indonesia	3,001	16.1	48,261
Iran	124	26.8	3,303
Japan	361	22.5	8,074
South Korea	840	19.2	16,148
North Korea	375	22.8	8,551
Thailand	926	21.0	19,451
Turkey	148	29.8	4,404
Vietnam	346	11.2	3,854
Asia	30,801	19.8	606,578
Australia	89	28.9	2,569
World totals	133,511	28.6	3,816,580

Source: Adapted from USDA Agricultural Statistics, 1992.

1. Acres do not add to continental or world totals because of rounding.
2. Acreage × average yield does not equal total production because of rounding.

TABLE 7.5   Some of the Over 200 Products Produced in the United States from Soybean and Soybean Fractions.

I. Whole Soybean Products

   A. Soy sprouts
   B. Full fat soy flour
      1. Bread
      2. Doughnut mix
      3. Frozen desserts
      4. Pancake flour
   C. Roasted soybeans
      1. Candy ingredient
      2. Confection
      3. Soy coffee

II. Soybean Meal Products

   A. Edible Uses
      1. Noodles
      2. Cereals
      3. Baby food
      4. Hypo-allergenic milk
      5. Special diet foods
      6. Meat extender
   B. Industrial Uses
      1. Adhesive
      2. Plywood
      3. Wallboard
      4. Tape joint cements
      5. Linoleum backing
      6. Texture paints
      7. Antibiotics
   C. Feed uses
      1. Calf milk replacers
      2. Livestock and pet feed
      3. Poultry feed

III. Oil Products

   A. Glycerol
   B. Fatty acids
   C. Sterol
   D. Refined soy oil
      1. Cooking oils
      2. Mayonnaise
      3. Margarine
      4. Pharmaceuticals
      5. Vegetable shortening
      6. Creamers
      7. Caulking compounds
      8. Disinfectants
      9. Insecticides
    10. Fungicides
    11. Herbicides
    12. Printing ink
    13. Putty
    14. Soap
    15. Waterproof cement
   E. Soybean lecithin
      1. Emulsifying agent
      2. Anti-foam agent
      3. Paint
      4. Ink
      5. Rubber
      6. Cosmetics

Source: Adapted from *Soybeans-The Miracle Crop.* American Soybean Association.

toasted and used in a number of products including soy meal and soy meal products (Table 7.5). The protein component of the soy flakes may be extracted for use in a number of food products.

MORPHOLOGY

Soybean, at least the cultivated type, is an erect, bushy annual that varies in height from 12 inches to 8 feet, depending upon cultivar, day length, temperature, moisture, and nutrition. The initial leaves are the seed leaves, or cotyledons. The first true leaves are simple and opposite, while all later occurring leaves are trifoliolate and alternate. Occasionally, leaves having four or more leaflets will occur. White or tan trichomes cover the leaves, stems, sepals, and pods of most cultivars, although glabrous types do exist.

The flowers of soybean develop from auxiliary buds along the main stem and branches. The inflorescence at each axil is a raceme consisting of 2 to 35 papilionaceous flowers. With indeterminate types, the inflorescences are axillary and at maturity the number of pods developing at each node tends to decrease toward the terminals of the main stem and branches. The determinate types, however, have both axillary and terminal inflorescences and at maturity should have a rather dense cluster of pods at the main stem and branch terminals. Flowering usually begins at about the fifth or sixth true leaf node, with vegetative branches arising from lower nodes.

The soybean flower is a typical papilionaceous flower—i.e., having a corolla composed of a standard, two wings, and two keel petals, typical of a legume. The stalk, or stem, of each inflorescence is morphologically typical of any stem or branch, sometimes referred to as a peduncle in other plant species. The inflorescence stalk gives rise to pedicled flowers. Each flower is subtended by two bracteoles and a fused or tubular calyx composed of five unequal lobes. The corolla has five petals, a large posterior banner petal, two lateral wing petals and two anterior keel petals that touch but are not fused. The stamens form a diadelphous androecium in which nine filaments are fused and elevated as a single structure with a posterior stamen remaining separate. The pistil is composed of a unicarpellate ovary with one to five (usually four) ovules and a style that is about half the length of the ovary and curves backward toward the posterior stamen and is terminated by a capitate stigma.

The soybean pod may contain up to five seeds but two to three is most common. Pods are straight to slightly curved, being 3/4-inch to 17 inches long with 2 to 3 inches being most common. The pod is made up of two halves of the single carpel connected by dorsal and ventral sutures. Pod color varies from yellow to yellow-gray, brown, or black.

SOYBEAN MATURITY CLASSES

Soybean is a photosensitive plant—i.e., it begins to produce flowers when a critical dark period is reached. (We usually refer to this as day sensitivity although the length of uninterrupted dark is the driving criteria.) Day length varies with latitude, and cultivars vary in their night length requirement for flower initiation. This situation means that most cultivars cannot be moved, generally, more than about 100 miles either north or south from their area of greatest adaptation. It also meant that prior to the 1920s, when scientists began to understand the photoperiodic response in soybeans, that producers were faced with trying to produce a crop that responded differently to location than those to which he or she was accustomed. Moving a high-yielding cultivar from one area to an area several miles north or south would result in a crop that could be considerably earlier or later maturing and perhaps much lower-yielding.

W. J. Morse was one of the first to make some sense of this confusing picture. In 1918, he reported that soybeans could be divided into three distinct types and that each of these types was suited to specific areas of the United States. Morse classified soybean cultivars as late, medium and medium late, and very early. The very early types were recommended for the northern states and the late group was recommended for southern states. In 1927, Morse classified a number of cultivars into five groups, with group one recommended for southern production and group five recommended for the northern tier of states. Scientists in Missouri in 1929 proposed that soybean cultivars be divided into four maturity groups, with group one being extremely early- and group five extremely late-maturing.

The maturity grouping that is used today in the United States was proposed by Morse in 1949 and was the culmination of 30 years of research into cultivar response to photoperiod. Morse's early-maturing group of cultivars were designated MG 0 and recommended for the northernmost areas of the United States, while cultivars adapted to the southernmost tier of states were categorized as MG VIII. Today this system places cultivars into 1 of 13 groups, 000, 00, 0, I, II to X. The lower numbers are adapted to more northern areas and classification progresses to X as one moves southward, with group X being adapted to the tropics. For a given location, a cultivar may be considered a full season, a midseason, or an early (or short) season cultivar. Earlier cultivars require fewer hours of continuous dark for flower initiation following a juvenile period.

A cultivar released for production in central Arkansas could be classified as a maturity group VI (Figure 7.2). Let us assume that this cultivar, then, is released for production in Arkansas from the 32nd parallel of latitude of the 35th parallel of latitude. Let us further assume that this determinate (indeterminate and determinate types will be discussed subsequently) cultivar begins to bloom on August 1, therefore requiring 605 minutes (10 hours and 5 minutes) of continuous dark (Figure 7.3). If we plant this cultivar in southeast Texas at the 30th parallel of latitude, then it will begin to flower on about July 12—i.e., it would be classified as an early-maturing cultivar. The opposite happens if we plant this cultivar at the 40th parallel of latitude, say in northern Missouri. At that latitude, this cultivar becomes later-maturing with flowering initiated about August 12. Therefore, early, mid, or late are relative terms applicable to a cultivar grown in a rather narrow geographical range relative to latitude.

One should also note that photosensitive crops like soybean respond to a complex of environmental cues, particularly the interaction of temperature, physiological age, and day length (Thomas and Raper, 1978). While soybean is sensitive to day length, temperature thresholds exist for soybean as well as for all other crops. Holmberg (1973) identified minimum and optimum temperatures required for the development of several stages of growth and reproduction (Table 7.6). Although photoperiod defines the latitudes of optimum adaptability of soybean cultivars, temperatures are not always adequate for their production. Figure 7.2 would suggest that group 00 or 000 could be grown in northwestern states as well as northeastern states and Canada. However, low night temperatures limit yields in those states although the frost-free period is sufficient.

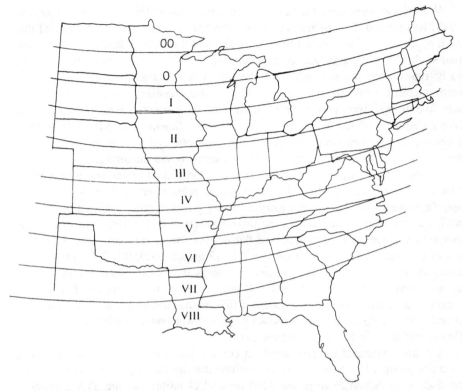

Figure 7.2 *General bands of adaptability of soybean groups in the United States. (Redrawn by Mike Hodnett)*

Soybean breeders, especially those with commercial companies, have developed cultivars having wider geographical adaption than earlier thought possible. Table 7.7 shows the seed yields of cultivars within maturity groups III through VII at Keiser, Arkansas, during 1987–88. Keiser lies between the 35th and 36th parallels and is considered suitable for maturity groups V and VI. The research reported in Tables 7.7 and 7.8 was aimed at determining if early-maturing groups III and IV, indeterminate types, could be produced in Arkansas and to determine if cultivars in these groups would mature a sufficient crop before late summer droughts occur that are normal for northeastern Arkansas. Regardless of the purpose of the experiment, their data clearly indicate wide adaptability for present-day cultivars in maturity groups III and IV.

GROWTH TYPES

Soybean is classified also by its growth and floral initiation. Indeterminate soybean types begin to flower and set pods early in the growing season while continuing to

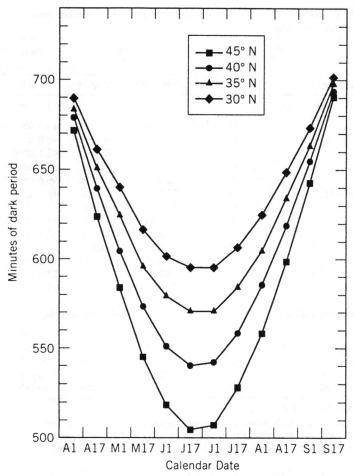

Figure 7.3 *Length of dark period as sundown to sunrise at four latitudes north of the equator for April 1 through September 17.*

TABLE 7.6 Temperature Thresholds for Soybean Growth and Reproduction.

	Range	
Stage of Development	Minimum	Optimum
	------------ (°F)-------------	
Germination	43	70
Emergence	46	70
Formation of floral buds	61	72
Flowering	63	75
Seed development/maturity	46	68

Source: Adapted from Raper and Kramer, 1987. Soybeans: Improvement, Production, and Uses. J. R. Wilcox (ed.), American Society of Agronomy, 1987.

TABLE 7.7 Yields and Maturity Dates of Several Soybean Cultivars from Four Maturity Groups at Keiser, Arizona.

Soybean Cultivar	Maturity Group	1987[1] Planting Dates		1988[2] Planting Dates	
		April 20	May 14	April 8	May 10
		------------------------(bu./ac.)------------------------			
Williams 82	III	45.8 bcd[3]	49.1 ab	44.2 abc	43.7 ab
Asgrow 3966	III	47.1 bc	44.0 bc	48.9 a	40.5 ab
FFR 441	IV	42.2 cd	42.3 c	46.1 abcd	41.6 ab
Egyptian	IV	40.6 de	52.3 a	37.4 bcd	47.6 a
Crawford	IV	45.8 bcd	48.5 ab	36.4 cd	42.5 ab
Ring Around 451	IV	56.8 a	49.3 ab	48.7 a	45.7 ab
Asgrow 5474	V	49.6 b	39.5 c	46.0 ab	47.3 a
Forrest	V	45.5 bcd	43.5 bc	44.6 abc	45.1 ab
Centennial	VI	45.2 bcd	33.6 d	36.4 cd	37.6 b
Leflore	VI	47.4 bc	40.3 c	33.7 d	48.0 a
Bragg	VII	35.3 e	22.8 e	39.6 bcd	41. ab
Hartz 7126	VII	48.8 b	25.6 e	33.9 d	29.4 c
Average Yield		45.8	40.9	40.9	42.5

Average Maturity Dates

		----------------------(month–day)----------------------			
Maturity group III		8–12	9–6	8–26	9–7
Maturity group IV		9–4	9–17	9–9	9–19
Maturity group V		9–23	9–28	9–20	9–29
Maturity group VI		10–23	10–27	10–27	10–29
Maturity group VII		10–24	10–28	10–30	10–31

Source: Adapted from May et al., 1989.

1. LSD$_{05}$ for comparing cultivar means between planting dates for 1987 = 5.8 bu/acre.
2. LSD$_{05}$ for comparing cultivar means between planting dates for 1988 = 10.2 bu/acre.
3. Means followed by the same letter(s) within a column are not statistically different.

TABLE 7.8 Soybean Seed Yields of Cultivars from Five Maturity Groups Grown at Keiser, Arizona, 1987–1988.

No. of Cultivars	Maturity Group	Planting Date	
		Early > Apr. 25	Normal
2	III	46.5	44.3
4	IV	43.7	46.2
2	V	46.4	43.9
2	VI	40.7	39.9
2	VII	39.4	29.8

Source: Adapted from May et al. 1989.

grow vegetatively—that is, adding new nodes from which floral buds are initiated. Determinate types add much less additional vegetative mass after flowering begins. Semideterminate cultivars that are intermediate in their vegetative and reproductive growth to the determinate and indeterminate types have been released during the last decade.

Traditionally, indeterminate types have been grown north of the 37th north parallel of latitude—that being the southern tip of Illinois. Groups 00 through IV are adapted to this northern tier of states and require fewer hours of continuous dark for flowering and maturity. Conversely, groups V through VIII are considered southern types and are determinate in growth and flowering habit, requiring more hours of continuous dark for initiation of reproductive development. Since 1978, some determinate cultivars have been released in maturity groups II, III, and IV. These cultivars are usually planted at very high plant densities, because yields of individual plants will be reduced because of their reduced plant size. Conversely, recent research has suggested that indeterminate cultivars of maturity group III can be planted early in many areas of the south and produce acceptable yields. The logic behind this production strategy is that flowering will be initiated in the spring, and therefore pod fill will occur while moisture and temperatures are more conducive to soybean growth, thus effectively avoiding the high temperatures of July and August in the Deep South. This production strategy also avoids summer droughts that are common in many southern states.

Determinate types were probably recommended originally for the south because of their ability to produce adequate forage. These cultivars would produce a vegetative structure on which seed could be set and mature late in the summer and early autumn after summer droughts in many years and so became established as the type to grow. Producers could plant earlier-maturing cultivars when planting early in the recommended planting date range, and these cultivars would produce vegetative mass early during the summer before flowering was initiated. If forced to plant later, then the southern producer planted a later-maturing cultivar that required more hours of dark for flower initiation. This meant that this cultivar would bloom later in the summer, thereby making more vegetative growth than the earlier-maturing cultivar would if it were planted late.

If a group V soybean is planted 30 days late, say June 20, at the 35th parallel of latitude, its maturity will be delayed by about 20 days. Yet, its yield can be substantially reduced, vegetative growth being reduced because of its photoperiodic response. Reduced number of nodes equates to fewer fruiting sites on the main stem and less branching that also reduces fruiting opportunity. Planting a later-maturing group VII cultivar 30 days late will result in less delay in maturity and possibly less reduction in yield, because the cultivar has a longer length of time for vegetative growth to occur before flower initiation. Therefore, producers are encouraged to pick their cultivars carefully, planting early-maturing cultivars early in the planting season and later-maturing groups later in the appropriate planting season.

However, the above scenario is not an absolute. Planting a late-maturing cultivar can result in low yields because of inclement weather during harvest or because of the onset of low temperatures before the crop has matured. Also, there is not a linear relationship between the number nodes on a soybean and yield. As noted above, there are up to 35 potential flowers at each node on the main stem and at branch nodes. Good management and especially irrigation will help the producer in the south to plant early-maturing cultivars and make maximum yields. Each producer must evaluate his or her situation carefully and choose the cultivar that fits their management strategies and capabilities.

Producers in the northern tier of states must be concerned with the relatively short duration of their growing season. Recommended planting dates for central Michigan, 45° north latitude, are early to mid-May. These producers must have cultivars that will grow vegetatively and reproductively, because they do not have enough growing season for determinate cultivars to make sufficient vegetative growth for maximum yields. Too, these producers must be concerned with an early frost. This situation mandates cultivars that require a relatively short period of continuous dark for flower initiation. Note in Table 7.9 that on June 21, northern Michigan, 46° north latitude, has only 495 minutes (8 hours and 15 minutes) of continuous dark, while southern Louisiana at 30° north latitude has 595 minutes (9 hours and 55 minutes). Some of the very early-maturing cultivars in maturity groups 00, 0, I, and II are insensitive to photoperiod and will flower after the completion of a short juvenile period.

Intermediate growth types (i.e., semideterminate) have been developed over the past decade. These types do not terminate vegetative growth as abruptly as do determinate types and tend to be intermediate in plant height, other variables being equal.

Table 7.10 compares determinate and indeterminate soybeans. The major difference in these types is that with determinate soybeans, reproductive growth occurs after the majority of the vegetative growth has taken place. With indeterminate types, less than half of the plant's final height has occurred when flowering begins —i.e., the plants grow vegetatively, adding new nodes, after flowering begins. In the latter type, vegetative growth, flowering, pod and seed growth, and maturity occur simultaneously during a considerable part of the plant's life.

There are pros and cons for each of the growth types of soybean. Determinate types could be penalized under stress conditions during vegetative stages if the stress was severe enough to reduce the number of main stem nodes and reduce branching. Thus, the number of fruiting sites could be reduced, thereby decreasing yield. This situation also could result in pods too close to the ground to harvest. Since determinate types flower for a shorter period of time, stress during flowering that causes shedding cannot be compensated for later in the season as in the case of indeterminates. However, flowering in determinate types does occur over several days and the plant can compensate for some flower shedding if the stress is relieved. Indeterminate types, on the other hand, mature over a longer time frame, which exposes early-set pods to predation by insects and diseases and to weathering that may cause shattering. Indeterminate types may grow taller and are therefore generally more susceptible to lodging.

TABLE 7.9 Duration of Daylight Defined as Sunrise to Sundown for Four North Parallels of Latitude.

Day of Month	Jan. h.	m.	Feb. h.	m.	Mar. h.	m.	Apr. h.	m.	May h.	m.	June h.	m.	July h.	m.	Aug. h.	m.	Sept. h.	m.	Oct. h.	m.	Nov. h.	m.	Dec. h.	m.
											Latitude 30° N.													
1	10	15	10	46	11	33	12	29	13	20	13	57	14	03	13	34	12	46	11	53	10	59	10	22
5	10	17	10	53	11	40	12	36	13	26	13	59	14	01	13	29	12	39	11	46	10	53	10	19
9	10	21	10	59	11	47	12	43	13	31	14	02	13	58	13	23	12	32	11	38	10	48	10	16
13	10	24	11	05	11	54	12	50	13	37	14	04	13	55	13	17	12	25	11	32	10	42	10	14
17	10	27	11	12	12	02	12	57	13	42	14	04	13	52	13	11	12	18	11	25	10	36	10	14
21	10	33	11	18	12	09	13	04	13	47	14	05	13	48	13	04	12	10	11	17	10	32	10	12
25	10	37	11	25	12	16	13	10	13	50	14	05	13	43	12	58	12	03	11	11	10	28	10	13
29	10	43	11	33	12	24	13	17	13	55	14	03	13	39	12	51	11	56	11	05	10	24	10	14
											Latitude 35° N.													
1	09	51	10	30	11	26	12	34	13	35	14	21	14	29	13	54	12	55	11	50	10	45	10	00
5	09	53	10	37	11	34	12	42	13	43	14	25	14	27	13	47	12	47	11	41	10	37	09	55
9	09	57	10	45	11	44	12	52	13	50	14	27	14	24	13	40	12	38	11	32	10	31	09	52
13	10	02	10	53	11	52	13	00	13	57	14	30	14	19	13	33	12	29	11	24	10	24	09	50
17	10	06	11	01	12	01	13	08	14	03	14	30	14	16	13	25	12	21	11	16	10	18	09	48
21	10	11	11	08	12	09	13	16	14	09	14	31	14	10	13	18	12	12	11	07	10	11	09	48
25	10	17	11	17	12	19	13	24	14	14	14	31	14	05	13	09	12	03	11	00	10	06	09	48
29	10	25	11	26	12	27	13	32	14	19	14	29	13	59	13	01	11	54	10	51	10	01	09	50

(continued)

TABLE 7.9 Duration of Daylight Defined as Sunrise to Sundown for Four North Parallels of Latitude.

Day of Month	Jan.		Feb.		Mar.		Apr.		May		June		July		Aug.		Sept.		Oct.		Nov.		Dec.	
	h.	m.	h.	m.	h.	m.	h.	m.	h.	m.	h.	m.	h.	m.	h.	m.	h.	m.	h.	m.	h.	m.	h.	m.
									Latitude 40° N.															
1	09	23	10	10	11	18	12	39	13	54	14	49	14	58	14	16	13	05	11	47	10	29	09	33
5	09	27	10	19	11	28	12	50	14	02	14	53	14	55	14	08	12	55	11	36	10	20	09	29
9	09	31	10	28	11	38	13	00	14	11	14	57	14	52	14	00	12	44	11	26	10	11	09	25
13	09	36	10	37	11	50	13	10	14	19	15	00	14	47	13	51	12	34	11	16	10	03	09	22
17	09	42	10	47	12	00	13	20	14	27	15	00	14	42	13	41	12	24	11	06	09	55	09	20
21	09	49	10	58	12	11	13	30	14	34	15	01	14	36	13	32	12	13	10	55	09	48	09	20
25	09	56	11	07	12	21	13	40	14	40	15	01	14	29	13	22	12	03	10	46	09	42	09	20
29	10	03	11	18	12	32	13	49	14	45	14	59	14	22	13	13	11	52	10	37	09	36	09	22
									Latitude 46° N.															
1	08	43	09	42	11	06	12	47	14	21	15	30	15	41	14	48	13	19	11	43	10	07	08	56
5	08	47	09	53	11	20	13	00	14	31	15	35	15	38	14	37	13	06	11	30	09	55	08	49
9	08	53	10	05	11	32	13	14	14	42	15	40	15	34	14	27	12	54	11	18	09	44	08	45
13	09	00	10	17	11	46	13	26	14	52	15	42	15	27	14	15	12	41	11	04	09	34	08	42
17	09	07	10	29	11	58	13	38	15	01	15	44	15	21	14	05	12	28	10	52	09	24	08	40
21	09	15	10	42	12	12	13	51	15	10	15	45	15	13	13	53	12	15	10	39	09	15	08	38
25	09	25	10	53	12	25	14	03	15	18	15	45	15	04	13	41	12	02	10	27	09	06	08	39
29	09	35	11	06	12	38	14	14	15	25	15	43	14	56	13	29	11	50	10	15	08	59	08	40

Source: Adapted from List, 1958.

TABLE 7.10 Comparison of Determinate and Indeterminate Types.

Determinate	Indeterminate
Little growth in height after flowering. Flowers occur about the same time throughout the plant.	Will usually double in height during flowering. Will flower at new nodes; flowering, pod and seed development occurring at same time.
Pods and seeds develop uniformly throughout the plant.	Pods and seeds develop from lower nodes upward.
Terminal leaves are approximately the same size as lower leaves.	Terminal leaves are smaller than lower leaves.
Terminal nodes on main stem usually have a long flowering stalk and several pods.	Few pods at terminal node.

STAGES OF GROWTH

Karl vol Linne (in Latin, *Carolus Linnaeus*) realized the need for a uniform method of naming plants so that learned persons could communicate about specific plants. This system, of course, was later extended to cover animals as well. Today we all benefit from the binomial (i.e., genus and species) classification of plants. In like manner, agricultural scientists realized the need for a system of describing the various stages throughout the life of crop plants. Describing when to apply herbicides, add or remove water from rice paddies, and describing when to scout for insects are but three common uses of growth stages among agricultural scientists. Fehr and Caviness (1977) described a system of stages of soybean development that can be applied equally well to both determinate and indeterminate types.

One must be able to identify the main stem, nodes, unifoliolate and trifoliolate leaves in order to use stage-of-growth descriptions. The main stem is the stem that has developed from the epicotyl of the seed. Secondary and tertiary branches can develop from axillary buds. These branches are not to be used in staging soybean because all branches will be later-maturing than the main stem.

A node is the location along the main stem or branch from which a leaf arises (Figure 7.4). Even if the leaf has been broken or shed from the plant, the node can still be identified by the remaining leaf scar. It is important to note that the soybean staging system is dependent on the number of main stem nodes and not on the number of main stem leaves.

As soybean is a dicot, the first leaves along the main stem are the seed leaves or cotyledons. The nodes from which the cotyledons arise are called the cotyledonary nodes and are opposite each other. Above the two cotyledonary nodes are two nodes where unifoliolate leaves are attached. These nodes also are opposite and give rise to the only unifoliolate leaves (i.e., only one leaflet) found on the soybean plant. The unifoliolate nodes are the first nodes counted when staging soybean plants, and since they are opposite each other, both are counted as one.

All nodes above the unifoliolate nodes support a trifoliolate leaf (i.e., a leaf composed of three leaflets). These trifoliolate leaves arise alternately along the

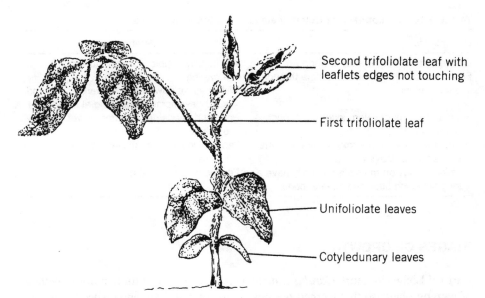

Second trifoliolate leaf with leaflets edges not touching

First trifoliolate leaf

Unifoliolate leaves

Cotyledunary leaves

Figure 7.4 *Illustration of the V2 stage of soybean development. The leaflets at the node above the first trifoliolate have unrolled sufficiently that the edges are not touching. (Drawing by Mike Hodnett)*

length of the stem. As the leaflets emerge, the edges are rolled inward, giving the leaflet a cylindrical shape (Figure 7.5). As growth continues, the leaflets unroll and flatten. Successful staging of a soybean plant requires one to identify fully developed leaves by examining the leaf at the node above the one in question. A leaf is considered fully developed (i.e., the node is counted) when the leaf at the node above has unrolled sufficiently such that the edges of each leaflet are not touching. The terminal node bears a trifoliolate leaf. This leaf is considered fully developed when the leaflets have flattened out and are similar to lower leaflets in size. However, the terminal leaf on indeterminate soybean is smaller than lower leaves while it is essentially equal in size to lower leaves on determinate types (Figures 7.6, 7.7, and 7.8).

Stages of soybean development are divided into vegetative (V) and reproductive (R) stages. Since indeterminate types continue to add new vegetative growth after reproductive growth has been initiated, the V and R stages overlap and plants or fields may be identified as V12 R3, for example. However, in most cases, once flowering begins, only R stages are considered. In determinate beans, identification as R3 stage would be sufficient to describe the stage of growth. Vegetative stages are described in Table 7.11.

Vegetative stages are based, as noted above, on node identification, while reproductive stages are based on flowering, pod development, seed development, and on plant maturation (Table 7.12). Again, only main stems should be used in determining reproductive stages of growth. Stages R1 and R2 may occur simultaneously in determinate cultivars because flowering begins at the upper nodes of the main stem.

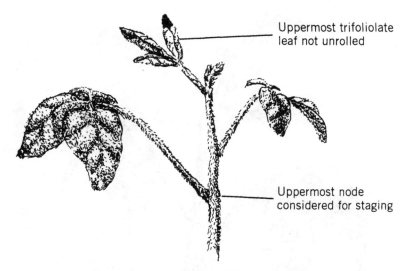

Figure 7.5 *Illustration of the uppermost soybean node with a fully developed leaf (uppermost node counted in staging). The uppermost trifoliolate leaf has not unrolled such that the edges are not touching. Therefore, the node that is penultimate to that leaf is not considered established for staging. (Drawing by Mike Hodnett)*

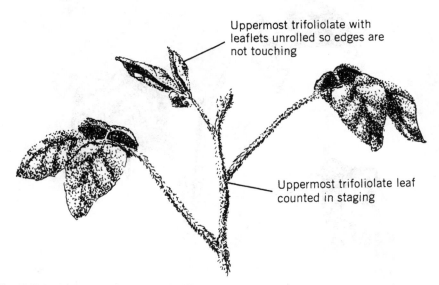

Figure 7.6 *Illustration of the uppermost soybean node with a fully developed leaf (uppermost node counted in staging). The uppermost trifoliolate leaf has unrolled such that the edges are not touching. Therefore, the node that is penultimate to that leaf is considered established for staging. (Drawing by Mike Hodnett)*

Figure 7.7 *Illustration of the terminal leaf of an indeterminate soybean plant showing that the terminal leaf has unrolled and flattened. The terminal leaf is smaller than lower leaves. (Drawing by Mike Hodnett)*

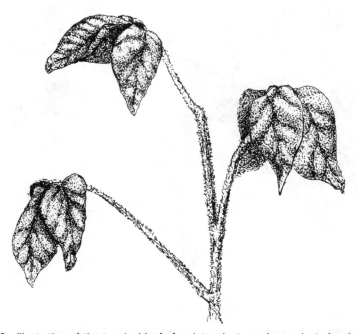

Figure 7.8 *Illustration of the terminal leaf of a determinate soybean plant showing that the terminal leaf has unrolled and flattened. The terminal leaf is only slightly reduced in size compared with lower leaves. (Drawing by Mike Hodnett)*

TABLE 7.11   Description of Vegetative Stages.

Stage	Abbreviated Title	Description
VE	Emergence	Cotyledons above the soil surface.
VC	Cotyledon	Unifoliolate leaves unrolled sufficiently so the leaf edges are not touching.
V1	First-node	Fully developed leaves at unifoliolate nodes (Figure 7.9).
V2	Second-node	Fully developed trifoliolate leaf at node above the unifoliolate nodes (Figure 7.4).
V3	Third-node	Three nodes on the main stem with fully developed leaves beginning with the unifoliolate nodes.
.		
.		
.		
V(n)	nth-node	n number of nodes on the main stem with fully developed leaves beginning with the unifoliolate nodes. n can be any number beginning with 1 for V1, first-node stage.

Source: Fehr, W. R. and C. E. Caviness; *Stages of Soybean Development*; Iowa State University.

R1 and R2 are approximately three days apart in indeterminate cultivars because flowering begins in the lower portion of the main stem and progresses upward.

Pods will reach near full size before seeds begin to develop rapidly. Pod measurements for stages R3 and R4 are made from the base of the leaf-like structure, the calyx, at the bottom of the pod to the tip of the pod. At stage R4, the seed cavities

TABLE 7.12   Description of Reproductive Stages in Soybeans.

Stage	Abbreviated Title	Description
R1	Beginning bloom	One open flower at any node on the main stem.
R2	Full bloom	Open flower at one of the two uppermost nodes on the main stem with a fully developed leaf (Figures 7.10 and 7.11).
R3	Beginning pod	Pod 5mm ($3/16$ inch) long at one of the four uppermost nodes on the main stem with a fully developed leaf (Figure 7.12).
R4	Full pod	Pod 2 cm ($3/4$ inch) long at one of the four uppermost nodes on the main stem with a fully developed leaf (Figure 7.13).
R5	Beginning seed	Seed 3mm ($1/8$ inch) long in pod at one of the four uppermost nodes on the main stem with a fully developed leaf (Figure 7.14).
R6	Full seed	Pod containing a green seed that fills the pod cavity at one of the four uppermost nodes on the main stem with a fully developed leaf (Figure 7.15).
R7	Beginning maturity	One normal pod on the main stem that has reached its mature pod color.
R8	Full maturity	Ninety-five percent of the pods have reached their mature pod color. Five to ten days of drying weather are required after R8 before the soybeans have less than 15 percent moisture.

Source: Fehr, W. R. and C. E. Caviness; *Stages of Soybean Development*; Iowa State University.

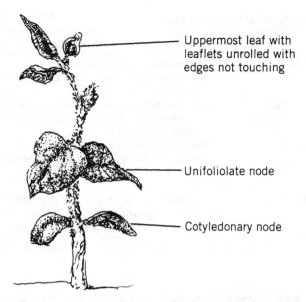

Uppermost leaf with
leaflets unrolled with
edges not touching

Unifoliolate node

Cotyledonary node

Figure 7.9   *Illustration of stage V1 with fully developed unifoliolate nodes. (Drawing by Mike Hodnett)*

inside the pods are outlined by a whitish membrane. By R6, the seeds nearly fill their seed cavities and will continue to enlarge through this stage of growth. Yellowing of leaves and pods usually occur together. However, in some situations, leaves may retain their green color after the pods have attained their mature color, usually brown or tan. Several reproductive growth stages are illustrated in Figures 7.10 to 7.15.

The producer, consultant, or pesticide applicator is usually more interested in the stage of growth of a field rather than that of individual plants. The stage of growth of a field of soybean is the stage attained by 50 percent or more of individual plants. This can be determined as illustrated in Tables 7.13 and 7.14. Since 50 percent of the plants staged had reached stage V5 in Table 7.13 and V8 or R1 in Table 7.14, the fields are so staged.

Persons staging fields of soybean should take certain precautions to ensure proper staging. These precautions are (1) obtain a representative sample by inspecting plants from several sites within the field; (2) stage at least ten plants at each site; (3) for fields of 25 acres or less, sample at least six sites, while large fields should be sampled at the rate of one site per 12 acres; and (4) carefully avoid plants with broken or damaged main stems and plants that are obviously different from the vast majority of surrounding plants—e.g., very tall or robust plants. If one wants both the vegetative stage and reproductive stage, then they must be evaluated independently, although the field data can be obtained from the same plants. This is illustrated in Table 7.14 where our field is at V8 R1 stage.

Soybean development, as with all plants, is influenced by temperature, soil moisture, cultivar, day length, and a multitude of other soil and environmental

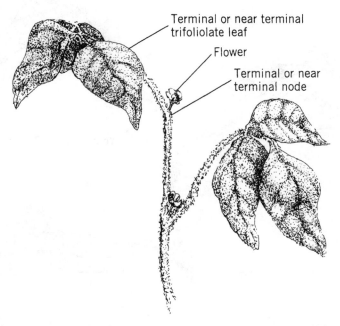

Terminal or near terminal
trifoliolate leaf

Flower

Terminal or near
terminal node

Figure 7.10 *Illustration of stage R2 in determinate type soybeans. (Drawing by Mike Hodnett)*

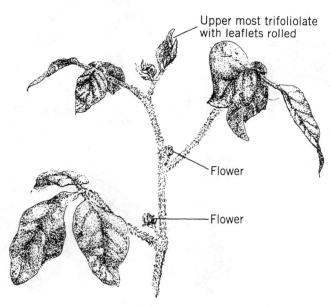

Upper most trifoliolate
with leaflets rolled

Flower

Flower

Figure 7.11 *Illustration of stage R2 in indeterminate type soybeans. (Drawing by Mike Hodnett)*

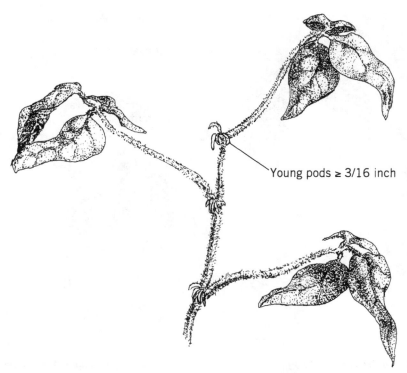

Young pods ≥ 3/16 inch

Figure 7.12 *Illustration of stage R3 in soybeans. (Drawing by Mike Hodnett)*

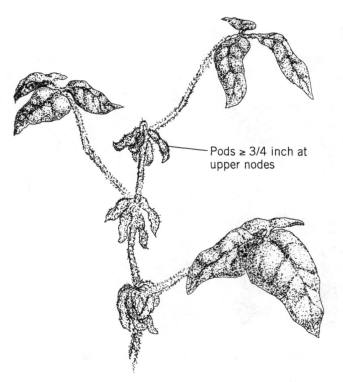

Pods ≥ 3/4 inch at
upper nodes

Figure 7.13 *Illustration of stage R4 in soybeans. (Drawing by Mike Hodnett)*

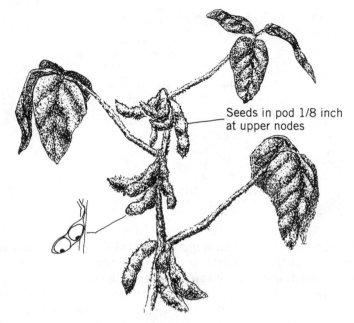

Seeds in pod 1/8 inch
at upper nodes

Figure 7.14   *Illustration of stage R5 in soybeans. (Drawing by Mike Hodnett)*

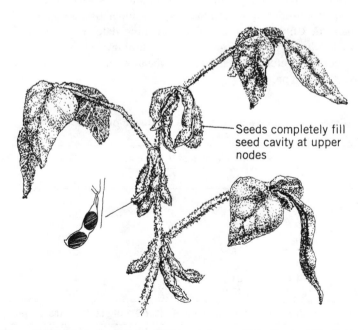

Seeds completely fill
seed cavity at upper
nodes

Figure 7.15   *Illustration of stage R6 in soybeans. (Drawing by Mike Hodnett)*

TABLE 7.13 Determining the Vegetative Stage of Growth of a 65 Acre Field of Soybean.

Site	No. of Plants at Stages		
	V3	V4	V5
1	——	5	5
2	1	4	5
3	——	4	6
4	1	3	6
5	——	7	3
Totals	2	23	25
Percentage	4.0	46.0	50.0
Stage of growth = V5 (≥50% of plants at this stage)			

factors. Since temperature is the most usual constraint during early plant development, one would anticipate greater variability in length of time between early growth stages than during mid-season growth stages (Table 7.15). Temperature, day length, and cultivar play major roles in flower initiation. High temperatures and short days (long nights) enhance flowering and reproductive development, while low temperatures and long days (short nights) have the opposite effect.

PRODUCTION PRACTICES

The objectives of individual soybean production practices vary little from state to state. Even when considering determinate versus indeterminate types, the goals remain unchanged. Cultivar choice, tillage, planting date, planting seed quality, rotations, row widths, fertility program, irrigation timing, and so on are all designed to maximize economic return from growing soybean whether in Michigan or Texas, Kansas or North Carolina.

TABLE 7.14 Determining the Vegetative Stage of Growth of a 20 Acre Field of Soybean.

Site	Number of Plants at V Stage or Without Flowers or at R1				
	V7	V8	V9	Without Flowers	at R1
1	2	4	4	3	7
2	3	5	2	5	5
3	2	3	5	6	4
4	4	3	3	3	7
5	1	6	3	2	8
6	2	3	5	4	6
Totals	1.4	24	22	23	37
Percentage	23.3	40.0	30.7	38.3	61.7
Stage = V8 (≥ 50% of plants at this stage) or R1 (≥ 50% of plants at this stage)					

TABLE 7.15   Number of Days Required for Soybean Development in the United States.

Stage	Average No. of Days	Range
Planting—VE	10	5–15
VE-VC	15	3–10
VC-V1	5	3–10
V1-V2	5	3–10
V2-V3	5	3–8
V3-V4	5	3–8
V4-V5	5	3–8
V5-V6	3	2–5
V6-Vn + 1	3	2–5
R1-R2	0*,3	0–7
R2-R3	10	5–15
R3-R4	9	5–15
R4-R5	9	4–26
R5-R6	15	11–20
R6-R7	18	9–30
R7-R8	9	7–18

Source: Fehr, W. R. and C. E. Caviness; *Stages of Soybean Development*; Iowa State University.

* R1 and R2 usually occur simultaneously in determinate cultivars. The time interval between R1 and R2 for indeterminate cultivars is about 3 days.

Cultivar Choice

Choosing a cultivar is a production decision that is hopefully made only once a season. Once a stand is established, the producer will live with his or her choice for the remainder of that season. Producers have a great number of cultivars to choose from, over 100 in some states if all maturity groups are included. Producers should consult cultivar performance guides provided by the cooperative extension service or their land grant university.

Producers should be careful to select a cultivar that will mature on time. Length of growing season, date of planting, and desired harvest time affect this choice. In addition to knowing the maturity grouping, cultivar developers rate their new releases relative to a known standard. For example, cultivar A may be marketed as three days earlier-maturing than Corsoy (−3), while cultivar B may be 3 days later-maturing—i.e., (+3) relative to Corsoy. Cultivars also vary in their degree of resistance to insects and diseases. Producers should be aware of reoccurring pests of soybean and choose their cultivars appropriately.

Date of Planting

The expected date of planting influences cultivar choice more than any other single factor. Early-maturing cultivars, whether determinate or indeterminate, should be

TABLE 7.16   Yields of Soybean Cultivars of Three Maturity Groups When Planted on Several Dates during 1981 and 1982 in Central Alabama.

Maturity Group and Cultivar	Bu./ac. at 13% Moisture When Planted					
	April		May		June	
	14	28	12	26	8	23
Group VI						
Coker 156	48	54	52	48	37	23
Davis	49	52	50	51	39	34
Average	48	53	51	50	38	29
Group VII						
Braxton	44	48	51	53	44	33
Ransom	41	45	50	56	44	31
Average	42	47	51	54	44	32
Group VIII						
Foster	48	50	50	56	41	31
Hutton	47	47	46	49	40	24
Average	48	49	48	53	40	27

Source: Adapted from *Soybeans,* 1986.

planted during the optimum planting period. These cultivars may not make full use of the growing season (i.e., they will mature early) but will help spread weather-related harvest-loss risks. Data from Auburn University in Table 7.16 illustrate this point. The optimum date of planting in 1981–82 for group VI cultivars was April 28, but May 26 for groups VII and VIII, which are considered full season in central Alabama. Within geographic limits, later maturing cultivars will remain vegetative longer and be more competitive with weeds. Mississippi producers are advised to plant between May 1 and June 10, while Arkansas producers aim for April 25 to June 10. Researchers in Arkansas recommend that group V be planted by May 30, followed by maturity groups VI, VII, and VIII.

Planting later than the optimum range for any maturity group will result in less yield in most years. Also, the later one plants, the greater the yield loss. Reports by Michigan State University Cooperative Extension Service indicates a one-half bushel/acre/day delayed decline in beans planted after the last recommended date, mid-May in southern Michigan and late May in central Michigan. Similar results were reported in Kentucky and Arkansas with determinate cultivars. Soybean planted after recommended dates in the mid-South will average a yield decline of 15 percent if planted June 15, 30 percent if planted July 1 and about 60 percent if not planted until July 15. One should be aware that these percentages apply to the average of all production. Producers with irrigation capabilities and who follow good management strategies may see no reduction from delayed, within reason, planting.

Planting early-maturing cultivars extremely early will result in yield loss because

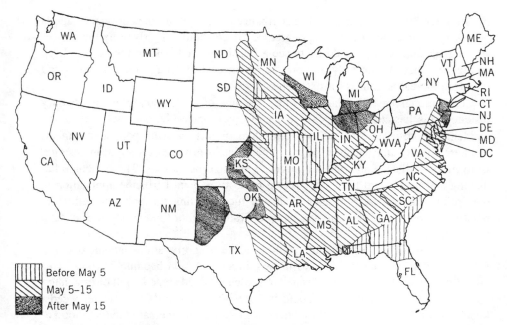

Figure 7.16 *Normal start of planting of soybean in the United States.* Source: *USDA-SRS, 1984. (Redrawn by Mike Hodnett)*

the night lengths are sufficiently long to induce flowering long before sufficient vegetative growth has been obtained for maximum yields. This is especially true in determinate types. However, very early planting of indeterminate group II cultivars has been advocated for producers in Texas and in other southern states. The logic behind this is that these soybeans will begin to flower in about three weeks, since the critical dark period exists at and/or just after planting. Therefore, these plants will grow vegetatively and reproductively during the spring months while moisture is normally available. This effectively avoids reproduction during likely periods of drought later in the growing season.

Soybean planting begins in more northern states and more arid states when soil temperature at planting depth is 55 to 60°F. In states with longer growing seasons, and more optimally distributed rainfall, planting is recommended when soil temperature at planting depth is about 65°F (Figure 7.16).

Tillage

The goals of seedbed preparation are the same for almost every crop: (1) control weeds, (2) conserve available moisture (especially true in more arid areas), (3) preserve or improve tilth, (4) prevent wind and water erosion, (5) provide for good seed-soil contact, (6) allow for proper placement of seed and fertilizer, and (7) aid in the control of plant diseases and insects. Producers should use only the tillage necessary to accomplish these goals. In some cases, these goals can be met by using reduced or minimum soil disturbance while full or conventional tillage is necessary in other locations. The degree of tillage should be driven by the above-stated goals and not by the desire to see a freshly plowed field or the need to keep employees busy. Using only appropriate and necessary tillage reduces costs and therefore increases returns to management.

Conventional or full tillage includes the use of moldboard plowing to bury residue. This hastens the decay of the previous crop residue and will reduce the incidence of diseases such as pod and stem blights, brown spot, stem canker, southern blight, and bacterial blight, which can overwinter on soybean residue. Conventional tillage can consist of eight to ten field operations. These include fall disking and moldboard plowing, disking prior to preplant herbicide application in the spring, disking and/or other means of incorporating the preplant herbicide, planting followed by rotary hoeing, and cultivating two or three times after a stand is established.

In some areas of the Mid-south, fall or early spring operations include disking, followed by subsoiling (sometimes in two directions), and bedding of rows. This allows producers to plant into a "stale" or settled seedbed some length of time later. Spring preplant operations then could include rebedding to destroy winter weeds, knocking off the top of the beds to reduce their height and increase the likelihood of planting seed into moist soil, applying preplant herbicides and incorporating twice with a harrow or bed conditioner, followed immediately by planting.

In some cases, producers may subsoil along the previous drill or row and rebed on the old bed or drill for four or more years. The effect of subsoiling in these systems is erratic, apparently depending on whether or not a compacted layer, plow, or traffic pan, develops. Data developed in Alabama illustrate this phenomenon (Table 7.17). In-row subsoiling within a conventional tillage system improved yields at the Tennessee Valley Substation, at Tallassee, and at the Wiregrass Substation, but not at six other sites in the state. These three sites have soil types susceptible to compaction. No-till yields exceeded those from conventional tillage at the Wiregrass Substation when in-row subsoiling was included. However, in-row subsoiling resulted in an increase as compared with no-till yields without in-row subsoiling at every location.

It is important to note that the use of beds has several distinct advantages. First, under full tillage, the soil at planting depth warms faster than with flat seedbeds. This is because the planting depth is closer to greater surface area than with flat seedbeds. Secondly, excess spring moisture drains away from the drill area, and this can result in faster reentry into some fields following spring rains. Once planted, the bed provides drainage away from the seedling, which promotes faster growth and prevents death of seedlings from excess water, which results in decreased oxygen available to the seedling roots. If a field is properly shaped with about a 0.4 percent slope, reentry into flat fields may be faster, because the producer would not have to wait for furrow bottoms to dry. This is especially true in clay soils. Third, in a season where little spring rain occurs and dryland farming is practiced, stale beds allow the producer to knock off the upper part of the bed and place the soybean seed into soil moistened by winter rains.

TABLE 7.17 Yields, as Percent of Conventional Tillage, of Soybean as Affected by Preplant Soil Preparation and In-Row Subsoiling on Eight Soils in Alabama. Individual Locations Evaluated for Two–Four Years between 1978–1981.

Tillage Treatment	Tennessee Valley Substa.	Sand Mountain Substa.	Upper Coastal Plain Substa.	Tallasssee	Prattville Expt. Field	Monroeville Experiment Field	Wiregrass Substa.	Gulf Coast Substa.
Conventional	100	100	100	100	100	100	100	100
Conventional plus in-row subsoil	113	92	106	115	105	99	156	98
No-till	114	71	87	84	66	85	85	79
No-till plus in-row subsoil	116	87	91	100	104	105	152	100

Source: Adapted from *Soybeans,* 1986. With permission from Dr. David Teem, Alabama Agricultural Experiment Station, Auburn University, Alabama.

The degree of tillage even under full tillage intent is often driven by soil type and the amount and distribution of rain during winter months. Fall operations such as moldboard plowing must be practiced in some northern states before the ground freezes. Where applicable, preplant herbicide can be applied in the fall following moldboard plowing or just ahead of bedding operations.

Minimum tillage as a concept should be practiced by every producer. Excess tillage can result in loss of seedbed moisture, soil structure, the creation of traffic pans that will inhibit moisture percolation and root penetration, and loss of top soil through wind and water erosion. Reduced tillage, especially no-till, has a number of advantages over maximum or full tillage systems: (1) reduced fuel and labor costs, (2) yield increase may result from reduced tillage in dry years through moisture conservation, (3) reduced dependence on weather at planting time as nondisturbed soil will dry faster than tilled soil, (4) reduced wind and water erosion, (5) production of soybean on fields with slopes too steep for conventional tillage because of erosion problems, and (6) dust pollution will be reduced. The Alabama Agriculture Experiment Station also reported that reduced tillage decreased the rate of buildup of cyst nematode during the growing season. Reduced tillage, again especially no-till, also has several disadvantages: (1) specialized equipment is required, such as planters with double disk soil openers and fluted coulters, (2) it requires a higher level of management, especially in the area of weed, insect, and disease awareness and control, and (3) increased herbicide costs.

No-tillage is usually practiced on soils that are fertile and well drained. These include the sandy and silt loams and some of the silty clay loams. Soils with large clay content and with high shrink-swell capacity are not usually suitable to no-tilled soybean. No-till planting should be done into killed sod of cool-season annual or perennial grasses or into killed stubble of winter annuals or of the previous crop. Planting is usually accomplished by means of special planter units designed with double disk openers and sometimes with front coulters that are notched, fluted, or waffled to help cut the plant debris and provide some loose soil in which to plant. Since planting is done in undisturbed soil that can be quite compacted and hard, especially if dry conditions exist, planters must be heavier than conventional planters. A large percentage of acres of soybean grown no-till are double-cropped with a winter grain, usually wheat, oats, or barley. For example, in Kentucky, 80 percent of the no-till soybean acreage was double-cropped in 1986.

TABLE 7.18   Effects of Rotation and Tillage System on Soybean Yields in Alabama.

Cropping System	Yield (bu./ac.) by Tillage System		
	Conventional	Strip	No-till
Continuous soybean	23	30	35
Soybean-corn	32	39	39
Wheat-soybean-corn	35	36	31

Source: Adapted from Thurlow, 1989. With permission from Dr. David Teem, Alabama Agricultural Experiment Station, Auburn University, Alabama.

Equipment is becoming available to till only a narrow band along the drill immediately ahead of planting. A powered rotary tiller is one way to accomplish this, tilling a four-inch wide band and leaving 80 percent or more of the previous crop's residue untouched. This approach, an extension of the fluted coulter technique, should become more popular as it accomplishes the goals of seedbed tillage while achieving the goals of minimum tillage (Table 7.18).

A modification of bedding rows in the fall or early spring was proposed by researchers at Purdue University in 1985. This system, termed ridge-tillage, prepares the field for next year's planting at the last cultivation of the current year. Ridges are formed at the last cultivation by the use of disk hillers (similar to hippers or bedders) or by the use of large sweeps with wings. The cultivator frame must be heavier than normal because the soil is usually dry and hard at this time of year. No additional tillage is performed until planting the next year, at which time the ridges are scraped just ahead of the planter unit to push off old crop residue and to lower the ridges. This accomplishes the same purposes as beds in other parts of the United States and allows the producer to plant into stale, moist soil. Normal cultivation and production practices can then be followed until the last cultivation when the cycle begins again. Although this is not a widely accepted production practice (only 9 percent of Indiana's soybean and corn acreage was ridge-tilled in 1987) it does hold promise similar to narrow-band preplant tillage—i.e., a semiprepared seedbed with many attributes of reduced tillage. Yields of soybean following corn were similar for conventional, ridge, and no-till over a seven-year study in Indiana.

Double-cropping

A large percentage of acreage planted to soybean in the United States is double-cropped with a winter grain. Fifty percent of soybean acres in Delaware were double-cropped in 1991 and 21 percent of Kentucky's total soybean acreage was double-cropped in 1986. Double-cropping is the practice of producing two crops at different times in the same field in a one-year period of time. A modification of true double-cropping is the production of three crops in two years on the same acre of land. Double-cropping increased in popularity during the 1970s as a result of food shortages in many areas of the world and the need of American farmers to maximize land utilization to offset relatively low commodity prices, high land prices, and high production costs. As noted earlier, double-cropping also may be appropriate for soil and water conservation in certain areas.

Soybean is often double-cropped with winter wheat and in some areas with barley. Barley will usually mature about two weeks earlier than wheat, and so in areas where a market exists, barley may be the better of the two winter grains for double-cropping with soybean. The most important factor, all other variables being equal, in determining success or failure of double-cropped soybean is planting date. Other cultural factors affecting double-cropped soybean are (1) winter crop stubble and residue management, (2) seeding method, (3) tillage, (4) row spacing/plant density, (5) irrigation, and (6) weed control.

Soybeans begin to flower when the night length becomes sufficiently long;

TABLE 7.19 Fourteen-Year (1967–68 and 1970–81) Average Soybean Yields When Planted on Different Dates in Alabama.

Maturity Group		No. of Cultivars	Yield and Avg. Loss/day from Previous Planting When Planted		
			May 22	June 5	June 26
			-------------------- (bu./ac.) ----------------------		
V	(v. early)	1–4	30.8	25.4(−.39)	17.0(−.40)
VI	(early)	4–7	34.0	28.0(−.43)	20.3(−.38)
VII	(full season)	2–5	31.8	28.1(−.26)	20.4(−.23)
VIII	(late)	2–4	28.2	25.5(−.19)	20.7(−.23)
Avg.		9–17	31.2	26.5(−.34)	19.6(−.33)

Source: Soybeans, 1986.

therefore, planting late may result in plants without sufficient leaf area to carry out photosynthesis for the development of pods and seeds. Based on 40 bushels/acre yield for beans planted during the optimum planting date range, Kentucky has reported a decrease in yield of 4 bushels/week after June 15; Michigan reports a 5 bushels/week decrease after July 1; Arkansas reports a 5 bushels/week decline after June 25; and Delaware producers can expect a decline of about one-half bushels/day after June 20. Long-term studies in central Alabama show yield declines averaging 0.33 bushels/day after May 22 (Table 7.19). Soybean planted late, whether following a winter grain crop or not, should be planted at higher seeding rates and in more narrow rows. This practice can cut losses in half or more, especially as beans are planted nearer the optimum planting date. Seeding rates should be increased although the number of seeds planted per linear foot of row decreases. Narrow rows (i.e., rows less 30 inches apart in most cases) will allow for better light energy capture by allowing for faster canopy closure. Recommended increases in plant densities for double-cropped (or late-planted mono-crop) soybean range from 10 percent in Michigan to 40 percent in Delaware.

Stubble management has probably the second greatest impact on double-cropped soybean yields because of its interaction with stand establishment. The greatest effect of old crop residue is interference with planter operation. Stubble from winter grain should be no higher than 12 inches and preferable lower. Higher stubble will cause soybean plants to etiolate or grow taller than would be the case without stubble. This can be good in that the lower set pods with double-cropped beans will be higher, or bad if severe enough that lodging is increased. Stubble and residue interfere with herbicide effectiveness by shielding areas of soil from herbicide spray. In some areas, the environmentally unsound practice of burning residue is still practiced. Other possible means of dealing with stubble and residue are to distribute it as evenly as possible over the entire field with a combine shredder or to bale when a market exists for the straw.

What works one year for stubble management may not work the next, primarily because of the amount of moisture at planting and during the season, assuming irrigation is not practiced. No-till into standing stubble and residue will conserve moisture and result in a yield advantage over tillage during dry years, whereas tillage may be better during wetter years because of better seed placement resulting in a more uniform and desirable stand, better weed control, and better fertilizer placement.

Seeding Rates

Soybean plants have the capacity to compensate for a wide range of plant densities. Producers should always aim for the optimum number of plants per acre, which is achievable by knowing the number of seeds planted per linear foot of row, time of year, seedbed conditions, and seed viability or germination. Producers should determine the number of seeds/pound of their cultivar of choice to determine the number of pounds/acre to plant. Soybean production guides, Cooperative Extension Service or State Agricultural Experiment Station cultivar test reports, local County/Parish Extension personnel, or seed company technical representatives are sources of information on the number of seeds per pound for cultivars recommended for specific areas. However, weather conditions during the year of seed production and screening procedures used during cleaning can affect seed size.

Recommended plant densities for soybean production for seed vary from 100,000 to 175,000 plants per acre (ppa). Stands as low as 75,000 ppa may result in maximum yields if those plants are evenly distributed and do not undergo additional stress. However, excessively high populations, especially over 200,000 ppa, and excessively low ppa should be avoided. High plant populations can result in yield declines because of increased lodging and interplant competition for available sunlight, moisture, and nutrients. Low plant populations may result in less than optimum yields from inadequate competition with weeds, failure to optimize the capture of incoming radiation, and from losses due to pods too low to be captured by the combine. Overplanting is more common than planting too few seeds, as producers have learned that it is better to err on the high side than the low, since yields usually do not drop off as rapidly unless severe lodging is a determining factor. Also, producers want to "buffer" themselves against poor seed germination and vigor, seedling diseases, herbicide injury, insects, hail, poor cultivation, rotary hoeing to break soil crusts, and other pests and hazards, all of which can result in lower plant populations than expected.

Table 7.20 is indicative of the kind of data on seed size and recommended seeding rates for various row widths. Producers usually should increase seeding rates if (1) planting very early or very late, (2) planting no-till, (3) planting in very rough seedbeds, (4) using a very short-season cultivar, (5) planting semidwarf cultivars, (6) planting broadcast, and (7) planting into soils that crust. Seeding rate per foot of row should be lowered when there is a history of lodging, if planting very high-quality seed, say over 90 percent germination, when planting under ideal conditions, or when planting in rows less than 30 inches apart.

Research from most states indicate that soybean should be planted in rows less than 40 inches wide. Data from Michigan suggest an average increase of 5 bushels/acre when growing soybean in rows 20 inches wide or less as compared with

TABLE 7.20 Planting Guide for Soybean.

	Seed/Ft. at Interrow Spacings of							
	7 Inches		15 Inches		20 Inches		30 Inches	
Seeds/lb.	2	3	4	6	6	8	8	10
	------------------------------(lb. seed/ac.)------------------------------							
2000	80	110	85	105	78	104	70	87
2400	66	93	71	88	65	87	58	73
2800	57	80	61	75	56	75	50	62
3200	50	70	53	66	49	65	44	54
3600	44	62	47	58	44	58	39	48
4000	40	56	43	53	39	52	35	44
4400	36	51	39	48	35	47	32	40
Expected plants per ft. of row[1]	1.6–2.4		4.0–4.8		4.8–6.4		6.4–8.0	
Expected plants per acre	119,000– 179,000		139,000– 167,000		125,000– 167,000		111,000– 139,000	

Source: Adapted from Herbek and Bitzer, 1988.

1. At 80% survival.

beans grown in rows 28 inches wide or wider; Kentucky reported similar data. Reasons for this increased yield potential include less lodging, higher pod height, weed suppression, decreased branching, less harvest loss from low branches, and better erosion control. Disadvantages include restricted opportunity for cultivation, possible equipment changes, and perhaps greater inputs of fertilizer, seed, and herbicides. Increased management input is usually required for any crop when the producer is operating at an increased production level.

Fertility

A bushel of soybean contains approximately 3.8 pounds of nitrogen, 0.9 pounds of phosphate (P_2O_5 equivalent), and about 1.4 pounds of potassium (K_2O equivalent). A yield of 40 bushels (2,240 pounds), therefore, removes about 152 pounds of nitrogen, 36 pounds of P_2O_5, and 52 pounds of K_2O. In addition to these levels of macronutrients found in mature soybean seed, approximately 1.5 pounds of nitrogen, 0.25 pounds of P_2O_5, and about 1 pound of K_2O are required for vegetative growth for each bushel produced. That is, again with a 40 bushels/acres yield, the vegetative portion will require 60 pounds of nitrogen, 10 pounds of P_2O_5, and 40 pounds of K_2O. However, there is not a linear relationship between seed yield and foliage produced. Depending on availability and amounts found in the soil, soybean may respond to applications of secondary and microelements such as calcium, magnesium, sulfur, boron, chlorine, copper, iron, manganese, zinc, and molybdenum.

Producers should have their soil tested and apply recommended fertility and pH corrections. Proper soil pH is essential for any crop, but especially so with a leguminous crop such as soybean. Proper pH for soybean production is between 5.8

and 7, with 6 or slightly higher being optimum. Maintaining an optimum pH accomplishes the following: (1) makes some micronutrients and both the native phosphorus in the soil and that applied to the soil more readily available to plants; (2) reduces the toxicity of manganese and aluminum; (3) enhances the activity of several microorganisms that decay plant residue and thus release plant nutrients; (4) improves root growth; and (5) enhances the activity of nodule-forming bacteria that are responsible for fixing atmospheric nitrogen.

The most common ways to apply commercial fertilizer to soybean fields is to broadcast and incorporate with a moldboard plow, disk, or similar implement before planting, or to band fertilizer at planting. Broadcast incorporated has a number of advantages, including (1) reduced salt injury to young plants, (2) less labor required, (3) speeds the planting process, (4) usually less expensive since fertilizer can be bought in bulk, and (5) nutrients are uniformly distributed throughout root zone and are therefore more readily available to the plant, especially during drought periods. The major disadvantage of broadcast is that it increases soil-to-fertilizer contact, thus making phosphorus fixation a greater problem in acid or alkaline soils. Banding fertilizer two to four inches below and to the side of the drill results in more efficient use of the applied fertilizer, especially in the first year of application. Fertilizer should be banded in highly acid or alkaline soils having low levels of nutrients already present.

A third method for applying fertilizer is by foliar application. This method has not been successful for the major and secondary nutrients, but some micronutrient deficiencies can be corrected by foliar application, especially if the deficiency is properly diagnosed early.

In a true no-till situation, the phosphorus and potassium are not likely to move below the top two inches of the soil, especially if the pH is not optimum for soybean production. If phosphorus and potassium test low, then these nutrients should be broadcast and incorporated by moldboard plowing if plowing will not cause excessive erosion (Figure 7.17). Nitrogen is a mobile nutrient and will move down into the soil profile with rain or irrigation water. However, some forms of nitrogen will volatilize into the atmosphere when applied to soil surfaces and others will further decrease the pH of the upper two inches of soil. Nitrogen is recommended in the United States only when soybean is planted into soil for the first time, or for the first time in three years or more. In no-till production, some states recommend that a starter fertilizer be banded at planting if a large amount of residue covers the soil surface sufficiently to keep the soil cool enough to suppress early plant development.

As a legume, a symbiotic relationship exists between the soybean plant and a soil bacterium, *Rhizobium japonicum,* where the bacterium derives its carbohydrate needs from the host plant and the host plant derives its nitrogen needs from the bacterium that has the capability to "fix" atmospheric nitrogen. The rhizobia form nodules, or swollen spherical structures, or soybean roots. *Rhizobium japonicum* is not native to the soils of the United States and therefore must be added to the soil by inoculating soybean seed with commercially available formulations just prior to planting, or it can be soil-applied just ahead of planting if the field has not been in

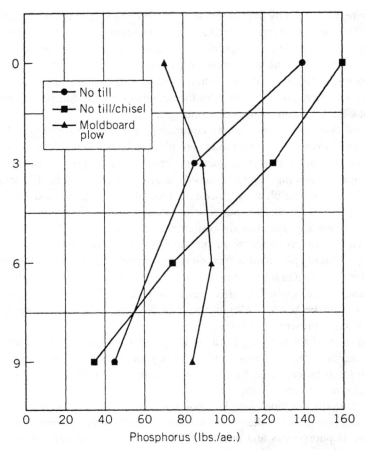

Figure 7.17 *Soil test values for phosphorus at different soil depths and for three tillage regimes.* Source: *Adapted from Illinois Growers Guide, 1982*

soybean within the previous five years. Some experts recommend adding rhizobia every three to five years just to ensure a healthy supply in the soil. When adding rhizobia, the producer should be careful to follow label directions, as the rhizobia are living organisms and can be killed by improper handling and storage.

While rhizobia fix atmospheric nitrogen in sufficient quantities to meet plant needs, several states recommend a starter fertilizer containing nitrogen to supply nitrogen needs until nodules have formed and begin to fix N_2. Commercial nitrogen may be needed also if beans are planted after a large amount of wheat residue has been freshly incorporated as the microorganisms involved in the decomposition of the straw will temporarily use up the available soil nitrogen. Other than this situation, essentially no data support the application of nitrogen except in situations where nodulation by rhizobia is reduced or eliminated. Application of fertilizer nitrogen will reduce the amount of N_2 fixed under normal conditions. Research in Illinois showed that the soybean-rhizobia complex fixed 48 percent of the nitrogen needs, the other 52 percent coming from soil organic matter, nitrification from soil nitrogen, or from previously applied nitrogen, when no commercial nitrogen was applied ahead of the soybean crop. However, only 10 percent was fixed by rhizobia when 400 pounds of nitrogen per acre was applied to the soil.

Research has been conducted aimed at isolating strains of rhizobium that are more efficient nitrogen fixers than present inoculants. Research at Iowa State University suggests that a new strain should be applied at a rate 1,000 times higher than the natural population if the new strain is going to have a chance at becoming the predominant strain. USDA-ARS scientists at Beltsville, Maryland, have succeeded in genetically altering rhizobium such that the altered strain produces 56 percent more nodules, increases plant nitrogen content by 50 percent, and increases plant weight by 25 percent. It is unclear at this time when genetically improved *Rhizobium japonicum* will be available to U.S. producers of soybean.

Harvest

Soybean is harvested over a long time frame in the United States because of the many different maturity groups grown. As a general rule, harvest in the upper Midwest begins after mid-September, while harvest in the remainder of the country may begin before or after this date (Figure 7.18).

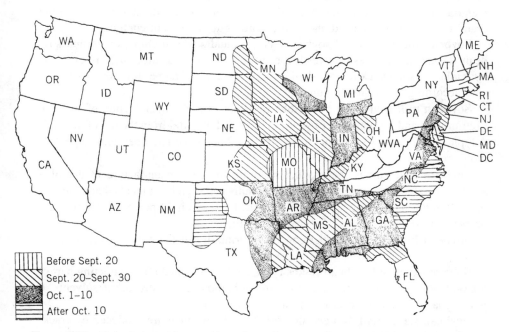

Figure 7.18 *Normal start of harvest dates for various soybean production areas across the United States.* Source: USDA-SRS, 1984. *(Redrawn by Mike Hodnett)*

BIOTIC PESTS

Four conditions must exist for plant disease or insect infestation to be detrimental: (1) a susceptible host plant, (2) a sufficient quantity of a virulent pathogen that is capable of rapid reproduction and spread, (3) an environment that is suitable to the pest, and (4) time. The most environmentally sound means of controlling biotic pests of the soybean plant is through integrated pest management or IPM. In soybean, IPM includes (1) the use of resistant cultivars, (2) crop rotation and clean tillage, (3) high-quality planting seed, (4) proper plant nutrition, and (5) judicial use of chemical pesticides.

Soybean producers should obtain publications with photos and descriptions of diseases and insects, and become familiar with the symptoms so that they can make timely decisions about crop protection or corrections. Such information can be obtained from the cooperative extension service in most counties/parish where soybean is grown.

SOYBEAN PROCESSING PRODUCTS

Product Standards have been established and defined by the National Soybean Processors Association as follows:

Soybean cake or soybean chips is the product after the extraction of part of the oil by pressure or solvents from soybeans. A term descriptive of the process of manufacture, such as expeller, hydraulic, or solvent extracted, shall be used in the brand name. It is designated and sold according to its protein content.

Soybean meal is ground soybean cake, ground soybean chips, or ground soybean flakes. A term descriptive of the process of manufacture, such as expeller, hydraulic, or solvent extracted, shall be used in the brand name. It is designated and sold according to its protein content.

Soybean mill feed is the by-product resulting from the manufacture of soybean flour or grits and is composed of soybean hulls and the offal from the tail of the mill. A typical analysis is 13 percent crude protein, 32 percent crude fiber, and 13 percent moisture.

Soybean mill run is the product resulting from the manufacture of dehulled soybean meal and is composed of soybean hulls and such bean meals that adhere to the hull in normal milling operations. A typical analysis is 11 percent crude protein, 35 percent crude fiber, and 13 percent moisture.

Soybean hulls is the product consisting primarily of the outer covering of the bean.

Solvent extracted soybean flakes is the product obtained after extracting part of the oil from soybeans by the use of hexane or homologous hydrocarbon solvents. It is designated and sold according to its protein content.

Soybean flakes and 44 percent protein soybean meal are produced by cracking, heating, and flaking soybeans and reducing the oil content of the conditioned

TABLE 7.21 Insect Pests of Soybean.

Insect	Description	Symptoms/Injury	Control
Soil Insects:			
Cutworms	Vary in appearance; see other chapters of this text for descriptions.	Plants are cut off at or just above ground level; many cutworms feed only at night; many remain in the soil and feed on roots while others feed on above ground portions cutting and/or consuming parts of leaves.	Good seedbed preparation two to three weeks before planting; cutworms can be severe in grassy fields early season; Chemical control may be necessary.
Grape Colaspis *Colaspis brunnea* (Fabricius)	Small white larvae with three pairs of true legs and a dark brown head.	Damage sometimes confused with cyst nematode; feeding occurs on stems and roots below the soil surface.	Cultural.
Seedling Soybean Insects:			
Yellow-striped Armyworm *Spodoptera ornithogalli* (Guenee)	Reddish brown to black velvety worm with bright yellow stripes running lengthwise along each side; may reach 1/2 inch or longer.	Foliage feeder on young plants.	Chemical control if over 50% of foliage is consumed.
Fall Armyworm *Spodoptera frugiperda* (Smith)	Four sets of prolegs; dark brown body with variable markings; head capsule black with an inverted white Y on front but not always well defined.	Foliage feeder.	Chemical control if over 50% of foliage is consumed.
Corn Earworm (Bollworm or Tobacco Budworm) *Helicoverpa zea* (Boddie) *Heliothis virescens* (Fabricius)	Four sets of prolegs; yellowish brown head capsule; body color varies widely.	Foliage feeders.	Chemical control if over 50% of foliage is consumed.

(continued)

TABLE 7.21 (Continued)

Insect	Description	Symptoms/Injury	Control
Thrips several species; see other chapters this text.	Very small winged or wingless insects; color varies from black to tan; white or silvery streaks or "thrips tracts" can be found on underside of leaves.	Leaves may be puckered; plants unthrifty.	Chemical control may be necessary if populations are high and poor soybean growing weather being experienced.
Three-cornered Alfalfa Hopper	Wedge-shaped, green insect about 1/4 inch long having a humped-back appearance.	Girdling of young stems resulting in dead plants; lower yields on surviving plants and lodging; damage after plants are 10 inches tall mostly confined to petioles.	Chemical.
Bean Leaf Beetle *Cerotoma trifurcata* (Forster)	Usually a light tan to reddish brown beetle; 4 to 6 black spots on back.	Damage is usually expressed as round holes eaten in leaves.	Chemical control if foliage loss exceeds 40%.
Foliage Feeding Insects: Green Cloverworm *Plathypera scabra* (Fabricius)	Light green body and head with three pairs of ventral prolegs; wiggles violently when disturbed.	Usually a minor foliage feeder.	Chemical.
Cabbage Loopers *Trichoplusia ni* (Hubner)	Light green with white strip on each side; two pairs of ventral prolegs and moves in a characteristic looping fashion.	Usually a minor foliage feeder but can require treatment some years.	Chemical.
Soybean Looper *Pseudoplusia includens* (Walker)	Light green with white strip on each side; two pairs of ventral prolegs and moves in a characteristic looping fashion; can be separated from the cabbage looper by its black true legs.	Usually a minor foliage feeder but can require treatment some years.	Chemical.

Insect	Description	Remarks	Control
Velvetbean Caterpillar *Anticarsia gemmatalis* (Hubner)	Light green to black body with a yellow to orange head; appearance of having a slightly forked tail and wiggles violently when disturbed; has 4 pairs of prolegs.	Usually a minor foliage feeder but can require treatment some years.	Chemical.
Beet Armyworm *Spodoptera exigua* (Hubner)	Olive drab to black body with 4 pairs of ventral prolegs; a black spot on each side of the body on the second segment; found grouped together in webbing on underside of leaves.	Occasionally a pest of soybean, especially following a period of dry weather.	Chemical; irrigation.
Fall Armyworm (see seedling insects) Corn Earworm (see seedling insects)			
Pod Feeding Insects: Corn Earworm and Tobacco Budworm (see seedling insects)	See seedling insects.	Neonate larvae begin feeding on young leaves and move to blooms and pods.	Probably the most important pod feeder of soybean and chemical control is required when larvae average 4 or more worms per row foot.
Stinkbugs (Green, Southern Green, and Brown) *Acrosternum hilare* (Say) *Nezara viridula* (Linne) *Euschistus servus* (Say)	Shield-shaped body and piercing-sucking mouth parts; adults are green or yellow-gray to brown; nymphs are light green with orange-yellow and black markings.	Beans are deteriorated tissue and therefore lowered seed quality from the digestive juices injected into the beans during feeding.	Chemical.

TABLE 7.22 Common Diseases of Soybeans in the United States.

Disease	Cause/Source	Symptom/Injury	Control
Planting to Bloom: Seed Decay and Seedling Blights or Rots	Various soil-borne and seed-borne fungi.	Seed decay; pre and post emergence seedling loss.	Plant high-quality, disease-free seed; plant in warm, well prepared seedbed; fungicide seed treatments.
Phytophthora Root and Stem Rot	Soil-borne fungus that can survive in the soil as resting spores for several years.	Pre- and post-emergence damping off; leaves on older plants turn yellow and wilt; roots rot; brown discoloration of stem from soil level.	Resistant cultivars; crop rotation; good field drainage; fungicide seed treatments.
Rhizoctonia and Fusarium Root and Stem Rot	Soil-borne fungus.	Red to reddish brown cankers on roots, hypocotyl, and stem below soil level (with fusarium the taproot shows brownish decay).	Same as for seedling blights; adequate phosphorus and potassium levels; cultivate soil into the drill to help secondary roots develop above canker.
Soybean Cyst Nematode (SCN)	Soil-borne nematodes.	Stunted and yellow plants; may resemble nitrogen or potassium deficiency; roots are smaller than normal with few Rhizobium nodules; some root decay; careful excavation and examination of roots shows pinhead-sized, pear-shaped, white to brown cysts on roots near surface.	Crop rotation; resistant cultivars; rotation of resistant and susceptible cultivars to prevent development of new races; good potassium nutrition and early planting.
Powdery Mildew	Air-and residue-borne fungus.	Thin, superficial, white to pale gray, powdery growth over the upper surface of leaves; lower surfaces may be reddened; leaves may turn brown and drop prematurely.	Resistant cultivars; rotation and clean plowdown may not be effective as fungus is air-borne.
Downy Mildew	Air-and residue-borne fungus.	Pale green to light yellow lesions with undefined margins on upper leaf surfaces; lower leaf surfaces may have grayish tufts of fungal growth.	Resistant cultivars; crop rotation; clean tillage; fungus favored by cool, moist weather and narrow rows that prevent rapid dew drying.
Bacterial Blight	Seed- and residue-borne bacteria.	Small water-soaked leaf spots which turn brown with a yellow border; leaves becoming ragged and torn.	High-quality, disease-free seed; crop rotation; avoid cultivating when plants are wet.
Bacterial Pustule	Seed- and residue-borne bacteria.	First symptoms are small, angular, yellowish green leaf spots with dark reddish brown centers; center of spot is raised to form a whitish pustule, especially on lower leaf surface.	Resistant cultivars plus control measures for bacterial blight.

Disease	Cause	Symptoms	Control
Wildfire	Residue-borne bacteria.	Light brown to black leaf spots up to ½-inch diameter bordered by a broad yellow band; usually found associated with bacterial pustule.	Same as for bacterial pustule.
Bloom to harvest:			
Charcoal Rot	Soil- or seed-borne fungus.	Plants turn yellow and slowly die, leaves staying attached; light gray areas in bark; speckled charcoal discoloration may be visible under bark near the soil line.	Rotation; clean till; good fertility; plant late-maturing cultivars.
Anthracnose	Seed- or residue-borne fungus.	Infection develops from ground upward; first and most characteristic is a dying or wilting of the trifoliolate feeding the inflorescences giving a characteristic "shepherd's crook"; pods and stems show irregularly shaped, reddish brown or dark brown areas; severely infected seed may be moldy, dark brown and shriveled.	Crop rotation; clean till; disease-free seed.
Stem Canker	Seed- or residue-borne fungus.	Sunken brownish canker with dark brown margin on lower nodes with green stems above.	Resistant cultivars; crop rotation; disease-free seed; clean till; seed treatment.
Pod and Stem Blight	Seed- or residue-borne fungus.	Rows of small black spots appear on stems and/or pods at maturity; seed may be cracked, shriveled or moldy.	Seed treatment; foliar sprays at stage R5 for seed production; crop rotation; disease-free seed; clean till.
Brown Spot	Residue- and seed-borne fungus.	Angular brown spots on leaves; leaves later turning yellow and dropping.	Crop residue plowdown; crop rotation; disease-free seed.
Frogeye Leaf Spot	Residue- and seed-borne fungus.	Enlarging reddish brown, circular to angular leaf spots with centers becoming olive gray or ash gray.	Resistant cultivars; crop rotation; clean tillage; disease-free seed.
Cercospora Purple Seed Stain and Leaf Blight	Seed- and residue-borne fungus.	Seed, pods, leaves and stems can be affected; characteristic pink to purple discoloration of seed coats; stem, petiole and pod infections characterized by sunken, irregular reddish purple spots.	Plant disease-free seed; crop rotation; clean plowdown.
Soybean Mosaic, Bean Yellow Mosaic, Bean Pod Mottle, Bud Blight (Tobacco Ringspot)	Seed-borne virus.	Leaves may be dwarfed, crinkled, puckered, ruffled, or mottled light green or yellow; plants may be stunted and growing points die; pods may be stunted, flattened or curved and have few or no seeds.	Disease free-seed; control perennial broad-leaf weeds near field.

product by the use of hexane or homologous hydrocarbon solvents. The extracted flakes are cooked and marketed as such or ground into meal. Typical analysis is:

Protein: minimum 44.0 percent

Fat: minimum 0.5 percent

Fiber: maximum 7.0 percent

Moisture: maximum 12.0 percent

Ground soybean is obtained by grinding whole soybeans without cooking or removing any of the oil.

Ground soybean hay is the ground soybean plant including the leaves and beans. It must be reasonably free of other crop plants and weeds and must contain not more than 33 percent crude fiber.

Soybean feed, solvent extracted, is the product remaining after the partial removal of protein and nitrogen-free extract from dehulled solvent extracted soybean flakes.

Heat processed soybeans (dry roasted soybeans) is the product resulting from heating whole soybeans without removing any of the component parts. It may be ground, pelleted, flaked, or powdered. It must be sold according to its crude protein content.

Soy protein concentrate is prepared from high-quality sound, clean, dehulled soybean seeds by removing most of the oil and water-soluble nonprotein constituents and must contain not less than 70 percent protein on a moisture-free basis.

Kibbled soybean meal is the product obtained by cooking ground, solvent-extracted soybean meal under pressure and extruding from an expeller or other mechanical pressure device. It must be designated and sold according to its protein content and shall contain not more than 7 percent crude fiber.

Ground extruded whole soybean is the meal product resulting from the extrusion by friction, heat and/or steam of whole soybeans without removing any of the component parts. It must be sold according to its crude protein, fat, and fiber content.

Soy grits is the granular material resulting from the screened and graded product after removal of most of the oil from selected, sound, clean, and dehulled soybeans by a mechanical or solvent extraction process.

Soy flour is the finely powdered material resulting from the screened and graded product after removal of most of the oil from selected, sound, cleaned, and dehulled soybeans by a mechanical or solvent-extraction process.

U.S. STANDARDS AND GRADES

Definition of Soybean

Grain that consists of 50 percent or more of whole or broken soybeans (*Glycine max* (L.) Merr.) that will not pass through an $^8/_{64}$ round-hole sieve and not more than

TABLE 7.23   Grades and Grade Requirements for Soybeans.

	Minimum Test Weight per Bushel	Maximum Limits of				
		Damaged Kernels				Soybeans of Other Colors
Grade		Heat Damaged	Total	Foreign Material	Splits	
	(lbs.)	(%)	(%)	(%)	(%)	(%)
U.S. No. 1	56.0	0.2	2.0	1.0	10.0	1.0
U.S. No. 2	54.0	0.5	3.0	2.0	20.0	2.0
U.S. No. 3[1]	52.0	1.0	5.0	3.0	30.0	5.0
U.S. No. 4[2]	49.0	3.0	8.0	5.0	40.0	10.0

U.S. Sample grade

U.S. sample grade is soybeans that

(a) Do not meet the requirements for U.S. Nos. 1, 2, 3, or 4, or;

(b) Contain 8 or more stones which have an aggregate weight in excess of 0.2 percent of the sample weight, 2 or more pieces of glass, 3 or more Crotalaria seeds (*Crotalaria* spp.), 2 or more castor beans (*Ricinus communis* L.), 4 or more particles of an unknown substance(s), 10 or more rodent pellets, bird droppings, or equivalent quantity of other animal filth per 1,000 grams of soybeans; or

(c) Have a musty, sour, or commercially objectionable foreign odor (except garlic odor[3]); or

(d) Are heating or otherwise of distinctly low quality.

Source: USDA-FGIS, 1988.

1. Soybeans that are purple or stained are graded not higher than U.S. No. 3.

2. Soybeans that are materially weathered are graded not higher than U.S. No. 4.

3. Garlicky soybeans contain five or more green garlic bulblets or an equivalent quantity of dry or partly dry bulblets in a 1,000 gram sample.

10.0 percent of other grains for which standards have been established under the United States Grain Standards Act (Table 7.23).

Definitions of Other Terms

(a) **Classes.** There are two classes for soybean: yellow soybeans and mixed soybeans.

(1) **Yellow soybean.** Soybeans that have yellow or green seed coats, and which in cross-section are yellow or have a yellow tinge, and may include not more than 10 percent of soybeans of other colors.

(2) **Mixed soybeans.** Soybeans that do not meet the requirements of the class yellow soybeans.

(b) **Damaged kernels.** Soybeans and pieces of soybeans that are badly ground-damaged, badly weather-damaged, diseased, frost-damaged, germ-damaged, heat damaged, insect-bored, mold-damaged, sprout-damaged, stinkbug-stung, or otherwise materially damaged. Stinkbug-stung kernels are considered damaged kernels at the rate of one-fourth of the actual percentage of the stung kernel.

(c) **Foreign material.** All matter that passes through an $^8/_{64}$ round-hole sieve and all matter other than soybeans remaining in the sieved sample after sieving according to procedures prescribed in FGIS instructions.

(d) **Heat-damaged kernels.** Soybeans and pieces of soybeans that are materially discolored and damaged by heat.

(e) **Purple mottled or stained.** Soybeans that are discolored by the growth of a fungus, or by dirt, or by a dirt-like substance(s), including nontoxic inoculants, or by other nontoxic substances.

(f) **Sieve—$^8/_{64}$ round hole sieve.** A metal sieve 0.032 inch thick perforated with round holes 0.125 inch in diameter.

(g) **Soybeans of other colors.** Soybeans that have green, black, brown, or bicolored seed coats. Soybeans that have green seed coats will also be green in cross-section. Bicolored soybeans will have seed coats of two colors, one of which is brown or black, and the brown or black color covers 50 percent of the seed coats. The hilum of a soybean is not considered a part of the seed coat for this determination.

(h) **Splits.** Soybeans with more than one-fourth of the bean removed and that are not damaged.

BIBLIOGRAPHY

Alabama Soybean Production Guide. 1991. Auburn Univ. Coop. Ext. Ser., Agron. Ser. Timely Info.

American Soybean Association. 1974. *Soybean Digest Blue Book.* Am. Soybean Assoc., Hudson, IA.

American Soybean Association. 1988. *'88 Soya Blue Book.* Soyatech, Inc., Bar Harbor, ME.

American Soybean Association. 1990. *'90 Soya Blue Book.* Soyatech, Inc., Bar Harbor, ME.

American Soybean Association. (not dated). *Soybeans: The Miracle Crop.* Am. Soybean Assoc., St. Louis, MO.

Board, J. E., and W. Hall. 1985. *Factors Involved in Soybean Yield Losses at Nonoptimal Planting Times in Louisiana.* Louisiana St. Univ. Agri. Expt. Sta. Bul 769.

Boquet, D. J., and D. M. Walker. 1984. *Wheat-soybean Double-cropping: Stubble Management, Tillage, Row Spacing and Irrigation.* Louisiana St. Univ. Agri. Expt. Sta. Bul. 760.

Boquet, D. J., K. L. Koonce, and D. M. Walker. 1983. *Row Spacing and Planting Date Effects on Yield and Growth Responses of Soybeans.* Louisiana St. Univ. Agri. Expt. Sta. Bul. 754.

Carroll, R. B., R. P. Mulrooney, R. W. Taylor, and T. A. Evans. 1991. *Soybean Facts: Seed and Seeding Diseases of Soybeans.* Univ. of Delaware. Coop. Ext. Ser. SF-19.

Carroll, R. B., R. W. Taylor, R. P. Mulrooney, and T. A. Evans. 1991. *Soybean Facts: Root and Leaf Diseases of Soybeans.* Univ. of Delaware Coop. Ext. Ser. SF-21.

Chang, K. C. 1977. "Ancient China." *In* K. C. Chang (ed.) *Food in Chinese Culture: Anthropological and Historical Perspectives.* Yale Univ. Press, New Haven, CT.

Criswell, J. G., and D. J. Hume. 1972. "Variation in sensitivity to photoperiod among early maturing soybean strains." *Crop Sci.* 12:657–660.

Delta Agricultural Digest. 1992. *Soybean Production.* G. Rutz (ed.) Farm Press Pub. Clarksdale, MS.

Dunigan, E. P., O. B. Sober, J. L. Rabb, and D. J. Boquet. 1980. *Effects of Various Inoculants on Nitrogen Fixation and Yields of Soybeans.* Louisiana St. Univ. Agri. Expt. Sta. Bul. 726.

Fehr, W. H., and C. E. Caviness. 1977. *Stages of Soybean Development.* Iowa St. Univ. Agri. Expt. Sta. Spec. Rep. 80.

Fuhrmann, J. J., and B. L. Vasilas. 1991. *Soybean Production.* G. Rutz (ed.) Farm Press Pub., Clarksdale, MS.

Garner, W. W., and H. A. Allard. 1920. "Effect of relative length of day and night and other factors of the environment in growth and reproduction on plants." *J. Agri. Res.* 18:553–606.

Griffith, D. R., J. V. Mannering, D. B. Mengel, S. D. Parsons, T. T. Bauman, D. H. Scott, C. R. Edwards, F. T. Turpin, and D. H. Doster. 1984. *A Guide to No-till Planting After Corn or Soybeans.* Purdue Univ. Coop. Ext. Ser. (Tillage) ID-154.

Helsel, Z. R. 1981. *Soybeans.* Michigan St. Univ. Coop. Ext. Ser. Ext. Bul. E-1528, File 22.22.

Helsel, Z. R., T. J. Johnson, and L. P. Hart. 1981. *Soybean Production in Michigan.* Michigan St. Univ. Coop. Ext. Ser. Bul. E-1549.

Henderson, J. 1991. *Alabama Soybean Production Guide.* Auburn Univ. Coop. Ext. Ser. Agron. Ser. Timely Info.

Henderson, J., J. Everest, T. Whitwell, and C. B. Ogburn. 1983. *Soybean Production Fact Sheet: Conservation Tillage for Soybeans.* Auburn Univ. Coop. Ext. Ser. Cir. ANR-39.

Herbek, J. H., and M. J. Bitzer. 1988. *Soybean Production in Kentucky.* Univ. of Kentucky Coop. Ext. Ser. AGR-128.

Herbek, J. H., and M. J. Bitzer. 1988. *Soybean Production in Kentucky Part II: Seed Selection, Variety Selection and Fertilization.* Univ. of Kentucky Coop. Ext. Ser. AGR.-129.

Herbek, J. H., and M. J. Bitzer. 1988. *Soybean Production in Kentucky. Part III: Planting Practices and Double-Cropping.* Univ. of Kentucky Coop. Ext. Ser. AGR-130.

Herbek, J. H., and M. J. Bitzer. 1988. *Soybean Production in Kentucky. Part IV: Weed, Disease and Insect Control.* Univ. of Kentucky Coop. Ext. Ser. AGR-131.

Hestermann, O. B., J. J. Kells, and M. L. Vitosh. 1987. *Producing Soybeans in Narrow Rows.* Michigan St. Univ. Coop. Ext. Ser. Bul. E-2080.

Hymowitz, T., and J. R. Harlan. 1983. "Introduction of the soybean to North America by Samuel Bowman in 1765." *Econ. Bot.* 37:371–379.

Hymowitz, T., and R. J. Singh. 1987. "Taxonomy and Speciation." *In* J. R. Wilcox (ed.) *Soybeans: Improvement, Production, and Uses.* Am. Soc. Agron., Madison, WI.

Hymowitz, T., and R. L. Bernard. 1991. "Origin of the Soybean and Germplasm Introduction and Development in North America." *In* H. L. Shands and L. E. Wiesner (eds.) *Use of Plant Introductions in Cultivar Development: Part I.* Crop Sci. Soc. Am., Madison, WI.

Illinois Growers Guide to Superior Soybean Production. 1982. Univ. of Illinois Coop. Ext. Ser. C. 1200.

Kuykendall, L. D. 1992. *Super bacteria: A boost for yields.* USDA-ARS Agri. Res.

List, R. J. 1958. *Smithsonian Meteorological Tables.* Smithsonian Misc. Col. Vol. 114, Pub. 4014.

May, M. L., C. E. Caviness, and I. L. Eldridge. 1989. *Soybean response to early planting in northeast Arkansas.* Ark. Farm Res. 4:5.

Morse, W. J. 1918. *The Soy Bean: Its Culture and Uses.* USDA Farmers' Bul. 973.

Morse, W. J. 1927. *Soy Beans Culture and Varieties.* USDA Farmers' Bul. 1520.

Morse, W. J., J. L. Carter, and L. F. Williams. 1949. *Soybeans Culture and Varieties.* USDA Farmers' Bul. 1520 (Rev.)

Morse, W. J. 1950. "History of Soybean Production." *In* K. S. Markley (ed.) *Soybeans and Soybean Products.* Interscience Pub. Inc., New York.

Mulrooney, R. P., R. W. Taylor, and B. L. Vasilas. 1991. *Soybean Facts: Soybean Cyst Nematode.* Univ. of Delaware Coop. Ext. Ser. SF-22.

Osborne, T. B., and L. B. Mendel. 1917. "The use of soybean as food." *J. Biol. Chem.* 32:369–387.

Raper, C. D., and P. J. Kramer. 1987. "Stress Physiology." *In* J. R. Wilcox (ed.) *Soybeans: Improvement, Production, and Uses.* Am. Soc. of Agron., Madison, WI.

Seddigh, M., and G. D. Jolliff. 1984. "Effects of night temperature on dry matter partitioning and seed growth of indeterminate field-grown soybean." *Crop Sci.* 24:704–710.

Shanmugasundaram, S. 1978. "Variation in the photoperiod response to flowering in soybean." *Soybean Genet. Newsl.* 5:91–94.

Sims, J. R., R. W. Taylor, and B. L. Vasilas. 1991. *Soybean Facts. Nutrient Management for Soybeans.* Univ. of Delaware Coop. Ext. Ser. SF-8.

Snyder, C. S., L. O. Aslock, and G. M. Lorenz. 1989. *Foliar Nitrogen Fertilization of Soybeans after Bloom.* Univ. of Arkansas Coop. Ext. Ser. Fact Sheet 2044.

Soybean Production Guide. 1980. Univ. of Florida Coop. Ext. Ser. Cir. 277E.

Soybean Production Handbook. 1991. R. P. Nester (ed.) Univ. of Arkansas Div. Coop. Ext. Ser. MP 197.

Soybean Production Handbook. 1988. Kansas St. Univ. Coop. Ext. Ser. C-5449.

Soybeans. 1986. Auburn Univ. Agri. Expt. Sta. Res. Rep. Ser. No. 4.

Taylor, R. W., B. L. Vasilas, and R. Uniatowski. 1991. *Soybean Facts: Double-cropping Soybeans in Delaware.* Univ. of Delaware Coop. Ext. Ser. SF-31.

Taylor, R. W., B. L. Vasilas, R. Uniatowski, and J. J. Fuhrmann. 1991. *Soybean Facts: Tillage and Soybean Production.* Univ. of Delaware Coop. Ext. Ser. SF-32.

Thomas, J. F., and C. D. Raper Jr. 1978. "Effect of day and night temperature during floral induction on morphology of soybean." *Agron. J.* 70:893–898.

Thurlow, D. L. 1989. *Research Update 1989 Soybeans.* Auburn Univ. Agri. Expt. Sta.

Uniatowski, R., R. W. Taylor, and B. L. Vasilas. 1991. *Soybean Cultivar Selection for Delaware.* Univ. of Delaware Coop. Ext. Ser. SF-5.

United States Department of Agriculture. 1936, 1942, 1952, 1961, 1972, 1982 and 1992. *Agricultural Statistics.* U.S. Government Printing Office, Washington, DC.

United States Department of Agriculture-Statistical Reporting Service. 1984. *Usual Planting and Harvest Dates for U.S. Field Crops.* USDA Agri. Handbook No. 628.

Vasilas, B. L., R. W. Taylor, and J. J. Fuhrmann. 1991. *Soybean Facts: Development and Growth Stages with Soybeans.* Univ. of Delaware Coop. Ext. Ser. SF-24.

Vasilas, B. L., R. W. Taylor, and R. Uniatowski. 1991. *Soybean Growth Types.* Univ. of Delaware Coop. Ext. Ser. SF-2.

Vasilas, B. L., R. W. Taylor, and R. Uniatowski. 1991. *Soybean Maturity Classification and Selection.* Univ. of Delaware Coop. Ext. Ser. SF-4.

Vasilas, B. L., R. W. Taylor, and R. Uniatowski. 1991. *Soybean Facts: Soybean Row Spacing.* Univ. of Delaware Coop. Ext. Ser. SF-9.

Vasilas, B. L., R. W. Taylor, and R. Uniatowski. 1991. *Soybean Plant Populations.* Univ. of Delaware Coop. Ext. Ser. SF-10.

Vasilas, B. L., R. W. Taylor, R. Uniatowski, and J. J. Fuhrmann. 1991. *Soybean Planting Date.* Univ. of Delaware Coop. Ext. Ser. SF-3.

Whalen, J. M., and R. W. Taylor. 1991. *Soybean Facts: Insect Control in Soybeans.* Univ. of Delaware Coop. Ext. Ser. SF-16.

Wood, J. M., C. Tutt, and T. Pfeiffer. 1991. *1990 Kentucky Soybean Performance Tests.* Univ. of Kentucky Agri. Expt. Sta. Prog. Rep. 332.

8

PEANUT
(Arachis hypogaea L.)

INTRODUCTION

The peanut is known also a groundnut, earth nut, monkey nut, goober (from the Bantu "nguber") pinder, pinda, and Manilla nut. Over 60 percent of U.S. production goes for human consumption with the remainder used as seed, animal feed, and oil. Peanuts are rich in energy with seeds continuing 42 to 52 percent oil and 25 to 32 percent protein. One pound of peanuts has the approximate equivalent energy value of 2 pounds of beef, 1.5 pounds of cheddar cheese, or 36 medium eggs.

The peanut is a tropical legume native to South America. It was being grown extensively throughout tropical and subtropical regions of the New World by the 15th century. The cultivated peanut was found by early Spanish and Portuguese explorers in present-day Brazil, Argentina, Paraguay, Bolivia, Peru, and several West Indies islands. Twenty-nine species of *Arachis* are found in South America, with no wild species of *Arachis* found on any other continent. Archaeological evience found in the Casma Valley of Peru dates the appearance of the peanut at 1800 to 1200 B.C.

A number of early explorers and colonists encountered and recorded facts and dates relative to the peanut. Hammons (1982) identified 24 such encounters that occurred before 1697. The peanut was cultivated by the Inca civilization of Peru and the peoples of the Antilles (present-day West Indies, including the Bahamas). The plant and nuts were called "ynchic" by the Inca Amerindians, but the Spanish preferred to call them "mani." That was the name used by the Arawak Amerindians who were native to the Antilles. The first European to encounter the peanut probably was Bartolome Las Cases, a missionary to the island of Hispaniola (present-day Haiti/Dominican Republic) in the early 1500s. Hammons (1982) quotes Las Cases concerning the peanut or mani: "They had another fruit which was sown and grew beneath the soil, which were not roots but which resembled the meat of the filbert nut . . . . These had thin shells in which they grew and . . . (they) were dried in a manner of the sweet pea or chick pea at the time they are ready for harvest. They are called mani."

MOVEMENT FROM ORIGIN

The peanut was introduced into. Europe, at least into Spain, by 1574. Evidence suggests that the peanut was grown by the Aztecs of central Mexico at the time of Spanish occupation. Reports returned to the King of Spain following the conquest of Mexico by Cortez (1518 to 1521 A.D.), and later its exploration and colonization provided no clear evidence that the peanut was used as a food. Higgins (1951) noted that a Mr. Sahagun, a school instructor and historian of Mexican culture from 1529 to 1550, wrote of *tlalcacauatl* as a poultice for swollen gums. *Tlalcacauatl* could have been made from the peanut since in the Nahuatl language spoken by the Aztecs, Toltecs, and other Amerindian tribes of central Mexico and parts of Central America, *tlale* means earth and *cacauatl* means cacao seed, cacao being the seed from which cocoa and chocolate are made. Sahagun did not list *tlalcacauatl* as a food of the Nahuatl-speaking people suggesting that it was not made from the cacao seed which was a food of many of these Amerindian tribes.

A Spanish priest who lived in Hispaniola from 1510 to 1530 and Oviedo, the official historian of New Spain from 1513 to 1524, both noted that the peanut was grown in the West Indies. Oviedo wrote: "A fruit which the Indians in this Isle Espanola have, they call mani and they plant it, and harvest it and it is a common plant in their gardens and fields, and it is the size of a pine nut with a shell, and they think it is healthful; the Christians pay little attention to it, unless they are common people, or children, or slaves or people who do not have a fine taste. It is of mediocre taste, and of little substance, and is a very common food of the Indians who use it in quantity."

Although archaeological evidence may place the oldest remains of peanut in Peru, authorities believe that it evolved in the boundary area of Brazil, Bolivia, and Paraguay. So, we are left with the same question as with most other New World crops: How and when did the peanut spread to the West Indies and maybe as far north as northern Mexico where it was found by early Spanish explorers? There is little if any definitive evidence on the movement of the peanut. It could be assumed that the peanut became an item of trade or a portable source of food between New World Amerindian nations. Surely the Amerindians of Peru and western South America traded with tribes and nations both north and south.

Distribution could have been by animals or during floods or by other natural means. And there is no doubt that plant habits do become established by these natural occurrences. The other possibility is that the peanut was carried and established in suitable areas by Amerindians before Europeans discovered the New World. But who moved the peanut to other areas of the world? Pedro Alvares

Calval, a Portuguese explorer sailing from India, "discovered" and claimed Brazil for the Portugal government on April 22, 1500. He departed thence, sailed around the Cape of Good Hope and on to the East Indies. The Portuguese soon established trade routes from Brazil, east to Africa, and west to the East Indies. They may have been responsible also for introducing maize, cassava, and tobacco as well as peanuts to the African continent. Peanut was so well adapted to the African climate that travelers in Africa a century later would described it as a native plant. Portugal probably spread the peanut up and down the African coasts (history records the presence of Portuguese colonies and trading posts along the east coast of Africa by 1522). The commonly accepted means of introduction into the United States was with slaves brought from Africa, beginning around 1600. However, introduction from the West Indies cannot be ruled out. The peanut was recorded in the Moluccas (Indonesia) Islands in 1690, perhaps introduced during trade between Acapulco, Mexico, and Manila, Philippines, as early as 1565. There is little additional solid evidence about the movement of the peanut from its origins.

TYPES OF COMMERCIAL PEANUTS

Runner (var. *hypogaea*)

The introduction into present-day United States may have occurred with slave introductions from Africa about 1707 to 1725, absolute proof being lacking. We do know, however, that the peanut was introduced into the southeastern United States and that it was a small-pod type with a spreading, or runner, vegetative growth habit. It was a full-season annual known by a number of cultivar names, including African, Wilmington, North Carolina, Georgia, and Southeastern Runner. Hammons (1982) noted, however, that the morphological description of this supposed African introduction was the same as the peanut found growing in the Caribbean Islands and described by a 17th century French naturalist.

(The American Peanut Research and Education Society and the Crop Science Society of America have suggested that the terms spanish, valencia, and virginia will begin with a lower-case letter when used to refer to types of peanuts. Although this is not an accepted rule of capitalization, it is consistent with other, commonly accepted lower-case spellings when referring to common names for plants such as johnsongrass or bermudagrass.)

Spanish (var. *vulgaris*)

The spanish type originated in northeastern Argentina, Paraguay, and southern Brazil. Hammons referenced Krapovickas (1968) who documented that this type of peanut was introduced to Europe by Don Jose Campus at Lisbon, Portugal, in 1784. The spread of the spanish type is credited to Tabares de Ulloa (who later developed a machine to shell peanuts) of Valencia, Spain, and Lucien Bonaparte in France around

1800. The Spanish cultivated peanuts for oil and for preparation of chocolate-covered peanuts.

The spanish-type peanut was introduced into the United States from Malaga, a city in southern Spain, in 1871, it requiring a shorter growing season than the runner type predominating United States acreage at that time. The 1871 consignment was to Thomas B. Roland of Norfolk, Virginia.

Virginia (var. *hypogaea*)

This large-seed type apparently originated in the Bolivian and Amazonian regions and was introduced into the United States around 1844, although some authorities place its introduction after 1871. The virginia-type peanut grown commercially in the United States may be the result of a natural hybridization of *A. hypogaea* var. *hypogaea* (i.e., a large-seed type) with the African runner type introduced in the early 1700s. The time of this natural hybridization, its selection, and early maintenance are unknown.

Valencia (var. *fastigiata*)

The valencia is believed to have evolved in Paraguay and central Brazil. Beattie identified this type of peanut as an introduction from Valencia, Spain, (thus the name) to the United States in 1911.

DEVELOPMENT OF U.S. INDUSTRY

The peanut, being introduced into the United States either with slaves from Africa or by Amerindians of the West Indies, was regarded early on as a food for the poor and unsophisticated. Commercialization was first as a feed, primarily for hogs, turkeys, and chickens. The biochemical properties of the peanut produced "soft pork" when fed to swine, resulting in the distinctive flavor associated with Virginia ham. Farmers in the Southeast would fatten hogs almost entirely on peanuts by "hogging off"—that is, penning swine in fields of peanuts and allowing the animals to root out the underground nuts. The peanut, although very high in protein, is notoriously low in minerals. Hogs that were fed peanuts exclusively would have weakened bones that would break easily. Lactating sows that were fed an exclusive diet of peanuts would contract a condition known as "down in the back." The high protein (and therefore high energy) content of the peanut resulted in an increase in milk production in these sows, and thereby an increased demand for phosphorus (P) and calcium (Ca). As the extra Ca was not supplied in the diet, it was mobilized from bone tissue, resulting in weakened skeletal structure. Dislocation and broken bones were therefore common.

Producers and meat packers tried to capitalize on the distinctive flavor of soft pork. They were successful to a degree with the acceptance of cured (i.e., preserved by salt and hickory smoke) hams and pork loins. These products were referred to as Virginia or Smithfield hams. However, the appearance and taste of the remainder of the pork carcass left much to be desired.

During the Civil War, Northern (and many Southern) soldiers were introduced to the peanut as a food. Demand increased rather rapidly afterwards and the size of the U.S. crop tripled from 1865 to 1870. But the peanut could be produced only on a limited scale because of the labor required to produce the crop. Plants were dug in the fall, nuts picked from the vine by hand, washed, spread on sheets to dry in the sun, and usually roasted and consumed during the winter months. Some nuts were shelled and pounded into a paste or butter. But, this laborious harvest and the lack of organized marketing meant that peanut remained a regional crop until the early 1900s. Four events occurred around the turn of the 20th century that vaulted peanut into the nationally known commodity that it is today.

HISTORICAL EVENTS

Peanut Butter

Dr. John H. Kellogg, Director of a Sanitorium in Battle Creek, Michigan, encouraged the use of peanut butter as a nutritious and easily digestible health food in the 1890s. Dr. Kellogg obtained the first two U.S. patents for this process in 1898. Joseph Lambert developed the first equipment for this process: a roaster, a blancher, and a hand-operated nut-grinding machine.

Penny-in-the-Slot Machines/Candy

F. V. Mills and H. S. Mills introduced the penny-in-the-slot peanut machine in 1901. The popularity of this device for selling roasted peanuts grew rapidly, and 30,000 such machines were in use within a few years. These machines gave great impetus to spanish-type peanut production in North Carolina. Why these machines were used predominately or wholly with spanish-type peanuts is not obvious. Confectioners began to produce peanut candy on a large scale in 1901.

Peanut Picking Machines

The production of peanuts in the United States was quickly reaching a plateau around the turn of the 20th century. Lack of available labor for picking the nuts from the vines was becoming the limiting factor in expansion. There were those who suggested that production would be discontinued if harvesting equipment was not developed to relieve this situation. F. F. Ferguson and J. T. Benthall patented the first successful peanut picker in 1905 after demonstrating its suitability in Richmond, Virginia in 1904. With this system of harvest, plants were dug, allowed to dry briefly on the ground, the soil was then shaken from the vines, and vines were then stacked around a tall pole, maybe seven feet high, to allow the nuts to dry.

Stacks were then transported to Ferguson's and Benthall's stationary picking machine.

Boll Weevil

Cotton was king in the southern United States because it was essentially the only cash crop produced. Producers of cotton had had license to produce this commodity with almost no threat of disease or major insect pests, with the exception of occasional outbreaks of leaf worms and boll worms during years of relatively high rainfall. But this situation changed abruptly in 1892 as the Mexican boll weevil, *Anthonomus grandis* Boh., moved across the Rio Grande River and into the cotton fields of south Texas. By 1933, the weevil had migrated to every cotton-producing region of the United States. Producers scrambled for control measures and for new crops that would be less hazardous to produce.

World War I

With the advent of the picking machine, and increased demand for peanuts for roasting and for use in confectionaries, farmers in North and South Carolina, Georgia, and Alabama turned to the peanut as an alternative to cotton. Acreage expanded rapidly from 1909 to 1919. World war brought a new need that was alleviated by the peanut, along with other crops, the need for increased production of plant oils. Spurred by this need as traditional sources of plant oils dried up because of political events surrounding World War I, 260,000,000 pounds of peanut oil were processed in the southern U.S. in 1916. Processing was completed mostly in converted cottonseed oil mills. The boll weevil and World War I helped establish peanut as a viable cash crop for the southeastern U.S. Producers and businessmen of Coffee County in southeast Alabama raised $3,000 to erect a monument, a statue of a pedicel and a boll weevil, to recognize the weevil as the driving force behind diversification of agriculture in the Southeast. That monument still stands today.

World War II

Two events surrounding World War II also gave impetus to increased peanut production in the United States. First, the government designated "peanut for oil" as an essential crop in 1941 because of the need for more plant oils to support the war effort. Area and production increased until 1948 when demand for the U.S. peanut declined to the point that acreage allotments were imposed to protect producers from bankruptcy. Allotments also ensured a stable domestic supply.

Peanut Combine

W. D. Kinney and J. L. Shepherd developed the first once-over mobile peanut combine. This machine meant the end of the laborious task of manually shaking vines free of soil and then stacking them onto a "peanut pole." The mobile picker was quickly accepted by the industry.

Other Industrial Events

Other events that were important in the growth of the U.S. peanut industry were as follows:

1. American Peanut Corporation was founded in 1849 by L. D. Bain, who bought and sold hand-picked peanuts.
2. The first mill committed to the cleaning, grading, and shelling of peanuts was established in an abandoned cotton mill near Suffolk, Virginia, in 1890.
3. Dawson Cotton Oil Company was established in Dawson, Georgia, in 1904. The company later changed its name to Stevens Industries and today is one of the largest peanut cleaning, shelling, crushing, and peanut butter operations in the industry.
4. The Lilliston Company marketed its first Lilliston Peanut Picker in 1910. Twenty-five machines were built in 1911.
5. Amadeo Ovici founded the Planters Nut and Chocolate Company in Suffolk, Virginia, in 1912. Planters continues a diverse and extensive operation today and primarily uses virginia-type peanuts. The company created "Mr. Peanut" in 1981, which is still their trademark. Mr. Peanut walks in front of peanut shops and gives away a reported $100,000 worth of peanuts each year.
6. Tom Houston Peanut Company began business in a wood-frame cabin in Columbus, Georgia, in 1925. Three employees toasted spanish-type peanuts and sold them in the streets of Columbus. Today, Tom's is a multimillion-dollar business. In 1987, several silos were built at the Columbus plant with a storage capacity of 28,000,000 pounds.

SYSTEMATICS

There have been a number of attempts to organize the classification of the numerous "types" of peanut. Gregory et al. (1951) suggested that peanuts should be classified on the basis of their branching patterns and recommended the use of the terms virginia, spanish, and valencia. However, this system failed to account for intermediate yet stable forms found in Paraguay and northeastern Argentina. A. H. Bunting (1955) proposed that peanuts be categorized on the basis of branching habit, since there are only two basic forms, those being alternative and sequential. He also suggested that those two categories should have subspecies status. Wynne and Coffelt (1982) note that Krapovickas and Rigoni agreed with Bunting and proposed in 1960 that *A. hypogaea* should be divided into *A. hypogaea* L. subsp. *hypogaea* and *A. hypogaea* subsp. *fastigiata*. This classification apparently fits the diversity found in the six primary and secondary centers of origin (or diversity):

Guarnian, Bolivian, Peruvian, Amazonian, Goias and Minas Gerais, and northeastern Brazil.

The accepted botanical classification of peanut grown in the United States today is *A. hypogaea* subsp. *hypogaea* var. *hypogaea* for virginia and runner types, and *A. hypogaea* subsp. *fastigiata* var. *vulgaris* and var. *fastigiata* for spanish and valencia, respectively. These four types grown in the United States are interfertile, each having 2n = 40 somatic chromosomes. Intermediate morphological progeny have been developed. Subspecies *hypogaea* also contains the var. *hirsuta* that is grown commercially in China and Mexico. Two botanical varieties, *peruviana* and *uequatoriana,* that are included in the subspecies *fastigiata* are grown commercially in South America.

MORPHOLOGY

The peanut is a perennial, dicotyledonous legume grown in all temperate and tropical climate zones. It's growth habit is so complicated that Gregory et al. (1951) stated that before 1949 only one description of the morphology of the seed and seedling had been attempted and that it was incorrect. They further stated that only two descriptions of the reproductive morphology existed and that only a few correct accounts of the relationship between the aerial flower and the subterranean fruit could be found.

Each peanut seed consists of two large seed leaves, or cotyledons, an epicotyl consisting of the apical meristem and six to eight differentiated leaves, a hypocotyl that is extremely distinctive in young seedlings, and a radicle or primary root. During the first few days following germination under optimum conditions, the root grows eight times as rapidly as the shoot, which will not produce any visible new initiatives for about two weeks. The root, on the other hand, may produce over 100 new lateral roots by the twelfth day after germination. An interesting morphological oddity of the peanut is that the roots are devoid of root hairs except in cases of root injury or high temperatures and high humidity.

Another unique feature of the peanut is that the cotyledons are neither epigeal nor hypogeal. In beans, for example, the cotyledons are pushed (and partly pulled) upward by the expanding hypocotyl until they are just above the soil surface. These seedlings are call epigeal. In the garden pea, the cotyledons remain where the seeds were placed when planted, thus hypogeal. The peanut cotyledons are pushed upward by an expanding hypocotyl until light is encountered, whereupon the hypocotyl stops growing and the cotyledons remain at the soil-atmosphere interface. The hypocotyl normally elongates only four to five inches. When seed are placed at this soil depth, the expanding hypocotyl consumes essentially all of the energy stored in the cotyledons and the epicotyl will emerge pale and yellow, and the seedling will be almost devoid of roots. However, peanut seedlings can emerge because of epicotyl elongation when seeds are planted more than four inches deep.

Apical dominance (i.e., the inhibition of lateral branching) occurs in most plants. However, very little apical dominance is demonstrated in peanuts, with two lateral branches arising from the cotyledonary node, equaling or exceeding the length of the primary or main shoot within a few days. This morphological feature provides the plant with essentially three initial shoot apices. Third and fourth laterals arise from the central or main stem from the axis of the first two true leaf nodes. Additional vegetative branches arise from the first two nodes of each of the cotyledonary lateral branches in virginia-type peanut. Reproductive branches arise from axillary buds in the axis of foliage leaves of the cotyledonary branches of virginia, spanish, and valencia types. Reproductive branches rarely occur on the main or central stem of virginia-type peanut, but do occur regularly on the main stem of both spanish and valencia types, although not before they appear in lateral vegetative branches. These arrangements will result in a more spreading phenotype for the virginia-type peanut (and therefore for runner also) and a more "upright" appearance for spanish and valencia.

The compound leaves, usually tetrafoliate, of the peanut plant occur alternately along the stem at $2/5$ phyllotaxy. The leaf is normally even-pinnate—i.e., has an equal number of leaflets on each side of a common stalk. Leaf shape varies from abovate, or inversely ovate, to elliptical to near linear lanceolate. Leaves on the main stem of the peanut plant are well formed, and all tend to be near equal in size under good growing conditions. On secondary and later branches, leaf size and completeness may be less at the proximal end relative to the distal end. In fact, the first and second nodes from the base are often represented by mere scales or stipule-like leaves called cataphylls. Most secondary and later branches are subtended by these modified leaves.

The peanut plant is morphologically a very complex plant. Multiple branching, specialized leaf structures, and aerial flowers but subterranean fruit all complicate its structure to the novice. Most, if not all, producers and agriculturists who have worked with an array of peanut types and an array of peanut plants can classify them to type—i.e., spanish, virginia, or valencia. Many can probably correctly classify them as being erect or runner types. Obviously then, there must be some tangible and morphologically observable traits that distinguish them.

A botanical key similar to that of Richter (1899) and described by Gregory et al. (1951) is found below. Richter used a system of n + 1...j to describe the branching habit while the author prefers 1°, 2°, etc. (Figure 8.1). Recall that the peanut plant has a primary or central stem, now called the 1° stem, that originates from the epicotyl. Branches arising from this central stem, including those arising from the cotyledonary axis, are 2° branches and limbs arising from those are 3°, etc.

A. Lateral buds of the central stem are vegetative;
  1. 2° branches consist of two laterals arising one each from cotyledonary axils and vegetative laterals arising from the axis of the first two true-leaf nodes of the main stem;
  2. 3° branches arising from the first two nodes of the 2° branches from the cotyledonary axis are vegetative but occasionally reproductive from the second node;
  3. Additional branches are usually vegetative but reproductive branches occasionally occur at higher nodes on central stem;

Central stem

(a)

Figure 8.1 *Illustration, half profile, of growth and fruiting of virginia (a), spanish (b), and valencia (c) type peanuts. (Drawing by Mike Hodnett and Linn White)*

    4. Branches arising from 3° branches, i.e. 4°, alternate two vegetative and two reproductive;

    5. Larger seeded .... virginia

    6. Smaller seeded .... runner

B. Lateral buds of the central stem are vegetative or reproductive;

    1. First and second nodes of 2° branches are reproductive;

       a. 3° branches are irregularly reproductive and vegetative; pods are two to three seeded .... spanish

       b. 3° branches all reproductive or sometimes vegetative until the eighth node; all 4° branches are reproductive; pods are three to six seeded .... valencia

Central stem

(*b*)

Figure 8.1 (*Continued*)

The inflorescence of the peanut plant is described as spicate or subpaniculate, occurring in leaf or cataphyll axil (Figure 8.2). The inflorescences are not terminal, and each will have three to several flowers. The rachis of this modified panicle has cataphylls with simple flowering branches arising from the axil of cataphyll and rachis. The flowering branch is extremely short and terminates in a cataphyll (may be bifid) and a flower bud. The flowers usually open one at a time, but two flowers within one inflorescence may appear simultaneously in spanish-type peanuts. Flowers within a single inflorescence may appear at intervals of one to several days.

The flower of the cultivated peanut is orange papilionate and sessile. Each flower

Main stem

(c)

Figure 8.1 (Continued)

contains an extremely long calyx composed of five fused sepals that separate at the tips. Four of the sepals remain fused to form a superior lip that arises behind the standard petal. The remaining sepal tip is linear and lies below the keel petal. The staminal column is composed of ten filaments, but only eight terminate in anthers, half of which will be two-lobed and half one-lobed. The filaments are fused through one-half to two-thirds of their length. The staminal column lies horizontal to the elongated calyx tube. The pistil is made up of a single, sessile carpel (ovary) with an elongated style extending through the calyx tube; it bends sharply to go through the staminal column and bends again upward with the reflexed filaments. The pistil terminates in a club-shaped, feather-like stigma.

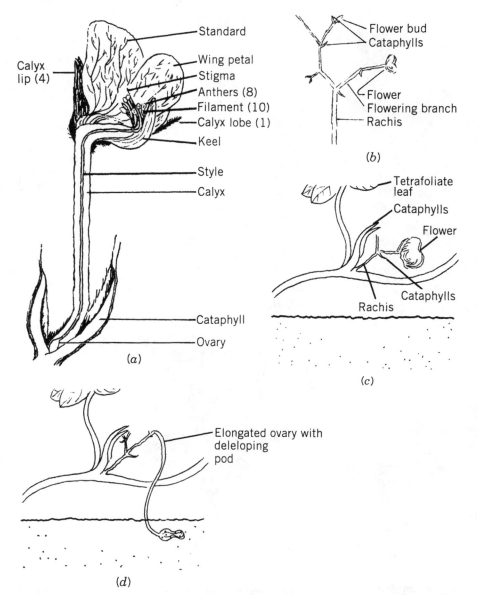

Figure 8.2 *Illustration of the inflorescence of peanut. (a) Peanut flower; (b) and (c) inflorescence elongated to show branching; and (d) elongated ovary with pod below soil surface. (Inflorescence size exaggerated to show details.) (Drawing by Amy Smith)*

The next feature of the peanut flower sets it apart from all other crop flowers and is responsible for the peanut fruit maturing underground. The peg has been referred to as a petalless flower or a gynophore—that is, a stalk that supports an ovary. It is obvious to today's student of botany that this elongated form called the "peg" is not an apetalous flower. Smith (1950) recognized that the peg is not a gynophore but is an elongation of the ovary itself. The loculus of the ovary is in the distil end—that is, the tip of the peg—and contains two to six ovules (Figure 8.1).

Once fertilization occurs, the peg that is positively geotropic at this time begins to elongate. The peg will grow into the soil if the flower is not more then six inches above the ground. If the peg fails to reach the ground, the tip dies without pod and seed development. There are occasional exceptions to this. In such exceptions, viable seed will develop in the tip to the aerial peg but pods do not develop. Numerous short aerial pegs without pods are frequently observed in fields with thin stands of spanish-type cultivars during late stages of growth under cool temperature conditions.

The peg will normally grow one to three inches into the soil. At some point after the peg has grown into the soil, it loses its positive geotropism and the tip turns horizontal and the pod begins to develop. The mature pod has a shell that is usually reticulate with ridges resulting from the atrophy of endocarp cells of the mesocarp layer that surrounded the ovules during development. The shell usually has constrictions between the seeds.

INTRODUCTION INTO THE UNITED STATES AND CULTIVAR DEVELOPMENT

As noted above, the most accepted scenario for the introduction of the peanut into the United States was through early slave trade. Thus associated, the peanut remained an isolated, little considered, regional crop until after the U.S. Civil War. These early plantings were from seed of unknown origins, and little if anything was known about their adaptability to U.S. conditions. We will consider climatic and agricultural requirements for peanut production later, but we must first consider plant introductions and development of adapted cultivars.

The collection of peanut accessions, breeding lines, and cultivars is fairly extensive. Over 8,000 accessions have been catalogued and numbered with a plant inventory (PI) number (PI originally was the abbreviation for plant introduction). The working collection is held at the Southern Regional Plant Introduction Station of Experiment, Georgia. Long-term storage of this material is at Fort Collins, Colorado. As with all crops, scientists are actively seeking to collect, identify, store for future use, and utilize the genetic variability that exist within *Arachis* and closely selected species before it is destroyed forever by encroachment of civilization.

The "type" issue is confounded by the descriptive growth-form terms "runner" and "bunch." These terms refer to growth habits and are used in cultivar descriptions regardless of botanical classification. The runner cultivars have a spreading habit of growth, while bunch types tend to be more erect. Therefore, virginia and runner-type peanut cultivars may be referred to as runner, and spanish and valencia-type peanut cultivars may be referred to as bunch types.

The spanish peanut was the most widely grown type in the United States until the late 1970s when its acreage was exceeded by runner type after the introduction of a new cultivar developed by the Florida Agricultural Experiment Station in 1969.

Florunner was selected from a cross of Early Runner, a popular early-maturing runner type, and Florispan. Florunner produces larger seed than its predecessor runner-type cultivars. Seed size classes of this cultivar overlap with spanish-type cultivars such that there is a movement in the industry to discontinue the use of "runner" as a market class. (Note that runner, virginia, spanish, and valencia are terms used as market classes in the industry and refer to peanuts of the runner type, virginia type, spanish type and valencia type.) As more cultivars are developed that do not fit the present market classification use of these terms, the industry could be forced to abandon this terminology.

Spanish-type peanut had occupied about 45 percent of U.S. acreage prior to the introduction of Florunner. By 1972, runner-type peanuts occupied 44 percent of U.S. acreage, spanish was grown on 32 percent, with virginia and valencia types occupying 23 percent and less than 1 percent, respectively. Even more startling documentation of the increased productivity and therefore popularity of the runner-type peanut is the fact that spanish type accounted for 36 percent of the total U.S. production in 1960 and the runner type accounted for 32 percent. By 1989, runner-, virginia-, and spanish-type cultivars accounted for 79, 16, and 5 percent, respectively (valencia production not included). Although runner-type peanuts have been grown traditionally in the southeastern states of Mississippi, Alabama, Florida, Georgia, and South Carolina, this type is becoming more popular in the Texas and Oklahoma area where spanish-type peanut has been traditionally grown.

U.S. PRODUCTION, TRADE, AND USES

Total U.S. acreage devoted to all types of peanut in 1990 was 1,840,000 (Table 8.1). The Virginia-North Carolina region planted 261,000 acres; the Southeast region composed of South Carolina, Georgia, Alabama, and Florida grew 1,134,000 acres; Texas and Oklahoma had 395,000 acres; and New Mexico grew 20,000 acres (Table 8.2). Total U.S. production of all types in 1990 was 3,603,000,000 pounds of farmer stock—that is, in-shell. As an average of 1988 to 1990, 694,886,660 pounds/year of runner and virginia were produced in the Virginia-North Carolina production area; 2,414,416,500 pounds of predominately runner-type peanuts were produced in South Carolina, Georgia, Alabama, and Florida; 706,055,000 pounds of spanish and runner types were produced per year in Texas and Oklahoma; and New Mexico averaged 42,077,332 pounds of valencia-type peanut.

These nine states accounted for practically all of the peanuts produced in the United States since acreage allotments or production quotas were imposed by Congress in 1954. Those states, their share as an annual average during 1988 to 1990, and types of peanuts grown are shown in Table 8.3.

The U.S. Congress established a price support program in 1934 (exempted for 1936). In 1941, Congress designated peanut as a basic commodity and authorized quotas and price supports. With this encouragement and to help supply plant oils for the war effort, the American peanut farmers increased the number of acres harvested for nuts from 2,052,000 in 1940 to 3,160,000 in 1945, a 54 percent increase. The

TABLE 8.1 United States Peanut Acreage, Yield, and Foreign Trade.

Year	Acreage (×1,000)				Average Yield (lbs./ac.)[3]	Total Production (× 1 mil.lb.)	Foreign Trade (×1,000 lbs.)	
	Total	Grown Alone[1]	Not Harvested[2]	Harvested			Exports	Imports
1910	730	—	—	464	827	384	5,447	22,788
1915	1,064	—	—	617	779	481	8,669	38,110
1920	1,715	—	720	995	699	696	13,149	69,303
1925	1,667	1,279	671	996	725	722	3,711	54,039
1930	1,881	1,433	808	1,073	650	697	2,645	14,040
1935	2,546	1,972	1,049	1,497	770	1,153	284	315
1940	3,108	2,599	1,056	2,052	861	1,767	637	6,409
1945	4,058	3,853	898	3,160	646	2,042	62,767	34,639
1950	2,734	2,633	469	2,262	898	2,035	67,441	13
1955	1,913	1,882	244	1,669	928	1,548	2,251	65,730
1960	1,546	1,526	151	1,395	1,232	1,718	84,602	66
1965	—	1,820	82	1,438	1,661	2,390	4,553	2
1970	—	1,518	49	1,469	2,030	2,983	6,436	32
1975	—	1,532	32	1,500	2,564	3,847	22,708[5]	1,850
1980	—	1,521	1	1,400	1,645[4]	2,303	33,071	3,597
1985	—	1,490	23	1,467	2,810	4,123	83,747	1,493
1990	—	1,840	31	1,810	1,991	3,603	118,664	9

Source: Adapted from USDA Agricultural Statistics, 1936 and following years.

1. Peanuts were often grown as a companion crop with corn for grazing by swine.
2. Acreage not harvested for nuts, mostly fed to swine without being dug, i.e. "hogged-off."
3. Farmer stock yield, i.e. in shell weights.
4. Severe drought year in Southeast.
5. Foreign trade reported by different classification by 1975 and years following.

TABLE 8.2 Harvested Acreage (Harvested for Nuts) (×1,000) of Peanuts and Average Yield (Pounds/Acre) by State and Decade, 1940–1990.

State	1940		1950		1960		1970		1980[1]		1990	
	(ac.)	(yd.)	(ac.)	(yd.)	(ac.)	(yd.)	(ac.)	(yd.)	(ac.)	(yd.)	(ac.)	(yd.)
VA	158	1,365	148	1,540	104	1,900	102	3,060	101	1,285	97	3,195
NC	262	1,400	232	1,090	178	1,750	167	2,670	166	1,755	164	2,900
SC	24	750	19	800	11	1,100	14	1,880	13	1,100	14	2,230
GA	705	825	728	935	472	1,275	507	2,220	514	1,935	770	1,750
FL	94	780	72	850	48	1,200	53	2,075	55	2,600	94	2,480
TN	6	750	4	800	2	900						
AL	310	735	335	970	195	1,100	190	1,660	200	1,325	256	1,510
MS	28	410	10	425	5	400	4	1,100	6	1,250		—
AR	22	350	8	475	2	450						—
LA	11	360	3	340		—						
OK	90	600	212	590	110	1,350	116	1,655	105	1,335	106	2,220
TX	330	560	490	675	275	775	306	1,405	230	1,275	289	1,850
NM		—	7	820	6	1,900	8	2,230	9	2,540	20	2,500
Total	2,040	858	2,268	898	1,408	1,259	1,467	2,031	1,399	1,645	1,810	1,991

Source: Adapted from USDA Statistics, 1941 and following years.

1. Severe drought in Southeast.

TABLE 8.3 Percent of Peanuts Produced and Types Grown by State during 1988–1990.

State	Percent of Production	Predominate Type(s) Grown
North Carolina	11	runner and virginia
Virgina	7	runner and virginia
Alabama	13	runner
Florida	6	runner
Georgia	43	runner
South Carolina	1	runner
Oklahoma	6	spanish and runner
Texas	12	spanish and runner
New Mexico	1	valencia

Source: Adapted from USDA Agricultural Statistics, 1992.

government's incentive programs for peanut oil ended in 1951. Since then, the government farm programs have been geared to encouraging peanut production for edible uses. Congress established an acreage allotment system in 1954 that set the minimum U.S. acreage at 1.61 million. In 1981, in response to the decreased production caused by the severe drought of 1980, the acreage allotment system was replaced by a poundage quota system that specifies the quantity of peanuts that a producer can sell at the maximum government loan price. The national poundage quota for 1984 was 1.1 million short tons. These quotas are adjusted each year by the Secretary of Agriculture according to estimated U.S. consumption for edible and seed purposes. The price support system, which included acreage allotments and now poundage quotas, was originally set to ensure the producer a return of 90 percent of parity, that being the price that gives the commodity the same purchasing power that it had in 1910 to 1914. Price supports were reduced to 75 percent of parity in 1970 and discontinued in 1978.

A producer was assigned an allotment based on his farm unit history relative to the national requirements and, beginning in 1981, on his historical production. In the early years of the quota system, a producer could plant his entire allotment. If his production exceeded his poundage quota, the poundage above that quota could be sold as "additional" peanuts. Legislation in 1985 suspended acreage allotments and set poundage quotas based on estimated domestic edible, seed, and related uses. Quota peanuts are supported at a minimum of $550/short ton, adjusted for changes in costs of production, excluding adjustments for return to land since 1982, but not exceeding 106 percent of the previous year's support price. Beginning in 1978, farmers without acreage or quota history could produce peanuts that were eligible for government support as additionals. The U.S. Government, through the Commodity Credit Cooperation, (CCC), buys or extends loans on quota peanuts (producers can sell directly to processors) and beginning in 1978, on additional peanuts. The loan rate for additional peanuts is announced by the Secretary of Agriculture no later than February preceding the crop year. The average price for quota peanuts from 1985 through 1988 was $588/ton and additionals were valued at $152/ton. While the peanut producer can sell (i.e., put in the loan program) additional peanuts to the government, most are sold as exports. Recall that the peanut support system is designed to ensure an adequate supply for domestic use, so additionals can serve as a backup supply for U.S. consumption or be sold through exports.

The producer sells his or her peanut crop to a "first handler." These are buying points operated mostly by shellers, independent commodity dealers, or independent warehouse owners. First handlers normally provide drying, cleaning, purchasing or arranging for price support loans from the CCC, and warehousing of peanut going into the loan. According to the USDA's Food Safety and Quality Service, there are 170 buying points in North Carolina, 72 in Virginia, 55 in Alabama, 10 in Florida, 160 in Georgia, 33 in Oklahoma, and 74 in Texas. There are 11 additional buying points in Arizona, Arkansas, Louisiana, Mississippi, South Carolina, and New Mexico.

The peanut price support system specifies that producers' peanut lots (a lot generally being a trailer or truckload as delivered to the first handler) be separated into one of three classes, those being segregation 1, 2, or 3, (to be discussed under grades) depending on the amount of and types of damage in each lot. First handlers can buy only segregation 1 peanut for use in edible products.

The United States Department of Agriculture, Agricultural Marketing Service (USDA-AMS) places inspectors at each buying point. Peanuts placed into storage must be cleaned if foreign matter exceeds 10 percent by weight of the sample. Foreign matter consists of dirt, sticks, stones, vines, insects, broken shells, etc. Cleaning takes place immediately after drying each lot to 8 to 10 percent moisture. Other factors considered in grading peanuts are sound mature kernels (SMK), sound splits (SS), other kernels (OK), loose shelled kernels (LSK), and damaged kernels. Peanuts are tested also for aflatoxin contamination. The following summary of movement from and marketing from first handler to consumer products and identification of product manufacturers was adapted from McArthur, Grise, Doty, and Hacklander (1983).

Producer to first handler: first handler removes foreign material; drys; segregates peanut with *Aspergillus flavus;* makes commercial purchases and arranges loans; takes delivery of contracted additionals; may warehouse; inspected by USDA-AMS.

First handler to sheller: sheller may be the first handler; screens in shell; shells; stores; ships to buyer; buys quota and additionals peanut from CCC for commercial use; chemical analyses determined before shipping domestic edible.

Sheller to edible product processor: edible product processor buys segregation 1 only; may buy additional from CCC for edible uses at quota prices or more.

Roaster: an in-shell processor.

Crusher: primary product is oil; uses segregation 2 and 3 grades; may use additionals via contract or from CCC at additionals price.

Crusher-restricted: monitored by grower associations to dispose of surplus peanut; price is less than additionals; may crush for CCC and CCC retains ownership of oil.

Exporter-shelled: may market quota peanut but at prices above quota loan, or use contracted or CCC additionals.

Exporter-crushed: uses peanut fragments from sheller at a price equal to or greater than additionals loan rate.

Edible product processor: retail trade includes such items as peanut butter, salted nuts, candy, etc.; may export.

Roaster to retail trade—ball park: in-shell market.

Crusher to retail: domestic oil and meal; export oil; retail fertilizer from peanut with *A. flavus* mold.

Crusher-restricted: oil restricted to domestic use and unrestricted meal.

Peanut is an oilseed crop that is consumed primarily in the United States as edible products, with relatively little consumed as oil or meal. A breakdown of uses for virginia-, spanish-, and runner-type peanuts from 1955 through 1990 is shown in Table 8.4. The majority (52 percent in 1990 and 57 percent in 1980) of peanuts consumed in this country is in the form of peanut butter, with consumption of shelled, salted peanuts accounting for 25 percent of consumption of shelled peanuts in 1990 and use in candy accounting for about 21 percent of consumption. Cleaned, unshelled peanuts predominately are consumed as roasted, ball-park peanuts.

Each of the four types of peanut has characteristics that make them desirable for use in different primary products. In recent years, the increased production of runner-type peanut has resulted in encroachment by runner type into end uses that traditionally had been filled by one of the other three types. The virginia type has the largest kernel of commercial cultivars in the United States, about 60 percent larger by weight than nuts produced by runner-type cultivars. Most of the shelled virginia type (50 percent in 1990) are marketed as salted nuts with most of the remaining shelled production going to peanut butter. Most of the peanuts roasted and sold in the shell are virginia-type peanuts.

Most runner-type peanuts are marketed as peanut butter, with 63 and 55 percent of shelled nuts going to peanut butter in 1980 and 1990, respectively. However, there has been considerable encroachment of runner type into the whole nut market with only 2 percent of runner type going to the salted nut market in 1970, but up to 18 percent in 1990. Seed of the Florunner cultivar varies in size and shape, which causes problems in size grading and therefore marketing of this cultivar as salted, shelled or unshelled roasted peanuts.

Relative to use in primary products, about 60 percent of spanish-type nut consumption in 1990 was as peanut butter, with use as shelled, salted nuts and candy each accounting for about 20 percent. Spanish has a higher oil content than the other three types, which is an advantage when crushing.

The valencia-type peanut usually has three or more kernels per pod. Valencias

TABLE 8.4 Peanuts—Shelled Raw Basis—Used in Primary Products and Cleaned Unshelled Used in United States since 1955.

	Peanut Butter		Shelled Uses				
			Cleaned Peanuts				Unshelled
Type/Year	Butter	Sandwiches	Salted	Candy	Other	Total	
			(×1000 lbs.)				
Virginia							
1955	35,647	—	100,534	36,559	2,053	174,793	—
1960	57,702	5,010	133,105	48,037	3,948	247,802	—
1965	69,871	7,565	150,476	51,389	5,506	284,807	—
1970	50,294	7,895	176,816	58,729	5,895	299,629	—
1975	62,244	3,068	147,933	52,335	2,746	268,326	—
1980	26,142	2,487	50,095	18,903	1,137	98,764	—
1985	52,885	1,939	113,577	32,900	5,286	206,587	—
1990	99,218	na	141,915	26,072	17,237	284,442	—
Runner							
1955	153,636	—	7,721	18,733	3,948	184,038	—
1960	219,727	9,009	2,225	17,995	6,594	255,550	—
1965	235,617	9,925	3,965	32,093	8,668	290,268	—
1970	232,567	9,941	5,820	95,373	7,743	351,444	—
1975	447,182	15,204	94,633	124,423	9,742	691,184	—

Year							
1980	529,923	21,289	127,430	177,098	15,327	871,067	—
1985	605,696	22,261	207,503	240,198	16,192	1,091,850	—
1990	582,069	na	188,939	260,014	19,464	1,050,486	—

Spanish

Year							
1955	136,378	—	38,072	63,305	3,510	241,265	—
1960	144,011	3,698	54,302	78,685	3,729	284,425	—
1965	194,002	4,012	58,918	96,577	4,139	357,648	—
1970	257,489	7,481	56,168	89,137	3,933	414,208	—
1975	138,631	2,685	59,069	62,918	3,509	266,812	—
1980	32,491	302	27,964	41,889	3,224	105,870	—
1985	42,735	424	37,463	40,739	2,039	123,400	—
1990	60,624	na	23,841	19,319	1,388	105,172	—

All Types

Year							
1955	325,661	—	146,327	118,597	9,511	600,096	61,650
1960	421,440	17,717	189,632	144,717	14,271	787,777	80,619
1965	499,490	21,502	213,359	180,059	18,313	932,723	94,648
1970	540,350	25,317	238,804	243,239	17,571	1,065,281	119,236
1975	648,057	20,957	301,635	239,676	15,997	1,226,322	142,029
1980	588,556	24,078	205,489	237,890	19,688	1,075,701	120,263
1985	701,316	24,624	358,543	313,837	23,517	1,421,837	234,601
1990	741,911	na	354,695	305,405	38,089	1,440,100	229,641

Source: Adapted from USDA Agricultural Statistics, 1962 and following years.

429

have a sweeter taste than either of the other three types and are sold primarily in the shell. There is a very small market for fresh (i.e., not dried) peanuts that is filled primarily by valencia type. These nuts are boiled in-shell and sold at road side stands or small grocery/service stations. Most valencia type are grown in New Mexico, but some are produced in other areas for the fresh peanut market.

As noted above, the peanut is used mostly in the edible market. Peanut oil accounted for only about 0.02 percent of edible vegetable oil produced in the United States from 1975 to 1981. Peanut meal, a by-product of oil production from peanuts, accounted annually for 0.34 to 0.66 percent of all high-protein edible meal for the same time period. The major reason that peanut in this country is not grown for oil and meal is its poor competitive position with oil and meal from soybean and oil from sunflower and corn. In 1981, the cost of producing oil from peanuts was $0.48/pound, compared with $0.15/pound for soybean oil. The cost of 50 percent protein peanut meal was twice as expensive as 50 percent protein soybean meal. A highly unsaturated fatty acid content (80 percent), relatively high smoke point (about 440°F) during deep-fat frying, and a slightly nutty flavor imported to fried foods make peanut oil a desirable product. However, its drawbacks are that it solidifies when refrigerated and cannot be "winterized," (i.e., made to remain liquid at low temperatures) without producing nonfilterable crystals. Peanut oil also is more susceptible to oxidation than competing vegetable oils, which means a relatively shortened shelf life. However, none of these pros or cons appreciably affect the production of peanut oil. The volume of peanut oil produced in this country is dictated by the volume of excess nuts and nuts considered undesirable for direct consumption. While the United States is an exporter of peanut oil, Argentina and Brazil are the largest exporting countries.

Three additional points need to be made about peanut meal. First, almost all U.S.-produced meal is sold domestically as animal feed, although it is relatively unimportant as a feed supplement. Second, peanut meal that is destined for the food market must be crushed under sanitary conditions. Such products are produced in extremely small quantities, although they (i.e., grits or flour) are excellent high-protein sources for meat extenders, and bakery and snack products. Last, peanuts contaminated with A. flavus can be crushed for oil. The fungal toxin remains in the meal, which is sold as fertilizer since the cost of removing the toxin is prohibitive.

WORLD PRODUCTION AND TRADE

The United States ranked a distant third in peanut production in 1990–91 behind India and the Peoples' Republic of China. Average production for 1988–89 through 1990–91 for countries producing at least 300 short tons (2,000 pounds) is shown in Table 8.5.

The United States is the world's leading exporter of peanuts. Exports of shelled peanuts have averaged 504,016,200 pounds/year since 1970, ranging from a low of 213,027,000 pounds for 1970 to a high of 745,593,000 pounds in 1978. As a percent of production of shelled peanuts, the United States exported from 11 to 31 per-

**TABLE 8.5   Peanut Production for Selected Countries
as an Average of 1988/89 through 1990/91.**

Country	In Shell Production
	(×1,000 lbs.)
India	17,918,000
People's Republic of China	12,802,000
United States	3,857,000
Indonesia	1,916,000
Senegal	1,598,000
Burma	1,030,000
Sudan	864,000
Zaire	838,000
Nigeria	786,000
Argentina	774,000
Vietnam	646,000
Other[1]	6,526,000
World Average	49,556,000

Source: Adapted from USDA Agricultural Statistics, 1990.

1. Split year to account for Northern and Southern Hemisphere countries.

2. Includes 21 countries.

cent of production during the 1970s and 1980s. The United States exports relatively little peanut as "processed" food—some peanut butter, a very small amount of peanut oil, and essentially no peanut meal. U.S. exports of shelled peanuts go primarily to Canada, France, and England.

GROWTH STAGES

A system of staging the peanut plant was proposed by Brite (1982) in Florida that was based on visually observable traits during the life of the plant. Brite proposed his staging system based on research with spanish- and virginia-type peanut. Vegetative stages are based on the number of nodes on the main stem with the cotyledonary node identified as node zero (Table 8.6). The next node is the node of the first true leaf, a tetrafoliate leaf, and when the leaflets at this node have developed to the point that they are unfolded and flat, then the plant is at stage V-1. Additional main stem nodes, and therefore V stages, are counted as soon as the leaflets at those nodes are unrolled and flat. Reproductive stages are based on visually observable events relative to flowering, pegging, and fruit growth and development. Reproductive stage 1 (i.e., R-1) is defined as the point in plant development when the plant has its first bloom or open flower at any node. Stage R-2 occurs when the plant has an elongated peg, and so on. A field is considered in a specific stage, vegetative or reproductive, when 50 percent or more of the plants staged are at a given stage of growth (see Chapter 7 section on soybean staging for example).

The purpose of defining growth stages is to specifically identify the physiological

TABLE 8.6   Growth Stages and Their Descriptions for Peanut.

Stage[1]	Abbreviated Title	Description
V-E	Emergence	Cotyledons near the soil surface with some part of the seedling plant visible.
V-O	——	Cotyledons are flat and open at or below the soil surface.
V-1	First tetrafoliate	One node on the main axis with its tetrafoliolate leaflets unfolded and flat.
V-n	(additional V stages)	n developed nodes on the main axis, a tetrafoliolate node is counted when its leaflets are unfolded and flat.
R-1	Beginning bloom	One open flower at any node on the plant.
R-2	Beginning peg	One elongated peg.
R-3	Beginning pod	One peg in the soil with turned, swollen ovary at least twice the width of the peg.
R-4	Full pod	One fully-expanded pod; size characteristic of the cultivar.
R-5	Beginning seed	One fully-expanded pod in which seed cotyledon growth is visible when the fruit is cut in cross-section.
R-6	Full seed	One pod with seed cavity apparently filled.
R-7	Beginning maturity	One pod showing visible, natural coloration or blotching of inner pericarp or testa.
R-8	Harvest maturity	Two-thirds to three-fourths of all developed pods have testa or pericarp coloration. Fraction is cultivar-dependent, lower for virginia types.
R-9	Over-mature pod	One undamaged pod showing orange-tan coloration of the testa and/or natural peg deterioration.

Source: Adapted from Boote, 1982.

1. Vegetative (V) and reproductive (R) stages for populations are considered established when at least 50% of the plants in the sample have reached that stage. The establishment of a R Stage for an individual plant is based on the first occurrence of the specific trait without regard to position on the plant.

and morphological degree of plant development and to avoid the use of days after planting and calendar date in describing plant or crop development. However, it remains of interest to have a general idea of the calendar age of a plant or crop as well as the growth stage so we have some idea of the length of season. These are shown for spanish and virginia peanuts used by Boote in developing his stages (Table 8.7).

PRODUCTION PRACTICES

Cultivar Choice

Peanut breeders, as do breeders of most crops, have to consider the needs of all segments of the industry in breeding and selecting improved peanut cultivars. The producer looks for higher yields of good quality nuts and tolerance to biotic and abiotic stresses. The peanut processors want more uniform maturity, thinner shells,

**TABLE 8.7 Days from Planting to Specific Stages
of Growth of Two Peanut Cultivars When Grown
at Gainesville, Florida, 1979.**

Stage	Days after Planting	
	Starr (Spanish)	Florunner (Runner)
V-4	20	20
V-10	40	40
V-16	60	58
V-20	80	80
V-21	100	——
V-22	——	100
V-23	120	——
V-24	——	120
R-1[1]	31	31
R-2	39	42
R-3	46	51
R-4	52	60
R-5	57	62
R-6	67	74
R-7	80	93
R-8	119	129
R-9	NA	123

Source: Adapted from Boote, 1982.

1. Days from planting to stage R-1 for 40 cultivars grown in Virginia in
1982 and 1983 averaged 42 days.

and easier-blanching peanuts, while the consumer looks for improved nutritional
properties, and better aesthetic and flavor properties. The producer usually receives
the first consideration because he or she is the primary user of the breeding technology, but the other segments of the consumptive chain cannot be ignored.

The decision of which peanut cultivar to plant is one that the producer will make
only once, and he or she will live with that decision throughout the growing season.
Choice of cultivar defines (1) the maturity or number of days from planting to digging, (2) seed size, which affects seeding rate, (3) degree of resistance or susceptibility to diseases, and (4) in a minor way, its market value. Even though the cultivar
affects all four of these issues, yield potential is the driving force behind a farmer's
cultivar choice, followed by quality characteristics, which are identified as item
four above.

Growers should plant more than one cultivar to spread their risk of crop failure
because of adverse weather and/or biotic pest epidemics. Planting cultivars with
different maturity dates will allow better use of limited harvest and on-farm drying
equipment. A comparison of four cultivars available to North Carolina growers in
1991 shows the variability in virginia-type peanuts adapted to North Carolina and
Virginia (Table 8.8).

The Southeastern states predominately grow runner-type cultivars, principally

TABLE 8.8 Comparison of Four Virginia-Type Peanut Cultivars.

		Cultivar		
Factor	NC-6	NC-7	NC-10C	VA81B
Growth type	Semibunch	Semibunch	Runner	Bunch
Maturity[1]	+5	−10	−2	−14
PR[2]	0	−	0	−
Seed coat color	Tan	Tan	Pink	Pink
STA[3]	M-L	L-M	L-M	M-H
Seeds/lb.	500	500	600	570
Seed calcium[4]	Low	Low	Moderate	Moderate
Seedling vigor[5]	−	−	+	+
Fertility[6]	Mg	Mn	N	S
Grade[5]				
% ELK[7]	+	+	−	+
% SMK[8]	+	+	0	+
% Fancy[9]	+	+	0	−
Acceptance[5]	−	0	0	0

Source: Adapted from Sullivan, 1991.

1. Days from planting to digging, Florigiant is standard reference cultivar and requires 160 days to maturity.
2. PR = pod retention; 0 = same as Florigiant, + = better, − = worse.
3. STA = soil type adaptation; L = light, M = moderate, H = heavy.
4. Ability to absorb soil calcium.
5. 0, −, + = same, worse or better than Florigiant.
6. S = sensitive to fertility level, N = non-sensitive, Mg = generally low in Magnesium, Mn = generally low in manganese.
7. Extra large kernel.
8. Sound mature kernel.
9. See section on grading.

Florunner released in 1969 by the Florida Agricultural Experiment Station. Florunner is widely adapted, being grown from Virginia to New Mexico. This cultivar and Florigiant, a virginia-type peanut released in 1961, revolutionized the production of peanut in the United States. The dramatic effect on yield has been documented earlier in the chapter. (Compare total production and per-acre production for 1960 with later years in Tables 8.1 and 8.2.) But in addition to yield potential, it has been the most widely accepted peanut cultivar in U.S. history and remains the standard to which all new cultivars are compared. When first released, its variable seed size was considered a detrimental character but since has proven to be one of it's strong points. This character makes it possible for processors to market different kernel sizes for varying market demands. Its shelling, processing, and flavor characteristics have added greatly to its acceptance in most peanut markets.

Four cultivars dominate production in the Southeast, those being Florunner, Sunrunner, GK-7, and Southern Runner. Sunrunner is similar to Florunner except it has more uniform seed size; Southern Runner has a moderate level of resistance to leaf spot diseases and requires fungicide spray on a three-week schedule rather than the more normal 10 to 14 days. All mature in 135 to 145 days except Southern Runner, which requires 150 to 170 days.

A number of spanish-type peanut cultivars are available to producers in the United States, although predominately grown in Texas and Oklahoma. Again, four cultivars dominate acreage; Tamspan 90, Starr, Spanco, and Pronto. Spanish cultivars mature in 110 to 140 days, except for Goldin I, which requires approximately two weeks longer.

New Mexico's Valencia A was released in 1971 by the New Mexico Agricultural Experiment Station. Valencia A was selected out of Tennessee Red, a red-seed cultivar introduced about 1925. Tennessee Red has been called the "typical" valencia-type peanut. Valencia production is limited mostly to the Portales Valley of New Mexico, which is characterized by little rainfall, high ($\sim$ 10,000 feet) elevation, and near full sun. Production is irrigated and valencia cultivars mature in 130 to 140 days.

Date of Planting

Optimum planting dates vary by geographical location across the United States. In southern Texas, producers may begin planting by April 10 while others may not plant until early June. Planting in the southeastern states, including Virginia, is recommended between April 1 and June 1. Planting of Valencia peanuts in the Portales Valley of New Mexico occurs typically during May. As with most crops, the ability to irrigate makes the decision of when to plant more fluid, as droughts during both planting and growing seasons can be overcome.

Producers should be cognizant of (1) the number of days required for maturation of their cultivar(s) of choice; (2) the average date of first frost for their location; (3) general weather patterns such as normal drought periods or periods of increased rain probability; and (4) their ability to dig and combine in an efficient manner. Planting should commence when the soil temperature at the four-inch depth is at least 65°F at midday for three consecutive days, with a good five-day forecast. Producers of peanuts should not plant if rain, especially a cold rain, is expected within 48 hours because this would slow germination, thereby making the seed more susceptible to seed rots and seedling disease pathogens. Heavy rain before emergence also can cause soil crusting, which can be a mechanical impediment to the young seedling and cause reduced or complete loss of stand.

Early planting within the optimum range for a given location will result usually in higher yields, more mature pods, and of course, earlier harvest. Comparison of two planting dates and two harvest dates at three locations in North Carolina in 1990 showed a decrease of 200 pounds/acres when planting was delayed at least 15 days (Table 8.9). Later planting mandated a later harvest for maximum production in these experiments.

Data from North Carolina and Virginia compared gross returns per acre when varying planting and harvesting dates (Table 8.10). As planting was delayed from April 23 to May 23, maximum returns were associated with later harvest. There were no real differences in gross returns between digging dates of October 2 and 12 when planted April 23 or between digging dates October 12 and 22 when planted May 23. That is, a month's delay in planting shifted the optimum digging date only

TABLE 8.9   Yields of NC7 Peanuts Averaged over Three Locations in North Carolina in 1990.

Planting Dates	Harvest Dates	Average Yield
		(lbs./ac.)
Before 5 May	Late September	3,907
	After 10 October	4,000
After 20 May	Late September	3,635
	After 10 October	3,881
Early Planted		3,954
Later Planted		3,758
Early Harvest		3,771
Later Harvest		3,941

Source: Adapted from Sullivan, 1991.

10 to 20 days, although we would need at least one more digging date, November 2, to be sure that October 22 was optimum when planted May 23. Under these assumptions, the average optimum number of days from planting to digging was 157 days. The average expected maturity for these four cultivars is 152 days, with a range of 146 to 160. It also is obvious from these data that the optimum date of planting is inextricably tied to the optimum date of digging and vice versa. Cultivars should be chosen and planted such that digging should occur when the threat of inclement weather is least. Physiological maturity of peanuts and the recognition of such will be discussed shortly.

Tillage

Peanuts should be planted into well-drained, sandy or sandy loam soils. Planting peanuts into soils having high levels of silt or clay is not advisable. Given the

TABLE 8.10   Gross Returns as Ranks[1] and Days after Planting (DAP) to Digging Averaged over Three Years and Four Cultivars[2] in North Carolina and Virginia.

	Planting Date							
	April 23		May 3		May 13		May 23	
Digging Date	Rank	(DAP)	Rank	(DAP)	Rank	(DAP)	Rank	(DAP)
September 12	5	(142)	5	(132)	5	(122)	5	(112)
September 22	4	(152)	4	(142)	4	(132)	4	(122)
October 2	2	(162)	2	(152)	3	(142)	3	(132)
October 12	1	(172)	1	(162)	1	(152)	2	(142)
October 22	3	(182)	3	(172)	2	(162)	1	(152)

Source: Adapted from Sullivan, 1991.

1. 1 = highest return and 5 = lowest return.
2. Cultivars and usual maturity requirements: Florigiant (160 days), NC 7 (150 days), NC 9 (255), and VA81B (146).

growth habit (i.e., the fertilized ovary growing into the soil and the pod developing underground), one can understand that soils that have a high-bulk density would, in most cases, inhibit such growth, thereby reducing yield. The most desirable soils are those that are loose, pliable and easily tilled, and have a moderately deep rhizosphere that is easily penetrable by air, water, roots, and pegs. Soils that do not meet these criteria can result in lowered production, lowered quality, and/or increased digging losses. Producers also should be concerned about topsoil depth, internal drainage, and erosion potential. Heavier-textured soils not only make digging more difficult but will lower quality by causing misshaped and/or off-colored shells. These soils will remain wetter for a longer period of time, which can delay timely field operations and decrease yields. Peanuts produced in heavy-textured soils with high organic matter may have off-colored, darkened, or stained shells that will affect the marketability of virginia type and valencia type for the roasted in-shell market. Discolored shells without off-colored kernels are not a problem with spanish- and runner-type peanuts intended for shelled uses. Pods, especially if headed for in-shell use, should be clean and have a bright appearance.

Because the fruit of the peanut plant must grow through the soil surface and develop within the top three inches of soil, it can come into contact with any fungal or bacterial agents that overwinter on old crop or weed residue. Southern blight or white mold caused by *Sclerotium rolfsii* can be especially troublesome. Because of southern blight and other disease pathogens that overwinter on plant debris, peanut producers practice clean tillage more so than producers of other agronomic crops. Land preparation begins with management of the previous crop residue. Operations may include shredding and disking of harrowing debris into the upper soil surface early in the autumn and well ahead of deep plowing. If subsoiling or chiseling is practiced (although it is not recommended in most states for peanut production), it should be done in the autumn to allow maximum benefit for subsoil accumulation of moisture from winter rains.

Deep plowing, six to eight inches, should occur near planting time. Moldboard plows may be equipped with cup coulters (which may be called jointers or cover-boards) to push or place residue at the bottom of the preceding furrow. This effectively places plant residue under anaerobic conditions that are detrimental to pod disease organisms. Subsequent tillage operations should be conducted in such a manner as to not alter this placement of plant debris.

Additional disking or harrowing operations are necessary to break up clods resulting from deep plowing. Producers strive for a smooth, uniform, and well-prepared seedbed. Some producers who utilize beds—i.e., raised area for rows— use a power rotor tiller or a rigid blade just ahead of the planting operation to obtain a "table top" seedbed.

The best response to applied fertilizer occurs when the fertilizer is applied according to soil tests to the preceding summer crop or to a winter cover crop. The next best response occurs from applying fertilizer ahead of disking and/or moldboard plowing.

Preplant tillage is aimed primarily at achieving a good seedbed for uniform seed placement. In addition, preplant tillage and post-plant tillage are designed to control weeds that decrease yield through competition for sunlight, moisture, and nutrients and by harboring disease and insect pathogens. Peanut fields must remain free of weeds for six to seven weeks after emergence if maximum yields are to be achieved. Peanut is a slow-growing annual that does not compete well with most broadleaf weeds or grasses.

Herbicides have greatly simplified weed control in peanuts. Herbicides are available as preplant-incorporated, pre-emergence, cracking stage, and over the top or layby. Peanut offers the unique cracking stage time frame of herbicide application. These herbicides are applied five to seven days after planting into a moist, well-prepared seedbed. Since the peanut plant is not strictly epigeal and since the taproot grows eight times faster than the shoot during the first 12 days following germination, herbicides that are foliage-active can be used to control small weeds that do have sufficient foliage to be controlled by such herbicides while very little peanut foliage is exposed. The peanut plant also is safe because the root system is deep enough at cracking to not be affected by leaching of these herbicides. The drawback of depending on cracking stage herbicides is that the plants grow through this stage rapidly and rainfall over a few days can cause producers to miss the appropriate window for application.

Mixing herbicides with fertilizer, either dry or liquid, is not an uncommon practice or consideration, because such combination would reduce the number of field operations, a goal for economic and agronomic reasons. However, C. W. Swim of the University of Georgia offers the following cautions:

1. Preplant herbicides should be incorporated not more than three inches to ensure that the peanut root will grow through the herbicide zone rapidly, preventing excessive crop damage. This also ensures that label rates are not diluted in too large a volume of soil that would decrease herbicide effectiveness. On the other hand, fertilizer should be incorporated more than three inches to ensure adequate amounts throughout the rhizosphere.

2. Pod rot problems are accentuated by high levels of fertility, a condition easily resulting from shallow incorporation of fertilizer. High levels of potassium in the pegging zone will decrease the uptake of calcium necessary for normal pod and seed development.

3. Fertilizer is often applied well ahead of planting, sometimes as much as five to six months. Many herbicides begin to lose activity when mixed with the soil for this length of time.

Decreased in-season tillage is desirable to prevent peanut damage, especially in runner types. If soil is thrown onto the lateral branches that support fruiting structures, then flowering, pegging, and pod development may be inhibited. Soil around fruiting branches also favors the development of stem rot diseases. Southern stem rot or white mold is more likely to be a problem when stem damage has occurred from mechanical cultivation and/or by excess soil around the crown of the plants.

As a final note, tillage should be employed to keep weeds under control when

TABLE 8.11   Effect of Bed Height on Peanut Yields in Texas.

Year	Elevation of Drill			
	Furrow	0–2 Inches	3–4 Inches	5–6 Inches
	-------------------------------(lbs./ac.)-------------------------------			
1	1,304 c[1]	1,883 b	2,126 a	1,983 ab
2	2,136 c	2,281 c	2,472 b	2,746 a
3	3,059 b	3,041 b	3,241 b	3,570 a
Avg.	2,166	2,402	2,613	2,766

Source: Adapted from Harrison et al., 1975.

1. Yields within rows are not statistically different if they are followed by the same letters.

necessary and to maintain good soil tilth to ensure good pegging conditions. Producers should be "minimum-till" conscious for maximum profits.

Before the advent of herbicides, peanut producers prepared a flat seedbed and at planting create a shallow furrow in which seeds were planted. This technique allowed producers to "dirt" the peanut plants (i.e., cultivate soil toward the plant), primarily to control weeds. In the 1950s, as production technology accelerated, scientists quickly learned that this practice was detrimental to maximum production. Today, many producers plant on beds—i.e., raised areas on which rows are established. In the Southeast, producers may prepare fields for planting and then establish a "bed effect" by simply driving a tractor over the field. The area under the tractor, or between the wheel tracks, then is called the bed. On the soil types suited to peanut production, this results in furrows—i.e., wheel tracks—perhaps four inches deep. The number of rows on each bed depends on row width and tractor tire width. In other parts of the United States, beds are established with some sort of bedding equipment—i.e., an implement designed to create a raised area for planting. Results shown below from research conducted at Yoakum, Texas, show the advisability of planting on beds (Table 8.11).

These data suggest that the higher the bed, the higher the production level; however, beds higher than six inches may favor development of certain soil pests. Producers generally end up with a final planted bed height of two to four inches.

From time to time investigators look into the possibility of producing peanuts with reduced, minimum, or no-till. This type of production holds promise for major reductions in the cost of production. Hartzog and Adams (1988) suggested that reduced tillage has been slow to be accepted because of the belief that surface residue plays a major role in causing diseases, a belief they suggest is not founded on convincing research data. Their data indicate that tillage for peanut should be reevaluated (Table 8.12).

Seeding Rate

Seeding rates depend on several factors: (1) seed size; (2) seed quality; (3) row configuration; (4) irrigated or dryland production; and, of course (5) desired final

TABLE 8.12   Effect of Reduced Tillage on Yield and Grade of Florunner Peanut in Alabama.

Site	Yield[1]		Grade	
	Conv. Till.	Red. Till.	Conv. Till.	Red. Till.
	---------- (lbs./ac.) ----------		---------- (% SMK) ----------	
1	3,940 a	4,170 a	75	74
2	4,200 b	4,860 a	72	71
3	5,100 a	3,650 b	74	73
4	3,570 a	3,620 a	70	70
5	2,930 a	2,870 a	72	74

Source: Adapted from Hartzog and Adams, 1988.

1. Yields within rows are not statistically different if they are followed by the same letter.

stand. Planting seed size is affected by cultural practices such as digging date, irrigation, cultivar, and environmental conditions during maturation and harvest. Data from Texas show the impact of digging dates on the distribution of seed size in Starr spanish-type peanut (Table 8.13). However, as with many crop species, size of planting seed does not affect yield potential, providing that stand and seedling vigor are not affected. This being the case, producers may save a few dollars in variable costs by planting medium-size seeds. Planting small seed probably should be avoided because of the potential for lower seedling vigor when grown under stress such as very early planting or production in areas where temperatures can remain cool for extended periods during early season. Care should be exercised to ensure that all other factors, such as percent germination, are equal between seed lots when faced with this choice. Spanish-type peanut cultivars average about 1,200 seeds/pound, runner cultivars have about 750 seeds/pound, and virginia-type cultivars will average between 400 and 500 seeds/pound (Table 8.14). Producers of runner and spanish types should strive for four to six seeds/linear foot of row, while those who plant virginia type should plant three to four seeds/linear foot of row.

TABLE 8.13   Effect of Digging Date on Seed Size[1] and Distribution in Starr Peanut.

Days after Planting	Date of Digging	Seed Yield	Percent of Seed in Each Size			
			Large	Regular	Medium	Small
		(lb./ac.)				
104	8-9	813	18	44	26	12
111	8-16	1,033	25	43	24	7
120	8-25	1,370	28	44	19	9
129	9-3	1,736	35	41	19	5
140	9-14	1,930	46	39	14	2
150	9-24	331	35	41	20	4

Source: Adapted from Clark, 1975.

1. Width of seed in each size: large > 19/64-inch; regular > 17/64; medium > 15/64; small > 13/64.

TABLE 8.14 Seed Size of Several Cultivars of Each Peanut Type, Seeding Rates in Pounds/Acre at Varying Seeds/Row-Foot and Final Stand When Planted at Either Four or Five Seeds/Row-Foot.

Variety	Seed Size	Seeding Rate When Planted at Seeds/Ft.				Seeds/Foot of Row	
		3	4	5	6	4	5
	(#/lb.)	---------------- (lbs./ac.) -----------------				---- (plants/ac.)[1] -----	
		Runner					
Florunner	779	——	75	93	112	52,583	65,202
GK-7	754	——	77	96	116	52,252	65,146
Sunrunner	750	——	77	97	116	51,975	65,475
		Spanish					
Tamnut 74	1,150	——	51	63	76	52,785	62,205
Pronto	900	——	65	81	97	52,650	65,610
		Virginia					
Florigiant	493	88	118	——	——	52,357	——
GK-3	429	102	135	——	——	52,124	——
NC-7	407	107	142	——	——	52,015	——
Early Bunch	482	90	120	——	——	52,056	——

Source: Adapted from Beasley, 1987.

1. Assumes 36-inch interrow spacing and 90% germination.

Producers in the Southeast plant about 100 pounds of seed/acre of runner peanuts, resulting in a desired final stand of about 60,000 plants/acre. The desired final stand in Texas also is about 45,000 plants for dryland production and up to 60,000 plants/acre if irrigated. Higher final stands, up to 100,000 plants/acre, are recommended for multiple rows per bed under irrigation.

Rotations

Rotation is essential for successful and sustained production of peanuts. Long-term rotation with peanuts grown in at least a three-year cycle is best, but a two-year rotation is better than monoculture. The crops included in the rotation will depend on profitability, but a rotation with at least two years of a grass crop such as corn or sorghum is desirable. Grass crops reduce nematode problems where they occur and allow for the use of a different spectrum of herbicides for control of weed species. Corn and grain sorghum also respond to high levels of fertilizer, leaving the soil with good nutrient supply for healthy peanut growth. Other benefits include:

1. Improved organic matter content, which increases the water-holding capacity, tilth, and structure;
2. Reduced soil erosion helps maintain top soil depth; and

3. Reduced peanut yield losses to stem, root, peg, and pod rots caused by *Sclerotium rolfsii, Rhizoctonia solani,* and *Pythium myriotylum.*

Peanut following soybean or other legumes and some vegetables will have higher losses from diseases. Cotton is a good rotational crop, but crop residue from the strong taproot and woody plant structure is difficult to manage.

Inoculation

The peanut is in the cowpea cross-inoculation group, and the *Rhizobium* species that cause nodulation of this group of legumes is quite widespread. Nitrogen fixation occurs in sufficient amounts to meet plant and rhizobia needs beginning about 30 days after planting. Pink or red coloration inside a nodule indicates that nitrogen is being fixed. Other legumes that are in the cowpea cross-inoculation group include alyce clover, bush clover, cowpea, crotalaria, guar, hairy indigo, kudzu, lespedeza, mung bean, pigeonpea, and velvet bean. These rhizobia are not very host-specific, and almost any cultivar of peanut can be infected by a number of different strains. Nodulation usually occurs worldwide without inoculant being added. While a number of rhizobia strains may infect peanut, only a limited number are highly effective in fixing nitrogen.

The *rhizobium* bacteria infect the roots of legumes, causing an enlargement or nodule to form. The bacteria inside these nodules convert (often referred to as "fixing") atmospheric N_2 into a form usable by the plant—i.e., NH_4^+. The bacteria derive their carbohydrate supply from the plant while supplying the plant with nitrogen, a true symbiotic relationship.

Cox et al. (1982) noted that the application of the appropriate rhizobia has rarely resulted in increased yields, as this bacterial group is so widely distributed in soils. Researchers in Alabama identified several fields that were isolated by distance and/or woodlands where peanut had not been grown for at least 15 years. They planted Florunner with and without inoculation with a highly effective rhizobium strain or with addition of 100 pounds/acres of ammonium nitrate fertilizer without adding inoculant. As shown in Table 8.15, no consistent beneficial effects were derived by the addition of either rhizobia or nitrogen fertilizer.

Research in North Carolina, on the other hand, showed a small increase in yield from the use of commercial inoculant, and adding the inoculant in-furrow was better than adding it as a seed treatment. Most states recommend applying inoculant, especially if fields have not been planted to peanuts within the past four or five years.

Fertility

The primary (N, P, and K) and secondary (Ca and Mg) nutrients removed in 3,000 pounds of farmer stock peanuts in Oklahoma are shown in Table 8.16. Research reported from Texas indicated that peanuts require about 18 pounds of sulfur for production of 2,000 lbs. of farmer stock, plus the accompanying 2,000 lbs. of hay.

TABLE 8.15 Effects of Added Nitrogen Fertilizer or Seed Inoculant on Peanut Yields when Grown at Five Sites in Alabama That Had Not Been Planted to Peanut for at Least Fifteen Years.

| Site | Yield | | | Grade[1] | | |
	None	Fert.	Inoculated	None	Fert.	Inoculated
	----------------- (lbs./ac.) -------------------			--------------------- (%) ---------------------		
1	2,910	2,600	3,030	71	69	71
2	2,600	2,620	3,130	74	71	72
3	4,190	3,990	4,210	74	71	72
4	3,690	3,470	3,420	73	72	72
5	4,040	3,930	3,760	75	74	75

Source: Adapted from Hartzog and Adams, 1988.

1. Grade = % sound mature kernels.

Minor elements (zinc, iron, manganese, copper, boron, molybdenum, and perhaps chlorine) required per ton of farmer stock ranges from 0.02 to 0.51 pounds.

Nitrogen deficiency rarely occurs in peanut fields, since peanut is a legume and rhizobia are so widespread throughout the United States. Occasionally, deficiency symptoms are seen but can be quickly alleviated by using 15 to 50 pounds N/acre. Even so, some states recommend 15 pounds/acre at or before planting, especially on deep sandy soils where soil nitrogen could have been leached out of the upper levels of the soil.

Most soils in the United States are adequate in phosphorus and potassium. Even when soils test as low as 4 pounds P/acre, yields may not be affected by applied phosphorus. Peanut does respond however to applied potassium when soil test levels are less than about 20 pounds K/acre. Deficiencies of these macro elements usually occur on deep sands where there is not a clayey subsoil. Production of a grass crop for several years prior to planting peanuts also will deplete the topsoil of potassium.

Some researchers believe that a high level of potassium in the pegging zone increases the incidence of pod rot and interferes with the uptake of calcium by pegs and pods. In fields where potassium levels are high, a higher rate of calcium should be applied through gypsum, calcium sulfate, or other liming source. Calcium is probably the most often yield-limiting nutrient in peanut production. Calcium that is

TABLE 8.16 Nutrients (lbs.) Removed by 2,000 Pounds of Farmer Stock Peanuts.

Crop Segment	Yield (lbs.)	Nitrogen	Phosphorus	Potassium	Calcium	Magnesium
Peanuts[1]	2,000	70	15	18	3	3
Hay	3,000	98	21	90	53	23
Totals	5,000	168	36	108	56	26

Source: Adapted from Sholar et al., 1991; Chapin et al., 1975.

1. Includes seed and shell.

taken up by peanut roots is transported throughout the plant, including the ovary, until the ovary (now the peg) enters the soil. Once the developing fruit enters the pegging phase and throughout pod development, no calcium is transported to it through the xylem. The peg and pod passively absorb water and nutrients directly from the soil solution; therefore calcium must be available in a water-soluble form in the upper three inches of soil. Calcium deficiency is expressed symptomatically as underdeveloped kernels in pods (known as "pops"), darkened plumules (i.e., the rudimentary terminal bud in seeds), poor seed germination, higher rates of aflatoxin contamination, and higher incidence of pod rot. With any or all of these symptoms, vegetative growth can be excessive but not to the point of appearing so. Profuse flowering has been reported in calcium-deficient peanuts with many of these flowers infertile. The large-seeded virginia-type peanut requires more careful calcium management than the smaller-seeded runner and spanish types.

The calcium required for economic returns is a separate issue from soil pH levels. Peanut plants performs best at pH levels between 5.8 and 6.2 in the coastal plain soils of the Southeast, and 6 to 7 on slightly heavier soils in the Southwest. When soil solution pH drops below 6, and especially below 5.5, aluminum, zinc, and manganese become soluble and are toxic to many plants; cations such as calcium, potassium, and magnesium are precipitated to insoluble forms and therefore become unavailable; and the symbiotic relationship between rhizobia and the peanut plant is adversely affected. Low soil pH should be corrected by use of limestone at soil test recommended rates. Many producers use dolomitic limestone ($MgCO_3$) rather than calcitic limestone ($CaCO_3$) because they are correcting pH and getting a cheap source of magnesium. Limestone is not water-soluble at a neutral pH, but is soluble under acidic conditions; therefore, lime is not an acceptable form of calcium for peg and pod development.

Calcium sulfate, called gypsum or land plaster, is the recommended form of calcium for peanut production. Gypsum is recommended as a top dressing at early flowering if a soil test shows less than 400 pounds Ca/acre and/or the ratio of calcium to potassium is less than 3:1—i.e., three times as much calcium as potassium, for virginia-type peanut production. When growing runner-type or spanish-type cultivars, gypsum is recommended for soils testing less than 220 pounds Ca/acre or a ratio of calcium to potassium of less than 3:1. Since calcium is absorbed passively into the peg or pod, moisture from either rain or irrigation must be present for calcium uptake. Producers who have irrigation capabilities should apply timely irrigation after gypsum application. If heavy rains, four to six inches, occur within three weeks after application, a second, lighter application of gypsum may be called for about four weeks after the initial application.

Digging

Peanut is unique in that its fruit develops underground from an aerial flower. This means that the crop must be "dug" in order for harvesting to proceed. This presents a small problem to the producer as there are no aerial indicators of maturity. Another confounding factor is that the peanut is botanically an indeterminate plant that puts on fruit over a period of time, meaning that the producer has to balance the maturing of late flowers and the loss of pods from early flowers. Lastly, the producer is interested in the maturity of a kernel that is inside a shell.

It is obvious from earlier comments in this chapter that the number of days from planting to maturity is cultivar-dependent. For example, Florunner cultivar has a maturity range of 135 to 150 days. Disease-free, well-maintained fields of Florunner, however, may require more than 150 days, while fields of diseased or otherwise unthrifty plants may need to be dug in less than 135 days. Unseasonably cool weather or late planting also will delay maturity. The trick is to commence the harvest process in such a manner as to maximize the number of sound, mature kernels within pods that will remain attached to the plant when dug. Diseased fields may need early digging because of peg deterioration that could result in the pod breaking free of the vine. The timeliness of digging peanuts has been called an art rather than an exact science.

There are a couple of ways that producers estimate the best time to dig their crop. First, all pods from four plants collected from different areas of a field are removed. All pods are opened and the percent of pods with the inside of their shells having brown spots is calculated. Harvest is recommended for spanish-type and runner-type peanuts when 75–80 percent of the pods are dark or darkening. Again, overall health of the plants should be taken into consideration. This is called the "shell out" method.

Cultivars of virginia-type peanut are assessed by examining the seedcoat or testa rather than the color of the hull. Again, at least four representative plants are dug and all pods shelled. Pods are separated into three groups, those with kernels having (1) seedcoats white, (2) seedcoats light pink, and (3) seedcoats dark pink with prominent veins. Normally, harvest takes place when 60 to 65 percent of the pods have kernels with dark pink seedcoats with prominent veins.

An alternative method for runner-type (not virginia-type) peanut is the hull-scrape method. With this method, the exocarp, or outside, of the peanut hull is removed by scraping with a knife or by blasting with sand and water at high pressure. Maturity is then based on the color of the mesocarp. As the pod matures, the mesocarp changes from white to yellow to dark yellow and then orange to brown, and finally to black. Pods are compared with a color chart to predict days to optimum harvest.

Peanuts are dug with a digger-shaker-windrower that removes the plants from the soil, shakes excess soil from pods and roots, and then inverts the plants so that pods are off the ground for maximum drying. Peanut should be dug at 35 to 50 percent kernel moisture and field-cured to 18 to 24 percent moisture. In some environments of the Southwest, producers may be tempted to allow kernels to dry to 10 percent moisture to avoid drying cost at the first handler. However, pod loss during threshing increases as moisture decreases below 18 percent. On average, loss of one Florunner peanut pod per square foot during harvest represents a loss of 150 pounds/acre. Obviously, decreased field loss more than offsets the cost of drying at the first handler point.

BIOTIC PESTS

Peanut is attacked by a number of disease pathogens and insect pests. White mold or southern blight caused by *Sclerotium rolfsii* and leafspots are the most serious disease problems, while the number-one insect pest is the lesser cornstalk borer. Most others are minor pests, occurring sporadically. These and other pests plus their scientific names, a brief description of symptoms or damage, and their control are listed in Tables 8.17 and 8.18.

U.S. STANDARDS AND GRADES

Farmers' Stock Virginia Type

U.S. No. 1 shall consist of unshelled virginia-type peanuts that are mature, dry, and practically free from foreign material. Not less than 45 percent, by weight, of any lot shall consist of jumbo and fancy hand-picks and of this amount not less than three-fourths, or 34 percent, of the total unshelled weight shall be jumbo hand-picks.

 U.S. No. 2 shall consist of unshelled virginia-type peanuts that are mature, dry, and practically free from foreign material. Not less than 35 percent, by weight, of any lot shall consist of jumbo and fancy hand-picks and of this amount not less than three-fifths, or 21 percent, of the total unshelled weight shall be jumbo hand-picks.

 U.S. No. 3 shall consist of unshelled virginia-type peanuts that are mature, dry, and practically free from foreign material.

 Unclassified shall consist of unshelled virginia-type peanuts that do not meet the requirements of the foregoing grades.

 Large kernels: Peanuts that will pass over a screen having $^{20}/_{64} \times$ 1-inch perforations.

 Medium kernels: Peanuts that will pass through a screen having $^{20}/_{64} \times$ 1-inch perforations and over a screen having $^{15}/_{64} \times$ 1-inch perforations.

 Small kernels: Peanuts that will pass through a screen having $^{15}/_{64} \times$ 1 inch perforations.

 Note: The above size terms, large, medium and small, are equivalent to extra large No. 1 and No. 2 now in common use to designate size of kernels. They have been provided to avoid confusion with the grade terms U.S. No. 1, U.S. No. 2, etc., which deal with hand-picks, shelled, and defective peanuts.

Farmers' Stock Runner Type

U.S. No. 1 shall consist of unshelled runner peanuts that are mature, dry, and practically free from damage. When shelled, the sound and mature kernels shall not be less than 65 percent of the total weight of the sample (Table 8.19).

 U.S. No. 2 shall consist of unshelled runner peanuts that are mature, dry, and practically free from damage. When shelled, the sound and mature kernels shall not be less than 60 percent of the total weight of the sample.

TABLE 8.17   Insect Pests of Peanuts in the United States.

Insect	Scientific Name	Description/Damage	Control
Whitefringed Beetle	*Graphognathus leucoloma*	White or cream-colored larvae, legless, up to ½ inch long; feed on roots, often cutting the tap root causing plant death or season-long stunting.	Rotation; chemical.
Bahiagrass Borer	*Derobrachus brevicollis*	Soil tunneling larvae do not feed on peanut but cut tap roots causing plant death.	Deep tillage with or without deep placement of insecticide; avoid following bahiagrass in rotation.
Lesser Cornstalk Borer	*Elasmopalpus lignosellus*	Dark, blue-green larvae, ½–¾ inch long with brown or purple bands; flips about rapidly when disturbed; will feed on any plant part in contact with ground or tunnel into soil to feed on pegs and pods.	Chemical; maintain good soil moisture if possible.
Southern Corn Rootworm	*Diabrotica undecimpunctata howardi*	Slender, white to cream colored larvae, ½–¾ inches long; fragile, wrinkled body with 3 pairs of inconspicuous legs; head and last body segment dark brown to black; feeds on roots, pegs, and pods; pegs and pods will have tiny, cylindrical holes.	Chemical.
Thrips	*Frankliniella fusca*	Tiny, slender insects; 1/16 inch long; adults jump or fly when disturbed; immatures similar without wings; yellow to black color.	Systemic insecticides.
Aphids (also called plant lice)	*Aphis* spp.	Small, soft bodied, sucking insects, about 1/15 inch long; secretes "honeydew" while feeding; pale yellowish green, dark green or black; winged and wingless forms; phloem feeders that can cause reduced plant vigor.	Systemic insecticides.
Three-cornered Alfalfa Hopper	*Spissistilus festinus*	Light green, wedge shaped; about ¼ inch long; immature similar but wingless; feeds in such a fashion as to girdle main stems; feeding area will gall over but remain weakened.	Chemical control as necessary.

(continued)

TABLE 8.17   (Continued)

Insect	Scientific Name	Description/Damage	Control
Granulate Cutworms	Feltia subterranea	Smooth, cylindrical worms, grey to brown; up to 1½ inches long; feed and cut young plants at soil surface.	Bait formulation of insecticide.
Corn Earworm	Helicoverpa zea	Larvae vary in color from light green to pink to brown to near black with alternating light and dark stripes running length of body; feed on plant foliage beginning with spiral meristems.	Chemical control.
Armyworms (Fall and Yellow Striped)	Spodoptera spp.	Feed in foliage, occasionally causing complete defoliation; feed during day and night.	Chemical control.
Velvetbean Caterpillar	Anticarsia gemmatalis	Rarely defoliates plants; late season pest; larvae have yellow head with light green to black body; yellowish-white stripes run length of body.	Chemical.
Green Cloverworm	Plathypena scabra	Larvae green with two narrow stripes on each side of body; three pairs of abdominal prolegs; foliage feeders.	Chemical.
Looper	Trichoplusia ni and T. includens	Foliage feeders; move in characteristic looping motion; light green with whitish stripes; up to 1½ inches long.	Chemical.
Rednecked Peanutworm	Stegasta bosquella	Larvae feed on apical meristems; early infestations may stunt plants; larvae are cream to green colored with brown head and plate just behind head.	Chemical.
Spider Mites	Tetranychus urticae	Tiny, insect type pests that colonize underside of leaves; upper leaf surface becomes speckled, may yellow and die.	Chemical applied to underside of leaves; maintain good soil moisture if possible.

U.S. No. 3 shall consist of unshelled runner peanuts that are mature, dry, and practically free from damage. When shelled, the sound and mature kernels shall not be less than 55 percent of the total weight of the sample.

U.S. sample grade shall consist of unshelled runner peanuts that do not meet any of the foregoing grades.

TABLE 8.18   Diseases of Peanuts in the United States.

Disease	Scientific Name	Symptom/Injury	Control
		Foliage Diseases	
Early Leafspot (also called Cercospora Leaf Spot)	*Cercospora arachidicola*	Small, light brown specks enlarging to as much as ½-inch diameter; may become black; usually surrounded by yellow halo.	Crop rotation; burial of plant residue before planting; fungicides beginning at first symptoms.
Late Leafspot (also called Cercospora Leaf Spot)	*Cercosporidium personatum*	Same as early leafspot except halo less obvious or absent.	Same as early leafspot.
Rust	*Puccinia arachidis*	Leaves in distinct areas of fields have many small, orange colored pustules.	Fungicides if occurs 3 weeks or more before crop maturity.
Web Blotch	*Phoma arachidicola*	Webbing or net-like pattern; tan to bronze color on top of leaf; larger, circular tan to brown blotches that may cover entire leaflet.	Runner types may be resistant but spanish likely susceptible; fungicides.
Pepper Spot	*Leptosphaerulina crassiasca*	Small, dark brown or black areas, sometimes sunken.	Fungicides.
Leaf Scorch	*Leptosphaerulina crassiasca*	Light tan to brown V-shaped lesions with bright yellow margins; commonly develops at leaflet tips.	Fungicides.
		Soil-Borne Diseases	
White Mold, (also called Southern Blight)	*Sclerotium rolfsii*	Yellowing and wilting of branches, followed by necrosis; browning and dry rot of lower stem; death of only one or two branches or of entire plant; infected stems may be covered with dense, white fungus that forms orange sclerotia; in dry conditions, pod rot may occur without foliage symptoms.	Resistant cultivars; rotation with grass crop or other non-host for two or more years; deep burial of plant debris; avoid mechanical injury; avoid cultivating soil onto vines; fungicides.

(continued)

TABLE 8.18 (*Continued*)

Disease	Scientific Name	Symptom/Injury	Control
Rhizoctonia diseases (Seedling Disease or Damping Off, Stem Rot, Limb Rot, Foliage Blight, Root Rot, Peg Rot and Pod Rot.	*Rhizoctonia solani* *Sclerotium rolfsii* *Pythium myriotylum*	Damping Off: dark brown to reddish-brown, sunken lesions just below ground level; Stem Rot: circular, sunken, brown lesions showing concentric ring pattern (called zonate); Limb Rot: limbs in contact with ground affected and will be blackened and rotten; Foliage Blight: rare; zonate patterns are light to dark brown.	Resistant cultivars; litter burial; rotation; reduced late season irrigation; calcium sulfate at pegging for pod rot; fungicides.
Cylindrocladium Black Rot (CBR)	*Cylindrocladium crotalariae*	Yellowing and wilting of leaves; some stunting of plants; roots are black, rotten and stunted; pods partially to completely black and rotten; seed coats light tan with small, reddish orange specks.	Rotation for 2 or more years with crops other than soybean; resistant cultivars.
Crown Rot (also called Black Mold)	*Aspergillus niger*	Sudden wilting; stem at or slightly below ground level is blackened.	Irrigation may slow disease.
Sclerotinia Blight	*Sclerotinia minor*	Almost identical to White Mold except sclerotia are black and form on or inside stems, pods, seed, and pegs; causes stem shredding and pod loss; fungus active below 82°F and soil moist.	Avoid frequent irrigation if temperature conducive to fungal growth; resistant cultivars; chemical.
		Nematodes	
Root-knot	*Meloidogyne arenaria*	Stunted, yellow plants; sporadic areas of field; pegs, pods, and roots covered with large galls.	Rotation; nematicides.
		Virus Diseases	
Peanut Mottle (PMV)	—	Young leaves have light and dark green mottling.	Seed-borne and aphid transmitted; no control measures.

TABLE 8.18 (*Continued*)

Disease	Scientific Name	Symptom/Injury	Control
Peanut Stunt (PSV)	—	Severe stunting of parts or entire plant; leaflet size greatly reduced; pods may have a distinct constriction between kernels.	Seed-borne and aphid transmitted; no control measures.
Peanut Stripe (PStV)	—	Stripping or banding along leaflet veins.	Seed-borne and aphid transmitted; does not significantly affect yield; no control measures.
Tomato Spotted Wilt (TSWV)	—	Symptoms range from stunting to mottled leaves to death of apical meristem; leaves always have ringed spot.	200 different plant hosts-includes pepper, tomato, peas, beans; spread by thrips and not seed-borne.

Sample means the total quantity of material taken for examination including all shelled and unshelled stock and foreign material.

Practically free from damage relates to unshelled peanuts. No appreciable amount of nuts with cracked shells shall be noticeable.

Damaged kernels are (1) kernels that are rancid or decayed; (2) moldy kernels; (3) kernels showing sprouts over 1/8 inch long (however, all sprouted kernels, the separated halves of which show decay, shall be classed as damaged.); (4) dirty kernels where the surface is distinctly dirty and the dirt is ground in; (5) wormy or worm-injured kernels; (6) kernels that show a yellow discoloration when the skin is

TABLE 8.19 Allowable Percentages for Farmers' Stock Runner Based on Total Weight of Sample.

Grade	Tolerance for Other Varieties	Sound Kernels	Tolerance for Damaged Kernels
U.S. No. 1	1%	65%	2%
		66% +	3%
U.S. No. 2	1%	60%	2%
		61%	3%
		62%	4%
		63% +	5%
U.S. No. 3	1%	55%	2%
		56%	3%
		57%	4%
		58%	5%
		59% +	6%

Source: USDA-FGIS, 1983.

removed; and (7) kernels having skins that show dark brown discoloration, usually netted and irregular, and affecting more than 25 percent of the skin. Kernels having skins that are paler or darker in color than is usually characteristic of the variety, but which are not actually discolored, shall not be classed as damaged.

Mature means that the kernels are fully developed.

Hand-picks are unshelled peanuts that are clean, fairly bright, mature, dry, and free from mildewed or speckled shells, paper ends, and free from cracked or broken shells. Hand-picks shall not pass through a slotted screen having $^{32}/_{64} \times$ 3-inch perforations.

Fancy hand-picks are hand-picks that will pass through a slotted screen having $^{37}/_{64} \times$ 3-inch perforations, but which will not pass through a screen having $^{32}/_{64} \times$ 3-inch perforations.

Jumbo hand-picks are hand-picks that will not pass through a slotted screen having $^{37}/_{64} \times$ 3-inch perforations.

Shelled Spanish Type

U.S. No. 1 Spanish consists of shelled spanish peanut kernels of similar varietal characteristics that are whole and free from foreign material, damage and minor defects, and which will not pass through a screen having $^{15}/_{64} \times$ $^{3}/_{4}$-inch openings.

U.S. spanish splits consist of shelled spanish-type peanut kernels of similar varietal characteristics that are split or broken, but which are free from foreign material, damage and minor defects, and which will not pass through a screen having $^{16}/_{64}$-inch round openings.

U.S. No. 2 Spanish consists of shelled spanish peanut kernels of similar varietal characteristics that may be split or broken, but which are free from foreign material, damage and minor defects, and which will not pass through a screen having $^{16}/_{64}$-inch round openings.

Shelled Virginia Type

U.S. extra large Virginia consists of shelled virginia-type peanut kernels of similar varietal characteristics that are whole and free from foreign material, damage, and minor defects, and which will not pass through a screen having $^{20}/_{64} \times$ 1-inch openings. Unless otherwise specified, the peanuts in any lot shall average not more than 512/pound.

U.S. medium Virginia consists of shelled virginia peanut kernels of similar varietal characteristics that are whole and free from foreign material, damage, and minor defects, and which will not pass through a screen having $^{18}/_{64} \times$ 1-inch openings. Unless otherwise specified, the peanuts in any lot shall average not more than 640/pound.

U.S. No. 1 Virginia consists of shelled virginia peanut kernels of similar varietal characteristics that are whole and free from foreign material, damage, and minor defects, and which will not pass through a screen having $^{15}/_{64} \times$ 1-inch openings. Unless otherwise specified, the peanuts in any lot shall average not more than 864/ pound.

U.S. virginia splits consist of shelled virginia-type peanut kernels of similar varietal characteristics that are free from foreign material, damage, and minor defects, and which will not pass through a screen having $^{20}/_{64}$-inch round openings. Not less than 90 percent, by weight, shall be splits.

U.S. No. 2 Virginia consists of shelled virginia-type peanut kernels of similar varietal characteristics that may be split or broken, but which are free from foreign material, damage, and minor defects, and which will not pass through a screen having $^{17}/_{64}$-inch round openings.

Shelled Runner-Type

U.S. No. 1 Runner consists of shelled runner-type peanut kernels of similar varietal characteristics that are whole and free from foreign material, damage, and minor defects, and which will not pass through a screen having $^{16}/_{64} \times {}^{3}/_{4}$-inch openings.

U.S. runner splits consists of shelled runner peanut kernels of similar varietal characteristics that are slit or broken, but which are free from foreign material, damage, and minor defects, and which will not pass through a screen having $^{17}/_{64}$-inch round openings.

U.S. No. 2 Runner consists of shelled runner-type peanut kernels of similar varietal characteristics that may be split or broken, but which are free from foreign material, damage, and minor defects, and which will not pass through a screen having $^{17}/_{64}$-inch round openings.

Similar varietal characteristics means that the peanut kernels in the lot are not of distinctly different varieties. For example, spanish type shall not be mixed with runners.

Whole means that the peanut kernel is not split or broken.

Split means the separated half of a peanut kernel.

Broken means that more than one-fourth of the peanut kernel is broken off.

Foreign material means pieces or loose particles of any substance other than peanut kernels or skins.

Unshelled means a peanut kernel with part or all of the hull (shell) attached.

Minor Defects means that the peanut kernel is not damaged but is affected by one or more of the following: (1) skin discoloration that is dark brown, dark gray, dark blue, or black and covers more than one-fourth of the surface; (2) flesh discoloration that is darker than a light yellow color or consists of more than a slight yellow pitting of the flesh; (3) sprout extending more than one-eighth of an inch from the tip of the kernel; and (4) dirt when the surface of the kernel is distinctly dirty, and its appearance is materially affected.

Damage means that the peanut kernel is affected by one or more of the following: (1) rancidity or decay; (2) mold; (3) insects, worm cuts, web, or frass; (4) freezing injury causing hard, translucent, or discolored flesh; and (5) dirt when the surface of the kernel is heavily smeared, thickly flecked, or coated with dirt, seriously affecting its appearance.

Cleaned Virginia Type

U.S. jumbo hand-picked shall consist of cleaned virginia-type peanuts in the shell that are mature (1), dry and free from loose peanut kernels, dirt, or other foreign material, pops (2), paper ends (3), and from damage (4) caused by cracked or broken shells (4a), discoloration (4b), or other means. The kernels shall be free from damage (4c, 4d, 4e, 4f, 4g, 4h) from any cause. In addition, the peanuts shall not pass through a screen having $^{37}/_{64}$ × 3-inch perforations. Unless otherwise specified, the unshelled peanuts in any lot shall not average more than 176 count per pound (5).

U.S. fancy hand-picked shall consist of cleaned Virginia peanuts in the shell meeting the same standards as U.S. jumbo hand-picks, except the peanuts shall not pass through a screen having $^{32}/_{64}$ × 3-inch perforations. Unless otherwise specified, the unshelled peanuts in any lot shall not average more than 225 count per pound (5).

Unclassified shall consist of cleaned Virginia-type peanuts in the shell that fail to meet the requirements of either of the foregoing grades. The term "unclassified" is not a grade within the meaning of these standards but is provided as a designation to show that no definite grade has been applied to the lot.

As used in these standards, (1) mature means that the shells are firm and well developed; (2) pops are fully developed shells that contain practically no kernel; (3) paper ends are peanut shells that have very soft and/or very thin ends; (4) damage means any injury or defect that materially affects the appearance, edible or shipping quality of the individual peanut or the lot as a whole. The following shall be considered as damage:

(a) Cracked or broken shells that have been broken to the extent that the kernel within is plainly visible without minute examination and with no application of pressure, or where the appearance of the individual peanut is materially affected.

(b) Discolored shells that have dark discoloration caused by mildew, staining, or other means affecting one-half or more of the shell surface. Talc powder or other similar material that may have been applied to the shells during the cleaning process shall not be removed to determine the amount of discoloration beneath, but the peanut shall be judged as it appears with the talc.

(c) Kernels that are rancid or decayed.

(d) Moldy kernels.

(e) Kernels showing sprouts extending more than one-eighth inch from the end of the kernel.

(f) Distinctly dirty kernels.

(g) Kernels that are wormy, or have worm frass adhering, or have worm cuts that are more than superficial.

(h) Kernels that have dark yellow color penetrating the flesh or yellow pitting extending deep into the kernel.

(5) Count per pound means the number of peanuts in a pound. When determining the count per pound, one single-kernel peanut shall be counted as one-half peanut.

BIBLIOGRAPHY

Auburn University. 1991. *1991 Peanut: Insect, Disease, Nematode, and Weed Control Recommendations.* Auburn Univ. Coop. Ext. Ser. Circ. ANR-360.

Baldwin, J. S. 1989. "Peanut Inoculation and Nitrogen Fixation." pp. 3a/1–3a/4. *In* W. C. Johnson III (ed.) *Georgia Peanut Production Guide.* Univ. of Georgia Coop. Ext. Ser. SB 23.

Beasley, J. P. 1987. "Varieties." pp. 5/1–5/4. *In* W. C. Johnson III (ed.) *Georgia Peanut Production Guide.* Univ. of Georgia Coop. Ext. Ser. SB 23.

Bird, J. B. 1948. "America's oldest farmers." *Natural Hist.* (New York) 57:296–303.

Boote, K. J. 1982. "Growth stages of peanut (*Arachis hypogaea* L.)." *Peanut Sci.* 9:35–40.

Boswell, T. E. 1975. "Land Selection and Management." pp. 19–21. *In Peanut Production in Texas.* Texas A&M Univ. Agri. Expt. Sta. and Coop. Ext. Ser. Rm 3.

Brooks, G. E. 1975. "Peanuts and colonialism: Consequences of the commercialization of peanuts in West Africa, 1830–70." *J. African Hist.* 16:29–54.

Bunting, A. H. 1955. "A classification of cultivated groundnut." *J. Exp. Agri.* 23:152–170.

Carley, D. H. 1983. "Production and Marketing of Peanuts in the United States." pp. 19–40. *In* J. G. Woodroff (ed.) *Peanuts: Production, Processing, Products.* AVI Pub. Co., Inc., Westport, CT.

Carver, A. A. 1969. "Registration of the Florigiant peanuts." *Crop Sci.* 9:849–850.

Chapin, J. C., C. Gray, and W. B. Anderson. 1975. "Nutritional Requirements." pp. 22–25. *In Peanut Production in Texas.* Texas A&M Univ. Agri. Expt. Sta. and Coop. Ext. Ser. Rm 3.

Clark, L. E. 1975. "Seed Quality." pp. 10–18. *In Peanut Production in Texas.* Texas A&M Univ. Agri. Exp. Sta. and Coop. Ext. Ser. Rm 3.

Cox, F. R., F. Adams, and B. B. Tucker. 1982. "Liming, Fertilizing, and Mineral Nutrition." pp. 139–163. *In* H. E. Pattee and C. T. Young (eds.) *Peanut Science and Technology.* Am. Peanut Res. and Edu. Soc., Inc., Yoakum, TX.

Gooden, D. T., C. E. Drye, J. W. Chapin, E. C. Murdock, and L. A. Stanton. 1984. *Peanut Production Guide for South Carolina.* Clemson Univ. Coop. Ext. Ser. Circ. 588.

Gregory, W. C., B. W. Smith, and J. A. Yarbrough. 1951. "Morphology, Genetics and Breeding." pp. 28–88. *In The Peanut—The Unpredictable Legume.* Natl. Fert. Assoc., Washington, DC.

Gregory, W. C., and M. P. Gregory. 1976. "Groundnut *Arachis hypogaea (Leguminosae-Papilionatae)*." pp. 151–154. *In* N. W. Simmonds (ed.) *Evolution of Crop Plants.* Longman, London, England.

Hammons, R. O. 1982. "Origin and Early History of the Peanut." pp. 1–20. *In Peanut Science and Technology.* Am. Peanut Res. and Edu. Soc., Inc., Yoakum, TX.

Harper, W. H., L. Stockard, and B. Smith. 1983. *The Peanut Production and Marketing System in New Mexico.* New Mexico St. Univ. Agri. Exp. Sta. Res. Rep. 503.

Harrison, A. L., L. E. Clark, and C. E. Simpson. 1975. "Bed Type and Plant Population." pp. 38–41. *In Peanut Production in Texas.* Texas A&M Univ. Agri. Exp. Sta. and Coop. Ext. Ser. Rm 3.

Hartzog, D. L., and J. F. Adams. 1988. *Soil Fertility Experiments with Peanuts in Alabama, 1973–1986.* Auburn Univ. Agri. Exp. Sta. Bul. 594.

Hartzog, D. L., J. O. Donald, A. K. Hagan, T. W. Tyson, L. Curtis, J. C. French, J. R. Weeks, J. Everest, A. Miller, J. R. Crews, and J. L. Johnson. 1990. *Peanut Production in Alabama*. Auburn Univ. Coop. Ext. Ser. Circ. ANR-207.

Higgins, B. B. 1951. "Origin and Early History of the Peanut." *In The Peanut—The Unpredictable Legume*. The Natl. Fert. Assoc., Washington, D.C.

Hsi, D. C. H., and R. E. Finkner. 1972. "Registration of New Mexico Valencia A peanut." *Crop Sci*. 12:256.

Johnson, W. C., III. 1987. *The Hull Scrape Method to Assess Peanut Maturity*. Univ. of Georgia Coop. Ext. Ser. Bul. 958.

Krapovickas, A. 1969. "The Origin, Variability and Spread of the Groundnut (*Arachis hypogaea*)." pp. 427–441. *In The Domestication and Exploration of Plants and Minerals*. Gerald Duckworth Co., LTD, London, England. (lang. = French)

Krapovickas, A. 1973. "Evolution of the Genus *Arachis*." pp. 135–151. *In* R. Moav (ed.) *Agricultural Genetics, Selected Topics*. John Wiley and Sons, New York, NY.

McArthur, W. C., V. Grise, H. O. Doty Jr., and D. Hacklander. 1983. *U.S. Peanut Industry*. USDA Eco. Res. Ser., Washington, DC.

Melouk, H. A. 1991. *Confronting a new fungal nightmare*. USDA-ARS *Agri. Res*. 39:21–23.

Mozingo, R. W., T. A. Coffelt, and J. C. Wynne. 1987. *Characteristics of Virginia-type Peanut Varieties Released from 1944–1985*. Virginia Agri. Expt. Sta. So. Coop. Ser. Bul. 326.

Mozingo, R. W., T. A. Coffelt, and J. C. Wynne. 1988. *Quality Evaluations of Virginia-type Peanut Varieties Released from 1944–1985*. Virginia Agri. Exp. Sta. So. Coop. Ser. Bul. 335.

Norden, A. J., O. D. Smith, and D. W. Gorbet. 1982. "Breeding of the Cultivated Peanut." pp. 95–122. *In* H. E. Pattee and C. T. Young (eds.) *Peanut Science and Technology*. Am. Peanut Res. and Edu. Soc., Inc., Yoakum, TX.

Norden, A. J. 1980. "Peanut." pp. 443–456. *In* W. R. Fehr and H. H. Hadley (eds.) *Hybridization of Crop Plants*. Crop Sci. Soc. Am., Madison, WI.

Norden, A. J., R. W. Lipscomb, and W. A. Carver. 1969. "Registration of the Florunner peanuts." *Crop Sci*. 9:850.

Planters LifeSavers Co. 1990. *Peanuts: A Grower's Guide to Quality*. Planters LifeSavers Co., Winston-Salem, NC.

Sholar, R., J. Damicone, H. Greer, M. Kizer, G. Johnson, and P. Mulder. 1992. *1992 Peanut Production Guide for Oklahoma*. Oklahoma St. Univ. Coop. Ext. Ser. Cir. E-608.

Smith, Ben W. 1950. "*Arachis hypogaea* L. Aerial flower and subterranean fruit." *Am. J. Bot*. 37:802–815.

Squier, E. G. 1877. *Peru, Incidents of Travel and Exploration in the Land of the Incas*. MacMillan Press, New York.

Sullivan, G. A. 1991. "Peanut Production Practices." pp. 9–22. *In* 1992 *Peanuts*. North Carolina St. Univ. Coop. Ext. Ser. AG-331.

Swann, C. W. 1987. "Weed Control." p. 12/1–12/6. *In* W. C. Johnson III (ed.) *Georgia Peanut Production Guide*. Univ. of Georgia Coop. Ext. Ser. SB 23.

United States Department of Agriculture. *Agricultural Statistics, 1936, 1942, 1952, 1961, 1972, 1982 and 1992*. U.S. Government Printing Office, Washington, DC.

Weaver, R. W. 1974. "Effectiveness of Rhizobia forming nodules in Texas grown peanuts." *Peanut Sci.* 1:23–25.

Wells, R., T. Bi, W. F. Anderson, and J. C. Wynne. 1991. "Peanut yield as a result of fifty years of breeding." *Agron. J.* 83:957–961.

Williams, D. E. 1989. "Exploration of Amazonian Bolivia yields rare peanut landraces." *Diversity* 5:12–13.

Wolt, J. D., and F. Adams. 1979. "Critical levels of soil- and nutrient-solution calcium for vegetative growth and fruit development of Florunner peanut." *J. Soil Sci. Soc. Am.* 43:1159–1164.

Womack, H., J. S. French, F. A. Johnson, S. S. Thompson, and C. W. Swann. 1981. *Peanut Pest Management in the Southeast.* Univ. of Georgia Coop. Ext. Ser. Bul. 850.

Woodroof, J. G. 1983. "Historical Background." pp. 1–18. *In* J. G. Woodroof (ed.) *Peanuts: Production, Processing, Products.* AVI Pub. Co., Inc., Westport, CT.

Woodroof, J. G. 1983. "The Culture of Peanuts." pp. 41–89. *In* J. G. Woodroof (ed.) *Peanuts: Production, Processing, Products.* AVI Pub. Co., Inc., Westport, CT.

Wynne, J. C., and T. A. Coffelt. 1982. "Genetics of *Arachis hypogaea* L." pp. 50–94. *In* H. E. Pattee and C. T. Young (eds.) *Peanut Science and Technology.* Am. Peanut Res. and Edu. Soc., Inc., Yoakum, TX.

Index

CPSIA information can be obtained
at www.ICGtesting.com
Printed in the USA
JSHW011119190822
29490JS00002B/3